The Physics of Interstellar Dust

Series in Astronomy and Astrophysics

Series Editors: **M Birkinshaw**, University of Bristol, UK
M Elvis, Harvard–Smithsonian Center for Astrophysics, USA
J Silk, University of Oxford, UK

The Series in Astronomy and Astrophysics includes books on all aspects of theoretical and experimental astronomy and astrophysics. Books in the series range in level from textbooks and handbooks to more advanced expositions of current research.

Other books in the series

Dark Sky, Dark Matter
J M Overduin and P S Wesson

Dust in the Galactic Environment, 2nd Edition
D C B Whittet

An Introduction to the Science of Cosmology
D J Raine and E G Thomas

The Origin and Evolution of the Solar System
M M Woolfson

The Physics of the Interstellar Medium
J E Dyson and D A Williams

Dust and Chemistry in Astronomy
T J Millar and D A Williams (eds)

Observational Astrophysics
R E White (ed)

Stellar Astrophysics
R J Tayler (ed)

Forthcoming titles

Very High Energy Gamma Ray Astronomy
T Weekes

Numerical Methods in Astrophysics
P Bodenheimer, G Laughlin, M Rozyczka and H W Yorker

Series in Astronomy and Astrophysics

The Physics of Interstellar Dust

Endrik Krügel

Max-Planck-Institut für Radioastronomie,
Bonn, Germany

Institute of Physics Publishing
Bristol and Philadelphia

© IOP Publishing Ltd 2003

British Library Cataloguing-in-Publication Data

A catalogue record for this book is available from the British Library.

ISBN 0 7503 0861 3

Library of Congress Cataloging-in-Publication Data are available

Series Editors: **M Birkinshaw**, University of Bristol, UK
M Elvis, Harvard–Smithsonian Center for Astrophysics, USA
J Silk, University of Oxford, UK

Commissioning Editor: John Navas
Production Editor: Simon Laurenson
Production Control: Sarah Plenty
Cover Design: Victoria Le Billon
Marketing: Nicola Newey and Verity Cooke

Published by Institute of Physics Publishing, wholly owned by The Institute of Physics, London

Institute of Physics Publishing, Dirac House, Temple Back, Bristol BS1 6BE, UK

US Office: Institute of Physics Publishing, The Public Ledger Building, Suite 929, 150 South Independence Mall West, Philadelphia, PA 19106, USA

Typeset in LaTeX 2_ε by Text 2 Text, Torquay, Devon
Printed in the UK by MPG Books Ltd, Bodmin, Cornwall

Für meine Frau

Contents

Preface

Dear reader

Before you is a compilation of lectures held at the University of Bonn all revolving around interstellar dust and the formation of stars.

From lecture notes to print

The incentive to turn my scribbled lecture notes into a book was twofold: the desire to reach a larger audience, and the wish to hand students a more polished and lasting description of the tools for future work. Lecture and written text, even when covering the same topic, should not be identical in their contents, but complementary: they are two independent didactical challenges. In a lecture, the student should be able to follow from beginning to end. The speaker stresses ideas and concepts and does not waste time in elaborating lengthy formulae. A good lecturer may be likened to a salesman at the front door. He is aggressive, his arguments are compelling and what he says sounds exciting which prevents us from slamming the door in his face.

A serious writer, however, can convince only by more subtle tones, most of all through thoroughness. He is like the unobtrusive shopkeeper whom we have been visiting for years. We know we can trust his goods, although he himself may be a bit boring. Whereas an opinion about a lecture is formed quickly and is not likely to change afterwards, we esteem a book only at second sight. Not every chapter has to be grasped at first reading. Instead, there is opportunity to contemplate a figure, formula or paragraph at leisure, over a steaming pot of tea or the curly smoke rings of a pipe.

The topic

The central theme of this book is cosmic dust. Its relevance for astronomy and for the evolution of the cosmos is not obvious. Unless we use special equipment, more sophisticated than binoculars, it does not catch our attention as does a variable star, a comet or a globular cluster. Dust only screens the light at optical wavelengths. Its constituents, the grains, are disappointingly small and would barely be visible under a microscope. The total dust mass in the Milky Way is negligible compared to that of the stars or even the interstellar gas. However, we believe that man is made of this very dust and that, by itself, is reason to study it in depth.

But there is more to it. Interstellar dust is not an isolated component of the universe, like pulsars or white dwarfs, which could be removed and everything else would stay the same. Instead, it is in intimate contact with the rest of the world, giving and taking, and this is best exemplified by the influence it has on star formation and on the appearance of young stars and galaxies.

The addressee

This text was conceived for students who have received an elementary but comprehensive introduction to physics—this is usually the case after two years of university studies—and who have taken a general course in astronomy. It is also aimed at PhD students who are starting research and have come across interstellar dust in one of its many manifestations. Hopefully, this book will also be of service to astronomers in general.

I admit that it contains hardly any exciting new results; not because a book is never fresh, nor for fear that excitement might be detrimental to the heart. Instead, the goal was to supply the student with those basic facts about small solid particles that passed the test of time. Only being acquainted with the old results can one fully enjoy the new. As many of the basic facts are scattered over the literature and are sometimes hard to dig up, a selected compilation was thought to be useful.

Another reason to concentrate on matters where there is consensus and to avoid being specific or touching upon controversial topics lies in the very nature of the dust itself. Hardly any two dust grains in the universe are alike and this immense diversity explains, to a large degree, why all numbers about interstellar dust are vague and insecure. When an astronomical number is certain, say, the mass of a planet or the distance to a star, one can happily apply it in further work without worrying how it was derived. But when the number is ill determined, one should know the physical and technical pillars upon which its derivation rests. Only then can one estimate how far it may, or should, be stretched, or come up with a new number, physically founded and adapted to the particular problem.

Astronomy is a branch of physics

This is a provocative statement and may arouse indignation. As if I had forgotten how the great discoveries of the past 30 years have come about: As a result of revolutionary technologies and grand enterprises! Indeed, when one recalls how astronomical satellites have widened our outlook on the universe, it seems justified to consider astronomy a branch of Space Project Management, and when one thinks of the progress achieved by new telescopes, astronomy appears as a subfield of Telescope Engineering or Receiver Development. It was new-technology instruments that have allowed us to peep into hitherto hidden realms. Even ADM (Advanced Data Manipulation) may be more important to astronomy than physics in view of the gigantic quantities of data that have to be crunched and transformed into convincing numbers and pictures.

So I freely acknowledge the priority of management and technology over physics. If one were to reduce the physics in astronomy courses to a minimum (one cannot do entirely without it) and teach instead the fields mentioned earlier, astronomy would continue to thrive for a decade or two, if one includes Science Marketing, even for three. Despite all this, out of sheer pleasure for the subject, this book stresses the link between astronomy and physics. It attempts to summarize the major physical topics with direct application to interstellar grains and wishes to encourage students to try the physical approach to an astronomical problem, without polemizing against higher resolution or higher sensitivity.

The language

It is obviously English. The obvious needs no words but there are lamentable aspects about using the modern *lingua franca*. I consider it a trifle that no sentence came easy. Indeed, it did me good learning some more of a foreign language while composing the text. Nor do I mind that one suspects behind simple phrases a simple mind, this supposition may be true.

A serious argument against writing in a tongue one has not fully mastered is that style and clarity are akin because improving the style usually means improving the thought, nothing else. After all, a textbook on physical sciences is not a railway timetable. A poignant style enhances the understanding, helps memorize and carries the reader over difficult stretches. *Ach*, in this respect, German would have been beneficial to the reader.

More important still is the obligation to preserve and develop one's language as an inherited gift and an attribute of culture of no less import than the collection of national wines. As English has become so pervasive in our daily scientific work, we, the majority of astronomers, tend to forget technical terms in our mother tongue or do not update them and this has the deplorable consequence that we speak and write about our favourite subject clumsily in two languages: in English and in our own.

But the strongest point in a plea to retain in science one's mother tongue in all its might, parallel to the *lingua franca*, is that each language imprints on the mind its own pattern of thinking. Pondering a problem in different languages means approaching it on different paths, and each path offers its specific outlook. It is erroneous to think that the findings of natural sciences are fully expressed in numbers or formulae. Words are needed, too. A formula lacks cross relations and does not sufficiently take into account the analogous character of what it asserts. For example, I solve equations containing time but do not very well know what time is. If words are needed to explain a formula, how many more are required to arrive at it? What would quantum mechanics be if it were reduced to equations without extensive accompanying text? Who would shorten R Feynman's *Lectures on Physics*? They are the work of a genius not because of the formulae, they are the usual ones but because of the way the story is *told*. And a successful struggle with an astronomical problem also needs a vivid, precise and powerful language to put all its facets into a fruitful perspective.

To whom I am indebted

I owe to my colleagues who bore with me, helped with their expertise and advice and encouraged me, in particular: *David Graham, Michaela Kraus, Antonella Natta, Richard Porcas, Johannes Schmid-Burgk* and *Alexandr Tutukov*. I am grateful to those who undertook the pains of critically reading parts of the manuscript: *Christian Henkel, Aigen Li, Armin Kirfel, Ralf Siebenmorgen, Werner Tscharnuter, Nikolaj Voshchinnikov, Malcolm Walmsley* and *Jan Martin Winters*.

Two books served as guides (*Vorbilder*) which I tried to follow, without pretending to match them. Each has, to my mind, one outstanding merit: L Spitzer's *Diffuse Matter in Space* is of dazzling perfection. It has been on my desk for decades (and I am still struggling with parts of it today). M Harwit pioneered in his *Astrophysical Concepts* to teach astronomy anew, with the eyes of a physicist, addressing the student and enlightening the professor.

The philosophical headline

A long scientific text is frequently preceded, one might even say embellished, by words from an authority outside the field, such as a philosopher or a poet. Although far from being an expert in the scientific subject itself, his words carry weight because they shed light on the topic from a different angle of cognition and reassure the natural scientist in his moments of doubt. I wish to follow this custom.

Dabbling in poetry and philosophical treatises, I found numerous aphorisms suitable for such a purpose but the most appropriate headlines for this book came to me as a birthday gift from my daughters. It is the following verse by the 19th century North-American poet Walt Whitman which they had calligraphically written onto cardboard. Here is what Whitman left us:

> *When I heard the learn'd astronomer,*
> *When the proofs, the figures, were ranged in columns before me,*
> *When I was shown the charts and diagrams, to add, divide, and measure*
> *   them,*
> *When I sitting heard the astronomer where he lectured with much*
> *   applause in the lecture-room,*
> *How soon unaccountable I became tired and sick,*
> *Till rising and gliding out I wandered off by myself*

In the mystical moist night-air, and from time to time,
Looked up in perfect silence at the stars.

Of course, any literary praise of these lines from my side is out of place, being a layman in literature. So I will not say a word about the magic beat that pervades the poem: How the rhythm starts from impatience, condenses into anger and transforms into serenity. I will not admire how irresistibly Whitman conjures the lure of the night sky and contrasts it to the unnerving ambition of scholars. Nor will I marvel at his prophetic power to foresee and congenially describe the feelings of a backbencher at an astronomical meeting more than a century after his time.

The reason for picking this poem as the philosophical headline is that it pays a wise tribute to the irrational. Reflected or not, irrationality, like *the mystical moist night-air*, is at the root of any sincere endeavour, including the quest of an astronomer to understand the cosmos. Some colleagues strongly disagree and regard with contempt those who let themselves be charmed by such a poem. I take their objections very serious but find the occasional vehemence of their arguments soothing, corroborating, at least, that they are not moved by logic and astronomical data alone.

At the end of this longish foreword, a line comes to mind by F M Dostojevskji from his novel *The Demons*. At a benefit party, Stepan Verchovenskji, the aging hero of the narrative, makes an ambitious opening speech which Dostojevskji laconically summarizes by the words

Но все бы это ничего, и кто не знает авторских предисловий?

After intensive consultations with linguists and psychologists, I venture in the present context the translation: *Hmm well, well hmm!*

Let this be the concluding remark.

Yours sincerely
EK

Bonn

Easter 2002

Chapter 1

The dielectric permeability

We begin by acquainting ourselves with the polarization of matter. The fundamental quantity describing how an interstellar grain responds to an electromagnetic wave is the dielectric permeability which relates the polarization of matter to the applied field. We recall the basic equations of electrodynamics and outline how plane waves travel in an infinite non-conducting (dielectric) medium and in a plasma. We summarize the properties of harmonic oscillators, including the absorption, scattering and emission of light by individual dipoles. Approximating a solid body by an ensemble of such dipoles (identical harmonic oscillators), we learn how its dielectric permeability changes with frequency. This study is carried out for three scenarios:

- a dielectric medium where the electron clouds oscillate about the atomic nuclei;
- a dielectric medium where the charge distribution in the atomic dipoles is fixed but the dipoles themselves may rotate; and
- a metal where the electrons are free.

1.1 Maxwell's equations

At the root of all phenomena of classical electrodynamics, such as the interaction of light with interstellar dust, are Maxwell's formulae. They can be written in different ways and the symbols, their names and meaning are not universal, far from it. Before we exploit Maxwell's equations, we, therefore, first define the quantities which describe the electromagnetic field.

1.1.1 Electric field and magnetic induction

A charge q traveling with velocity $\mathbf{v}$ in a fixed electric field $\mathbf{E}$ and a fixed magnetic field of flux density $\mathbf{B}$ experiences a force

$$\mathbf{F} = q \left[\mathbf{E} + \frac{1}{c} \mathbf{v} \times \mathbf{B} \right] \tag{1.1}$$

1

called the *Lorentz* force; the cross $\times$ denotes the vector product. **B** is also called the magnetic induction. Equation (1.1) shows what happens mechanically to a charge in an electromagnetic field and we use it to define **E** and **B**.

The force **F** has an electric part, $q\mathbf{E}$, which pulls a positive charge in the direction of **E**, and a magnetic component, $(q/c)\mathbf{v} \times \mathbf{B}$, working perpendicular to **v** and **B**. When the moving charges are electrons in an atom driven by an electromagnetic wave, their velocities are small. A typical value is the velocity of the electron in the ground state of the hydrogen atom for which classically v/c equals the fine structure constant

$$\alpha = \frac{e^2}{\hbar c} \simeq \frac{1}{137} \ll 1.$$

Protons move still more slowly because of their greater inertia. For an electromagnetic wave in vacuum, $|\mathbf{E}| = |\mathbf{B}|$ and, therefore, the term $(\mathbf{v} \times \mathbf{B})/c$ in formula (1.1) is much smaller than **E** and usually irrelevant for the motion of the charge.

1.1.2 Electric polarization of the medium

1.1.2.1 *Dielectric permeability and electric susceptibility*

In an electrically neutral medium, the integral of the local charge density $\rho(\mathbf{x})$ at locus **x** over any volume V vanishes,

$$\int_V \rho(\mathbf{x})\, dV = 0.$$

However, a neutral body of volume V may have a dipole moment **p** (small letter) given by

$$\mathbf{p} = \int_V \mathbf{x}\rho(\mathbf{x})\, dV. \tag{1.2}$$

The dipole moment per unit volume is called the polarization **P** (capital letter),

$$\mathbf{P} = \frac{\mathbf{p}}{V}. \tag{1.3}$$

It may be interpreted as the number density N of molecules multiplied by the average electric dipole moment $\mathbf{p}_{\mathrm{mol}}$ per molecule:

$$\mathbf{P} = N\mathbf{p}_{\mathrm{mol}}.$$

E and **P** being defined, we introduce as an additional quantity the displacement **D**. It adds nothing new and is just another way of expressing the polarization of matter, **P**, through an electric field **E**:

$$\mathbf{D} = \mathbf{E} + 4\pi\mathbf{P}. \tag{1.4}$$

The local polarization **P** and the local average electric field **E** in the dielectric medium are much weaker than the fields on the atomic level and constitute just a small perturbation. Therefore, **P**, **D** and **E** are proportional to each other and we can write

$$\mathbf{D} = \varepsilon \mathbf{E} \tag{1.5}$$
$$\mathbf{P} = \chi \mathbf{E} \tag{1.6}$$

with

$$\chi = \frac{\varepsilon - 1}{4\pi}. \tag{1.7}$$

The proportionality factors ε and χ are material constants, they mean more or less the same; ε is called the *dielectric permeability* or *dielectric constant* or *permittivity*, χ bears the name *electric susceptibility*. In the trivial case of a vacuum, the polarization **P** vanishes, **E** = **D** and $\varepsilon = 1$. In a dielectric medium, however, a constant field **E** induces a dipole moment so that $\varepsilon > 1$ and $\mathbf{P} \neq 0$.

We have, altogether, three quantities describing the electric field: **D**, **E** and **P**, although two would suffice. For example, we could replace **D** in all equations by $\varepsilon\mathbf{E}$. Equation (1.5) is the first of two constitutive relations complementing the set of Maxwell's formulae.

1.1.2.2 The electric polarizability

Another quantity we will need is the *electric polarizability* α_e. If we place a small grain of volume V into a constant electric field **E**, it acquires a dipole moment

$$\mathbf{p} = \alpha_e V \mathbf{E}. \tag{1.8}$$

Because of the dipole moment, the field in the vicinity of the grain becomes distorted. It no has longer the value **E** but some other value, say $\mathbf{E}^e$ (*e* for *e*xternal). As one recedes from the grain, $\mathbf{E}^e$ approaches **E** asymptotically. There is also a field $\mathbf{E}^i$ *i*nside the grain. This differs from the constant outer field **E**. The relation between them is described in detail in section 3.4. The polarization **P** (capital letter) depends linearly both on $\mathbf{E}^i$ (from (1.6)) and on **E** (from (1.8)), the proportionality factors being α_e and χ, respectively,

$$\mathbf{P} = \chi \mathbf{E}^i = \alpha_e \mathbf{E}. \tag{1.9}$$

In the general case, α_e and χ are tensors, and the dipole moment and the fields do not have the same direction.

1.1.3 The dependence of the dielectric permeability on direction and frequency

Unless specified otherwise, we always assume (on a macroscopic level with dimensions much greater than an atom) homogeneity so that the dielectric permeability ε does not depend on the position in the grain. The relation (1.5), $\mathbf{D} = \varepsilon \mathbf{E}$, is linear, which means that ε is independent of the field strength. However, it may depend on the direction of the fields because, on a microscopic level, a grain is made up of atoms and is thus not homogeneous. Equation (1.5) must then be generalized to

$$D_i = \sum_i \varepsilon_{ik} E_i + D_{0i}.$$

The constant term D_{0i} implies a frozen-in dipole moment even in the absence of an outer field. This happens only in special cases known as pyroelectricity; for mono-crystals with a cubic structure (see section 7.1), D_{0i} is always zero. The tensor ε_{ik} is symmetric, $\varepsilon_{ik} = \varepsilon_{ki}$, and can always be brought into a diagonal form by the appropriate choice of coordinate system. It then has, at most, three independent components. If the three diagonal elements are equal, ε reduces to a scalar and the grain is said to be isotropic. Crystals of the cubic system have this pleasant property but in interstellar grains we will also encounter anisotropy.

It is an essential feature for a discussion of interstellar dust that ε is a function of frequency:

$$\omega = 2\pi \nu$$

or, as one says, that ε shows dispersion. The functional form of $\varepsilon(\omega)$ is called the dispersion relation and will be used extensively in what follows. So the term dielectric constant, which is the other frequently used name for $\varepsilon(\omega)$, is misleading when taken at its face value.

In electrostatics, only one parameter is needed to specify the polarization and χ or ε are real. However, in an alternating field, one needs for each frequency ω two independent numbers which, out of mathematical convenience, are written as *one* complex variable:

$$\chi(\omega) = \chi_1(\omega) + i\chi_2(\omega) \tag{1.10}$$
$$\varepsilon(\omega) = \varepsilon_1(\omega) + i\varepsilon_2(\omega). \tag{1.11}$$

1.1.4 The physical meaning of the electric susceptibility χ

We explain the physical meaning of χ in the case of a harmonically and slowly varying field,

$$E = E_0 e^{-i\omega t}.$$

We use scalar notation for the moment and let E_0 be real. The vector representing the complex field E then rotates in the complex plane clockwise, as depicted in figure 1.1. For a slowly varying field with small ω, the susceptibility χ is close to its static value:

$$\chi_1 \simeq \chi(0) \qquad \chi_2 \simeq \chi_2(0) = 0.$$

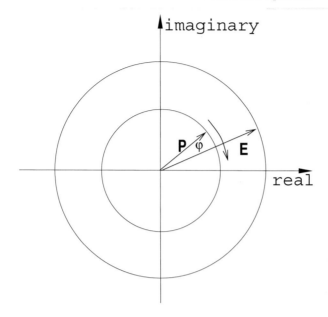

Figure 1.1. The electric field vector, **E**, in the medium is presented as a complex quantity of length E, which rotates at frequency ω. It induces a complex polarization **P** in the material of length P. When damping is weak, **P** lags behind **E** by a small angle φ. The real part of the complex electric susceptibility is equal to the ratio of the radii of the two circles, $\chi_1 = P/E$, whereas the imaginary part gives the phase lag, $\chi_2 = \chi_1 \varphi$. Only the real components of **P** and **E** have a physical meaning, the complex representation is for mathematical convenience.

The static value $\chi_1(0)$ is positive because the polarization must point in the direction of the electric field. Obviously the real part χ_1 determines the maximum polarization during one cycle via the relation $P_{max} \simeq \chi_1 E_0$.

As the field changes, E and P are not in phase because the electrons always need a little while to adjust. The polarization P, therefore, lags behind by some time $\Delta t \ll \omega^{-1}$ corresponding to a small angle

$$\varphi = \omega \Delta t = \tan(\chi_2/\chi_1) \simeq \chi_2/\chi_1 \ll 1. \qquad (1.12)$$

The interval Δt must be positive for reasons of causality: sensible people squeal only *after* they have been pinched. Therefore, the imaginary part χ_2 is positive and determines, for a given χ_1, the time lag. While adapting to the field, the electrons have to overcome internal friction, which implies dissipational losses. Some of the field energy is inevitably drained into the dielectric medium. If χ_2 and Δt were negative, the field would draw energy from the dielectric body, decreasing its entropy which contradicts the second law of thermodynamics. Even $\chi_2 = 0$ is impossible as it would mean that E and P are absolutely synchronous and dissipation is completely absent.

If the frequency ω is not small but arbitrary, the angle $\omega \Delta t$ is also not necessarily small. Nevertheless, we can exclude $\chi_2 \leq 0$ for the same reasons as before. The real part of the susceptibility, χ_1, however, may become zero as well as negative. The maximum of P is now given by $P_{\max} = \sqrt{\chi_1^2 + \chi_2^2} E_0$.

Of course, we could also choose a time dependence $e^{i\omega t}$ for the field and nothing would change. The sign of i is just a convention, although one that must be followed consistently. When $E = E_0 e^{i\omega t}$, the field rotates anticlockwise in figure 1.1 and the susceptibility, instead of (1.10), would be $\chi(\omega) = \chi_1(\omega) - i\chi_2(\omega)$, again with positive χ_2.

1.1.5   Magnetic polarization of the medium

In a macroscopic picture, a stationary motion of charges with density ρ and velocity $\mathbf{v}$, corresponding to a current density $\mathbf{J} = \rho\mathbf{v}$, produces a magnetic moment

$$\mathbf{m} = \frac{1}{2c} \int (\mathbf{x} \times \mathbf{J}) \, dV. \qquad (1.13)$$

When the integral extends over a unit volume, it is called the magnetization $\mathbf{M}$,

$$\mathbf{M} = \frac{\mathbf{m}}{V}. \qquad (1.14)$$

For example, a charge q traveling with velocity v in a circular orbit of radius r constitutes a current $I = qv/2\pi r$. The magnetic moment of this moving charge is

$$m = \frac{qvr}{2c} = \frac{AI}{c}$$

where $A = \pi r^2$ signifies the area of the loop. If A is small, the accompanying field is that of a dipole.

Without the motion of the macroscopic charge, the magnetization of matter comes, in the classical picture, from the atomic currents (electron orbits). The magnetic moment per unit volume $\mathbf{M}$ may be interpreted as the number density N of molecules in the substance multiplied by the total magnetic moment $\mathbf{m}_{\mathrm{mol}}$ per molecule:

$$\mathbf{M} = N\mathbf{m}_{\mathrm{mol}}.$$

The magnetic field $\mathbf{H}$ is defined by

$$\mathbf{H} = \mathbf{B} - 4\pi\mathbf{M}. \qquad (1.15)$$

Similar to the situation of the electric field, we could replace $\mathbf{H}$ by $\mathbf{B}$ in all equations and retain two field quantities, $\mathbf{B}$ and $\mathbf{M}$, instead of three. We would prefer the pairs $(\mathbf{E}, \mathbf{P})$ and $(\mathbf{B}, \mathbf{M})$, because the electric field $\mathbf{E}$, the magnetic flux density $\mathbf{B}$ as well as the polarizations $\mathbf{P}$ and $\mathbf{M}$ allow a direct physical interpretation; however, conventions urge us to drag $\mathbf{D}$ and $\mathbf{H}$ along.

1.1.6 The magnetic susceptibility

The second constitutive equation concerns the magnetic field **H** and the induction **B**. If the substance is not ferromagnetic, a linear relation holds as in (1.5):

$$\mathbf{B} = \mu\mathbf{H}. \tag{1.16}$$

μ is the magnetic permeability and also generally complex,

$$\mu = \mu_1 + i\mu_2.$$

It also shows dispersion. In vacuo, $\mathbf{B} = \mathbf{H}$ and $\mu = 1$. For diamagnetic substances, μ is a little smaller than one, for paramagnetic substances a little bigger, μ is large only for ferromagnets. Similar to (1.7), we define the magnetic susceptibility by

$$\chi = \frac{\mu - 1}{4\pi} \tag{1.17}$$

so that $\mathbf{M} = \chi\mathbf{H}$. If there is any chance of confusion, one must explicitly write χ_m or χ_e for the magnetic or electric case. In analogy to (1.8), a constant *outer* magnetic field **H** induces in a small body of volume V a magnetic dipole moment

$$\mathbf{m} = V\mathbf{M} = \alpha_m V\mathbf{H} \tag{1.18}$$

where α_m is the magnetic polarizability and **M** the magnetization *within* the body. For interstellar grains, which are weakly magnetic substances with $|\mu|$ close to one, the magnetic induction and magnetic field are practically the same inside and outside, and $\chi \simeq \alpha$. If **B** denotes the interstellar magnetic field and **M** the magnetization of the grain, one usually writes $\mathbf{M} = \chi\mathbf{B}$.

1.1.7 Dielectrics and metals

We will be dealing with two kinds of substances for which Maxwell's equations are often formulated separately:

- *Dielectrics.* These are insulators and no constant current can be sustained within them. Nevertheless, alternating currents produced by a time-variable electric field are possible. In these currents, the charges do not travel far from their equilibrium positions.
- *Metals.* This is a synonym for conductors and, in this sense, they are the opposite of dielectrics. When a piece of metal is connected at its ends to the poles of a battery, a steady current flows under the influence of an electric field. When this piece of metal is placed in a static electric field, the charges accumulate at its surface and arrange themselves in such a way that the electric field inside vanishes and then there is no internal current. However, time-varying electric fields and currents are possible.

In the interstellar medium, one finds both dielectric and metallic particles but the latter are probably far from being perfect conductors.

1.1.8 Free charges and polarization charges

We are generally interested in the electromagnetic field averaged over regions containing many atoms. So we usually picture the grain material to be a continuous medium. However, sometimes we have to dive into the atomic world. On a scale of 1 Å or less, the medium becomes structured. Atoms appear with positive nuclei and negative electrons spinning and moving in complicated orbits and the electric and magnetic fields, which are smooth on a macroscopic scale, have huge gradients.

Consider a microscopic volume δV in some piece of matter. It contains protons and electrons and the net charge divided by δV gives the total local charge density ρ_{tot}. For an electrically neutral body, electrons and protons exactly cancel and ρ_{tot} integrated over the body is zero. Often ρ_{tot} is written as the sum of two terms:

$$\rho_{\text{tot}} = \rho_{\text{pol}} + \rho_{\text{free}}. \tag{1.19}$$

The first arises from polarization. In a dielectric medium, the electric field polarizes the atoms and thus produces aligned dipoles. If the polarization within the particle is uniform, ρ_{pol} is zero because the positive and negative charges of the neighboring dipoles exactly balance; the only exception is the surface of the particle where charges of one kind accumulate. However, if the polarization $\mathbf{P}$ is spatially non-uniform, ρ_{pol} does not vanish. Then the separation of charges is inhomogeneous and leads to a charge density

$$\rho_{\text{pol}} = - \operatorname{div} \mathbf{P}. \tag{1.20}$$

The second term in (1.19), ρ_{free}, comprises all other charges besides ρ_{pol}. For example, when the polarization is constant, $\operatorname{div} \mathbf{P} = 0$ and ρ_{free} stands for *all* charges, positive and negative. In an uncharged body, their sum is zero. When charges are brought in from outside, $\rho_{\text{free}} \neq 0$. Within a metal, $\mathbf{P} = 0$, and polarization charges can only appear on the surface.

Moving charges constitute a current. The total current density $\mathbf{J}_{\text{tot}}$ may also be split into parts similarly to (1.19),

$$\mathbf{J}_{\text{tot}} = \mathbf{J}_{\text{free}} + \mathbf{J}_{\text{pol}} + \mathbf{J}_{\text{mag}}. \tag{1.21}$$

The first term on the right-hand side is associated with the motion of the free charges, the second with the time-varying polarization of the medium and the third represents the current that gives rise to its magnetization:

$$\mathbf{J}_{\text{free}} = \mathbf{v} \rho_{\text{free}} \qquad \mathbf{J}_{\text{pol}} = \dot{\mathbf{P}} \qquad \mathbf{J}_{\text{mag}} = c \cdot \operatorname{rot} \mathbf{M}. \tag{1.22}$$

1.1.8.1 Sign conventions

We have to make a remark about the sign convention. An electron, of course, has a negative charge, its absolute value being $e = 4.803 \times 10^{-10}$ esu (electrostatic units). After Maxwell's equation (1.27), the electric field of an isolated positive charge (proton) is directed away from it and repels other positive charges according to (1.1). The moment **p** of a dipole created by two opposite but equal charges is, from (1.2), directed from the minus to the plus charge and anti-parallel to the electric field along the line connecting the two charges. Hence a dielectric grain in a constant field **E** has a polarization **P** parallel to **E** and surface charges as depicted in figure 3.4 of chapter 3. With the help of this figure, we can explain the minus sign in equation (1.20). Going from left to right, the polarization jumps at the left edge of the grain from zero to its value inside. The gradient there is positive and thus a negative charge appears on the surface. In the case of non-uniform polarization within the grain, at a place where div **P** > 0, the electric field pulls a small charge δq out of a tiny volume δV leaving behind an unbalanced negative charge $-\delta q$.

1.1.9   The field equations

We have now defined all quantities that appear in Maxwell's equations and we write them down, first, for a neutral dielectric medium:

$$\operatorname{div} \mathbf{D} = 0 \tag{1.23}$$

$$\operatorname{div} \mathbf{B} = 0 \tag{1.24}$$

$$\operatorname{rot} \mathbf{E} = -\frac{1}{c}\dot{\mathbf{B}} \tag{1.25}$$

$$\operatorname{rot} \mathbf{H} = \frac{1}{c}\dot{\mathbf{D}} \tag{1.26}$$

second, for a medium with free charges and currents:

$$\operatorname{div} \mathbf{D} = 4\pi \rho_{\text{free}} \tag{1.27}$$

$$\operatorname{div} \mathbf{B} = 0 \tag{1.28}$$

$$\operatorname{rot} \mathbf{E} = -\frac{1}{c}\dot{\mathbf{B}} \tag{1.29}$$

$$\operatorname{rot} \mathbf{H} = \frac{1}{c}\dot{\mathbf{D}} + \frac{4\pi}{c}\mathbf{J}_{\text{free}}. \tag{1.30}$$

A dot above a letter stands for partial time derivative, for instance, $\dot{\mathbf{B}} = \partial\mathbf{B}/\partial t$. Applying the operator div to (1.30) gives the expression for charge conservation:

$$\dot{\rho}_{\text{free}} + \operatorname{div} \mathbf{J}_{\text{free}} = 0. \tag{1.31}$$

E denotes the electric and **H** the magnetic field, **D** the electric displacement, **B** the magnetic induction, **J** the current density and c, of course, the velocity of light. Our choice of mathematical symbols is summarized in appendix A together with some common relations of vector analysis.

1.2   Waves in a dielectric medium

We derive from Maxwell's equations how a plane wave propagates in an infinite medium, define the optical constant, or refractive index and recall a few formulae concerning energy, flux and momentum of the electromagnetic field.

1.2.1   The wave equation

In harmonic fields, which have a sinusoidal time dependence proportional to $e^{-i\omega t}$, Maxwell's equations (1.25) and (1.26) become simpler:

$$\text{rot } \mathbf{E} = i\frac{\omega\mu}{c}\mathbf{H} \tag{1.32}$$

$$\text{rot } \mathbf{H} = -i\frac{\omega\varepsilon}{c}\mathbf{E}. \tag{1.33}$$

Applying the operator rot to (1.32) and (1.33), we arrive, with the help of formula (A.6) of vector analysis, at

$$\Delta\mathbf{E} + \frac{\omega^2}{c^2}\mu\varepsilon\mathbf{E} = 0 \qquad \Delta\mathbf{H} + \frac{\omega^2}{c^2}\mu\varepsilon\mathbf{H} = 0 \tag{1.34}$$

or

$$\Delta\mathbf{E} - \frac{\mu\varepsilon}{c^2}\ddot{\mathbf{E}} = 0 \qquad \Delta\mathbf{H} - \frac{\mu\varepsilon}{c^2}\ddot{\mathbf{H}} = 0. \tag{1.35}$$

These are the wave equations which describe the change of the electromagnetic field in space and time. In an infinite medium, one solution to (1.34) is a plane harmonic wave,

$$\mathbf{E}(\mathbf{x}, t) = \mathbf{E}_0 \cdot e^{i(\mathbf{k}\cdot\mathbf{x}-\omega t)} \tag{1.36}$$

$$\mathbf{H}(\mathbf{x}, t) = \mathbf{H}_0 \cdot e^{i(\mathbf{k}\cdot\mathbf{x}-\omega t)} \tag{1.37}$$

where $\mathbf{k}$ is a vector with $\mathbf{k}^2 = \omega^2\mu\varepsilon/c^2$. The characteristics of a plane harmonic wave are an $e^{i\mathbf{k}\cdot\mathbf{x}}$ space variation and an $e^{-i\omega t}$ time variation. All waves in interstellar space that interact with interstellar matter are planar because the sources from which they arise are very distant.

1.2.1.1   *Flux and momentum of the electromagnetic field*

An electromagnetic wave carries energy. The flux, which is the energy per unit time and area, is given by the Poynting vector $\mathbf{S}$. For *real* fields, its *momentary* value is

$$\mathbf{S} = \frac{c}{4\pi}\mathbf{E} \times \mathbf{H}. \tag{1.38}$$

In a vacuum, when E_0 denotes the amplitude of the electric field vector, the *time average* is

$$\langle S \rangle = \frac{c}{8\pi}E_0^2. \tag{1.39}$$

When we use *complex* fields, the time average of the Poynting vector is a real quantity and, from (A.33), is given by

$$\langle \mathbf{S} \rangle = \frac{c}{8\pi} \, \text{Re}\{\mathbf{E} \times \mathbf{H}^*\}. \tag{1.40}$$

$\mathbf{H}^*$ is the complex conjugate of $\mathbf{H}$. The wave also transports momentum, through a unit area at a rate $\mathbf{S}/c$. In the corpuscular picture, an individual photon travels at the speed of light, has an energy $h\nu$, a momentum $h\nu/c$ and an angular momentum $\hbar$.

1.2.2 The wavenumber

The vector $\mathbf{k} = (k_x, k_y, k_z)$ in (1.36) is called the wavenumber and is, in the most general case, complex:

$$\mathbf{k} = \mathbf{k}_1 + i\mathbf{k}_2$$

with real $\mathbf{k}_1$, $\mathbf{k}_2$ and obeys the relation

$$\mathbf{k}^2 = \mathbf{k}_1^2 - \mathbf{k}_2^2 + 2i\mathbf{k}_1 \cdot \mathbf{k}_2 = \frac{\omega^2}{c^2}\mu\varepsilon. \tag{1.41}$$

Of course, $\mathbf{k}^2 = \mathbf{k} \cdot \mathbf{k} = k_x^2 + k_y^2 + k_z^2$. Inserting the field of a plane wave given by (1.36), (1.37) into equations (1.32) and (1.33) yields

$$\frac{\omega\mu}{c}\mathbf{H} = \mathbf{k} \times \mathbf{E} \qquad \text{and} \qquad \frac{\omega\varepsilon}{c}\mathbf{E} = -\mathbf{k} \times \mathbf{H} \tag{1.42}$$

and after scalar multiplication with $\mathbf{k}$

$$\mathbf{k} \cdot \mathbf{E} = \mathbf{k} \cdot \mathbf{H} = 0. \tag{1.43}$$

- First, we consider the standard case in which $\mathbf{k}$ is real. Its magnitude is then

$$k = \frac{\omega}{c}\sqrt{\varepsilon\mu}$$

and the imaginary parts ε_2 and μ_2, which are responsible for the dissipation of energy, must be zero. This is strictly possible only in a vacuum. Any other medium is never fully transparent and an electromagnetic wave always suffers some losses. As $\mathbf{k}$ specifies the direction of wave propagation, it follows from (1.42) and (1.43) that the vectors of the electric and magnetic field are perpendicular to each other and to $\mathbf{k}$. The wave is planar and travels with undiminished amplitude at a phase velocity

$$v_{\text{ph}} = \frac{\omega}{k} = \frac{c}{\sqrt{\varepsilon\mu}}. \tag{1.44}$$

The real amplitude of the magnetic and electric field, E_0 and H_0, are related through

$$H_0 = \sqrt{\frac{\varepsilon}{\mu}} E_0. \qquad (1.45)$$

In particular, in a vacuum

$$v_{\text{ph}} = c \qquad H_0 = E_0 \qquad k = \frac{\omega}{c} = \frac{2\pi}{\lambda}.$$

- Next, let **k** be complex. Then ε_2 and μ_2 are complex, too, and the electric field vector equals

$$\mathbf{E}(\mathbf{x}, t) = \mathbf{E}_0 \cdot e^{-\mathbf{k}_2 \cdot \mathbf{x}} \cdot e^{i(\mathbf{k}_1 \cdot \mathbf{x} - \omega t)}.$$

As the wave propagates, light is absorbed and the wave amplitude suffers damping proportional to $e^{-\mathbf{k}_2 \cdot \mathbf{x}}$. The phase of the wave is constant in a plane perpendicular to $\mathbf{k}_1$ and the amplitude is constant in a plane perpendicular to $\mathbf{k}_2$.

If the vectors $\mathbf{k}_1$ and $\mathbf{k}_2$ are parallel, the planes of constant phase and constant amplitude coincide. There is a unit vector **e** parallel to **k** defining the direction of wave propagation such that $\mathbf{k} = (k_1 + ik_2) \cdot \mathbf{e}$, where k_1, k_2 are the lengths of $\mathbf{k}_1$, $\mathbf{k}_2$. Such a wave is called homogeneous.

If $\mathbf{k}_1$ and $\mathbf{k}_2$ are not parallel, one speaks of an inhomogeneous wave and the surface where the field is constant is not a plane. The geometrical relations (1.42), (1.43) then lose their obvious interpretation.

1.2.3   The optical constant or refractive index

For the propagation of light, the electromagnetic properties of matter may be described either by the wavevector (from (1.41)) or by the optical constant

$$m = \sqrt{\varepsilon \mu}. \qquad (1.46)$$

m is also called the (complex) refractive index. When the medium is metallic and has a conductivity σ, one must use the dielectric permeability ε as defined in (1.115), rather than (1.5). It is clear from the definition of the optical constant (1.46) that m does not contain any new information, in fact less than all material constants ε, μ and σ taken together. The name *optical constant* is unfortunate as m is not constant but varies with frequency. It is a complex dimensionless number,

$$m(\omega) = n(\omega) + ik(\omega) \qquad (1.47)$$

with real part n and imaginary part k. In the common case of a non-magnetic medium, where

$$\mu = 1$$

the real and imaginary parts of the optical constant follow from ε_1 and ε_2,

$$n = \frac{1}{\sqrt{2}}\sqrt{\sqrt{\varepsilon_1^2 + \varepsilon_2^2} + \varepsilon_1} \qquad (1.48)$$

$$k = \frac{1}{\sqrt{2}}\sqrt{\sqrt{\varepsilon_1^2 + \varepsilon_2^2} - \varepsilon_1} \qquad (1.49)$$

and, vice versa,

$$\varepsilon_1 = n^2 - k^2 \qquad (1.50)$$

$$\varepsilon_2 = 2nk. \qquad (1.51)$$

(1.46) has two solutions and we pick the one with positive n and non-negative k[1].

1.2.4 Energy dissipation of a grain in a variable field

The total energy of an *electrostatic field* **E** in vacuum, produced by a fixed distribution of charges, is

$$U = \frac{1}{8\pi}\int \mathbf{E}\cdot\mathbf{E}\,dV$$

where the integration extends over all space. This formula follows readily from the potential energy between the charges, which is due to their Coulomb attraction. It suggests an energy density

$$u = \frac{1}{8\pi}\mathbf{E}\cdot\mathbf{E}. \qquad (1.52)$$

If the space between the charges is filled with a dielectric,

$$u = \frac{1}{8\pi}\mathbf{E}\cdot\mathbf{D}. \qquad (1.53)$$

When we compare (1.52) with (1.53), we find that the energy density of the dielectric medium contains an extra term $\frac{1}{2}\mathbf{E}\mathbf{P}$ which accounts for the work needed to produce in the medium a polarization **P**. Therefore, if we place, into a constant field in a vacuum, some small dielectric object of volume V, the total field energy changes by an amount

$$w = -\tfrac{1}{2}V\mathbf{P}\cdot\mathbf{E}. \qquad (1.54)$$

It becomes smaller because some work is expended on the polarization of the grain. If the outer field slowly oscillates, $\mathbf{E} = \mathbf{E}_0 e^{-i\omega t}$, so does the polarization and for small energy changes

$$dw = -\tfrac{1}{2}V\,d(\mathbf{P}\cdot\mathbf{E}) = -V\mathbf{P}\cdot d\mathbf{E}.$$

---

[1] In order to abide by the customary nomenclature, we use the same letter for the imaginary part of the optical constant (italic type k) and for the wavenumber (Roman type k).

The time derivative $\dot{w}$, which we denote by W, gives the dissipated power and its mean is

$$\langle W \rangle = -V \langle \mathbf{P} \dot{\mathbf{E}} \rangle = \tfrac{1}{2} V \omega \, \text{Im}\{\alpha_e\} |\mathbf{E}_0|^2 > 0. \tag{1.55}$$

It is positive and proportional to frequency ω and volume V, which makes sense. The bracket $\langle \ldots \rangle$ denotes the time average, and we have used $\mathbf{P} = \alpha_e \mathbf{E}$ from (1.8). Because of the mathematical relation (A.33), the real part of the dielectric polarizability $\text{Re}\{\alpha_e\}$ disappears in the product $\langle \mathbf{P} \dot{\mathbf{E}} \rangle$ confirming that only the imaginary part $\text{Im}\{\alpha_e\}$ is responsible for heating the grain.

Formally similar expressions but based on different physics, hold for magnetism. We first look for the formula of the magnetic energy density. Whereas static electric fields are produced by fixed charges and the total energy of the *electric* field is found by bringing the charges to infinity (which determines their potential energy), static magnetic fields are produced by constant currents. These currents also exert a force $\mathbf{F}$ on a charge but from (1.1) do no work because $\mathbf{F} \cdot \mathbf{v} = (q/c)\mathbf{v} \cdot (\mathbf{v} \times \mathbf{B}) = 0$. Therefore, in the case of a magnetic field, one has to evaluate its energy by *changing* $\mathbf{H}$. This produces according to (1.25) an electric field $\mathbf{E}$ and thus a loss rate $\mathbf{E} \cdot \mathbf{J}$. After some elementary vector analysis, one gets, for the total energy of the magnetostatic field,

$$U = \frac{1}{8\pi} \int \mathbf{H} \cdot \mathbf{B} \, dV$$

which implies an energy density of the magnetic field

$$u = \frac{1}{8\pi} \mathbf{H} \cdot \mathbf{B}. \tag{1.56}$$

In complete analogy to the electric field, a time-variable magnetic field, $\mathbf{H} = \mathbf{H}_0 e^{-i\omega t}$, leads in a small body of volume V and magnetic polarizability α_m (see (1.18)) to a heat dissipation rate

$$\langle W \rangle = \tfrac{1}{2} V \omega \, \text{Im}\{\alpha_m\} |\mathbf{H}_0|^2. \tag{1.57}$$

(1.55) and (1.57) are the basic equations for understanding the absorption of radiation by interstellar grains.

1.2.4.1   *The symmetry of the polarizability tensor*

We have already mentioned that, for an anisotropic substance, the dielectric permeability ε is a tensor and symmetric. The same applies to the polarizability. So instead of (1.9), we write more generally

$$P_i = \sum_j \alpha_{ij} E_j.$$

The symmetry of the tensor α_{ij} can be shown by computing after (1.54) the total energy W expended on polarizing a particle in a cycle consisting of four steps. At the beginning, the particle is unpolarized. In the first step, one applies an electric field in the x-direction, in the second step in the y-direction. Then one takes back the component E_x and finally the component E_y. If the cycle is loss-free, the condition $W = 0$ requires $\alpha_{xy} = \alpha_{yx}$. In a similar way, one finds $\alpha_{xz} = \alpha_{zx}$ and $\alpha_{yz} = \alpha_{zy}$.

1.3 The harmonic oscillator

In the early history of atomic theory, H A Lorentz applied the harmonic oscillator model to the motion of an electron in an atom. Despite its simplicity and the fact that electrons 'move' in reality in complicated paths, the Lorentz model is quite successful in a quantitative description of many phenomena that occur on the atomic level and reveal themselves in the macroscopic world. The oscillator concept is very fruitful, although a precise idea of what we mean by it is often missing. Usually we have in mind the electrons in an atom but occasionally the oppositely charged atomic nuclei of a crystal or some other kind of dipoles are meant. But generally we assume that a grain is built up of a system of oscillators. Using this concept, we later derive the dispersion relation $\varepsilon = \varepsilon(\omega)$ of the dielectric permeability around a resonance at frequency ω_0.

1.3.1 The Lorentz model

We imagine the following idealized situation: An electron of mass m_e and charge e is attached to a spring of force constant κ. A harmonic wave with electric field

$$E = E_0 e^{-i\omega t}$$

exerts a force $F = eE$ which causes the electron to move, say, in the x-direction. Its motion is governed by the equation

$$m_e\ddot{x} + b\dot{x} + \kappa x = F. \tag{1.58}$$

On the left-hand side, there is (a) an inertia term $m_e\ddot{x}$; (b) a frictional force, $-b\dot{x}$, which is proportional to velocity and leads to damping of the system unless it is powered from outside; and (c) a restoring force, $-\kappa x$, that grows linearly with the displacement from the equilibrium position. Putting

$$\omega_0^2 = \frac{\kappa}{m_e} \qquad \gamma = \frac{b}{m_e}$$

we get

$$\ddot{x} + \gamma\dot{x} + \omega_0^2 x = \frac{eE}{m_e}. \tag{1.59}$$

The oscillator has only three properties:

- charge-to-mass ratio e/m_e,
- damping constant γ and
- resonant frequency ω_0 (if the electron is unbound, $\omega_0 = 0$).

In the following, we discuss only the one-dimensional oscillator, where the electron moves along the x-axis but one may readily generalize this to three dimensions. It is assumed that the electric field is spatially constant over the displacement x of the electron. This is correct as long as the velocity v of the electron is small, $v \ll c$, because then $x \simeq v/\omega$ can be neglected in comparison to the wavelength c/ω of the field. Indeed, the velocity of an electron in an atom is typically of order $v/c \sim 1\%$.

1.3.2   Free oscillations

In the simplest case, when there is no friction ($\gamma = 0$) and no perturbation from outside ($E = 0$), (1.59) reduces to

$$\ddot{x} + \omega_0^2 x = 0 \tag{1.60}$$

and the electron oscillates harmonically forever at its natural frequency ω_0. If there is friction ($\gamma \neq 0$) but no external force,

$$\ddot{x} + \gamma\dot{x} + \omega_0^2 x = 0 \tag{1.61}$$

the solution is called a *transient*. It has the general form

$$x = e^{-\gamma t/2} \cdot (Ae^{-i\omega_\gamma t} + Be^{i\omega_\gamma t}) \tag{1.62}$$

where A and B are complex constants and

$$\omega_\gamma = \sqrt{\omega_0^2 - \gamma^2/4}.$$

For x to be real, it is required that B be the complex conjugate of A,

$$B = A^*$$

which means $\text{Re}\{A\} = \text{Re}\{B\}$ and $\text{Im}\{A\} = -\text{Im}\{B\}$. If $\gamma^2 < 4\omega_0^2$, i.e. when damping is weak, ω_γ is real. The electron then oscillates at a frequency ω_γ somewhat smaller than the natural frequency ω_0, with an exponentially decaying amplitude. If friction is strong, $\gamma^2 > 4\omega_0^2$ and ω_γ is imaginary. The amplitude then subsides exponentially without any oscillations. Critical damping occurs for

$$\gamma = 2\omega_0. \tag{1.63}$$

1.3.3 The general solution to the oscillator equation

In the most general case of equation (1.59), there is friction plus an external field that acts on the electron. It is now convenient to write the displacement and the electric field as complex variables; sometimes for clarity we mark the complexity of a quantity explicitly by a bar. So if we assume an harmonic field,

$$\bar{E} = \bar{E}_0\, e^{-i\omega t},$$

$\bar{E}$ and $\bar{E}_0$ are complex and

$$\ddot{\bar{x}} + \gamma\dot{\bar{x}} + \omega_0^2\bar{x} = \frac{e}{m_e}\bar{E}_0 e^{-i\omega t}. \tag{1.64}$$

One obtains the general solution of this inhomogeneous equation by adding to the general solution (1.62) of the associated homogeneous equation (1.61) one particular solution of (1.64), for example

$$\bar{x} = \bar{x}_0\, e^{-i\omega t}. \tag{1.65}$$

In such a sum, the transient (1.62) eventually dies out and only (1.65) remains. The electron then oscillates at frequency ω and not ω_0. Equation (1.65) describes the steady-state solution into which any particular solution, satisfying certain initial values for x and $\dot{x}$ at some earlier time t_0, evolves. For the complex amplitude of the displacement of the electron one finds

$$\bar{x}_0 = \frac{e\bar{E}_0}{m_e(\omega_0^2 - \omega^2 - i\omega\gamma)}. \tag{1.66}$$

Putting $E_0 = |\bar{E}_0|$ and $x_0 = |\bar{x}_0|$, the real amplitude of the electron's displacement becomes

$$x_0 = \frac{e E_0}{m_e\sqrt{(\omega_0^2 - \omega^2)^2 + \omega^2\gamma^2}}. \tag{1.67}$$

If damping (γ) is small and the incoming wave vibrates with the natural frequency of the electron ($\omega = \omega_0$), the amplitude x_0 is proportional to γ^{-1} and can become very large.

A static field E_0 induces in the oscillator a permanent dipole moment

$$p = e x_0 = \frac{e^2 E_0}{m_e\omega_0^2}.$$

One can apply this equation to atoms to obtain a rough estimate for the dipole moment induced by a field E_0 if one chooses for the characteristic frequency ω_0 a value such that $\hbar\omega_0$ is of order of the ionization potential of the atom.

1.3.3.1 The phase shift

Equation (1.65) tells us that although the electron has a natural frequency ω_0, it moves in the forced oscillation with the frequency ω of the external field. Without friction ($\gamma = 0$), the electron and the field are synchronous. With friction, the complex variables for position and field, $\bar{x}$ and $\bar{E}$, are out of phase because of the imaginary term $i\omega\gamma$ in (1.66). The electron always lags behind the field by a positive angle

$$\Phi = \begin{cases} \tan^{-1}(x) & \text{if } \omega < \omega_0 \\ \tan^{-1}(x) + \pi & \text{if } \omega > \omega_0 \end{cases} \qquad \text{where } x = \frac{\omega\gamma}{\omega_0^2 - \omega^2}.$$

- At low frequencies ($\omega \ll \omega_0$), the lag is small.
- Around the resonance frequency, the phase shift changes continuously and amounts to 90° at $\omega = \omega_0$.
- For $\omega \gg \omega_0$, it approaches 180°. Then the electron moves opposite to the direction in which it is being pushed by the external force eE. This is not a miracle but reflects the *steady-state* response of the electron. Initially, when the field E was switched on, the acceleration vector of the electron pointed, of course, in the same direction as the electric field.

1.3.4 Dissipation of energy in a forced oscillation

The total energy of the oscillator is the sum of kinetic energy T plus potential energy V:

$$T + V = \tfrac{1}{2}m_e(\dot{x}^2 + \omega_0^2 x^2). \tag{1.68}$$

The total energy declines when there is friction. If we think of a grain as being composed of many oscillators (atoms), friction results from collisions of the electrons with the lattice and this leads to heating. The damping constant γ, which has the dimension s^{-1}, is then interpreted as the collisional frequency. Because the mechanical power W, which is converted in a forced oscillation into heat, equals force multiplied by velocity:

$$W = F\dot{x}$$

multiplication of formula (1.58) with $\dot{x}$ gives

$$W = F\dot{x} = \frac{d}{dt}\left(\tfrac{1}{2}m_e\dot{x}^2 + \tfrac{1}{2}m_e\omega_0^2 x^2\right) + \gamma m_e\dot{x}^2.$$

The terms in the brackets represent the total energy. If γ is small, the time derivative of the total energy almost vanishes and only the term $\gamma m_e\dot{x}^2$ remains. With $\dot{x} = -\omega x$, the *time-averaged* dissipation rate is

$$W = \frac{1}{2}\gamma m_e\omega^2 x_0^2 = \frac{\gamma e^2 E_0^2}{2m_e} \frac{\omega^2}{(\omega_0^2 - \omega^2)^2 + \omega^2\gamma^2}. \tag{1.69}$$

(In fact, we should denote the time average by $\langle W \rangle$ but for simplicity we write W.)

- When the frequency is very high or very low, the heating rate W goes to zero because the velocities are small.
- Near the resonance frequency ($\omega \simeq \omega_0$), however, and especially when damping is weak, the power W becomes large.

1.3.5 Dissipation of energy in a free oscillation

If the system is not driven by an external force and if damping is small, the electron swings almost freely near its natural frequency ω_0, while its amplitude gradually declines like $e^{-\gamma t/2}$. If the motion of the electron is described by

$$x(t) = \begin{cases} x_0 e^{-\gamma t/2} e^{-i\omega_0 t} & \text{for } t \geq 0 \\ 0 & \text{for } t < 0 \end{cases} \tag{1.70}$$

the total energy E (unfortunately, the same letter as for the electric field) at time $t = 0$ is

$$E_0 = \tfrac{1}{2} m_e \omega_0^2 x_0^2$$

and afterwards it falls due to dissipational losses like

$$E = E_0 e^{-\gamma t}. \tag{1.71}$$

The energy of the system drops by a factor e in a time $\tau = 1/\gamma$ or after $\omega_0/2\pi\gamma$ cycles; the initial loss rate is

$$W = \tfrac{1}{2} \gamma m_e \omega_0^2 x_0^2.$$

The Fourier transform of $x(t)$ in equation (1.70) is, by definition,

$$f(\omega) = \int_{-\infty}^{\infty} x(t) e^{i\omega t} \, dt \tag{1.72}$$

or, in view of the reciprocity relation,

$$x(t) = \frac{1}{2\pi} \int_{-\infty}^{\infty} f(\omega) e^{-i\omega t} \, d\omega. \tag{1.73}$$

The two functions $x(t)$ and $f(\omega)$ form the Fourier transform pair. The integral (1.73) can be regarded as an infinite expansion of the motion of the electron, $x(t)$, into harmonic functions $e^{-i\omega t}$ of amplitude $f(\omega)$ and with continuously varying frequencies ω. For the amplitudes in the Fourier expansion of the free oscillator of (1.70), we find

$$f(\omega) = x_0 \int_0^{\infty} e^{-\gamma t/2} e^{-i(\omega_0 - \omega)t} \, dt = \frac{x_0}{i(\omega_0 - \omega) + \gamma/2}. \tag{1.74}$$

In the decomposition of $x(t)$ into harmonics, only frequencies with a substantial amplitude $f(\omega)$ are relevant. As we assume weak damping, they all cluster around ω_0. The loss rate corresponding to each frequency component is then proportional to the square of the amplitude:

$$W(\omega) \propto \gamma |f(\omega)|^2 \propto \frac{\gamma}{(\omega - \omega_0)^2 + (\gamma/2)^2}.$$

Equation (1.74) is the basis for the Lorentz profile to be discussed in section 1.4.

1.3.6 The plasma frequency

The plasma frequency is defined by

$$\omega_p = \sqrt{\frac{4\pi N e^2}{m_e}} \tag{1.75}$$

where N is the number density of free electrons. The inverse, ω_p^{-1}, gives the relaxation time of the electron density. To prove it, we apply a perturbation to a plasma in equilibrium which consists of heavy, immobile positive ions and unbound light electrons. Prior to the disturbance, the mean charge density ρ over any macroscopic volume ΔV is zero: positive and negative charges balance. Now consider in the plasma a slab S_1 with sides Δx, Δy, Δz and $\Delta y \cdot \Delta z = 1$, so that the volume of the slab $\Delta V = \Delta x$. All electrons inside S_1 are suddenly displaced along the x-axis by the infinitesimal length Δx. After this disturbance, they are in the neighboring slab S_2, which has the same shape and volume as S_1, and the net charge density (ions plus electrons) in S_2 is no longer zero but equal to Ne. The electric field E arising from the charge separation as a result of shifting electrons from S_1 to S_2 exerts a force F on the displaced electrons and accelerates them:

$$F = Ne\Delta x \cdot E = N\Delta x m_e \ddot{x}.$$

The field $E = 4\pi Nex$ is found by integrating the basic equation $4\pi\rho = \operatorname{div} \mathbf{D}$; therefore,

$$4\pi Ne^2 x = m_e \ddot{x}.$$

Comparison with (1.60) shows that the electrons will oscillate around their equilibrium position with the frequency ω_p of (1.75).

1.3.7 Dispersion relation of the dielectric permeability

Because the restoring force, $-\kappa x$, in (1.58) is due to electrostatic attraction between the electron and a proton, we may consider the oscillating electron as an alternating dipole of strength $\bar{x}e$. A real grain contains many such oscillators (electrons) that are driven by the incoming wave up and down. If they swing in

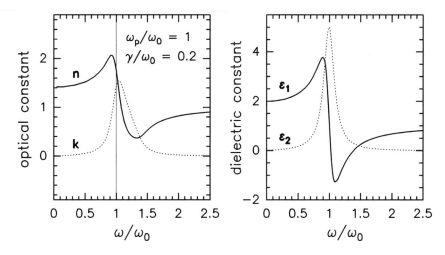

Figure 1.2. The dispersion relation specifies how the dielectric permeability $\varepsilon = \varepsilon_1 + i\varepsilon_2$ or, equivalently, the optical constant $m = n + ik$, change with frequency; here for a harmonic oscillator after (1.77). The vertical line helps us to locate the maxima with respect to the resonance frequency.

phase and their volume density equals N, the dipole moment per unit volume becomes, from (1.66),

$$\bar{P} = N\bar{x}e = \frac{Ne^2}{m_e} \frac{\bar{E}}{\omega_0^2 - \omega^2 - i\omega\gamma} = \frac{\varepsilon - 1}{4\pi}\bar{E} \qquad (1.76)$$

where the equals sign on the far right comes from (1.6). The bar designates complex quantities. Equation (1.76) represents the so called dispersion relation of the complex dielectric permeability and specifies how ε varies with frequency. If we use the plasma frequency of (1.75), where N is the number density of *free* electrons, we get

$$\varepsilon = \varepsilon_1 + i\varepsilon_2 = 1 + \frac{\omega_p^2(\omega_0^2 - \omega^2)}{(\omega_0^2 - \omega^2)^2 + \gamma^2\omega^2} + i\frac{\omega_p^2\gamma\omega}{(\omega_0^2 - \omega^2)^2 + \gamma^2\omega^2}. \qquad (1.77)$$

Figure 1.2 presents an example of the dispersion relation $\varepsilon(\omega)$ calculated with $\gamma/\omega_0 = 0.2$ and $\omega_p = \omega_0$. Despite this particular choice, it shows the characteristic behavior of the dielectric permeability at a resonance. Also of interest are the limiting values in very rapidly and very slowly changing fields. As we can compute m from ε after (1.48) and (1.49), figure 1.2 also gives the dispersion relation for n and k.

- In a constant field E, all quantities are real. The induced dipole moment per unit volume and the static permeability are:

$$P = \frac{Ne^2 E}{m_e \omega_0^2}$$

$$\varepsilon(0) = 1 + \frac{\omega_p^2}{\omega_0^2}.$$

- In a slowly varying field ($\omega \ll \omega_0$),

$$\varepsilon = 1 + \frac{\omega_p^2}{\omega_0^2} + i\frac{\omega_p^2 \gamma}{\omega_0^4}\omega. \tag{1.78}$$

  The real part ε_1 approaches the electrostatic value $\varepsilon(0)$, whereas ε_2 becomes proportional to frequency ω times the damping constant γ and is thus small. Correspondingly, n goes towards a constant value and k falls to zero.

- Around the resonance frequency, ε_2 is quite symmetric and has a prominent peak. But note in figure 1.2 the significant shifts of the extrema in n, k and ε_1 with respect to ω_0. At exactly the resonance frequency, $\varepsilon_1(\omega_0) = 1$ and $\varepsilon_2(\omega_0) = \omega_p^2/\gamma\omega_0$.

- The imaginary part ε_2 goes to zero far away from the resonance on either side but always remains positive. ε_1 increases from its static value as ω nears ω_0, then sharply drops and rises again. It may become zero or negative, as displayed in figure 1.2.

- At high frequencies ($\omega \gg \omega_0$) and far from the resonance, there is very little polarization as the electrons cannot follow the field due to their inertia and ε can be approximated by

$$\varepsilon = 1 - \frac{\omega_p^2}{\omega^2} + i\frac{\omega_p^2 \gamma}{\omega^3}. \tag{1.79}$$

  ε is then an essentially real quantity asymptotically approaching unity. This is a necessary condition for the fulfilment of the Kramers–Kronig relations (see section 2.5). Refraction disappears ($n \rightarrow 1$), and the material becomes transparent because there are no dissipational losses (k, $\varepsilon_2 \rightarrow 0$).

- At *very high* frequencies, when the wavelength is reduced to the size of an atom, the concept of a continuous medium breaks down and modifications are necessary.

1.4   The harmonic oscillator and light

We continue to discuss the optical constant from the viewpoint that matter is made of harmonic oscillators. We derive the emission of an accelerated charge, compute how an oscillator is damped by its own radiation and evaluate the cross section for absorption and scattering of a single oscillator.

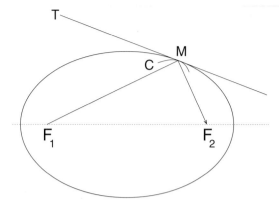

Figure 1.3. For light traveling from focus F_1 of an ellipse to focus F_2 via the reflection point M on the circumference of the ellipse, the total path length $\overline{F_1 M F_2}$ is indifferent to variations around M. If the reflection point M lies on the tangent T, the actual path is a minimum, if it is on the curve C which has a curvature radius smaller than the ellipse at M, the path is a maximum.

1.4.1 Attenuation and refraction of light

The attenuation of light is due to the dissipation of energy. In the harmonic oscillator model, the electrons transfer kinetic energy in collisions to the lattice. The loss rate depends foremost on the imaginary part of the dielectric constant ε_2, which itself is proportional to the frictional parameter γ. In a resonance, ε_2 has its maximum value close to but not coinciding with the natural frequency ω_0 (see figure 1.2). ε_2 or k can never be negative as this would imply amplification of light, which is impossible for a dust grain in thermodynamic equilibrium. In an idealized loss-free medium ε_2 and k vanish.

The real part of the dielectric constant, ε_1, may take up positive and negative values. The real part of the optical constant, n, however, is always positive. The index n determines the phase velocity v_{ph} of the wave. Without damping one finds, from (1.44) and (1.46),

$$v_{ph} = \frac{c}{n}.$$

n is responsible for the phenomenon of refraction. According to Fermat's principle, light traveling in a medium of varying n chooses, from among all possible paths, the quickest. More precisely, when infinitesimally varying the actual path, to first order, the travel time does not change. So the chosen route might be a local minimum or a local maximum (see figure 1.3).

Fermat's principle *formally* explains how spectacles and binoculars work and why a light ray changes its direction when it passes from a medium with index n_1 to one with n_2. The *physical* reason that, for example, the phase velocity in glass is different from that in vacuum is contained in the concept of superposition which states that the field at any given point is the sum of the fields from all oscillators anywhere in the world.

Consider, for this purpose, a plane wave that encounters a flat, thin sheet of glass oriented perpendicular to the direction of wave propagation. The electric field of the wave accelerates the electrons in the glass and they start to radiate themselves. The field E_a far *a*head of the sheet is, according to the principle of superposition, the sum of the fields created by all electrons plus the field of the incident wave. When one computes E_a (it is not too difficult), one finds that it has a phase shift relative to the incident wave, i.e. to the wave if there were no sheet. One gets exactly the same phase shift if one neglects the fields from the electrons altogether and assumes the wave has traveled in the sheet at a reduced phase velocity c/n. So the phase velocity c/n follows, in a subtle way, from the interference of the incoming wave with waves generated by the excited oscillators.

For a medium that is almost transparent, i.e. when the damping is weak and $\gamma \ll \omega_0$, the expression for n reduces, from (1.48) and (1.77), to

$$n \simeq 1 + \frac{\omega_p^2}{2(\omega_0^2 - \omega^2)}. \tag{1.80}$$

So for small attenuation, n is nearly constant at low frequencies and equal to $1 + \omega_p^2/2\omega_0^2$. As one approaches the resonance (see figure 1.2), n rises, reaches its maximum shortward of ω_0, dips afterwards below one and goes asymptotically to unity at high frequencies. In terms of actual frequencies, it all depends, of course, on the value of ω_0. When n rises with ω, i.e. to the left of ω_0 in figure 1.2, one speaks of *normal dispersion*. This is true, for example, for glass in the visible range: blue light entering a prism is bent more than red light. However, if $dn/d\omega < 0$, the dispersion is called *anomalous*, for historical reasons. When n is less than one, the phase velocity exceeds the velocity of light in vacuum. This does not violate special relativity as it is not possible to transmit information with a monochromatic wave. A wave package, which is composed of waves of different frequencies and is capable of carrying a message, cannot travel faster than c (see (1.131)).

If a substance has several resonances arising from different oscillators with number density N_j and natural frequency ω_j, one has instead of (1.80)

$$n(\omega) \simeq 1 + \frac{2\pi e^2}{m_e} \sum_j \frac{N_j}{\omega_j^2 - \omega^2}.$$

1.4.2   Retarded potentials of a moving charge

An oscillating electron represents a current that is surrounded by a time-variable magnetic field which, in turn, induces an electric field and so on; in the end, an electromagnetic wave is emitted. To calculate the emission, we first summarize what the electric and magnetic fields produced by arbitrarily moving charges of density ρ look like. We start with Maxwell's equations in the so-called microscopic form where all contributions to charge and current are included, not only the 'free' parts:

$$\text{div}\,\mathbf{E} = 4\pi\rho_{\text{tot}} \qquad \text{div}\,\mathbf{B} = 0 \qquad\qquad (1.81)$$

$$\text{rot}\,\mathbf{E} = -\frac{1}{c}\dot{\mathbf{B}} \qquad \text{rot}\,\mathbf{B} = \frac{1}{c}\dot{\mathbf{E}} + \frac{4\pi}{c}\mathbf{J}_{\text{tot}}. \qquad (1.82)$$

They follow immediately from the equation set (1.27)–(1.30) with the help of (1.19)–(1.22). Again there is charge conservation, $\dot{\rho}_{\text{tot}} + \text{div}\,\mathbf{J}_{\text{tot}} = 0$. We now drop the suffix 'tot'. Because div $\mathbf{B} = 0$, there exists a vector potential $\mathbf{A}$ such that

$$\mathbf{B} = \text{rot}\,\mathbf{A} \qquad\qquad (1.83)$$

and because $\text{rot}(\mathbf{E} + \dot{\mathbf{A}}/c) = 0$, there exists a scalar potential ϕ with

$$\mathbf{E} + \frac{1}{c}\dot{\mathbf{A}} = -\nabla\phi. \qquad\qquad (1.84)$$

These potentials are gauged by imposing on them the *Lorentz condition*,

$$\text{div}\,\mathbf{A} + \frac{1}{c}\dot{\phi} = 0 \qquad\qquad (1.85)$$

which leaves the fields $\mathbf{E}$ and $\mathbf{B}$ untouched. Then ϕ and $\mathbf{A}$ obey the relations

$$\Delta\mathbf{A} - \frac{1}{c^2}\ddot{\mathbf{A}} = -\frac{4\pi}{c}\mathbf{J} \qquad\qquad (1.86)$$

$$\Delta\phi - \frac{1}{c^2}\ddot{\phi} = -4\pi\rho. \qquad\qquad (1.87)$$

(1.86) and (1.87) are equivalent to Maxwell's equations. In vacuum, they become wave equations and their right-hand sides vanish. If the charges are localized around the center of the coordinate sytem and if their position is denoted by $\mathbf{x}_1$, the potentials at the point $\mathbf{x}_2$ of the observer are (see any textbook on electrodynamics):

$$\phi(\mathbf{x}_2, t) = \int \frac{\rho(\mathbf{x}_1, t')}{|\mathbf{x}_2 - \mathbf{x}_1|}\, dV \qquad\qquad (1.88)$$

$$\mathbf{A}(\mathbf{x}_2, t) = \int \frac{\mathbf{J}(\mathbf{x}_1, t')}{c\,|\mathbf{x}_2 - \mathbf{x}_1|}\, dV. \qquad\qquad (1.89)$$

The potentials $\mathbf{A}$ and ϕ are called retarded as they refer to the *present time* t but are determined by the configuration at an earlier epoch t'. The delay corresponds to the time it takes light to travel from $\mathbf{x}_1$ to $\mathbf{x}_2$:

$$t' = t - \frac{|\mathbf{x}_2 - \mathbf{x}_1|}{c}. \qquad\qquad (1.90)$$

If the size $d \sim |\mathbf{x}_1|$ of the region over which the charges are spread is small compared with the distance $r = |\mathbf{x}_2|$ to the observer,

$$d \ll r$$

(astronomers are always at a safe distance from the action), we may put $r \simeq |\mathbf{x}_2 - \mathbf{x}_1|$ and take r out from under the integrals. Moreover, $|\mathbf{x}_2 - \mathbf{x}_1| \simeq |\mathbf{x}_2| - \mathbf{x}_1 \cdot \mathbf{e}$, where $\mathbf{e}$ is the unit vector pointing from the atom to the observer, so

$$t' \simeq t - (r - \mathbf{x}_1 \cdot \mathbf{e})/c. \tag{1.91}$$

If the d is also small compared with the wavelength,

$$d \ll \lambda$$

there will be no phase shift among the waves emitted from different parts of the source and the emission is coherent, so we can write

$$t' \simeq t - r/c$$

and

$$\mathbf{A}(\mathbf{x}_2, t) = \frac{1}{cr} \int \mathbf{J}(\mathbf{x}_1, t - r/c) \, dV. \tag{1.92}$$

This expression applies, for example, to atoms where a charge is oscillating at a frequency ν and the linear dimension of the system $d \sim \nu^{-1} v = \lambda v/c$, where v is the non-relativistic charge velocity.

1.4.3   Emission of an harmonic oscillator

Under such simplifications, one obtains the *dipole field*. The vector potential $\mathbf{A}$ follows from a volume integral over the current density $\mathbf{J}$ at the earlier epoch $t' = t - r/c$. If there is only one small charge q of space density $\rho(\mathbf{x})$ that is oscillating about an opposite charge at rest, equation (1.92) yields

$$\mathbf{A}(\mathbf{x}_2, t) = \frac{1}{cr} \dot{\mathbf{p}}(\mathbf{x}_1, t - r/c) \tag{1.93}$$

because the current density is $\mathbf{J} = \rho \mathbf{v}$ and the wiggling charge q constitutes a dipole $\mathbf{p} = q\mathbf{x}_1$ whose derivative is given by the integral over the current density. If there are many charges q_i shaking at velocity $\mathbf{v}_i$, the result is the same with $\mathbf{p} = \sum q_i \mathbf{x}_i$.

We do not discuss the near field here. Far away from the charge, the wave is planar and

$$\mathbf{H} = \mathbf{e} \times \mathbf{E} \qquad \mathbf{E} = \mathbf{H} \times \mathbf{e}.$$

Far away, $\nabla \phi$ in (1.84) is negligible because ϕ represents the potential of the charge (see the comment after equation (6.10)) and the electric field of a charge falls off like $1/r^2$, whereas the field of a light source falls off more slowly like $1/r$. We thus have $\mathbf{E} = -\dot{\mathbf{A}}/c$ and

$$\mathbf{H} = \frac{1}{c^2 r}\dddot{\mathbf{p}} \times \mathbf{e} = \frac{\omega^2}{c^2 r}\mathbf{e} \times \mathbf{p} \tag{1.94}$$

$$\mathbf{E} = \frac{\omega^2}{c^2 r}(\mathbf{e} \times \mathbf{p}) \times \mathbf{e} \tag{1.95}$$

where we have assumed that the dipole varies harmonically at frequency ω. Having determined the fields $\mathbf{E}$ and $\mathbf{H}$ via the dipole moment $\mathbf{p}$, we can compute the flux dW carried in a solid angle $d\Omega = \sin\theta\, d\theta\, d\phi$ into the direction $\mathbf{e}$ which forms an angle θ with the dipole moment $\mathbf{p}$. This flux is given by the Poynting vector of (1.38), so

$$dW = \frac{|\ddot{\mathbf{p}}|^2}{4\pi c^3}\sin^2\theta\, d\Omega. \tag{1.96}$$

The emission is zero in the direction of the motion of the charge and maximum perpendicular to it. Integration over all directions yields for the total *momentary power* radiated by a dipole:

$$W = \frac{2}{3c^3}|\ddot{\mathbf{p}}|^2. \tag{1.97}$$

If the dipole oscillates harmonically proportional to $\cos\omega t$, the *time-averaged power* equals half the maximum value of W in (1.97).

1.4.4   Radiation of higher order

More generally, if one does not make the simplification $t' = t - r/c$ but sticks to equation (1.91) and expands $\mathbf{J}(\mathbf{x}_1, t')$ in (1.89) for small $\mathbf{x}_1 \cdot \mathbf{e}$, equation (1.93) is modified and at first order one has to add two correction terms:

$$\mathbf{A} = \frac{1}{cr}\dot{\mathbf{p}} + \frac{1}{cr}\dot{\mathbf{m}} \times \mathbf{e} + \frac{1}{6c^2 r}\ddot{\mathbf{Q}}. \tag{1.98}$$

$\mathbf{m}$ is the magnetic dipole moment of equation (1.13) and

$$Q_{ij} = \int \rho(\mathbf{x})[3x_i x_j - x^2 \delta_{ij}]\, dV \tag{1.99}$$

the electric quadrupole moment, a traceless tensor.   The vector $\mathbf{x}$ has the components x_i and the length x; δ_{ij} denotes the Kronecker symbol. From Q_{ij} one defines the vector $\mathbf{Q}$,

$$\mathbf{Q} = (Q_1, Q_2, Q_3) \qquad \text{with } Q_i = \sum_j Q_{ij}e_j \qquad (i = 1, 2, 3)$$

and this $\mathbf{Q}$ is used in (1.98). The vector potential $\mathbf{A}$ in formula (1.98) is also evaluated at the time $t' = t - r/c$. Again, one finds the Poynting vector $\mathbf{S} = (c/4\pi)\mathbf{E} \times \mathbf{H}$ from $\mathbf{E} = -\dot{\mathbf{A}}/c$ and $\mathbf{H} = \mathbf{e} \times \mathbf{E}$.

Even when the electric dipole moment vanishes, there may still be electric quadrupole or magnetic dipole radiation, associated with $\ddot{Q}$ and $\dot{m}$, respectively but they are several orders of magnitude weaker. One can easily show that if all particles have the same charge-to-mass ratio as, for instance, the two atoms in the hydrogen molecule H_2, the time derivatives $\dddot{p}$ and $\ddot{m}$ are zero so that only the electric quadrupole radiation remains.

1.4.5 Radiation damping

An accelerating electron radiates and the emitted light reacts on the electron because of the conservation of energy and momentum. Although theory (quantum electrodynamics) and experiment agree extremely well, there is currently no strict, self-consistent description of the feedback. We describe the feedback in classical terms here.

When a force F accelerates a free and otherwise undamped electron, we write the equation of motion in the form

$$F + F_{\text{rad}} = m_e \dot{u}.$$

The additional force F_{rad} accounts for the retardation caused by the radiative loss and $u = \dot{x}$ is the velocity. F_{rad} acts oppositely to F. Averaged over some time interval Δt, we assume that the emitted power due to the radiative deceleration (see (1.97)) is equal to the work that the force F_{rad} has done on the electron:

$$-\int_{\Delta t} F_{\text{rad}} u \, dt = \frac{2e^2}{3c^3} \int_{\Delta t} \dot{u}^2 \, dt.$$

When we integrate the right-hand side by parts and assume a periodic motion where $u\dot{u} = 0$ at the beginning and at the end of the time interval Δt, we find

$$\int_{\Delta t} \left(F_{\text{rad}} - \frac{2e^2}{3c^3} \ddot{u}^2 \right) u \, dt = 0.$$

So the force becomes

$$F_{\text{rad}} = m_e \tau \ddot{u}$$

with

$$\tau = \frac{2e^2}{3m_e c^3} = 6.27 \times 10^{-24} \text{ s}.$$

F_{rad} has to be added in the equation of motion of the harmonic oscillator, so (1.59) becomes

$$\omega_0^2 x + \gamma u + \dot{u} - \tau \ddot{u} = \frac{eE_0}{m_e} e^{-i\omega t}.$$

It contains time derivatives in x up to third order but, because $\ddot{u} = -\omega^2 u$ for a harmonic oscillation, the steady-state solution is again

$$\bar{x} = \bar{x}_0 e^{-i\omega t}$$

(a bar denotes complexity) and the complex amplitude becomes

$$\bar{x}_0 = \frac{e\bar{E}_0}{m_e} \frac{1}{\omega_0^2 - \omega^2 - i\omega(\gamma + \tau\omega^2)}.$$

This is the same expression for $\bar{x}_0$ as in (1.66) if one replaces there the dissipation constant γ by

$$\gamma + \tau\omega^2 = \gamma + \gamma_{rad}.$$

The term

$$\gamma_{rad} = \frac{2e^2}{3m_e c^3} \omega^2 \qquad (1.100)$$

is the damping constant due to radiation alone and follows from equating the time-averaged losses $\frac{1}{2}\gamma_{rad} m_e \omega^2 x_0^2$ of the harmonic oscillator after (1.69) to the time-averaged radiated power $p_0^2 \omega^4 / 3c^3$ of (1.97). Note that we always assume the radiative losses over one cycle to be small in comparison to the energy of the oscillator.

1.4.6 The cross section of an harmonic oscillator

1.4.6.1 Scattering

An oscillating electron with dipole moment

$$\bar{p} = \bar{p}_0 e^{-i\omega t} = e\bar{x}_0 e^{-i\omega t}$$

scatters, from (1.67) and (1.97), the power

$$W^{sca} = \frac{2}{3c^3} \omega^4 p_0^2 = \frac{2}{3c^3} \frac{e^4 E_0^2}{m_e^2} \frac{\omega^4}{(\omega_0^2 - \omega^2)^2 + \omega^2 \gamma^2} \qquad (1.101)$$

where we put $p_0 = |\bar{p}_0|$. So its cross section σ^{sca}, defined as the scattered power divided by the incident flux $S = (c/4\pi)E_0^2$ of (1.38), becomes

$$\sigma^{sca}(\omega) = \frac{W^{sca}}{S} = \sigma_T \frac{\omega^4}{(\omega_0^2 - \omega^2)^2 + \omega^2 \gamma^2} \qquad (1.102)$$

where

$$\sigma_T = \frac{8\pi r_0^2}{3} = 6.65 \times 10^{-25} \text{ cm}^2 \qquad (1.103)$$

is the *Thomson* scattering cross section of a single electron; it is frequency independent. r_0 is the classical electron radius and follows from equalling the rest mass energy $m_e c^2$ to the electrostatic energy e^2/r_0:

$$r_0 = \frac{e^2}{m_e c^2} = 2.82 \times 10^{-13} \text{ cm.} \qquad (1.104)$$

If there is only radiative damping, we have to set $\gamma = \gamma_{rad}$ after (1.100). From equation (1.102) we find the following approximations for the scattering cross section at the resonance, and at low and high frequencies:

$$\sigma^{sca}(\omega) = \begin{cases} \sigma_T \left(\dfrac{\omega}{\omega_0}\right)^4 & \text{if } \omega \ll \omega_0 \\[2ex] \dfrac{\sigma_T}{4} \dfrac{\omega_0^2}{(\omega - \omega_0)^2 + (\gamma/2)^2} & \text{if } \omega \simeq \omega_0 \\[2ex] \sigma_T & \text{if } \omega \gg \omega_0. \end{cases} \tag{1.105}$$

When the frequency is high, the electron is essentially free and the cross section constant and equal to σ_T. When the frequency is small, σ^{sca} falls with the fourth power of ω (Rayleigh scattering). Near the resonance, we have put $\omega^2 - \omega_0^2 \simeq 2\omega_0(\omega - \omega_0)$.

1.4.6.2 Absorption

In a similar way, we get from (1.69) for the absorption cross section of the harmonic oscillator

$$\sigma^{abs}(\omega) = \frac{W^{abs}}{S} = \frac{4\pi e^2}{m_e c} \frac{\omega^2 \gamma}{(\omega_0^2 - \omega^2)^2 + \omega^2 \gamma^2} \tag{1.106}$$

with the approximations

$$\sigma^{abs}(\omega) = \frac{\pi e^2}{m_e c} \begin{cases} 4\gamma \omega_0^{-4} \omega^2 & \text{if } \omega \ll \omega_0 \\[2ex] \dfrac{\gamma}{(\omega - \omega_0)^2 + (\gamma/2)^2} & \text{if } \omega \simeq \omega_0 \\[2ex] 4\gamma \omega^{-2} & \text{if } \omega \gg \omega_0. \end{cases} \tag{1.107}$$

A frequency dependence according to the middle line of (1.107) or (1.105) results in a Lorentz profile. It has the characteristic feature that the intensity over an emission or absorption line changes proportional to $[(\omega - \omega_0)^2 + (\gamma/2)^2]^{-1}$. As we have seen in the Fourier analysis of the motion of a free oscillator after (1.74), at the root of such a profile is the exponential decline $e^{-\gamma t/2}$ of the amplitude of the electron.

1.4.7 The oscillator strength

Integrating $\sigma(\omega)$ after (1.106) over frequency yields the total cross section for absorption,

$$\sigma_{tot} = \int \sigma(\omega)\, d\omega = \frac{2\pi^2 e^2}{m_e c}. \tag{1.108}$$

The integrand has significant values only around ω_0, so to evaluate σ_{tot} we could also use the approximate second formula of (1.107). We note that

$$\int_0^\infty \frac{\gamma}{(\omega - \omega_0)^2 + (\gamma/2)^2}\, d\omega = 2 \int_{-2\omega_0/\gamma}^\infty \frac{dx}{1 + x^2} \simeq 2\pi \qquad (\gamma \ll \omega_0).$$

When the cross section σ is a function of $\nu = \omega/2\pi$, we get

$$\sigma_{\text{tot}} = \int \sigma(\nu)\, d\nu = \frac{\pi e^2}{m_e c}.$$

The true quantum mechanical cross section for a downward transition $j \to i$ integrated over frequency is often written as

$$\sigma_{\text{tot}} = \frac{\pi e^2}{m_e c} f_{ji} \tag{1.109}$$

where f_{ji} is called oscillator strength. It is of order unity for strong lines, otherwise smaller. f_{ji} is related to the Einstein coefficients for induced and spontaneous emission, B_{ji} and A_{ji} (see section 6.3), through

$$\frac{\pi e^2}{m_e c} f_{ji} = \frac{h\nu}{c} B_{ji} = \frac{c^2}{8\pi \nu^2} A_{ji} \tag{1.110}$$

where $h\nu$ is the energy difference between level j and i.

1.4.8 The natural linewidth

If the oscillator is not driven by an external field but swings freely near the resonance frequency ω_0, its energy decays exponentially like $e^{-\gamma_{\text{rad}}(\omega_0)t}$ (see (1.71) and (1.100)). The intensity of emission, $I(\omega)$, at a frequency ω is proportional to $x_0^2(\omega)$, the square of the elongation given in (1.67). One may call $x_0^2(\omega)$ the resonance curve (see figure 1.4). Because all relevant frequencies for emission are close to ω_0, we have $\omega^2 - \omega_0^2 \simeq 2\omega_0(\omega - \omega_0)$ and the full width $\Delta\omega$ of the resonance curve $x_0^2(\omega)$ taken at half maximum becomes

$$\Delta\omega = \gamma_{\text{rad}} = \frac{2r_0}{3c}\omega_0^2$$

or, expressed in wavelength,

$$\Delta\lambda = 2\pi c \frac{\Delta\omega}{\omega_0^2} = \frac{4\pi}{3} r_0 = 1.18 \times 10^{-4} \text{ Å}. \tag{1.111}$$

This is the *natural linewidth* as derived from classical physics. It does not depend on the frequency of the transition, all intrinsic properties of the oscillator have canceled out.

In quantum mechanics, the Einstein coefficient A_{ji} is the counterpart to the radiative damping constant γ_{rad} of classical electrodynamics. A_{ji} specifies the rate at which a system in level j spontaneously decays to a lower level i (see section 6.3). The probability of finding the system in state j decreases with time like $e^{-A_{ji}t}$, in analogy to the classical formula.

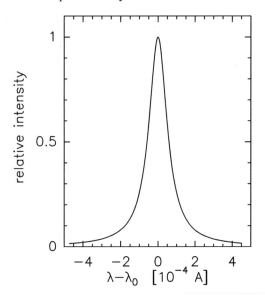

Figure 1.4. The emitted intensity of a classical oscillator with a resonance at λ_0 that is damped only radiatively with a damping constant γ_{rad} after (1.100). Irrespective of λ_0, the width of the line is 1.18×10^{-4} Å in wavelengths or γ_{rad} in frequency.

Furthermore, in quantum mechanics, the energy E_j of state j is fuzzy because of the uncertainty principle, $\Delta E \Delta t \geq \hbar$. Here $\Delta t \simeq A_{ji}^{-1}$ is the average time that the atom stays in level j. Therefore, E_j is undetermined by an amount $\Delta E = \hbar A_{ji}$ resulting in a frequency uncertainty $\Delta \omega = A_{ji}$. (More accurately, when calculating $\Delta \omega$, one has to sum the A coefficients of all possible downward transitions starting in j and something similar must be done for the lower level i, unless it is the ground state.)

The time A_{ji}^{-1} that the atom resides in state j also gives the duration of the emitted pulse. As the pulse travels with the velocity of light, the length of the wave train is cA_{ji}^{-1}. The emission thus cannot be strictly monochromatic but has a frequency spectrum centered at ω_0.

1.5   Waves in a conducting medium

We modify the expression for the dielectric permeability or the complex wavenumber for the case that some of the electrons in the material are not bound to individual atoms but are free so that they can conduct a current. The medium is then a plasma and the dispersion relation is known as the Drude profile. We show that, in the presence of a magnetic field, the optical constant n of a plasma depends for circularly polarized plane waves on the sense of rotation. This leads to the phenomenon of Faraday rotation. We also derive the group velocity of a wave package in a plasma.

1.5.1 The dielectric permeability of a conductor

A conductor contains free charges that can support a constant current, as we know from everyday experience with electricity. Ohm's law asserts a proportionality between the electric field **E** and the current density **J** (omitting the suffix 'free'),

$$\mathbf{J} = \sigma \mathbf{E}. \tag{1.112}$$

σ is called the conductivity and is usually defined for a quasi-stationary electric field; in a vacuum, $\sigma = 0$. When a conductor is placed in a *static field*, no current can flow because the charges arrange themselves on its surface in such a way that the electric field inside it cancels out. In a *dielectric medium*, however, a static electric field is, of course, possible. An electric field may exist in a metal, either because it varies rapidly and the charges do not have time to reach their equilibrium positions or it is produced by a changing magnetic field according to (1.25).

We treat conductors because some interstellar grains have metallic properties and because the interstellar medium is a plasma. Let us start with the Maxwell equations (1.29) and (1.30) which are appropriate for conductors,

$$\operatorname{rot} \mathbf{E} = -\frac{1}{c}\dot{\mathbf{B}} \qquad \operatorname{rot} \mathbf{H} = \frac{1}{c}\dot{\mathbf{D}} + \frac{4\pi}{c}\mathbf{J}.$$

When we insert harmonic fields into these formulae, we get for a homogeneous medium with the help of (1.112)

$$\operatorname{rot} \mathbf{E} = i\frac{\omega\mu}{c}\mathbf{H}$$

$$\operatorname{rot} \mathbf{H} = -i\frac{\omega}{c}\left(\varepsilon + i\frac{4\pi\sigma}{\omega}\right)\mathbf{E}. \tag{1.113}$$

We may look at (1.113) as an expansion of rot **H** into powers of ω. The term with the conductivity σ dominates at low frequencies, more precisely when

$$|\varepsilon| \ll \frac{4\pi\sigma}{\omega}.$$

Then equation (1.113) reduces to

$$\operatorname{rot} \mathbf{H} = \frac{4\pi\sigma}{c}\mathbf{E} \tag{1.114}$$

which describes the magnetic field produced by a quasi-stationary current. Formally, we can retain the field equation

$$\operatorname{rot} \mathbf{H} = \frac{\varepsilon}{c}\dot{\mathbf{E}},$$

as formulated in (1.33) for a dielectric medium, also for a conducting medium. However, ε must then denote the sum of the dielectric permeability of (1.5), written here as ε_d to indicate that it refers to a *dielectric*, plus a term relating to the conductivity:

$$\varepsilon = \varepsilon_d + i\frac{4\pi\sigma}{\omega}. \tag{1.115}$$

The complex wavenumber of (1.41), which appears in the wave equations (1.34), is in a conducting medium

$$\mathbf{k}^2 = \frac{\omega^2\mu}{c^2}\left(\varepsilon_d + i\frac{4\pi\sigma}{\omega}\right). \tag{1.116}$$

The optical constant m follows from ε of (1.115) via $m = \sqrt{\varepsilon\mu}$ (see (1.46)).

At low frequencies, the dielectric permeability of a metal is approximately (see (1.123) for an exact expression)

$$\varepsilon(\omega) = i\frac{4\pi\sigma}{\omega}.$$

ε is then purely imaginary ($\varepsilon_1 = 0$), much greater than one ($|\varepsilon| \gg 1$) and has a singularity at $\omega = 0$. From (1.48) and (1.49) the optical constants n and k are, therefore, also large and roughly equal:

$$n \simeq k \rightarrow \infty \qquad \text{for } \omega \rightarrow 0. \tag{1.117}$$

The difference between dielectrics and metals vanishes when the electromagnetic field changes so rapidly that the electrons make elongations that are small compared to the atomic radius. Then the restoring force, $-\kappa x$, in (1.58) is negligible and all electrons are essentially free. This happens around $\omega \simeq 10^{17}\,\mathrm{s}^{-1}$.

1.5.2 Conductivity and the Drude profile

We repeat the analysis of the harmonic oscillator in section 1.3 for a plasma. The electrons are then free, they experience no restoring force after an elongation and, therefore, $\omega_0 = 0$ in (1.59). With the same ansatz $\bar{x} = \bar{x}_0 e^{-i\omega t}$ as before, we find the velocity of a free electron in an harmonic electric field:

$$\bar{v} = \frac{e}{m_e(\gamma - i\omega)}\bar{E}.$$

The conductivity follows from the current density $\mathbf{J} = \sigma\mathbf{E} = Ne\mathbf{v}$. If there are N free electrons per cm^3,

$$\sigma = \frac{Ne^2}{m_e(\gamma - i\omega)} = \frac{\omega_p^2}{4\pi(\gamma - i\omega)} \tag{1.118}$$

where $\omega_p = \sqrt{4\pi N e^2/m_e}$ is the plasma frequency of (1.75). So generally speaking, the conductivity is a complex function and depends on ω,

$$\sigma(\omega) = \sigma_1(\omega) + i\sigma_2(\omega) = \frac{\omega_p^2}{4\pi}\frac{\gamma}{\gamma^2 + \omega^2} + i\frac{\omega_p^2}{4\pi}\frac{\omega}{\gamma^2 + \omega^2}. \tag{1.119}$$

- At long wavelengths ($\omega \ll \gamma$), σ becomes frequency independent. It is then an essentially real quantity approaching the direct-current limit

$$\sigma(\omega = 0) = \frac{\omega_p^2}{4\pi\gamma} \tag{1.120}$$

  which appears in Ohm's law. The conductivity depends on the density of free electrons through ω_p and on the collision time through γ.
- If the frequency is high or damping weak ($\gamma \ll \omega$), the conductivity is purely imaginary and inversely proportional to frequency. There is then no dissipation of energy in the current.

When we put $\varepsilon_d = 0$, $\mu = 1$, and substitute, for the conductivity σ, the expression from (1.118) into (1.116), we find for the wavenumber of a plasma

$$k^2 = \frac{\omega^2}{c^2}\left(1 - \frac{\omega_p^2}{\gamma^2 + \omega^2} + i\frac{\gamma}{\omega}\frac{\omega_p^2}{\gamma^2 + \omega^2}\right). \tag{1.121}$$

Without damping,

$$k^2 = \frac{\omega^2}{c^2}\left(1 - \frac{\omega_p^2}{\omega^2}\right). \tag{1.122}$$

In this case, when ω is greater than the plasma frequency, the real part n of the optical constant is, from (1.48), smaller than one, so the phase velocity is greater than the velocity of light, $v_{ph} > c$. For $\omega = \omega_p$, n becomes zero. A wave with $\omega \le \omega_p$ cannot penetrate into the medium and is totally reflected. The bending of short-wavelength radio waves in the ionosphere or the reflection on metals (see (3.60)) are illustrations.

The dielectric permeability $\varepsilon(\omega)$ corresponding to (1.121) is known as the *Drude profile* (see (1.41)):

$$\varepsilon(\omega) = 1 - \frac{\omega_p^2}{\gamma^2 + \omega^2} + i\frac{\gamma}{\omega}\frac{\omega_p^2}{\gamma^2 + \omega^2}. \tag{1.123}$$

To get some feeling for the numbers associated with the conductivity, here are two examples:

- *Copper* is a pure metal with a free electron density $N \simeq 8 \times 10^{22}$ cm^{-3}, a plasma frequency $\omega_p = 1.6 \times 10^{16}$ s^{-1} and a damping constant $\gamma \simeq$

3×10^{13} s^{-1}. It has a high direct-current conductivity $\sigma \sim 7 \times 10^{17}$ s^{-1}. Therefore copper cables are ideal for carrying electricity.

- *Graphite*, which is found in interstellar space, has $N \simeq 1.4 \times 10^{20}$ cm^{-3} and $\gamma \simeq 5 \times 10^{12}$ s^{-1}, so the direct-current conductivity becomes $\sigma \simeq 7 \times 10^{15}$ s^{-1}; it is two orders of magnitude smaller than for copper. Graphite is a reasonable conductor only when the electrons move in the basal plane.

1.5.3   Electromagnetic waves in a plasma with a magnetic field

In the presence of a magnetic field, wave propagation is generally complicated. We consider the fairly simple situation when a plane wave with electric vector $\mathbf{E} = (E_x, E_y, 0)$ travels without attenuation through a plasma parallel to a constant magnetic field $\mathbf{B} = (0, 0, B)$ oriented in the z-direction. The trajectory $\mathbf{r}(t) = (x, y, 0)(t)$ of an electron follows from integrating the equation of motion, $\mathbf{F} = m_e \ddot{\mathbf{r}}$, where $\mathbf{F}$ is the Lorentz force of (1.1),

$$\ddot{x} = \frac{e}{m_e} E_x + \frac{eB}{m_e c} \dot{y} \qquad \ddot{y} = \frac{e}{m_e} E_y - \frac{eB}{m_e c} \dot{x}. \qquad (1.124)$$

The variable magnetic field of the wave is neglected. If the wave is linearly polarized, it can be regarded as a superposition of two oppositely circularly polarized waves. For example, if the electric vector stays along the x-axis,

$$\mathbf{E} = (1, 0, 0) E_0 e^{-i\omega t} = \frac{1}{2}[(1, i, 0) + (1, -i, 0)] E_0 e^{-i\omega t}.$$

We, therefore, put

$$r_{\pm} = x \pm iy \qquad E_{\pm} = E_x \pm i E_y$$

and interpret the (x, y)-plane as a complex plane, and E_+ and E_- as the electric vectors of two circularly polarized waves associated with the electron trajectories $r_+(t)$ and $r_-(t)$. It follows from (1.124) that

$$\ddot{r}_{\pm} = \frac{e}{m_e} E_{\pm} \mp i \frac{eB}{m_e c} \dot{r}_{\pm}.$$

The solution to this differential equation is

$$r_{\pm} = \frac{e/m_e}{\omega^2 \pm \omega\omega_{\mathrm{cyc}}} E_{\pm}$$

with the cyclotron frequency

$$\omega_{\mathrm{cyc}} = \frac{eB}{m_e c}. \qquad (1.125)$$

A circularly polarized wave makes an electron rotate. The presence of the magnetic field $\mathbf{B}$ modifies the orbit, and the modification is different for clockwise and anti-clockwise rotation. If there are N electrons per unit volume, their motion implies a current $J_\pm = \sigma_\pm E_\pm = Ne\dot{r}_\pm$, where the conductivity

$$\sigma_\pm = i\,\frac{Ne^2/m_e}{\omega \mp \omega_{cyc}}.$$

Without damping, the optical constant $n^2 = \varepsilon_1$. The dielectric permeability ε is given in (1.115) with $\varepsilon_d = 1$, therefore,

$$n_\pm^2 = 1 - \frac{\omega_p^2}{\omega^2}\left(1 \pm \frac{\omega_{cyc}}{\omega}\right)^{-1}. \tag{1.126}$$

We conclude that the two circularly polarized waves travel with different phase velocities $v_{ph} = c/n_\pm$. This leads to a change in the direction of the polarization vector of the linearly polarized wave. This effect is called Faraday rotation.

1.5.4 Group velocity of electromagnetic waves in a plasma

A one-dimensional wave package is a superposition of monochromatic waves (see (1.36)):

$$u(x, t) = \int_{-\infty}^{\infty} A(k) \cdot e^{i(kx - \omega t)}\, dk \tag{1.127}$$

in which the amplitude $A(k)$ has a sharp peak at some wavenumber $k = k_0$. We therefore develop the dispersion relation $\omega = \omega(k)$ around k_0,

$$\omega(k) = \omega_0 + \frac{d\omega}{dk}(k - k_0). \tag{1.128}$$

Here $\omega_0 = \omega(k_0)$ and the derivative $d\omega/dk$ is evaluated at $k = k_0$. At the time $t = 0$, the wave package has the form

$$u(x, 0) = \int_{-\infty}^{\infty} A(k)e^{ikx}\, dk.$$

Inserting (1.128) into (1.127) gives

$$u(x, t) = \underbrace{\exp\left[i\left(\frac{d\omega}{dk}k_0 - \omega_0\right)t\right]}_{=C} \int_{-\infty}^{\infty} A(k)\exp\left[ik\left(x - \frac{d\omega}{dk}t\right)\right] dk$$

$$= C \cdot u\left(x', 0\right) \tag{1.129}$$

with

$$x' = x - \frac{d\omega}{dk}t.$$

Therefore, as the factor C is purely imaginary and irrelevant, the wave package travels with the *group velocity*

$$v_g = \frac{d\omega}{dk}. \tag{1.130}$$

According to (1.122), for a plasma without damping

$$v_g = \frac{d\omega}{dk} = c \cdot \sqrt{1 - \omega_p^2/\omega^2} \le c. \tag{1.131}$$

For the product of group and phase velocity $v_{ph} = \omega/k$ (see (1.44)),

$$v_{ph} \cdot v_g = c^2. \tag{1.132}$$

1.6   Polarization through orientation

The polarization which we have modeled with the help of the harmonic oscillator of section 1.3 is due to the deformation of atoms in the sense that their internal charges are shifted relative to each other. In this section, we consider molecules with an intrinsic dipole moment and a *rigid* charge distribution that are allowed to rotate. This process is principally relevant to gases and liquids and only marginally to solids, provided the molecules are sufficiently 'round' so that they can turn and are not hooked too strongly to their neighbors. However, the analysis which we develop here also applies to the alignment of atomic magnets and we will use the results later in chapter 11 on grain alignment.

1.6.1   Polarization in a constant field

An atom or molecule with a dipole moment $\mathbf{p}$ in an electric field $\mathbf{E}$ has a potential energy U that depends on orientation,

$$U = -\mathbf{p} \cdot \mathbf{E} = -pE \cos\theta \tag{1.133}$$

where θ is the angle between the vectors $\mathbf{p}$ and $\mathbf{E}$, their lengths being p and E. The potential energy is at its minimum when the dipole moment lines up with the field and $\theta = 0$.

The atoms are never perfectly aligned because they are tossed around by the motions of their neighbors. In thermal equilibrium at temperature T, the distribution of orientations is expressed by the Boltzmann factor $e^{-U/kT}$. If $f(\theta)\, d\theta$ denotes the number of atoms that have angles between $\theta \ldots \theta + d\theta$, then

$$f(\theta) = f_0 e^{-U/kT} = f_0 e^{pE \cos\theta/kT}. \tag{1.134}$$

The average value of $\cos\theta$ follows from integrating $f(\theta)$ over all directions,

$$\langle \cos\theta \rangle = \frac{\int f(\theta) \cos\theta \, d\Omega}{\int f(\theta)\, d\Omega} \tag{1.135}$$

where $d\Omega$ is an element of the solid angle. If there are N dipoles per unit volume, the polarization P in the direction of **E** equals

$$P = Np\langle\cos\theta\rangle. \tag{1.136}$$

In our particular case, where $f(\theta) \propto e^{pE\cos\theta/kT}$, the mean cosine is given by

$$\langle\cos\theta\rangle = L(x) \qquad \text{with } x = \frac{pE}{kT}$$

where $L(x)$ denotes the *Langevin function*:

$$L(x) = \coth(x) - \frac{1}{x} = \frac{x}{3} - \frac{x^3}{45} + \cdots. \tag{1.137}$$

We see that the strength of the net polarization P depends on temperature. Easy to handle and relevant to our applications are weak fields ($pE \ll kT$) for which

$$f(\theta) = \frac{N}{4\pi}\left[1 + \frac{pE}{kT}\cos\theta\right]. \tag{1.138}$$

$f(\theta)$ is now almost constant, all directions are almost equally likely and the average angle is

$$\langle\cos\theta\rangle = \frac{pE}{3kT}.$$

An identical formula holds for magnetic dipoles (dipole moment m) in a magnetic field. One just has to replace pE by mB.

1.6.2 Polarization in a time-variable field

The distribution function f for the orientation of dipoles in the case of a time-variable field is considerably more complicated. It now depends not only on the angle θ between dipole moment and field but also on time, so we have to write $f(\theta, t)$ instead of $f(\theta)$. The time enters into f for two reasons. First, the field exerts on each dipole a torque

$$\tau = pE\sin\theta \tag{1.139}$$

which makes it rotate. The rotation speed $\dot\theta$ is assumed to be proportional to the torque τ, so

$$\tau = \zeta\dot\theta \tag{1.140}$$

where ζ is a friction coefficient. Second, there is Brownian motion because the dipoles are jostled about in collisions. We now sketch how one derives an expression for the distribution function leaving aside the mathematical details. Let $f(\theta, t)\, d\Omega$ be the number of dipoles whose axes fall into a small solid angle $d\Omega$ and make an angle θ with respect to the field **E**. The directions of the individual dipoles vary smoothly with time due to Brownian motion and the alternating electric field. One computes the change

$$\Delta = \dot{f}\, d\Omega\, \delta t$$

in the number of dipoles $f(\theta, t)\, d\Omega$ over the time interval δt. The interval δt is chosen long enough so that the dipole axes manage to escape from $d\Omega$ but also sufficiently short so that their directions are modified only by a small angle. The change Δ is split into a contribution due to Brownian motion and one due to the field

$$\Delta = \Delta_{\text{Brown}} + \Delta_{\text{field}}.$$

- To evaluate Δ_{Brown}, one can restrict the discussion to the dipoles in the vicinity of $d\Omega$, because δt is small. This leads to an expression for Δ_{Brown} that depends on f and its spatial derivatives times the mean square angular distance $\langle \phi^2 \rangle$ which the axis of an individual dipole travels during the time δt. But $\langle \phi^2 \rangle$ is known: for Brownian motion $\langle \phi^2 \rangle = kT\delta t/4\zeta$.
- The other term Δ_{field} is much easier to calculate. The axes are turned at an angular velocity $\dot{\theta}$ by a variable torque τ (see (1.140)) that tries to align the moments with the field. The difference in the number of axes that leave and enter the solid angle $d\Omega$ is Δ_{field}.

Altogether, Debye [Deb29] found for the time derivative of the distribution function

$$\zeta \dot{f} = \frac{1}{\sin\theta} \frac{\partial}{\partial\theta} \left[\sin\theta \left(kT \frac{\partial f}{\partial\theta} + \tau f \right) \right]. \tag{1.141}$$

The first term in brackets on the right-hand side of (1.141) is due to Brownian motion, the second term due to the rotation of the dipoles by the field.

1.6.3 Relaxation after switching off the field

The average orientation in a *constant time field* is described by the Maxwell–Boltzmann equation (1.134). It does of course not depend on time, so $\dot{f}$ vanishes. We may convince ourselves that (1.134) is, indeed, a solution to (1.141) for $\dot{f} = 0$.

Suppose now that the field is suddenly switched off at time $t = 0$. Then for $t > 0$, the torque τ is absent and the orientations begin to randomize; after a while, complete disorder is established. This process is called relaxation and governed by the equation

$$\frac{\zeta}{kT} \dot{f} = \frac{1}{\sin\theta} \frac{\partial}{\partial\theta} \left(\sin\theta \frac{\partial f}{\partial\theta} \right) \tag{1.142}$$

which follows from (1.141) for $\tau = 0$. To solve (1.142), we try a distribution function

$$f(\theta, t) = 1 + \varphi(t) \frac{pE_0}{kT} \cos\theta$$

which is similar to (1.138) but for the relaxation function $\varphi(t)$ and the factor $N/4\pi$ which has been dropped. Substituting $f(\theta, t)$ into (1.142) yields

$$\varphi(t) = e^{-\frac{2kT}{\zeta}t}. \tag{1.143}$$

The form of the relaxation function implies exponential decay. The system relaxes from statistical alignment in a constant field to random orientation on a time scale

$$t_{\text{rel}} = \frac{\zeta}{2kT}. \tag{1.144}$$

1.6.4 The dielectric permeability in Debye relaxation

In the presence of an harmonic field $E = E_0 e^{-i\omega t}$, the distribution function f depends explicitly on time. The Maxwell–Boltzmann formula (1.134) is then no longer applicable and needs to be generalized. It is easy to find a solution to Debye's equation (1.141) when the variable field is weak, i.e. when

$$pE_0 \ll kT.$$

Naturally, we try an ansatz for f in the spirit of (1.138), again without the factor $N/4\pi$,

$$f(\theta, t) = 1 + Ae^{-i\omega t}\frac{pE_0}{kT}\cos\theta.$$

When inserted into (1.141) and terms with E_0^2 are neglected, it yields

$$A = \frac{1}{1 - i\omega t_{\text{rel}}}.$$

So the full expression for the distribution function in a variable weak field is

$$f(\theta, t) = 1 + \frac{pE_0}{kT}\cos\theta\frac{e^{-i\omega t}}{1 - i\omega t_{\text{rel}}}. \tag{1.145}$$

For a static field ($\omega = 0$), we are back to Maxwell–Boltzmann. At very high frequencies, f becomes constant and is independent of the direction of the field.

We can now compute the polarization P of such a medium according to (1.136). For $\langle\cos\theta\rangle$ in (1.135) we have to use the distribution function from (1.145). Because $P = \chi E$, we obtain for the susceptibility χ (or the dielectric permeability ε) in the case of rotational polarization:

$$\chi = \frac{\varepsilon - 1}{4\pi} = \frac{Np^2}{3kT}\frac{1}{1 - i\omega t_{\text{rel}}} = \frac{Np^2}{3kT}\left[\frac{1}{1 + \omega^2 t_{\text{rel}}^2} + i\frac{\omega t_{\text{rel}}}{1 + \omega^2 t_{\text{rel}}^2}\right]. \tag{1.146}$$

This function is plotted in figure 1.5. It should be compared with $\varepsilon(\omega)$ for a dielectric and a metal, i.e. with (1.77) and (1.123). We have here the novelty

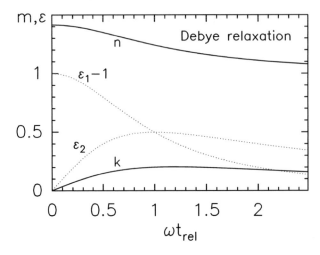

Figure 1.5. The dielectric permeability $\varepsilon = \varepsilon_1 + i\varepsilon_2$ after (1.146) and the optical constant $m = n + ik$ for the case that polarization is due to the alignment of molecules with a permanent dipole moment p; here we put $4\pi N p^2 / 3kT = 1$. Shown are n, k and $\varepsilon_1 - 1$ and ε_2.

that ε depends on temperature. For small frequencies, one may approximate the imaginary part of χ by

$$\chi_2 = \chi_0 \frac{\omega t_{\text{rel}}}{1 + \omega^2 t_{\text{rel}}^2} \simeq \omega \chi_0 t_{\text{rel}} \qquad \left(\chi_0 = \frac{N p^2}{3kT} \right). \qquad (1.147)$$

χ_0 is the static value of $\chi(\omega)$ and refers to a time constant field. The dissipation rate is proportional to ω, inversely proportional to temperature and otherwise determined by the relaxation time t_{rel}.

Let us consider water as an example of rotational polarization. The water molecule has an intrinsic electric dipole moment as the three nuclei in H_2O form a triangle with an obtuse angle of $105°$. Therefore the center of positive charge, which lies in the midst between the hydrogen atoms, and the center of negative charge, located near the oxygen atom, do not coincide, resulting in a permanent dipole moment $p = 1.9$ Debye.

Liquid water has, at room temperature, a viscosity $\eta \simeq 0.01$ g cm^{-1} s^{-1}. If the water molecules are approximated by spheres of radius a, the friction constant ζ follows from the viscosity using the equation for the torque (1.139) and the force F of Stokes' law (see (9.41)),

$$\zeta = \frac{\tau}{\dot{\theta}} \sim \frac{F a}{\dot{\theta}} \sim 6\pi \eta a^3. \qquad (1.148)$$

For $a \sim 1$ Å, $t_{\text{rel}} \simeq 2 \times 10^{-12}$ s approximately. The associated frequency t_{rel}^{-1} lies in the microwave region. With the support of experimental data, one can improve the estimate to $t_{rel} \simeq 8 \times 10^{-12}$ s but the first guess was not bad. As the density of water is $N = 3.3 \times 10^{22}$ cm^{-3}, at room temperature we have roughly $Np^2/3kT \sim 1$. When water freezes, the molecules find it hard to adjust their orientation because they cannot rotate freely in a solid. Then the viscosity and the relaxation time jump by a large factor and the frequency at which ε_2 has its maximum drops and polarization due to Debye relaxation becomes unimportant.

Chapter 2

How to evaluate grain cross sections

In section 2.1, we define cross sections, the most important quantity describing the interaction between light and interstellar grains. Section 2.2 deals with the optical theorem which relates the intensity of light that is scattered by a particle into exactly the forward direction to its extinction cross section. In sections 2.3 and 2.4, we learn how to compute the scattering and absorption coefficients of particles. The problem was first solved in the general case for spheres by G Mie [Mie08] and the underlying theory bears his name. Section 2.5 is concerned with a strange but important property of the material constants that appear in Maxwell's equations, such as ε or μ. They are complex quantities and Kramers and Kronig discovered a dependence between the real and imaginary parts. In the final section, we approximate the material constants of matter that is a mixture of different substances.

2.1 Defining cross sections

2.1.1 Cross section for scattering, absorption and extinction

For a single particle, the scattering cross section is defined as follows. Consider a plane monochromatic electromagnetic wave at frequency ν and with flux F_0. The flux is the energy carried per unit time through a unit area and given in (1.39) as the absolute value of the Poynting vector. When the wave hits the particle, some light is scattered into the direction specified by the angles (θ, ϕ) as depicted in figure 2.1. The flux from this scattered light, $F(\theta, \phi)$, which is received at a large distance r from the particle is obviously proportional to F_0/r^2; we, therefore, write

$$F(\theta, \phi) = \frac{F_0}{k^2 r^2} \mathcal{L}(\theta, \phi). \tag{2.1}$$

The function $\mathcal{L}(\theta, \phi)$ does not depend on r nor on F_0. We have included in the denominator the wavenumber

$$k = \frac{2\pi}{\lambda}$$

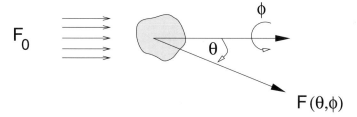

Figure 2.1. A grain scatters light from a plane wave with flux F_0 into the direction (θ, ϕ). In this direction, the scattered flux is $F(\theta, \phi)$.

to make $\mathcal{L}(\theta, \phi)$ dimensionless; the wavelength λ is then the natural length unit with which to measure the distance r.

- The cross section for scattering of the particle, C^{sca}, follows from the condition that $F_0 C^{\mathrm{sca}}$ equals the total energy scattered in *all* directions per unit time. Consequently,

$$C^{\mathrm{sca}} = \frac{1}{k^2} \int_{4\pi} \mathcal{L}(\theta, \phi) \, d\Omega = \frac{1}{k^2} \int_0^{2\pi} d\phi \int_0^{\pi} d\theta \, \mathcal{L}(\theta, \phi) \sin \theta$$

with the element of solid angle

$$d\Omega = \sin \theta \, d\theta \, d\phi. \tag{2.2}$$

The scattering cross section C^{sca} has the dimension of an area. It is assumed that the frequency of radiation is not changed in the scattering process.
- Besides scattering, a particle inevitably absorbs some light. The corresponding cross section C^{abs} is defined by the condition that $F_0 C^{\mathrm{abs}}$ equals the energy absorbed by the particle per unit time.
- The sum of absorption plus scattering is called extinction. The extinction cross section,

$$C^{\mathrm{ext}} = C^{\mathrm{abs}} + C^{\mathrm{sca}} \tag{2.3}$$

determines the *total* amount of energy removed from the impinging beam of light.
- The albedo is defined as the ratio of scattering over extinction,

$$A = \frac{C^{\mathrm{sca}}}{C^{\mathrm{ext}}}. \tag{2.4}$$

It lies between 0 and 1; an object with a high albedo scatters a lot of light.

All these various cross sections do not generally depend on the radiation field, the temperature or density of the dust material (which only changes very little anyway). In this respect, it is much simpler to calculate cross sections of grains than of gas atoms.

2.1.2   Cross section for radiation pressure

2.1.2.1   *Phase function and asymmetry factor*

In equation (2.1), $\mathcal{L}(\theta, \phi)$ specifies how the intensity of the scattered radiation changes with direction. We form a new function $\tilde{f}(\theta, \phi)$, which is proportional to $\mathcal{L}(\theta, \phi)$ but normalized so that the integral of $\tilde{f}$ over all directions equals 4π:

$$\int_{4\pi} \tilde{f}(\theta, \phi) \, d\Omega = 4\pi$$

and call it the phase function. An isotropic scatterer has $\tilde{f} = 1$.

For spheres, there is, for reasons of symmetry, no dependence on the angle ϕ, only on θ. It is then convenient to have a phase function f which has $\cos \theta$ as the argument, so we put $f(\cos \theta) = \tilde{f}(\theta)$. Again $f = 1$ for isotropic scattering and the normalization condition is

$$1 = \tfrac{1}{2} \int_{-1}^{+1} f(x) \, dx. \tag{2.5}$$

When one does not know or does not need the full information contained in $f(\cos \theta)$, one sometimes uses just one number to characterize the scattering pattern. This number is the asymmetry factor g, the mean of $\cos \theta$ over all directions weighted by the phase function $f(\cos \theta)$,

$$g = \langle \cos \theta \rangle = \tfrac{1}{2} \int_{-1}^{+1} f(x) \, x \, dx. \tag{2.6}$$

- It is easy to verify that g lies between -1 and 1.
- When scattering is isotropic, and thus independent of direction, $g = 0$.
- When there is mainly forward scattering, g is positive, otherwise it is negative. In the limit of pure forward scattering, $g = 1$; for pure backscattering $g = -1$.

2.1.2.2   *The momentum imparted on a grain by radiation*

Electromagnetic radiation also exerts a pressure on a grain. A photon that is absorbed deposits its full momentum $h\nu/c$. If it is scattered at an angle θ (see figure 2.1), the grain receives only the fraction $(1 - \cos \theta)$. Therefore, the cross section for radiation pressure, C^{rp}, can be written as

$$C^{\mathrm{rp}} = C^{\mathrm{ext}} - g \cdot C^{\mathrm{sca}}. \tag{2.7}$$

As g can be negative, although this case is unusual, C^{rp} may be greater than C^{ext}. To obtain the momentum transmitted per second to the grain by a flux F, we have to divide by the velocity of light c, so the transmitted momentum is

$$\frac{F \cdot C^{\mathrm{rp}}}{c}. \tag{2.8}$$

2.1.3 Efficiencies, mass and volume coefficients

The definition of the cross section of a single particle can be extended to 1 g of interstellar matter or 1 g of dust, or 1 cm^3 of space. We use the following nomenclature (omitting the dependence on frequency):

- The cross section of a single particle is denoted by the letter C.
- *Efficiency*, Q, is the ratio of C over the projected geometrical surface area σ_{geo}:

$$Q = \frac{C}{\sigma_{geo}}. \tag{2.9}$$

For spheres, $\sigma_{geo} = \pi a^2$, where a is the grain radius. There are efficiencies for absorption, scattering, extinction and radiation pressure, again

$$Q^{ext} = Q^{abs} + Q^{sca} \tag{2.10}$$
$$Q^{rp} = Q^{ext} - g \cdot Q^{sca}. \tag{2.11}$$

With the exception of spheres, σ_{geo} as well as the Cs and Qs change with the direction of the incoming light.

- *Mass coefficient*, K, is the cross section per unit mass. It refers either to 1 g of dust or to 1 g of interstellar matter. The latter quantity is some hundred times smaller.
- *Volume coefficient* is the cross section per unit volume and also denoted by the letter K. It refers either to 1 cm^3 in space (which typically contains 10^{-23} g of interstellar matter and 10^{-25} g of dust) or to 1 cm^3 of dust material with a mass of about 1 g.

2.2 The optical theorem

When a beam of light falls on a particle, some light is absorbed by the grain, heating it, and some is scattered. The optical theorem asserts that the reduction of intensity in the forward direction fully determines the particle's extinction cross section.

2.2.1 The intensity of forward scattered light

Consider a plane electromagnetic wave of wavelength λ propagating in a vacuum in the z-direction with electric field

$$E_i = e^{i(kz - \omega t)}. \tag{2.12}$$

For easier writing, we neglect the vector character of the field and assume a unit amplitude. When the wave encounters a grain located at the origin of the coordinate system, some light is absorbed and the rest is scattered into all

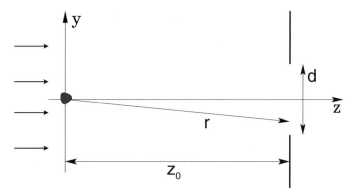

Figure 2.2. Light enters from the left, is scattered by a particle and falls at a large distance r through a disk of diameter d. The x-axis is perpendicular to the (y, z)-plane.

directions (θ, ϕ). At a distance r which is large when measured in wavelength units ($kr = 2\pi r/\lambda \gg 1$), the scattered field can be presented as

$$E_s = S(\theta, \phi)\frac{e^{i(kr-\omega t)}}{-i k r}.$$

(2.13)

The information about the amplitude of the scattered wave lies in the complex function $S(\theta, \phi)$, the exponential term contains the phase. Because $|E_s|^2$ is proportional to the scattered intensity, comparison of (2.13) with (2.1) tells us that $|S(\theta, \phi)|^2$ corresponds to $\mathcal{L}(\theta, \phi)$. As before, conservation of energy requires $E_s \propto 1/r$. The wavenumber $k = 2\pi/\lambda$ is introduced to make $S(\theta, \phi)$ dimensionless, the factor $-i$ in the denominator is just a convention.

Let us now determine the flux F through a disk of area A far behind the grain. The disk has a diameter d, so $d^2 \sim A$. Its center coordinates are ($x = 0$, $y = 0$, z_0). It lies in the (x, y)-plane and is thus oriented perpendicular to the z-axis. All points (x, y) in the disk fulfil the inequalities

$$|x| \ll z_0 \qquad |y| \ll z_0$$

and their distance to the particle is, approximately,

$$r \simeq z_0 + \frac{x^2 + y^2}{2z_0}.$$

The radiation that goes through the disk consists of the incident light (E_i) and the light scattered by the particle (E_s). The two fields interfere and to obtain the flux through the disk, they have to be added:

$$F = \int_A |E_i + E_s|^2 \, dA.$$

(2.14)

Dividing (2.13) by (2.12), we get

$$E_{\mathrm{s}} = E_{\mathrm{i}} S(\theta, \phi) \frac{e^{ik(r-z)}}{-ikr} \tag{2.15}$$

and for the sum of the incident and scattered field

$$E_{\mathrm{i}} + E_{\mathrm{s}} = E_{\mathrm{i}} \left\{ 1 - \frac{S(0)}{ikz_0} \exp\left(ik\frac{x^2+y^2}{2z_0}\right) \right\}. \tag{2.16}$$

Here we put

$$S(0) = S(\theta \simeq 0, \phi)$$

because $d \ll z_0$, so the disk as viewed from the grain subtends a very small angle. Therefore, in this expression for $|E_{\mathrm{i}} + E_{\mathrm{s}}|^2$, we may neglect terms with z_0^{-2}. In this way, we find

$$|E_{\mathrm{i}} + E_{\mathrm{s}}|^2 = |E_{\mathrm{i}}|^2 - \frac{2|E_{\mathrm{i}}|^2}{kz_0} \operatorname{Re}\left\{ \frac{S(0)}{i} \exp\left(ik\frac{x^2+y^2}{2z_0}\right) \right\}.$$

To extract the flux F from (2.14), we have to evaluate the integral (see (A.27))

$$\int_{-\infty}^{\infty} e^{\frac{ik(x^2+y^2)}{2z_0}} \, dx \, dy = \left[\int_{-\infty}^{\infty} e^{\frac{ikx^2}{2z_0}} \, dx \right]^2 = i\frac{2\pi z_0}{k}. \tag{2.17}$$

Of course, the disk does not really extend to infinity. But the integral still gives more or less the correct value as long as the disk diameter is much greater than $\sqrt{z_0\lambda} = \sqrt{2\pi z_0/k}$, i.e. it is required that

$$\lambda \ll d \simeq \sqrt{A} \ll z_0.$$

We, therefore, obtain the flux

$$F = |E_{\mathrm{i}}|^2 \left(A - \frac{4\pi}{k^2} \operatorname{Re}\{S(0)\} \right).$$

Now without an obstacle, the flux through the disk would obviously be $A|E_{\mathrm{i}}|^2$ and thus greater. The light that has been removed by the particle determines its cross section for extinction:

$$C^{\mathrm{ext}} = \frac{\lambda^2}{\pi} \operatorname{Re}\{S(0)\}. \tag{2.18}$$

This is the grand extinction formula, also known as the optical theorem. It is baffling because it asserts that C^{ext}, which includes absorption plus scattering into *all directions*, is fixed by the scattering amplitude in the *forward direction* alone. The purely mathematical derivation of (2.18) may not be satisfying. But

Figure 2.3. Light passes through a plane-parallel slab of thickness l with one face lying in the (x, y)-plane. The slab is uniformly filled with identically scattering particles. We determine the field at point P with coordinates $(0, 0, z)$.

obviously C^{ext} must depend on $S(0)$: the extinction cross section of an obstacle specifies how much light is removed from a beam if one observes at a large distance, no matter how and where it is removed (by absorption or scattering). If we let the disk of figure 2.2 serve as a detector and place it far away from the particle, it receives only the forward scattered light and, therefore, $S(0)$ contains the information about C^{ext}.

2.2.2 The refractive index of a dusty medium

One may also assign a refractive index to a dusty medium, like a cloud of grains. Let a plane wave traveling in the z-direction pass through a slab as depicted in figure 2.3. The slab is of thickness l and uniformly filled with identically scattering grains of number density N. When we compute the field at point P on the z-axis resulting from interference of all waves scattered by the grains, we have to sum over all particles. This leads to an integral in the (x, y)-plane like the one in (2.17) of the preceding subsection where we considered only one particle. But there is now another integration necessary in the z-direction extending from 0 to l. Altogether the field at P is:

$$E_i + E_s = E_i \left\{ 1 - N \frac{S(0)}{ikz} \int_0^l dz \, \frac{1}{z} \int_{-\infty}^{\infty} dx \, dy \, e^{\frac{ik(x^2+y^2)}{2z}} \right\}$$

$$= E_i \left(1 - S(0) \frac{\lambda^2}{2\pi} Nl \right). \tag{2.19}$$

The main contribution to the double integral again comes from a region of area $z_0\lambda$ (see figure 2.2 for a definition of z_0). We see from (2.19) that the total field at P is different from the incident field E_i and it is obtained by multiplying E_i by the factor

$$1 - S(0) \frac{\lambda^2}{2\pi} Nl. \tag{2.20}$$

Not only do the grains reduce the intensity of radiation in the forward direction but they also change the phase of the wave because the function $S(0)$ is complex. We can, therefore, assign to the slab an optical constant (with a bar on top)

$$\overline{m} = \overline{n} + i\overline{k}.$$

The field within the slab then varies according to (1.41), like $e^{i(k\overline{m}z - \omega t)}$. If the slab were empty containing only a vacuum, it would vary like $e^{i(kz - \omega t)}$. Therefore, the presence of the grains causes a *change* in the field at P by a factor $e^{ikl(\overline{m} - 1)}$. When $\overline{m}$ is close to one ($|\overline{m} - 1| \ll 1$), which is certainly true for any interstellar dust cloud,

$$e^{ikl(\overline{m} - 1)} \simeq 1 - ikl(\overline{m} - 1). \tag{2.21}$$

Setting (2.20) equal to (2.21) yields

$$\overline{n} - 1 = \frac{2\pi N}{k^3} \, \text{Im}\{S(0)\} \tag{2.22}$$

$$\overline{k} = -\frac{2\pi N}{k^3} \, \text{Re}\{S(0)\}. \tag{2.23}$$

The refractive index of the slab, $\overline{m} = \overline{n} + i\overline{k}$, refers to a medium that consists of a vacuum plus uniformly distributed particles. The way $\overline{m}$ is defined, it has the property that if the particles are pure scatterers, $\overline{k}$ is nevertheless positive. This follows from comparing (2.18) and (2.23) because the extinction coefficient C^{ext} of a single grain does not vanish. A positive $\overline{k}$ implies some kind of dissipation, which seems unphysical as no light is absorbed. One may, therefore, wonder whether $\overline{n}$ and $\overline{k}$ obey the Kramers–Kronig relation of section 2.5. However, they do, as one can show by studying the frequency dependence of $S(0)$.

To find the refractive index $\overline{m} = \overline{n} + i\overline{k}$ from (2.22) and (2.23) for a cloud filled with identical spheres of size parameter $x = 2\pi a/\lambda$ and refractive index $m = n + ik$, one can calculate $S(0)$, with m and x given, from (2.65).

2.3   Mie theory for a sphere

Scattering and absorption of light by spheres is a problem of classical electrodynamics. Its full derivation is lengthy and we present only the sequence of the main steps. Missing links may be filled in from the special literature (for instance, [Boh83, Hul57, Ker69]). The reader not interested in the mathematics may skip the next pages, or even cut them out carefully, and resume the text with the last subsection entitled *Absorption and scattering efficiencies*. What is in between can be condensed into one sentence: In Mie theory, one finds the scattered electromagnetic field and the field inside the particle by expanding both into an infinite series of independent solutions to the wave equation; the series coefficients are determined from the boundary conditions on the particle surface.

2.3.1 The generating function

Consider a spherical particle in vacuum illuminated by a linearly polarized monochromatic plane wave of frequency $\nu = \omega/2\pi$. Let $\mathbf{E}_i$ and $\mathbf{H}_i$ describe the *i*ncident field. We denote the field within the particle by $\mathbf{E}_1, \mathbf{H}_1$ and outside of it by $\mathbf{E}_2, \mathbf{H}_2$. The field outside is the superposition of the incident and the *s*cattered field (subscript s),

$$\mathbf{E}_2 = \mathbf{E}_i + \mathbf{E}_s \tag{2.24}$$

$$\mathbf{H}_2 = \mathbf{H}_i + \mathbf{H}_s. \tag{2.25}$$

The further calculations are greatly simplified by the following relations. Let $\mathbf{c}$ be an arbitrary constant vector and ψ a solution to the *scalar wave equation*

$$\Delta\psi + k^2\psi = 0 \tag{2.26}$$

where

$$k^2 = \frac{\omega^2\mu\varepsilon}{c^2}$$

from (1.41) (or (1.116) in the case of conductivity). All material properties are taken into account by the wavenumber k. Then the vector function $\mathbf{M}$, defined by

$$\mathbf{M} = \mathrm{rot}(\mathbf{c}\psi) \tag{2.27}$$

is divergence-free (div $\mathbf{M} = 0$) and a solution to the *vector equation*

$$\Delta\mathbf{M} + k^2\mathbf{M} = 0. \tag{2.28}$$

This is easy to prove either by the standard formulae of vector analysis or component–wise. The vector function $\mathbf{N}$ given by

$$\mathbf{N} = \frac{1}{k}\,\mathrm{rot}\,\mathbf{M} \tag{2.29}$$

also obeys the wave equation

$$\Delta\mathbf{N} + k^2\mathbf{N} = 0 \tag{2.30}$$

and $\mathbf{M}$ and $\mathbf{N}$ are related through

$$\mathrm{rot}\,\mathbf{N} = k\mathbf{M}. \tag{2.31}$$

2.3.2 Separation of variables

In this way, the problem of finding a solution to the vector wave equation reduces to finding one for the scalar wave equation. We start with the vector function

$$\mathbf{M} = \mathrm{rot}(\mathbf{r}\psi). \tag{2.32}$$

M is tangential to the sphere of radius $|\mathbf{r}|$ because the scalar product of $\mathbf{r}$ and **M** vanishes: $\mathbf{r} \cdot \mathbf{M} = 0$. In spherical polar coordinates (r, θ, ϕ), the wave equation for ψ reads:

$$\frac{1}{r^2} \frac{\partial}{\partial r}\left(r^2 \frac{\partial \psi}{\partial r}\right) + \frac{1}{r^2 \sin\theta} \frac{\partial}{\partial \theta}\left(\sin\theta \frac{\partial \psi}{\partial \theta}\right) + \frac{1}{r^2 \sin^2\theta} \frac{\partial^2 \psi}{\partial \phi^2} + k^2 \psi = 0. \quad (2.33)$$

We make an ansatz of separated variables

$$\psi(r, \theta, \phi) = R(r) \cdot T(\theta) \cdot P(\phi). \quad (2.34)$$

Arranging (2.33) in such a way that its left-hand side depends only on r and its right-hand side only on θ and ϕ, both sides must equal a constant value, which we write as $n(n + 1)$. In the same spirit, we can separate θ and ϕ. This leads to the three equations:

$$\frac{d^2 P}{d\phi^2} + m^2 P = 0 \quad (2.35)$$

$$\frac{1}{\sin\theta} \frac{d}{d\theta}\left(\sin\theta \frac{dT}{d\theta}\right) + \left[n(n+1) - \frac{m^2}{\sin^2\theta}\right] T = 0 \quad (2.36)$$

$$\frac{d}{dr}\left(r^2 \frac{dR}{dr}\right) + \left[k^2 r - n(n+1)\right] R = 0. \quad (2.37)$$

The linearly independent solutions to (2.35) are $\sin m\phi$ and $\cos m\phi$. Because they must be single-valued ($P(\phi) = P(\phi + 2\pi)$) it follows that $m = 0, \pm 1, \pm 2, \ldots$. Equation (2.36) is satisfied by the Legendre functions of the first kind $P_n^m(\cos\theta)$, where n and m are integer and $m \in [-n, n]$. Formula (2.37) has, as solutions, the spherical Bessel functions of the first (j_n) and second (y_n) kind

$$j_n(\rho) = \sqrt{\frac{\pi}{2\rho}} J_{n+\frac{1}{2}}(\rho) \quad (2.38)$$

$$y_n(\rho) = \sqrt{\frac{\pi}{2\rho}} Y_{n+\frac{1}{2}}(\rho) \quad (2.39)$$

where $\rho = kr$ and $n + \frac{1}{2}$ is half-integer. Altogether we obtain

$$\psi_{emn} = \cos(m\phi) \cdot P_n^m(\cos\theta) \cdot z_n(kr) \quad (2.40)$$

$$\psi_{omn} = \sin(m\phi) \cdot P_n^m(\cos\theta) \cdot z_n(kr). \quad (2.41)$$

Here z_n may either equal j_n or y_n. In the subindices of ψ, e stands for *even* (associated with cosine terms) and o for *odd* (sine terms). As ψ is the generating function for **M**, we get, from (2.27),

$$\mathbf{M}_{emn} = \text{rot}(\mathbf{r}\psi_{emn}) \quad (2.42)$$

$$\mathbf{M}_{omn} = \text{rot}(\mathbf{r}\psi_{omn}). \quad (2.43)$$

$\mathbf{N}_{emn}$ and $\mathbf{N}_{omn}$ then follow from (2.29).

2.3.3 Series expansion of waves

In the next step, one expands the incident, internal and scattered waves into the spherical harmonics $\mathbf{M}_{emn}$, $\mathbf{M}_{omn}$, $\mathbf{N}_{emn}$ and $\mathbf{N}_{omn}$. Starting with the incident plane wave $\mathbf{E}_i$, the not straightforward result is

$$\mathbf{E}_i = \sum_{n=1}^{\infty} E_n \left(\mathbf{M}_{o1n}^{(1)} - i\mathbf{N}_{e1n}^{(1)} \right) \qquad \text{with } E_n = E_0 i^n \frac{2n+1}{n(n+1)} \qquad (2.44)$$

where the scalar E_0 denotes the amplitude of the incident wave. The awkward expression (2.44) for the simple incident wave is the consequence of using spherical coordinates which are not suited for planar geometry; but, of course, these coordinates are the right ones for the scattered light. The superscript (1) at $\mathbf{M}$ and $\mathbf{N}$ signifies that the radial dependence of the generating function ψ is given by j_n and not by y_n; the latter can be excluded because of its behavior at the origin. Note that all coefficients with $m \neq 1$ have disappeared. For the rest of this paragraph, the symbol m is reserved for the optical constant.

For the internal and scattered field one obtains

$$\mathbf{E}_i = \sum_{n=1}^{\infty} E_n (c_n \mathbf{M}_{o1n}^{(1)} - i d_n \mathbf{N}_{e1n}^{(1)}) \qquad (2.45)$$

$$\mathbf{E}_s = \sum_{n=1}^{\infty} E_n (a_n \mathbf{N}_{e1n}^{(3)} - b_n \mathbf{M}_{o1n}^{(3)}). \qquad (2.46)$$

Superscript (3) denotes that the dependence of the generating function is given by the spherical Hankel function $h_n^{(1)}(z) = j_n(z) + i y_n(z)$ of order n. At large distances ($kr \gg n^2$), it behaves like

$$h_n^{(1)}(kr) \simeq \frac{(-i)^n e^{ikr}}{ikr}. \qquad (2.47)$$

The magnetic fields $\mathbf{H}_i$, $\mathbf{H}_s$ follow from the corresponding electric fields by applying the curl after Maxwell's equation (1.25).

2.3.4 Expansion coefficients

The expansion coefficients in (2.45) and (2.46) follow from the boundary conditions of the electromagnetic field at the surface of the grain. To derive the latter, consider a small loop in the shape of a rectangle with two long and two very much shorter sides. One long side lies just outside the particle in vacuum and runs parallel to the surface S of the grain, the other is immediately below the boundary within the particle. When we integrate $\mathrm{rot}\,\mathbf{E} = (i\omega\mu/c)\mathbf{H}$ from (1.32) over the rectangular loop employing Stokes' theorem (A.15) and make the loop infinitesimally small, the integral vanishes so that the tangential components of $\mathbf{E}$

must be continuous at the boundary of the grain. Applying the same procedure to equation (1.33), or to (1.113) for metals, we get the same result for the tangential component of **H**. Therefore, at all points $\mathbf{x} \in S$

$$[\mathbf{E}_2(\mathbf{x}) - \mathbf{E}_1(\mathbf{x})] \times \mathbf{e} = 0 \tag{2.48}$$

$$[\mathbf{H}_2(\mathbf{x}) - \mathbf{H}_1(\mathbf{x})] \times \mathbf{e} = 0 \tag{2.49}$$

where $\mathbf{e}$ is the outward directed normal to the surface S. Substituting for $\mathbf{E}_2, \mathbf{H}_2$ from (2.24) and (2.25) yields for the components of the fields

$$E_{1\theta} = E_{i\theta} + E_{s\theta}$$
$$E_{1\phi} = E_{i\phi} + E_{s\phi}$$
$$H_{1\theta} = H_{i\theta} + H_{s\theta}$$
$$H_{1\phi} = H_{i\phi} + H_{s\phi}.$$

This set of equations leads to four linear equations for the expansion coefficients a_n, b_n, c_n and d_n of the internal and scattered field (see (2.45) to (2.46)). If λ is the wavelength of the incident radiation, m the complex optical constant of the sphere, a its radius, and

$$x = \frac{2\pi a}{\lambda}$$

the size parameter, then a_n and b_n are given by

$$a_n = \frac{\psi_n(x) \cdot \psi_n'(mx) - m\psi_n(mx) \cdot \psi_n'(x)}{\zeta_n(x) \cdot \psi_n'(mx) - m\psi_n(mx) \cdot \zeta_n'(x)} \tag{2.50}$$

$$b_n = \frac{m\psi_n(x) \cdot \psi_n'(mx) - \psi_n(mx) \cdot \psi_n'(x)}{m\zeta_n(x) \cdot \psi_n'(mx) - \psi_n(mx) \cdot \zeta_n'(x)}. \tag{2.51}$$

In the trivial case, when the optical constant $m = 1$, the scattered field disappears as $a_n = b_n = 0$. The complex functions

$$\psi_n(z) = z j_n(z)$$
$$\psi_n'(z) = z j_{n-1}(z) - n j_n(z)$$
$$\zeta_n(z) = z[j_n(z) + i y_n(z)] = z h_n^{(1)}(z)$$
$$\zeta_n'(z) = z[j_{n-1}(z) + i y_{n-1}(z)] - n[j_n(z) + i y_n(z)]$$

may be calculated from the recurrence relations [Abr70, section 10.1.19]

$$j_n(z) = -j_{n-2}(z) + \frac{2n-1}{z} j_{n-1}(z)$$

$$y_n(z) = -y_{n-2}(z) + \frac{2n-1}{z} y_{n-1}(z)$$

starting with [Abr70, sections 10.1.11 and 10.1.12]

$$j_0(z) = \frac{\sin z}{z} \qquad j_1(z) = \frac{\sin z}{z^2} - \frac{\cos z}{z}$$

$$y_0(z) = -\frac{\cos z}{z} \qquad y_1(z) = -\frac{\cos z}{z^2} - \frac{\sin z}{z}.$$

When $|mx|$ is of order 100 or bigger, one encounters numerical difficulties because the complex Bessel functions contain the term e^z which becomes excessively large. A numerically much superior algorithm is printed in appendix A of [Boh83].

2.3.5 Scattered and absorbed power

If we imagine the particle to be surrounded by a large spherical and totally transparent surface A, the energy absorbed by the grain, W_a, is given by the difference between the flux which enters and which leaves the sphere. If $\mathbf{S}$ is the Poynting vector of the electromagnetic field outside the particle,

$$\mathbf{S} = \frac{c}{8\pi} \operatorname{Re}\{\mathbf{E}_2 \times \mathbf{H}_2^*\}$$

where $\mathbf{E}_2$ or $\mathbf{H}_2$ is the sum of the incident and scattered field from (2.24) and (2.25), W_a is determined by the integral

$$W_a = -\int_A \mathbf{S} \cdot \mathbf{e}_r \, dA. \tag{2.52}$$

Here $\mathbf{e}_r$ is the outward normal of the surface, and the minus sign ensures that W_a is positive. The Poynting vector $\mathbf{S}$ can be considered to consist of three parts:

$$\mathbf{S} = \mathbf{S}_i + \mathbf{S}_s + \mathbf{S}_{\text{ext}}$$

with

$$\mathbf{S}_i = (c/8\pi) \operatorname{Re}\{\mathbf{E}_i \times \mathbf{H}_i^*\}$$
$$\mathbf{S}_s = (c/8\pi) \operatorname{Re}\{\mathbf{E}_s \times \mathbf{H}_s^*\}$$
$$\mathbf{S}_{\text{ext}} = (c/8\pi) \operatorname{Re}\{\mathbf{E}_i \times \mathbf{H}_s^* + \mathbf{E}_s \times \mathbf{H}_i^*\}$$

and, therefore,

$$-W_a = \int_A \mathbf{S}_i \cdot \mathbf{e}_r \, dA + \int_A \mathbf{S}_s \cdot \mathbf{e}_r \, dA + \int_A \mathbf{S}_{\text{ext}} \cdot \mathbf{e}_r \, dA.$$

The first integral vanishes because the *incident* field (subscript i) enters and leaves the sphere without modification. The second integral obviously describes the scattered energy,

$$W_s = \frac{c}{8\pi} \operatorname{Re} \int_0^{2\pi} \int_0^{\pi} \left(E_{s\theta} H_{s\phi}^* - E_{s\phi} H_{s\theta}^* \right) r^2 \sin\theta \, d\theta \, d\phi. \tag{2.53}$$

The negative of the third integral, which we denote by $-W_{\text{ext}}$, is, therefore, the sum of absorbed plus scattered energy:

$$
\begin{aligned}
W_{\text{ext}} &= W_{\text{a}} + W_{\text{s}} \\
&= \frac{c}{8\pi} \operatorname{Re} \int_0^{2\pi} \int_0^{\pi} \left(E_{i\phi} H_{s\theta}^* - E_{i\theta} H_{s\phi}^* - E_{s\theta} H_{i\phi}^* + E_{s\phi} H_{i\theta}^* \right) r^2 \sin\theta \, d\theta \, d\phi
\end{aligned}
$$

$$(2.54)$$

and thus the total energy removed from the beam at a large distance. According to our definition in section 2.1, W_{a} is related to the absorption coefficient of the particle, C^{abs}, through $W_{\text{a}} = S_i C^{\text{abs}}$, where S_i is the time-averaged Poynting vector of the incident field. Likewise for the scattering and extinction coefficient, $W_{\text{s}} = S_i C^{\text{sca}}$ and $W_{\text{ext}} = S_i C^{\text{ext}}$.

2.3.6 Absorption and scattering efficiencies

As the fields $\mathbf{E}_i$ and $\mathbf{E}_s$ have been evaluated in (2.44)–(2.46), one obtains from (2.53) and (2.54) after some algebra the following formulae for the efficiencies of extinction and scattering:

$$
Q^{\text{ext}} = \frac{2}{x^2} \sum_{n=1}^{\infty} (2n+1) \cdot \operatorname{Re}\{a_n + b_n\}
$$

$$(2.55)$$

$$
Q^{\text{sca}} = \frac{2}{x^2} \sum_{n=1}^{\infty} (2n+1) \cdot \left[|a_n|^2 + |b_n|^2 \right].
$$

$$(2.56)$$

They must usually be evaluated with the help of a computer. The expansion coefficients a_n, b_n are given in (2.50) and (2.51). The asymmetry factor of (2.6) becomes

$$
g = \frac{4}{x^2 Q^{\text{sca}}} \sum_{n=1}^{\infty} \left[\frac{n(n+2)}{n+1} \operatorname{Re}\{a_n^* a_{n+1} + b_n^* b_{n+1}\} + \frac{2n+1}{n(n+1)} \operatorname{Re}\{a_n^* b_n\} \right].
$$

$$(2.57)$$

2.4 Polarization and scattering

We introduce the amplitude scattering matrix, define the Stokes parameters and compute the radiation field scattered into a certain direction. A number of results in this section are merely summarized, not fully derived but the missing links concern only the mathematics.

2.4.1 The amplitude scattering matrix

Consider a plane harmonic wave propagating in the z-direction and a particle at the origin of the coordinate system. Some light is scattered by the particle into a certain direction given by the unit vector **e**. The z-axis together with **e** define what is called the *scattering plane*.

The amplitude of the incident electric field $\mathbf{E}_i$ may be decomposed into two vectors, one parallel, the other perpendicular to the scattering plane; their lengths are denoted by $E_{i\parallel}$ and $E_{i\perp}$, respectively. Likewise, one may decompose the scattered electric field $\mathbf{E}_s$, the components being $E_{s\parallel}$ and $E_{s\perp}$. A three-dimensional model would help visualize the geometrical configuration but a two-dimensional drawing is no better than words and, therefore, missing.

At a large distance r from the particle, in the far field where $r \gg \lambda$, there is a linear relation between $(E_{s\perp}, E_{s\parallel})$ and $(E_{i\perp}, E_{i\parallel})$ described by the amplitude scattering matrix,

$$\begin{pmatrix} E_{s\parallel} \\ E_{s\perp} \end{pmatrix} = \frac{e^{ik(r-z)}}{-ikr} \begin{pmatrix} S_2 & S_3 \\ S_4 & S_1 \end{pmatrix} \begin{pmatrix} E_{i\parallel} \\ E_{i\perp} \end{pmatrix}. \tag{2.58}$$

The factor before the amplitude scattering matrix is the same as in (2.15) where we considered a scalar field, with only one function S. Here we deal with vectors and there is an amplitude scattering matrix consisting of four elements S_j. They depend, of course, on the scattering direction, which is specified by the unit vector **e** or by two angles, θ and ϕ.

As an example, we work out the scattering matrix of a small grain of unit volume. If its polarizability α is isotropic, its dipole, **p**, due to the incident wave is

$$\mathbf{p} = \alpha \mathbf{E}_i.$$

The scattered electric field is given by equation (1.95),

$$\mathbf{E}_s = \frac{\omega^2 \alpha}{c^2 r} (\mathbf{e} \times \mathbf{E}_i) \times \mathbf{e}.$$

It is always transverse to **e** and depends only on **e** and **p** (or $\mathbf{E}_i$) and not on the direction **k** from which the incident wave is coming. With the help of figure 2.4 we can easily figure out that $E_{s\parallel} = \cos\theta\, E_{i\parallel}$ and $E_{i\perp} = E_{s\perp}$, so for a dipole

$$\begin{pmatrix} S_2 & S_3 \\ S_4 & S_1 \end{pmatrix} = -ik^3\alpha \begin{pmatrix} \cos\theta & 0 \\ 0 & 1 \end{pmatrix}. \tag{2.59}$$

2.4.2 Angle-dependence of scattering

In the case of a sphere, the amplitude scattering matrix (2.58) also acquires a diagonal structure as the elements S_3 and S_4 vanish:

$$\begin{pmatrix} E_{s\parallel} \\ E_{s\perp} \end{pmatrix} = \frac{e^{ik(r-z)}}{-ikr} \begin{pmatrix} S_2 & 0 \\ 0 & S_1 \end{pmatrix} \begin{pmatrix} E_{i\parallel} \\ E_{i\perp} \end{pmatrix}. \tag{2.60}$$

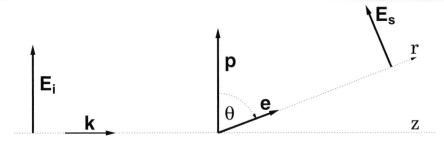

Figure 2.4. An incident wave with electric field $\mathbf{E}_i$ excites a dipole $\mathbf{p}$ and is scattered into the direction of the unit vector $\mathbf{e}$. The scattered wave has the electric field $\mathbf{E}_s$. The incident wave travels in the direction of the wavenumber $\mathbf{k}$. Both $\mathbf{E}_i$ and $\mathbf{E}_s$ lie in the scattering plane which is defined by $\mathbf{e}$ and $\mathbf{k}$. Alternatively (this scenario is not shown in the figure), $\mathbf{E}_i$ and $\mathbf{p}$ may be perpendicular to the scattering plane given by the same vectors $\mathbf{k}$ and $\mathbf{e}$. Then $\mathbf{E}_s$ is also perpendicular to the scattering plane.

S_1, S_2 depend only on $\mu = \cos\theta$, where $\theta = 0$ denotes the forward direction. They can again be expressed with the help of the expansion coefficients a_n, b_n from (2.50) and (2.51),

$$S_1 = \sum_n \frac{2n+1}{n(n+1)} (a_n \pi_n + b_n \tau_n) \qquad (2.61)$$

$$S_2 = \sum_n \frac{2n+1}{n(n+1)} (a_n \tau_n + b_n \pi_n) \qquad (2.62)$$

with

$$\pi_n(\cos\theta) = \frac{P_n^1(\cos\theta)}{\sin\theta} \qquad \tau_n(\cos\theta) = \frac{dP_n^1}{d\theta}.$$

The functions π_n and τ_n are computed from the recurrence relations

$$\pi_n(\mu) = \frac{2n-1}{n-1} \mu \pi_{n-1} - \frac{n}{n-1} \pi_{n-2}$$
$$\tau_n(\mu) = n\mu\pi_n - (n+1)\pi_{n-1}$$

beginning with

$$\pi_0 = 0 \qquad \pi_1 = 1.$$

- For unpolarized incident light ($E_{i\|} = E_{i\perp}$), the intensity of the radiation scattered into the direction θ is given by

$$S_{11}(\cos\theta) = \frac{1}{2}\left[|S_1|^2 + |S_2|^2\right]. \qquad (2.63)$$

The factor before the matrix in (2.60) has been negelected. The notation S_{11} comes from equation (2.76); in (2.1) the same quantity was denoted $\mathcal{L}$.

When integrated over all directions,

$$\int_0^\pi S_{11}(\cos\theta)\sin\theta\,d\theta = \tfrac{1}{2}x^2 Q^{\text{sca}}.$$

- The angle-dependence of the normalized phase function $f(\cos\theta)$ (see (2.5)) is related to S_{11} through

$$S_{11}(\cos\theta) = \frac{x^2 Q^{\text{sca}}}{4} f(\cos\theta). \tag{2.64}$$

- We now have two ways to determine the extinction cross section $C^{\text{ext}} = \pi a^2 Q^{\text{ext}}$ of a grain. Either from (2.55) or by inserting $S_1(0) = S_2(0)$ of (2.61) into the general extinction formula (2.18). Comparison of the two formulae yields

$$S_1(0) = S_2(0) = \tfrac{1}{2}\sum_n (2n+1)\cdot \text{Re}\{a_n + b_n\}. \tag{2.65}$$

- We add for completeness the formula for the *backscattering* efficiency. It is defined by

$$Q^{\text{back}} = 4\frac{S_{11}(180°)}{x^2} \tag{2.66}$$

and it is a useful quantity in radar measurements or whenever a particle is illuminated by a source located between the particle and the observer. Its series expansion is

$$Q^{\text{back}} = \frac{1}{x^2}\left|\sum_{n=1}^\infty (-1)^n (2n+1)\cdot(a_n - b_n)\right|^2. \tag{2.67}$$

For a large and perfectly reflecting sphere, $Q^{\text{back}} = 1$.

2.4.3 The polarization ellipse

Consider a plane harmonic wave of angular frequency ω, wavenumber $\mathbf{k}$ and electric field

$$\mathbf{E}(\mathbf{x}, t) = \mathbf{E}_0 \cdot e^{i(\mathbf{k}\cdot\mathbf{x} - \omega t)}$$

(see (1.36) and (1.37)) that travels in the z-direction of a Cartesian coordinate system. The amplitude $\mathbf{E}_0$ is, in the general case, complex,

$$\mathbf{E}_0 = \mathbf{E}_1 + i\mathbf{E}_2$$

with real $\mathbf{E}_1$ and $\mathbf{E}_2$. At any fixed z, the real part of the electric vector $\mathbf{E}$ rotates at frequency ω and the tip of the vector describes in the (x, y)-plane, which is perpendicular to the z-axis, an ellipse,

$$\text{Re}\{\mathbf{E}\} = \mathbf{E}_1 \cos\omega t + \mathbf{E}_2 \sin\omega t. \tag{2.68}$$

The sense of rotation changes when the plus sign in (2.68) is altered into a minus sign, i.e. when $\mathbf{E}_2$ flips direction by an angle π. There are two special cases:

- *Linear polarization.* For $\mathbf{E}_1 = 0$ or $\mathbf{E}_2 = 0$, or when $\mathbf{E}_1, \mathbf{E}_2$ are linearly dependent, the ellipse degenerates into a line. By adjusting the time t, it is always possible to make $\mathbf{E}_0$ real. At a fixed location z, the electric vector swings up and down along a straight line whereby its length changes; it vanishes when Re{$\mathbf{E}$} switches direction. The field can drive a linear oscillator in the (x, y)-plane.
- *Circular polarization.* The vectors are of equal length, $|\mathbf{E}_1| = |\mathbf{E}_2|$, and perpendicular to each other, $\mathbf{E}_1 \cdot \mathbf{E}_2 = 0$. The ellipse is a circle. At a fixed location z, the electric vector never disappears but rotates retaining its full length. The circularly polarized wave sets a two-dimensional harmonic oscillator in the (x, y)-plane (with equal properties in the x- and y-directions) into a circular motion; it transmits angular momentum.

One can combine two linearly polarized waves to obtain circular polarization and two circularly polarized waves to obtain linear polarization. The magnetic field has the same polarization as the electric field because $\mathbf{E}$ and $\mathbf{H}$ are in phase and have a constant ratio.

2.4.4   Stokes parameters

The polarization ellipse is completely determined by the length of its major and minor axes, a and b, plus some specification of its orientation in the (x, y)-plane. This could be the angle γ between the major axis and the x-coordinate. Instead of these geometrical quantities (a, b, γ), polarization is usually described by the Stokes parameters I, Q, U and V. They are equivalent to (a, b, γ) but have the practical advantage that they can be measured directly. We omit the underlying simple mathematical relations as well as the description of the experimental setup.

When a plane harmonic wave is scattered by a grain, the Stokes parameters of the *i*ncident (subscript i) and the *s*cattered wave (subscript s) are linearly related through

$$
\begin{pmatrix} I_s \\ Q_s \\ U_s \\ V_s \end{pmatrix} = \frac{1}{k^2 r^2} \begin{pmatrix} S_{11} & S_{12} & S_{13} & S_{14} \\ S_{21} & S_{22} & S_{23} & S_{24} \\ S_{31} & S_{32} & S_{33} & S_{34} \\ S_{41} & S_{42} & S_{43} & S_{44} \end{pmatrix} \begin{pmatrix} I_i \\ Q_i \\ U_i \\ V_i \end{pmatrix}. \tag{2.69}
$$

The Stokes parameters of the scattered light as well as the matrix elements refer to a particular scattering direction (θ, ϕ); r is the distance from the particle. The matrix S_{ij} contains no more information than the matrix in (2.58) and, therefore, only seven of its 16 elements are independent, which corresponds to the fact that the four elements S_j in the matrix of (2.58) have four absolute values $|S_j|$ and three phase differences between them.

- Only three of the four Stokes parameters are independent as for fully polarized light

$$I^2 = Q^2 + U^2 + V^2. \tag{2.70}$$

- Unpolarized light has

$$Q = U = V = 0$$

and for partially polarized light, one has, instead of the strict equality $I^2 = Q^2 + U^2 + V^2$, an inequality

$$I^2 > Q^2 + U^2 + V^2.$$

- For unit intensity ($I = 1$), linearly polarized light has

$$V = 0 \qquad Q = \cos 2\gamma \qquad U = \sin 2\gamma$$

so that

$$Q^2 + U^2 = 1.$$

The degree of linear polarization is defined by

$$\frac{\sqrt{Q^2 + U^2}}{I} \leq 1. \tag{2.71}$$

It can vary between 0 and 1.
- Circular polarization implies

$$Q = U = 0 \qquad V = \pm 1,$$

the sign determines the sense of rotation of the electric vector. The degree of circular polarization is

$$-1 \leq \frac{V}{I} \leq 1. \tag{2.72}$$

- Even when the incident light is unpolarized, i.e. when

$$Q_i = U_i = V_i = 0 \tag{2.73}$$

it becomes partially polarized after scattering if S_{21}, S_{31} or S_{41} are non-zero. Indeed, dropping the factor $1/k^2 r^2$ in (2.69), we get

$$I_s = S_{11} I_i \qquad Q_s = S_{21} I_i \qquad U_s = S_{31} I_i \qquad V_s = S_{41} I_i. \tag{2.74}$$

2.4.5 Stokes parameters of scattered light for a sphere

In the case of a sphere, the transformation matrix (2.69) between incident and scattered Stokes parameters simplifies to

$$\begin{pmatrix} I_s \\ Q_s \\ U_s \\ V_s \end{pmatrix} = \frac{1}{k^2 r^2} \begin{pmatrix} S_{11} & S_{12} & 0 & 0 \\ S_{12} & S_{11} & 0 & 0 \\ 0 & 0 & S_{33} & S_{34} \\ 0 & 0 & -S_{34} & S_{33} \end{pmatrix} \begin{pmatrix} I_i \\ Q_i \\ U_i \\ V_i \end{pmatrix}. \tag{2.75}$$

Out of the 16 matrix elements, eight are non-trivial and they have only four significantly different values:

$$S_{11} = \tfrac{1}{2}(|S_1|^2 + |S_2|^2) \qquad (2.76)$$
$$S_{12} = \tfrac{1}{2}(|S_2|^2 - |S_1|^2)$$
$$S_{33} = \tfrac{1}{2}(S_2^* S_1 + S_2 S_1^*)$$
$$S_{34} = \tfrac{i}{2}(S_1 S_2^* - S_2 S_1^*).$$

Only three of them are independent because

$$S_{11}^2 = S_{12}^2 + S_{33}^2 + S_{34}^2.$$

- If the incident light is 100% polarized and its electric vector *parallel* to the scattering plane so that $I_i = Q_i$, $U_i = V_i = 0$, we get, dropping the factor $1/k^2 r^2$ in (2.75),

$$I_s = (S_{11} + S_{12}) I_i \qquad Q_s = I_s \qquad U_s = V_s = 0.$$

  So the scattered light is also 100% polarized parallel to the scattering plane.
- Likewise, if the incident light is 100% polarized *perpendicular* to the scattering plane ($I_i = -Q_i$, $U_i = V_i = 0$), so is the scattered light and

$$I_s = (S_{11} - S_{12}) I_i \qquad Q_s = -I_s \qquad U_s = V_s = 0.$$

- If the incident wave is unpolarized ($Q_i = U_i = V_i = 0$), the scattered light is nevertheless usually polarized; in this case,

$$I_s = S_{11} I_i \qquad Q_s = S_{12} I_i \qquad U_s = V_s = 0.$$

When one defines the quantity p as the difference of the intensities $|S_1|^2$ and $|S_2|^2$ divided by their sum,

$$p = \frac{|S_1|^2 - |S_2|^2}{|S_1|^2 + |S_2|^2} = -\frac{S_{12}}{S_{11}} \qquad (2.77)$$

the absolute value $|p|$ is equal to the degree of linear polarization of (2.71). We will not compose a new name for p but also call it degree of polarization, although it contains, via the sign, additional information. In the forward direction, $S_1(0) = S_2(0)$ and $p = 0$.
- The sign of p, or of S_{12}, specifies the direction of polarization. Usually, S_{12} is negative and then linear polarization is *perpendicular* to the scattering plane. But S_{12} can, from (2.76), also be positive, as happens for big spheres (see figures 4.7 and 4.8). Then polarization is *parallel* to the scattering plane.

In the general case, when the particles are not spherical (anisotropic) and which is outside the scope of what we calculate here, none of the matrix elements in (2.69) vanishes. Then the polarization vector can have any inclination towards the scattering plane and p may also be non-zero in the forward direction.

2.5   The Kramers–Kronig relations

The formulae of *Kramers* and *Kronig* establish a link between the real and imaginary parts of the material constants, like χ, ε, μ or σ (see section 1.1). Their deduction is very abstract and the result baffling. The relations have some fundamental consequences, however, their practical value is, at present, moderate.

2.5.1   Mathematical formulation of the relations

We have to digress briefly into the theory of complex functions. Let

$$f(z) = u(z) + iv(z)$$

be a function of the complex variable $z = x + iy$. The integral over a path p in the z-plane, described through the parametrization $z(t) = x(t) + iy(t)$ with $\alpha \le t \le \beta$, is defined by

$$\int_p f(z)\,dz = \int_\alpha^\beta f(z(t)) \cdot z'(t)\,dt. \tag{2.78}$$

The fundamental theorem of complex functions states that if $f(z)$ is regular in a region G, i.e. it is single-valued and has a derivative, then the integral over any closed path p in G vanishes:

$$\int_p f(z)\,dz = 0. \tag{2.79}$$

For the function $f(z)$ we choose

$$f(z) = \frac{g(z)}{z - x_0} \tag{2.80}$$

where

$$g(z) = g_1(z) + ig_2(z)$$

is also a complex function and x_0 a *real* and positive number. If $f(z)$ is regular in the upper half of the z-plane, the integral along the closed path p, which is depicted in figure 2.5, is zero according to (2.79). The path p runs along the x-axis from left to right, makes a small semicircle of radius δ around x_0, and returns in a big half–circle of radius R to the starting position; we have divided it into four segments p_1 to p_4.

Now assume that R is very big and that $g(z)$ has the property to vanish for $|z| \to \infty$; the integral along the big half–circle p_4 then vanishes too. The small semicircle p_2 may be parametrized by $z(t) = x_0 - \delta e^{-it}$ with $0 \le t \le \pi$. The integral over p_2 gives

$$-i \int_0^\pi g(x_0 - \delta e^{it})\,dt = -i\pi g(x_0) \qquad \text{for } \delta \to 0. \tag{2.81}$$

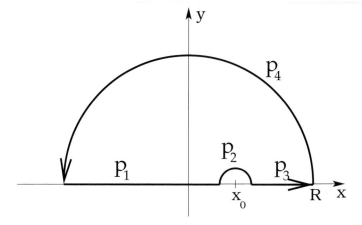

Figure 2.5. To derive the Kramers–Kronig relations, we integrate a function f that is related to the dielectric susceptibility χ, or permeability ε, over a closed path (thick line) in the complex (x, y)-plane. The path consists of four segments, p_1 to p_4. The points in this plane are identified with complex frequencies ω. Only the positive x-axis has a physical meaning but equation (2.88) allows us to define χ for any complex argument ω.

We thus have for the whole path p, from (2.79),

$$\fint_{-\infty}^{\infty} \frac{g(x)}{x - x_0}\, dx - i\pi g(x_0) = 0. \tag{2.82}$$

The integral with the bar is the Cauchy principal value defined by

$$\fint_{-\infty}^{\infty} \frac{g(x)}{x - x_0}\, dx = \lim_{\delta \to 0} \left\{ \int_{-\infty}^{x_0 - \delta} \frac{g(x)}{x - x_0}\, dx + \int_{x_0 + \delta}^{\infty} \frac{g(x)}{x - x_0}\, dx \right\}. \tag{2.83}$$

Equation (2.82) holds separately for the real and imaginary parts. Let the function g be symmetric (for real x) such that $g(-x) = g^*(x)$ or

$$g_1(-x) = g_1(x) \qquad \text{and} \qquad g_2(-x) = -g_2(x).$$

Writing the integral in (2.82) over the whole x-axis as the sum of two integrals with limits from $-\infty$ to 0 and from 0 to $+\infty$ and exploiting the symmetry of g, one obtains the Kramers–Kronig relations after small manipulations

$$g_1(x_0) = \frac{2}{\pi} \fint_0^{\infty} \frac{x g_2(x)}{x^2 - x_0^2}\, dx \tag{2.84}$$

$$g_2(x_0) = -\frac{2x_0}{\pi} \fint_0^{\infty} \frac{g_1(x)}{x^2 - x_0^2}\, dx. \tag{2.85}$$

Putting in (2.84) $x_0 = 0$ yields the special case (*no* Cauchy principal value)

$$g_1(0) = \frac{2}{\pi} \int_0^\infty \frac{g_2(x)}{x} \, dx. \tag{2.86}$$

In formula (2.83), both integrands within the big brackets go at x_0 to infinity like $1/x$ at $x = 0$. Therefore, each of the integrals alone diverges; however, their sum is finite.

2.5.2 The electric susceptibility and causality

In the application of equations (2.84)–(2.86) to a physical situation, the complex function g is identified with the electric susceptibility χ of (1.6) and the variable x with the frequency ω. Likewise, we could use the dielectric permeability ε, with the small modification that we have to put $g_1 = \varepsilon_1 - 1$ and $g_2 = \varepsilon_2$. We reiterate that, in the constitutive relation (1.6),

$$\mathbf{P} = \chi(\omega)\, \mathbf{E}$$

the polarization $\mathbf{P}$ depends linearly on the electric field $\mathbf{E}$ because $\mathbf{E}$ is so much weaker than the fields on the atomic level. In the most general case, the vector $\mathbf{P}$ at time t is not determined by the present value of $\mathbf{E}$ alone but by the whole preceding history. We, therefore, write the linear relation in the form

$$\mathbf{P}(t) = \int_0^\infty F(\tau)\, \mathbf{E}(t - \tau)\, d\tau. \tag{2.87}$$

This equation specifies how $\mathbf{P}$ responds to the application of an electric field $\mathbf{E}$. Besides linearity, formula (2.87) also expresses causality because $\mathbf{P}(t)$ results from an integration over the past. The function $F(\tau)$ depends on time and on the properties of the medium. It has values substantially different from zero only in the immediate past over an interval $\Delta\tau$ corresponding to the time scale for polarizing the molecules. At times much larger than $\Delta\tau$, the function $F(\tau)$ vanishes because the distant past does not influence the present.

When we consider monochromatic electric fields, $\mathbf{E}(t) = \mathbf{E}_0 e^{-i\omega t}$, the polarization $\mathbf{P} = \chi\mathbf{E}$ has exactly the form of (2.87), if the dielectric permeability is given by

$$\chi(\omega) = \int_0^\infty e^{i\omega\tau} F(\tau)\, d\tau. \tag{2.88}$$

Although only a real and positive frequency has a physical meaning, this equation formally extends χ to any complex argument ω of the upper half of figure 2.5; in the lower half where $\text{Im}\{\omega\} < 0$, the integral (2.88) diverges. Such an extension is necessary to perform the integration in (2.79). We summarize here some of the properties of the electric susceptibility of (2.88):

- If $|\omega| \rightarrow \infty$, χ goes to zero. For large and real frequencies, it is physically understandable because the electrons cannot follow the field, so the polarization is zero; for large imaginary ω, it is mathematically clear because of the factor $e^{-|\omega|\tau}$ in (2.88).

- The function $F(\tau)$ in (2.87) must for physical reasons, and to guarantee causality, be finite over the integration interval $[0, \infty]$, i.e. over the past. As a consequence, the susceptibility χ of (2.88) has no singularities in the upper half of the complex ω-plane and this is a necessary condition for obtaining the Kramers–Kronig relations (see (2.79)).
- χ fulfils the symmetry relations

$$\chi(-\omega) = \chi^*(\omega) \qquad \text{for real } \omega \qquad (2.89)$$
$$\chi(-\omega^*) = \chi^*(\omega) \qquad \text{for complex } \omega.$$

The last equation expresses the fact that a real field $\mathbf{E}$ produces a real polarization $\mathbf{P}$.

When, in (2.80), we replace $g(z)$ by $\chi(\omega)$, we may convince ourselves that $\chi(\omega)$ has all the desired mathematical properties to obey the formulae (2.84)–(2.86), if ω_0 is a real and positive frequency.

2.5.3 The Kramers–Kronig relation for the dielectric permeability

Here is the final result formulated explicitly for the dielectric permeability. For an arbitrary medium, ε_1 and ε_2 are not completely independent of each other but for any frequency ω_0,

$$\varepsilon_1(\omega_0) - 1 = \frac{2}{\pi} \int_0^\infty \frac{\omega \varepsilon_2(\omega)}{\omega^2 - \omega_0^2} \, d\omega \qquad (2.90)$$

$$\varepsilon_2(\omega_0) = -\frac{2\omega_0}{\pi} \int_0^\infty \frac{\varepsilon_1(\omega)}{\omega^2 - \omega_0^2} \, d\omega. \qquad (2.91)$$

For the static limit of ε_1 at zero frequency,

$$\varepsilon_1(0) - 1 = \frac{2}{\pi} \int_0^\infty \frac{\varepsilon_2(\omega)}{\omega} \, d\omega. \qquad (2.92)$$

Similar relations hold for the electric susceptibility χ, the electric polarizability α_e or the optical constant $m = n + ik$. Whenever the vacuum value of the material constant is one (as for ε), the -1 appears on the left-hand side (see (2.90) and (2.92)), when the vacuum value is zero (as for χ), the -1 is missing.

2.5.4 Extension to metals

In a metallic medium, the conductivity σ of (1.119) also follows the Kramers–Kronig (KK) relations. We may either put $g(\omega) = \sigma(\omega)$ or

$$g(\omega) = i\frac{4\pi\sigma(\omega)}{\omega}$$

which is the expression that appears in the generalized permeability of (1.115). However, for a metallic medium we have a pole not only at ω_0 (see figure 2.5) but also at $\omega = 0$. When, on itegrating the function $g(\omega)/(\omega - \omega_0)$ along the real frequency axis from $-\infty$ to $+\infty$, we circumvent this additional pole at $\omega = 0$ into the upper half of the complex plane in the same way as ω_0 (see (2.81)), we get an additional term $-4\pi^2\sigma(0)/\omega_0$ on the left-hand side of equation (2.82) where $\sigma(0)$ is the direct-current conductivity.

When we consider a conducting medium with a dielectric constant

$$\varepsilon = \varepsilon_d + i\frac{4\pi\sigma}{\omega} = \varepsilon_1 + i\varepsilon_2$$

from (1.115), equation (2.91) for $\varepsilon_2(\omega_0)$ has to be replaced by

$$\varepsilon_2(\omega_0) = -\frac{2\omega_0}{\pi} \int_0^\infty \frac{\varepsilon_1(\omega)}{\omega^2 - \omega_0^2} d\omega + \frac{4\pi\sigma(0)}{\omega_0}. \tag{2.93}$$

Equation (2.90) for $\varepsilon_1(\omega_0)$ stays in force as it is but the long wavelength limit $\varepsilon_1(0)$ must be adapted because $\varepsilon_2(\omega)$ has a singularity at $\omega = 0$. Going back to the more basic formula (2.84) for $\varepsilon_1(\omega_0)$, there now appears a term

$$\int_0^\infty \frac{d\omega}{\omega^2 - \omega_0^2}$$

which is always zero. What remains is the following modification of (2.92):

$$\varepsilon_1(0) - 1 = \frac{2}{\pi} \int_0^\infty \frac{\varepsilon_2(\omega) - 4\pi\sigma(0)/\omega}{\omega} d\omega.$$

2.5.5 Dispersion of the magnetic susceptibility

The magnetic susceptibility χ_m is defined in equations (1.16) and (1.17) and connects the field $\mathbf{H}$ with the magnetization $\mathbf{M}$ through

$$\mathbf{M} = \chi_m\mathbf{H}.$$

$\mathbf{M}$ has, from (1.13) and (1.14) the physical meaning of a volume density of magnetic moments. To speak of a magnetic susceptibility χ_m makes sense only if, neglecting free charges, in the expression (1.21) for the total current density $\mathbf{J}_{tot}$ the magnetic current dominates over the polarization current:

$$\mathbf{J}_{mag} = c\,\text{rot}\,\mathbf{M} \gg \mathbf{J}_{pol} = \dot{\mathbf{P}}.$$

With the help of Maxwell's equation, $\text{rot}\,\mathbf{E} = -\dot{\mathbf{B}}/c$, we get the order of magnitude estimate

$$\dot{P} = \chi_e\dot{E} \sim \chi_e\omega E \sim \chi_e\omega^2\mu Hl/c.$$

Assuming χ_e, $\mu \sim 1$ and a typical length l that must be much greater than an atomic radius, the condition $J_{mag} \gg J_{pol}$ requires frequencies

$$\omega \ll c\chi_m/l.$$

Therefore $\chi_m(\omega)$ becomes constant and real above some critical value ω_{cr}. In practice, magnetic dispersion stops well below optical frequencies, so $\omega_{cr} \ll 10^{15}$ s^{-1}. Whereas at high frequencies the dielectric permeability ε approaches one, the magnetic permeability μ goes to a value $\mu_{cr} = \mu(\omega_{cr})$ which may be different from unity. This necessitates the following modification in the KK relations:

$$\mu_1(\omega_0) - \mu_{cr} = \frac{2}{\pi} \int_0^\infty \frac{\omega\mu_2(\omega)}{\omega^2 - \omega_0^2}\, d\omega.$$

2.5.6   Three corollaries of the KK relation

2.5.6.1   *The dependence between ε_1 and ε_2*

Any set of physically possible values $\varepsilon_1(\omega)$ and $\varepsilon_2(\omega)$ for any grain material must obey the KK relations (2.90) and (2.91). They thus serve as a check for the internal consistency of data measured, for example, in the laboratory or derived otherwise. It is even sufficient to know one of them over the entire wavelength range, either $\varepsilon_1(\omega)$ or $\varepsilon_2(\omega)$, to compute the other. Whereas $\varepsilon_1(\omega)$ is not restricted at all, $\varepsilon_2(\omega)$ is associated with the entropy and must be positive everywhere. A data set for $\varepsilon_1(\omega)$ is wrong if it yields at just one frequency a negative value for $\varepsilon_2(\omega)$.

As an example of a dispersion formula that obeys the KK relations we may take equation (1.77) or (1.121). They apply to the harmonic oscillator or a metal, respectively. That they fulfil the KK relations may be verified from general mathematical considerations for the function $\varepsilon(\omega)$, which is the smart way; or by doing explicitly the KK integrals for (1.77), that is the hard way; or numerically, which is the brute way. Even the last method requires some delicacy when handling the Cauchy principal value. Numerical integration is inevitable when $\varepsilon(\omega)$ is available only in tabulated form.

When we look at the dispersion relation (1.77) and realize that ε_1 and ε_2 have the same denominator and contain the same quantities e, m_e, γ, ω and ω_0, equations (2.90)–(2.92) which link ε_1 with ε_2 are no longer perplexing. However, when we discussed the physics associated with the optical constants n and k, which determine the phase velocity and the extinction, the two parameters appeared very distinct and independent so that any general connection between them comes, at first glance, as a surprise.

2.5.6.2   *Dust absorption at very long wavelengths*

In a slowly varying electromagnetic field, a dielectric grain of arbitrary shape and composition absorbs, from (1.55), the power

$$W = \tfrac{1}{2}V\omega\,\mathrm{Im}\{\alpha_e\}E_0^2 = C^{abs}(\omega) \cdot S$$

where V is the volume of the grain, C^{abs} its cross section and $S = (c/8\pi)E_0^2$ the Poynting vector. The particle has to be small, which means that the frequency of the wave must stay below some critical value, say ω_1. We now apply equation (2.86) to the polarizability α of the grain and split the integral into two:

$$\alpha_1(\omega = 0) = \frac{2}{\pi} \int_0^\infty \frac{\alpha_2(\omega)}{\omega} \, d\omega = \frac{2}{\pi} \int_0^\infty \frac{\alpha_2(\lambda)}{\lambda} \, d\lambda$$

$$= \frac{2}{\pi} \int_0^{\lambda_1} \frac{\alpha_2(\lambda)}{\lambda} \, d\lambda + \frac{2}{\pi} \int_{\lambda_1}^\infty \frac{\alpha_2(\lambda)}{\lambda} \, d\lambda.$$

Note that we have swapped the integration variable from frequency to wavelength. Because α_2 is positive and the integral over α_2/λ in the total interval $[0, \infty]$ finite, the last integral must also be finite. If we make in the range $\lambda > \lambda_1 = 2\pi c/\omega_1$ the replacement

$$\frac{\alpha_2(\lambda)}{\lambda} = \frac{C^{abs}(\lambda)}{8\pi^2 V}$$

and write $C^{abs} = \sigma_{geo} Q^{abs}$, where σ_{geo} is the geometrical cross section, we get the following convergence condition:

$$\int_0^\infty Q^{abs}(\lambda) \, d\lambda < \infty.$$

Therefore, the absorption efficiency of any grain must, at long wavelengths, fall off more steeply than λ^{-1}. But this last constraint attains practical importance only when we know the threshold wavelength after which it is valid.

We could also have derived the result for $Q^{abs}(\lambda)$ at long wavelengths using the dielectric permeability ε instead of α_e but then metals would have required some extra remarks because in conductors $\varepsilon_2 \to \infty$ for $\omega \to 0$, whereas their polarizability α_e stays finite (see, for instance, (3.9)). A similar discussion to the one we carried out for α_e holds for the magnetic polarizability α_m.

2.5.6.3 Total grain volume

One may apply the KK relation not only to grain material but also to the interstellar medium as such [Pur69]. Because it is so tenuous, its optical constant $m = n + ik$ is very close to one. A small volume V in the interstellar medium of size d may, therefore, be considered to represent a Rayleigh–Gans particle. This class of grains, for which $|m - 1| \ll 1$ and $d|m - 1|/\lambda \ll 1$, is discussed in section 3.5. Of course, the particle is inhomogeneous as it consists of a mixture of gas, dust and mostly vacuum but for its absorption cross section we may nevertheless use (3.42)

$$C^{abs}(\lambda) = \frac{4\pi V}{\lambda} k(\lambda) = \frac{2\pi V}{\lambda} \varepsilon_2(\lambda).$$

The static limit (2.92) of the KK relation then gives

$$\varepsilon_1(0) - 1 = \frac{2}{\pi} \int_0^\infty \frac{\varepsilon_2(\lambda)}{\lambda} \, d\lambda = \frac{1}{\pi^2 V} \int_0^\infty C^{\text{abs}}(\lambda) \, d\lambda. \qquad (2.94)$$

In a *static* electric field **E**, a volume V of interstellar medium with dielectric permeability $\varepsilon(0)$ has, from equations (1.3), (1.6) and (2.94), a dipole moment

$$\mathbf{p} = \frac{\varepsilon(0) - 1}{4\pi} \mathbf{E} V = \frac{\mathbf{E}}{4\pi^3} \int_0^\infty C^{\text{abs}}(\lambda) \, d\lambda.$$

Alternatively, we may express **p** as resulting from the polarization of the individual grains within V. They have a total volume V_g and at zero frequency a dielectric constant $\varepsilon_g(0)$, so, from (3.8),

$$\mathbf{p} = \frac{3}{4\pi} V_g \frac{\varepsilon_g(0) - 1}{\varepsilon_g(0) + 2} \mathbf{E}.$$

When we crudely evaluate the long wavelength limit of this ratio from figure 7.19, we obtain a value of order one,

$$\frac{\varepsilon_g(0) - 1}{\varepsilon_g(0) + 2} \sim 1.$$

Along a line of sight of length L, the optical depth for absorption is $\tau(\lambda) = LC^{\text{abs}}$. Therefore, the total grain volume V_{dust} in a column of length L that produces an optical depth $\tau(\lambda)$ and has a cross section of 1 cm^2 is

$$V_{\text{dust}} \simeq \frac{1}{3\pi^2} \int_0^\infty \tau(\lambda) \, d\lambda.$$

When we take the standard normalized interstellar extinction curve of figure 7.8 for $\tau(\lambda)$, which refers to a visual extinction $A_V = 1$ mag, we obtain for the previous integral in the interval 0.1–10 μm a value of about 2×10^{-4}. Consequently, the total dust volume in a column of 1 cm^2 cross section with $A_V = 1$ mag is

$$V_{\text{dust}} \simeq 6 \times 10^{-6} \text{ cm}^{-3}. \qquad (2.95)$$

Such an estimate is of principal value, although its precision does not allow one to discriminate between grain models.

2.6 Composite grains

Interstellar grains are probably not solid blocks made of one kind of material but are more likely to be inhomogeneous so that the dielectric function within them varies from place to place. There are many ways in which inhomogeneity can come about. For instance, if particles coagulate at some stage during their evolution, the result will be a bigger particle with voids inside. The new big grain then has a fluffy structure and for its description, even if the grains before coagulation were homogeneous and chemically identical, at least one additional dielectric function is needed, namely that of vacuum ($\varepsilon = 1$). If chemically diverse particles stick together, one gets a heterogenous mixture. We note that purely thermal Brownian motion of grains (see section 9.3) is too small to make encounters between them significant but omnipresent turbulent velocities of 10 m s^{-1} are sufficient to ensure coagulation in dense clouds. Another way to produce inhomogeneities is to freeze out gas molecules on the surface of grains; this happens in cold clouds. A third possibility is that during grain formation tiny solid subparticles, like PAHs (section 12.1.1) or metal atoms, are built into the bulk material and contaminate it chemically. The chemical composition of interstellar dust is discussed in section 7.4 and in chapter 12.

2.6.1 Effective medium theories

The cross section of composite particles can be computed exactly in those few cases where the components are homogeneous and the geometrical structure is simple; examples are spherical shells, cylinders with mantles or coated ellipsoids. For any real composite particle, where different media are intermixed in a most complicated manner, such computations are out of the question.

To estimate the optical behavior of composite particles, one has to derive an average dielectric function ε_{av} representing the mixture as a whole. Once determined, ε_{av} is then used in Mie theory, usually assuming a spherical shape for the total composite grain. The starting point is the constitutive relation

$$\langle \mathbf{D} \rangle = \varepsilon_{av} \langle \mathbf{E} \rangle. \tag{2.96}$$

Here $\langle \mathbf{E} \rangle$ is the average internal field and $\langle \mathbf{D} \rangle$ the average displacement defined as

$$\langle \mathbf{E} \rangle = \frac{1}{V} \int \mathbf{E}(\mathbf{x}) \, dV \qquad \langle \mathbf{D} \rangle = \frac{1}{V} \int \varepsilon(\mathbf{x}) \, \mathbf{E}(\mathbf{x}) \, dV. \tag{2.97}$$

The integration extends over the whole grain volume V. If we envisage the grain to consist of a finite number of homogeneous components (subscript j), each with its own dielectric function ε_j and volume fraction f_j, we can write

$$\langle \mathbf{E} \rangle = \sum_j f_j \mathbf{E}_j \qquad \langle \mathbf{D} \rangle = \sum_j f_j \varepsilon_j \mathbf{E}_j. \tag{2.98}$$

The $\mathbf{E}_j$ are averages themselves (from (2.97)) over the subvolume $f_j V$; we have just dropped the brackets. The constitutive relation (2.96) is thus replaced by

$$\sum_j \varepsilon_j f_j \mathbf{E}_j = \varepsilon_{av} \sum_j f_j \mathbf{E}_j. \tag{2.99}$$

We envisage the components to be present in the form of many identical subparticles that are much smaller than the wavelength. When such a subparticle is placed into an extended medium with a spatially constant but time-variable field $\mathbf{E}'$, there is a linear relation between the field in the subparticle and the field in the outer medium (see sections 3.1 and 3.3),

$$\mathbf{E}_j = \beta \mathbf{E}'. \tag{2.100}$$

When one assumes that such a large-scale average field $\mathbf{E}'$ in the grain exists, one can remove the local fields $\mathbf{E}_j$ and find the average $\varepsilon_{\mathrm{av}}$.

2.6.2   Garnett's mixing rule

To exploit (2.99), we first imagine the grain to consist of a *matrix* (subscript m) containing *inclusions* (subscript i). For simplicity, let there be only one kind of inclusions with a total volume fraction f_i so that $f_i + f_m = 1$. For the constant large-scale field $\mathbf{E}'$ in (2.100), we take the field in the matrix and obtain

$$\varepsilon_i f_i \beta + \varepsilon_m f_m = \varepsilon_{\mathrm{av}} f_i \beta + \varepsilon_{\mathrm{av}} f_m.$$

For spherical inclusions, the proportionality factor β in equation (2.100) has the form

$$\beta = \frac{3\varepsilon_m}{\varepsilon_i + 2\varepsilon_m}$$

which is a generalization of equation (3.7) when the medium surrounding the sphere is not a vacuum but has some permeability ε_m. We thus arrive at the *Garnett* formula:

$$\varepsilon_{\mathrm{av}} = \varepsilon_m \frac{1 + 2 f_i (\varepsilon_i - \varepsilon_m)/(\varepsilon_i + 2\varepsilon_m)}{1 - f_i (\varepsilon_i - \varepsilon_m)/(\varepsilon_i + 2\varepsilon_m)}. \tag{2.101}$$

The expression is evidently not symmetric with respect to the inclusion and matrix. One has to make up one's mind which component should be regarded as the inclusion that pollutes the matrix. If the concentration of the inclusions, f_i, is small, equation (2.101) simplifies to

$$\varepsilon_{\mathrm{av}} = \varepsilon_m \left(1 + 3 f_i \frac{\varepsilon_i - \varepsilon_m}{\varepsilon_i + 2\varepsilon_m} \right).$$

The Garnett mixing rule is very similar to the Clausius–Mossotti law (see (3.55)). Indeed, the latter follows almost immediately from (2.101). The Clausius–Mossotti law gives the dielectric constant of an inhomogeneous medium that consists of a vacuum matrix ($\varepsilon_m = 1$) with embedded spherical inclusions, the latter being atoms of polarizability $\alpha = (3/4\pi)(\varepsilon_i - 1)/(\varepsilon_i + 2)$.

The Garnett rule can be extended to an arbitrary number of components. One still has to make the distinction between the matrix (subscript m) and the inclusions (index $i = 1, 2, 3, \ldots$). Putting as before

$$\mathbf{E}_i = \beta_i \mathbf{E}_m$$

we get the following generalization of (2.101),

$$\varepsilon_{av} = \frac{f_m \varepsilon_m + \sum f_i \beta_i \varepsilon_i}{f_m + \sum f_i \beta_i} \qquad (2.102)$$

where the sum of the volume fractions is unity,

$$f_m + \sum_i f_i = 1.$$

Written in the form of (2.102), the restriction that the inclusions are spheres has been lifted. They may be of any shape, for example ellipsoids, for which β_i can be calculated as outlined in Section 3.3. One can also take a mean over randomly oriented ellipsoids.

2.6.3 The mixing rule of Bruggeman

Next we suppose that none of the components of the grain is special, like the matrix in the Garnett theory. Then the components distinguish themselves only through their permeability and volume fraction; no assumption is made about the average field $\mathbf{E}'$. Inserting (2.100) into (2.99) yields the *Bruggeman* rule:

$$0 = \sum_j (\varepsilon_j - \varepsilon_{av}) f_j \beta_j \qquad (2.103)$$

with $\sum f_j = 1$. Contrary to the Garnett rule, this formula is symmetric in all components j. If they consist of spherical entities,

$$0 = \sum_j f_j \frac{\varepsilon_j - \varepsilon_{av}}{\varepsilon_j + 2\varepsilon_{av}}. \qquad (2.104)$$

Thus for n components, ε_{av} is determined from a complex polynomial of nth degree. When we imagine the interstellar dust to be a democratic compound of silicate, amorphous carbon, ice and vacuum, the Bruggeman mixing rule is preferred. However, when ice becomes dirty through contamination by tiny impurities (metal atoms or PAHs) that amount to only a small volume fraction ($f_i \ll f_m$), the Garnett rule is more appropriate.

An illustration of how the dielectric permeabilities of two components combine to an avarage ε_{av} is shown in figure 2.6 for the two mixing rules and for a very idealized situation. The materials have the optical properties of harmonic oscillators with different resonance frequencies. In the mixture, the resonances are damped, broadened and shifted.

2.6.4 Composition of grains in protostellar cores

Likely and astronomically relevant candidates for composite grains are the solid particles in cold and dense protostellar cloud cores (see section 15.3). Such grains

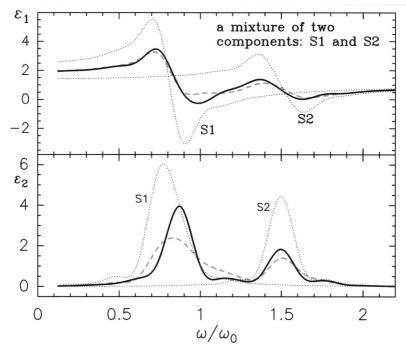

Figure 2.6. Let the dielectric permeability of two substances, S1 and S2, be represented by harmonic oscillators (see figure 1.2) and given by the dotted curves (top, real part ε_1; bottom, imaginary part ε_2). The *average* permeability of a mixture of these two substances calculated from the Bruggeman mixing rule is shown by the broken curve, that calculated from the Garnett mixing rule, by the full line. Both components have equal volume and both averaged permeabilities obey KK's relations, as do, of course, the dielectric constants of S1 and S2.

are also expected to be found in the cooler parts of stellar disks or in comets. We figure them as fluffy aggregates, probably substantially bigger than normal interstellar grains. They are composed of refractory (resistant to heating) and compact subparticles made of silicates or amorphous carbon. The subparticles are enshrouded by a thick sheet of ice as a result of molecules that have frozen out. The volatile (easy to evaporate) ice balls with their compact cores are loosely bound together and form, as a whole, the porous grain. In this picture, the frosting of molecules in ice layers on the surface of the subparticles precedes the process of coagulation.

Altogether, the grains in protostellar cores have four distinct components:

- compact silicate subparticles,
- compact carbon subparticles,
- ice sheets and

- vacuum.

The possible diversity of such grains is infinite and their structure can be extremely complex. The volume fractions of the components, f, add up to one,

$$f^{\text{Si}} + f^{\text{C}} + f^{\text{ice}} + f^{\text{vac}} = 1.$$

To be specific, we adopt for the volume ratio of silicate-to-carbon material in the grain

$$\frac{f^{\text{Si}}}{f^{\text{C}}} = 1.4\ldots2$$

compatible with the cosmic abundances in solids, and for the specific weight of the refractory subparticles

$$\rho_{\text{ref}} \approx 2.5 \text{ g cm}^{-3} \qquad \rho_{\text{ice}} \approx 1 \text{ g cm}^{-3}.$$

The available mass of condensable material is determined by the gas phase abundances of C, N and O and their hydrides. Standard gas abundances imply a ratio of volatile-to-refractory mass of

$$\frac{M_{\text{ice}}}{M_{\text{ref}}} = 1\ldots2.$$

Consequently, the ice volume is 2.5 to 5 times bigger than the volume of the refractories.

2.6.5 How size, ice and porosity change the absorption coefficient

These composite grains differ in three fundamental aspects from their compact silicate and carbon subparticles:

- they are bigger because of coagulation,
- they contain ices because of frosting and
- they are porous, again because of coagulation.

Coagulation, deposition of ice and porosity each affect the absorption coefficient and we now illustrate how. In the following examples, we fix λ to 1 mm, which is a wavelength where protostellar cores are frequently observed.

2.6.5.1 Grain size

We first investigate the influence of grain size. The mass absorption coefficient K^{abs} (defined in subsection 2.1.3) is for a fixed wavelength a function of grain radius a, so $K^{\text{abs}} = K^{\text{abs}}(a)$. To demonstrate the size effect, we normalize K^{abs} to its value when the grains are small ($a \ll \lambda$); it is then insensitive to a. We adopt $a = 0.1 \ \mu$m, which is a typical radius of an interstellar grain and certainly

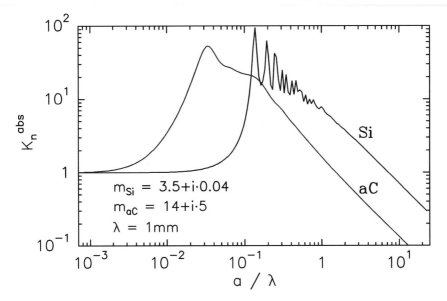

Figure 2.7. The influence of grain growth on the mass absorption coefficient K^{abs} for silicate (Si) and amorphous carbon spheres (aC) at a wavelength of 1 mm. The curves show the normalized coefficient K_n^{abs} defined in the text. For example, the mass absorption coefficient K^{abs} of carbon grains increases by a factor $K_n^{abs} \sim 50$ when the radius of the particles grows from 0.1 μm ($a/\lambda = 10^{-4}$) to 30 μm ($a/\lambda = 0.03$).

smaller than the wavelength (1 mm). The normalized coefficient is denoted K_n^{abs} and defined as

$$K_n^{abs}(a) = \frac{K^{abs}(a)}{K^{abs}(0.1\,\mu m)}.$$

It gives the relative change with respect to ordinary interstellar dust particles and it is plotted in figure 2.7 for silicate and carbon spheres. The spikes in figure 2.7 for silicates are resonances that disappear when the grains are not all of the same size but have a size distribution. When $a \ll \lambda$, the normalized coefficient K_n^{abs} equals one and is constant because this is the Rayleigh limit. However, when $a \gg \lambda$, the normalized coefficient $K_n^{abs} \propto a^{-1}$ because $Q^{abs} \approx 1$; big lumps are not efficacious in blocking light. For sizes in between, one has to do proper calculations. They reveal an enhancement in the mass absorption coefficient K^{abs} which can be very significant (> 10) and which would strongly boost millimeter dust emission because the latter is proportional to K^{abs} (see section 8.1.1).

One can create plots like those in figure 2.7 for other wavelengths and optical constants. The *qualitative* features stay the same but some details are quite interesting. For example, if all particles in the diffuse interstellar medium had a radius of 1 μm (without changing the total dust mass), the extinction optical

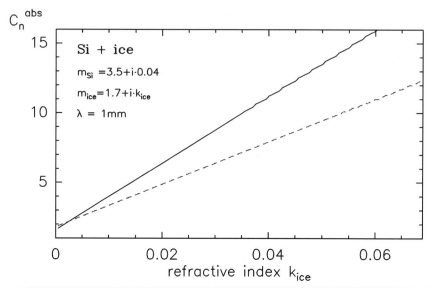

Figure 2.8. The influence of k_{ice}, the imaginary part of the optical constant of ice, on the absorption coefficient C^{abs} of a grain that consists of silicate and ice with a mass ratio 1:1. The grain radius a is much smaller than the wavelength; here $\lambda = 1$ mm. The ordinate C_n^{abs} gives the increase in the cross section with respect to the bare silicate core. The solid plot refers to a *coated sphere* with a silicate core and an ice mantle, the broken line to a *homogeneous sphere* where ice and silicate are mixed and the optical constant m_{av} of the mixture is computed after the Bruggeman theory. Both curves are based on Mie calculations.

depth at 2.2 μm (K-band) would be almost ten times larger than it really is.

2.6.5.2   Ice mantles

Next we illustrate the influence of ice in the grain material, again for a wavelength $\lambda = 1$ mm. We take a silicate sphere of arbitrary radius $a \ll \lambda$ and deposit an ice mantle of the same mass on it; such a mass ratio of ice to refractory core is suggested by the cosmic abundances in the case of complete freeze–out. The total grain's volume is then three and a half times bigger and the mass twice that of the bare silicate core. The relevant parameter for the absorption coefficient C^{abs} is k_{ice}, the imaginary part of the optical constant of ice. It depends on how much the ice is polluted by impurities. Estimates for k_{ice} at this wavelength are around 0.01 but uncertain. Defining the normalized absorption coefficient of a single grain by

$$C_n^{abs} = \frac{C^{abs}(\text{core} + \text{ice})}{C^{abs}(\text{core})}$$

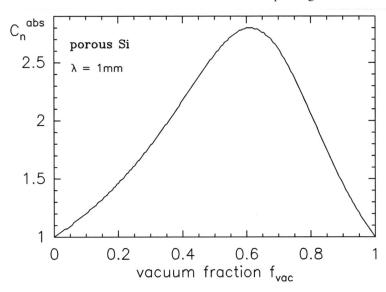

Figure 2.9. The influence of porosity on the grain cross section. The normalized cross section C_n^{abs} is defined in the text. The mass of the grain is kept constant and does not vary with the vacuum fraction f_{vac}, while the optical constant of the grain, which is calculated here after Bruggeman, changes with fluffines. The compact particle has $m_{Si} = 3.5 + i0.04$.

we learn from figure 2.8 that an ice mantle enhances C^{abs} by a factor $C_n^{abs} \sim 3$ if $k_{ice} = 0.01$. Note that an ice mantle increases C^{abs} even when $k_{ice} = 0$ because the mantle grain is larger than the refractory core and collects more light. We also show in figure 2.8 the value of C_n^{abs} when the ice is not in a mantle but mixed throughout the grain.

2.6.5.3 Fluffiness

A porous grain also has a greater absorption cross section C^{abs} than a compact one of the same mass. We consider, in figure 2.9, silicates with vacuum inclusions; the normalized cross section C_n^{abs} is defined by

$$C_n^{abs} = \frac{C^{abs}(\text{fluffy grain})}{C^{abs}(\text{compact grain of same mass})}.$$

When the volume fraction of vacuum, f^{vac}, equals zero, the grain is compact. Fluffy grains are obviously better absorbers because they are bigger. A porosity parameter f^{vac} between 0.4 and 0.8, which may or may not be a reasonable estimate for interstellar conditions, suggests an increase in the absorption coefficient by a factor of two.

Chapter 3

Very small and very big particles

In sections 2.3 and 2.4, we presented general solutions of the field equations in the case when a plane wave interacts with a particle with simple geometry; there we stressed the mathematical aspects. Usually, when we want to extract numbers from the theory for a specific astronomical application, we have to deliver ourselves to the mercy of a computer. No matter how efficient such a machine is, it is wise to retain some mental independence and bring to one's mind the physical aspects: computing must not be confused with understanding. In a few simple configurations, analytical solutions are possible and we turn to these in this chapter. They illuminate the problem and are useful for checking the correctness of a computer program.

3.1 Tiny spheres

We derive the efficiencies for small spheres of dielectric material. This is the basic section for understanding how and why interstellar dust absorbs and scatters light.

3.1.1 When is a particle in the Rayleigh limit?

When a sphere has a radius, a, which is small compared with the wavelength λ, i.e. when the size parameter

$$x = \frac{2\pi a}{\lambda} \ll 1 \tag{3.1}$$

the calculation of cross sections becomes easy. The particle itself is not required to be small, only the ratio a/λ. In fact, a may even be big. So the heading of this section is suggestive but not precise. With regard to the thermal emission of interstellar grains, which occurs at wavelengths where the condition $\lambda \gg a$ is usually very well fulfilled, one may use for the computation of cross sections, the approximations given here. If we additionally stipulate that the product of the size parameter multiplied by optical constant, $m = n + ik$, be small,

$$|m|x \ll 1 \tag{3.2}$$

80

we ensure two things:

- because $kx \ll 1$, the field is only weakly attenuated in the particle; and
- because $nx \ll 1$, the wave traverses the particle with the phase velocity $v_{ph} = c/n$ in a time $\tau \simeq nx\omega^{-1}$ which is much shorter than the inverse circular frequency ω^{-1}.

Grains for which conditions (3.1) and (3.2) hold are said to be in the Rayleigh limit. This concept can be applied to any particle, not just spheres, if one understands by the size parameter the ratio of typical dimension over wavelength.

3.1.2 Efficiencies of small spheres from Mie theory

A purely mathematical approach to finding simple expressions for the efficiencies of small spheres is to develop the first coefficients a_1, b_1 in the series expansion (2.50) and (2.51) for Q^{ext} and Q^{sca} into powers of x and retain only terms up to x^5:

$$a_1 = -i\frac{2x^3}{3}\frac{m^2-1}{m^2+2} - i\frac{2x^5}{5}\frac{(m^2-2)(m^2-1)}{(m^2+2)^2} + \frac{4x^6}{9}\left(\frac{m^2-1}{m^2+2}\right)^2 + O(x^7)$$

$$b_1 = -i\frac{x^5}{45}(m^2-1) + O(x^7).$$

- Usually non-magnetic materials ($\mu = 1$) are considered. Then the term with x^3 in the coefficient a_1 yields the electric dipole absorption, the one with x^6 electric dipole scattering:

$$Q^{ext} \simeq \frac{6}{x^2}\,\text{Re}\{a_1\} \simeq 4x\,\text{Im}\left\{\frac{m^2-1}{m^2+2}\right\} = 4x\,\text{Im}\left\{\frac{\varepsilon-1}{\varepsilon+2}\right\}$$

$$= \frac{8\pi a}{\lambda}\frac{6nk}{(n^2-k^2+2)^2+4n^2k^2} = \frac{8\pi a}{\lambda}\frac{3\varepsilon_2}{|\varepsilon+2|^2} \qquad (3.3)$$

$$Q^{sca} \simeq \frac{6}{x^2}|a_1|^2 \simeq \frac{8}{3}x^4\left|\frac{m^2-1}{m^2+2}\right|^2. \qquad (3.4)$$

If x is small, both Q^{ext} and Q^{sca} approach zero. Because

$$Q^{abs} \propto x \qquad \text{and} \qquad Q^{sca} \propto x^4$$

scattering is negligible at long wavelengths and extinction is reduced to absorption:

$$Q^{abs} \simeq Q^{ext}.$$

With respect to the wavelength behavior, this suggests the frequently cited dependences

$$Q_\lambda^{sca} \propto \lambda^{-4} \qquad \text{and} \qquad Q_\lambda^{abs} \propto \lambda^{-1}$$

for scattering and absorption. However, they are true only if $m(\lambda)$ is more or less constant.

- Because the optical constant m is, from (1.46), symmetric in ε and μ,

$$m^2 = (n + ik)^2 = \varepsilon \mu \qquad (k > 0)$$

equation (3.3), which contains only the coefficient a_1, is for purely magnetic material ($\mu \neq 1$, $\varepsilon = 1$) replaced by

$$Q_\lambda^{abs} = \frac{8\pi a}{\lambda} \cdot \frac{3\mu_2}{|\mu + 2|^2}. \qquad (3.5)$$

The dissipation process refers to magnetic dipole oscillations and is relevant only at frequencies $\omega < 10^{12}$ s^{-1}.

- The coefficient b_1 is discussed in section 3.2 for a non-magnetic conductor and presents magnetic dipole absorption (see (3.26)).

3.1.3 A dielectric sphere in a constant electric field

Besides cutting off the series expansion in Mie theory after the first term, there is another approach to obtaining Q^{sca} and Q^{ext} for small particles that gives physical insight. We now restrict the discussion to a dielectric medium. When $x \ll 1$ and $|mx| \ll 1$, the electric field in the grain changes in a quasi-stationary fashion. When we want to calculate the field in such a configuration, we are reduced to an exercise in electrostatics. The basic equations valid everywhere are

$$\text{rot } \mathbf{E} = 0 \qquad \text{div } \mathbf{D} = 0.$$

They impose the boundary conditions that the tangential component of the electric field $\mathbf{E}$ and the normal component of the displacement $\mathbf{D}$ are continuous on the grain surface. For a homogeneous medium ($\varepsilon = $ constant), this leads to the Laplace equation

$$\Delta\varphi = 0$$

where φ is the potential related to the field through $\mathbf{E} = -\nabla\varphi$.

When we place a sphere of radius a, volume V and dielectric constant ε into a constant field $\mathbf{E}$, the field becomes deformed. Let the sphere be at the center of the coordinate system. We label the field *inside* the sphere by $\mathbf{E}^i$ and outside it (*external*) by $\mathbf{E}^e$. For the potential φ of the deformed field we make the ansatz

$$\varphi = \begin{cases} -\mathbf{E} \cdot \mathbf{r} + c_1 \dfrac{\mathbf{E} \cdot \mathbf{r}}{r^3} & \text{outside sphere} \\ -c_2\mathbf{E} \cdot \mathbf{r} & \text{inside sphere} \end{cases} \qquad (3.6)$$

where $\mathbf{r}$ is the position coordinate of length $r = |\mathbf{r}|$, and c_1, c_2 are constants. In the absence of the sphere, the potential is simply

$$\varphi_0 = -\mathbf{E} \cdot \mathbf{r}.$$

Equation (3.6) reflects the expected behavior of the field. Inside the sphere, E^i is constant and parallel to E; outside, E^e is the sum of a dipole, which is induced by E and goes to zero at large distances, plus the constant field E. We determine c_1 and c_2 from the conditions on the grain surface. Continuity of the tangential component of the electric field implies continuity of the tangential derivative of φ, from which it follows that φ is itself continuous (parallel and perpendicular to the surface). This gives $c_2 = 1 - c_1/a^3$. Continuity of the normal component of the displacement D yields $\varepsilon c_2 = 1 + 2c_1/a^3$. Hence we derive

$$c_1 = a^3 \frac{\varepsilon - 1}{\varepsilon + 2} \qquad c_2 = \frac{3}{\varepsilon + 2}$$

and, therefore,

$$E^i = \frac{3}{\varepsilon + 2} E \qquad \text{for } r \le a. \tag{3.7}$$

Note that the field is smaller inside the body than outside, $E^i < E$. Because the polarization is given by (see (1.6))

$$P = \frac{\varepsilon - 1}{4\pi} E^i$$

the induced dipole moment of the grain is

$$p = PV = a^3 \frac{\varepsilon - 1}{\varepsilon + 2} E = \alpha_e V E \tag{3.8}$$

and the electric polarizability α_e becomes (see (1.8))

$$\alpha_e = \frac{3}{4\pi} \frac{\varepsilon - 1}{\varepsilon + 2}. \tag{3.9}$$

For other grain geometries there would be other dependences of α_e on ε. Equation (3.7) gives the internal field. If e denotes the unit vector in the direction r, the outer field E^e is the sum of E plus a dipole field:

$$E_{dip} = \frac{3e(e \cdot p) - p}{r^3} \tag{3.10}$$

so

$$E^e = E - \alpha_e V \, \text{grad} \left(\frac{E \cdot r}{r^3} \right) = E + E_{dip}. \tag{3.11}$$

3.1.3.1 A coated sphere in a constant electric field

It is not difficult to repeat the previous exercise for a sphere covered by a homogeneous shell. There is then a core (index 1) enveloped by a mantle (index 2) of a different substance. One now has boundary conditions at the interface between the two materials and, as before, on the outside. If f denotes the volume fraction of the inner sphere relative to the total sphere,

$$f = \left(\frac{a_1}{a_2}\right)^3 \le 1$$

expression (3.9) for the polarizability becomes a bit lengthy:

$$\alpha_e = \frac{3}{4\pi} \cdot \frac{(\varepsilon_2 - 1)(\varepsilon_1 + 2\varepsilon_2) + f(\varepsilon_1 - \varepsilon_2)(1 + 2\varepsilon_2)}{(\varepsilon_2 + 2)(\varepsilon_1 + 2\varepsilon_2) + f(2\varepsilon_2 - 2)(\varepsilon_1 - \varepsilon_2)}.$$

3.1.4 Scattering and absorption in the electrostatic approximation

3.1.4.1 Scattering

The field $\mathbf{E}$ is not really static but oscillates proportionally to $E_0 e^{-i\omega t}$ and so, synchronously, does the dipole moment $\mathbf{p}$. The oscillating dipole, which is now the grain as a whole, emits radiation. Its average power integrated over all directions is $W = |\ddot{\mathbf{p}}|^2/3c^3$ and follows from (1.97). However, W must also equal the total power scattered from the incident wave by the particle, therefore,

$$W = \frac{1}{3c^3}|\ddot{\mathbf{p}}|^2 = S\,C^{\mathrm{sca}}. \tag{3.12}$$

For the dipole moment, we have to insert $\mathbf{p} = \alpha_e V \mathbf{E}$ and for the time-averaged Poynting vector $S = (c/8\pi)E_0^2$ from (1.39), therefore

$$C^{\mathrm{sca}} = \frac{8\pi}{3}\left(\frac{\omega}{c}\right)^4 V^2|\alpha_e|^2. \tag{3.13}$$

The electric polarizability α_e for spheres is given by (3.9). It was obtained in the electrostatic approximation. Can we use it in the case when $\omega \ne 0$? Yes, we can because the variations of the field are assumed to be slow, so the electron configuration is always relaxed. This means the electrons always have sufficient time to adjust to the momentary field, just as in the static case. Of course, the polarization of the medium is not that of a static field. Instead, the dielectric permeability must be taken at the actual frequency ω of the outer field. Because $\omega \ne 0$, the permeability $\varepsilon(\omega)$ is complex, which automatically takes care of the time lag between $\mathbf{P}$ and $\mathbf{E}$. We thus find

$$C^{\mathrm{sca}} = \pi a^2 Q^{\mathrm{sca}} = \frac{24\pi^3 V^2}{\lambda^4}\left|\frac{\varepsilon - 1}{\varepsilon + 2}\right|^2 \tag{3.14}$$

which agrees with equation (3.4) for the scattering efficiency Q^{sca}.

3.1.4.2 Absorption

We can find the absorption coefficient of a small dielectric or magnetic grain if we use (1.55) or (1.57) and set the absorbed power equal to SC^{abs}. This gives

$$C^{abs} = \frac{4\pi}{c}\omega V \operatorname{Im}\{\alpha\}. \tag{3.15}$$

But let us be more elaborate and assume that the grain consists of NV harmonic oscillators excited in phase, V being the volume and N the oscillator density. The power W absorbed by the oscillators can be expressed in two ways, either by the dissipation losses of the harmonic oscillators given in (1.69) or via the Poynting vector S multiplied by the absorption cross section C^{abs}, so

$$W = VN\frac{\gamma E_0^2 e^2}{2m_e}\frac{\omega^2}{(\omega_0^2 - \omega^2)^2 + \omega^2\gamma^2} = SC^{abs}.$$

When we substitute the dielectric permeability of the harmonic oscillator after (1.77), we can transform this equation into

$$\frac{Vc}{4\lambda}\varepsilon_2 E_0^2 = SC^{abs}. \tag{3.16}$$

For the electric field E_0 which drives the oscillators, we must insert the field $\mathbf{E}^i$ inside the grain according to (3.7), and not the outer field $\mathbf{E}$. When we do this, we recover the efficiency for electric dipole absorption of (3.3) but now we understand the physics:

$$C^{abs} = \frac{6\pi V}{\lambda}\operatorname{Im}\left\{\frac{\varepsilon - 1}{\varepsilon + 2}\right\} = \frac{V\omega}{c}\frac{9\varepsilon_2}{|\varepsilon + 2|^2}. \tag{3.17}$$

3.1.5 Polarization and angle-dependent scattering

For small spheres, the scattering matrix given in (2.75) reduces further (we drop the factor in front of the matrix):

$$\begin{pmatrix} I_s \\ Q_s \\ U_s \\ V_s \end{pmatrix} = \begin{pmatrix} \frac{1}{2}(1 + \cos^2\theta) & -\frac{1}{2}\sin^2\theta & 0 & 0 \\ -\frac{1}{2}\sin^2\theta & \frac{1}{2}(1 + \cos^2\theta) & 0 & 0 \\ 0 & 0 & \cos\theta & 0 \\ 0 & 0 & 0 & \cos\theta \end{pmatrix}\begin{pmatrix} I_i \\ Q_i \\ U_i \\ V_i \end{pmatrix} \tag{3.18}$$

$\theta = 0$ gives the forward direction. There are several noteworthy facts:

- The scattering pattern no longer depends on wavelength, as the matrix elements contain only the angle θ.
- Scattering is symmetrical in θ about $\pi/2$ and has two peaks, one in the forward, the other in the backward ($\theta = \pi$) direction.

- The matrix element $S_{12} < 0$, so the polarization is perpendicular to the scattering plane.
- If the incident radiation has unit intensity and is unpolarized ($I_i = 1$, $Q_i = U_i = V_i = 0$), we have

$$I_s = \tfrac{1}{2}(1 + \cos^2 \theta) \qquad Q_s = -\tfrac{1}{2}\sin^2 \theta \qquad U_s = V_s = 0.$$

The degree of linear polarization becomes

$$p = \frac{\sin^2 \theta}{1 + \cos^2 \theta}. \tag{3.19}$$

- Consequently, the light is completely polarized at a scattering angle of 90°.
- The intensity of the scattered light, and thus the phase function $f(\theta, \phi)$, is proportional to $1 + \cos^2 \theta$. As a result, the integral in (2.6) vanishes and the asymmetry factor becomes zero, $g = 0$, although scattering by a small sphere is *not* isotropic.

3.1.6 Small-size effects beyond Mie theory

A real grain is not a homogeneous continuum but a crystal built up from atoms, rather regularly spaced and separated by a distance r_0, the lattice constant. Mie theory, which is based on the classical electrodynamics of a continuous medium, fails when the structure of matter or quantum effects become important. A more general theory is then needed to describe the optical behavior of particles. We will consider some quantum effects when we discuss PAHs, a specific kind of very small carbon grains. Here we only remark on the influence of the surface in the case of small grains.

Whereas atoms inside the particle are surrounded from all sides, the situation is different for those on the surface which have bonds only towards the particle's interior. This has consequences, for example, for their ability to bind to gas atoms or for the specific heat of the grain. If the particle has a diameter a, the ratio of the number of surface atoms N_{surf} to all atoms N in the grain is roughly

$$\frac{N_{surf}}{N} \simeq 6\frac{r_0}{a}.$$

As the lattice constant is of order 2 Å, a substantial fraction of atoms is on the surface only when the particle is small. One way in which the surface affects the optical grain properties can be understood when we interpret the damping constant γ in the motion of an electron (see (1.59)) as a collision frequency with atoms of the crystal. If the particle is small, an additional term must be added to the damping constant that comes from collisions (reflections) at the surface. Free electrons in a metal move with the Fermi velocity

$$v_F = \frac{\hbar}{m_e} \sqrt[3]{3\pi^2 n_e}$$

which is the threshold speed of fully degenerate (non-relativistic) electrons and follows from the Fermi energy of (6.54). For example, in graphite with an electron density $n_e \simeq 10^{20}$ cm^{-3} and $\gamma = 5 \times 10^{12}$ s^{-1}, one estimates that noticeable changes in γ, and thus in the dielectric permeability, occur for sizes $a \leq 50$ Å.

3.2 A small metallic sphere in a magnetic field

The case of a small metallic sphere in an *electric field* is included in the previous derivation for the dipole moment **p** by making ε in equations (3.9) very large, then

$$\mathbf{p} = a^3 \mathbf{E}.$$

The charges on the metal surface become polarized in the outer field but the electric field does not penetrate into the particle. One might, therefore, think there would be no absorption and just scattering with an efficiency $Q^{sca} = 8x^4/3$ (because Im$\{(\varepsilon - 1)/(\varepsilon + 2)\} = 0$, one has to use the x^6 term in the coefficient a_1 of equation (3.3)). But not quite because we have neglected in our small-size approximation the magnetic field which is also present in a wave. Let us now include it; Mie calculations automatically do.

When we place a particle in a constant magnetic field **H**, there is a formal identity with electrostatics. There are the same types of equations,

$$\text{rot } \mathbf{H} = 0 \qquad \text{div } \mathbf{B} = 0$$

and the same boundary conditions on the surface: the tangential component of **H** and the normal of $\mathbf{B} = \mu\mathbf{H}$ are continuous. So there is nothing new in magnetostatics.

3.2.1 Slowly varying field

The situation becomes interesting when the magnetic field **H** is slowly alternating, say, proportionally to $e^{-i\omega t}$. Slowly means that **H** is spatially uniform over the dimension of the body or, in other words, that the particle is small. The magnetic field $\mathbf{H}^i$ *inside the sphere* is then also changing and induces an electric field **E**. The presence of **E** in the conductor implies a current and, therefore, ohmic losses; their time average is $\sigma\mathbf{E}^2$. When one knows $\mathbf{H}^i$, one can derive the induced field **E** either from the Maxwell equation (1.25),

$$\text{rot } \mathbf{E} = i\frac{\mu\omega}{c}\mathbf{H}^i$$

or for a slowly changing field from (1.114),

$$\frac{4\pi\sigma}{c}\mathbf{E} = \text{rot } \mathbf{H}^i.$$

But one cannot use the static solution $\mathbf{H}_{\text{stat}}$ because it fulfils $\text{div}\,\mathbf{H}_{\text{stat}} = 0$ and $\text{rot}\,\mathbf{H}_{\text{stat}} = 0$ and thus gives no electric field at all. Instead, to determine the internal field $\mathbf{H}^i$, we start with the wave equation (1.34),

$$\Delta \mathbf{H}^i + k^2 \mathbf{H}^i = 0 \qquad (3.20)$$

together with $\text{div}\,\mathbf{H}^i = 0$. The square of the wavenumber is given by (1.116). When metallicity dominates ($\omega|\varepsilon_d| \ll 4\pi\sigma$),

$$k^2 = i\,\frac{4\pi\sigma\omega\mu}{c^2}.$$

We solve (3.20) by noting that the scalar function

$$f(r) = \frac{\sin(kr)}{r}$$

is a spherically symmetric solution of

$$\Delta f + k^2 f = 0$$

(see the Laplace operator in (2.33)). Because $\mathbf{H}$ is constant, the vector potential, $\mathbf{A}$, defined by the function

$$\mathbf{A} = \beta\,\text{rot}(f\mathbf{H})$$

also fulfils the wave equation

$$\Delta \mathbf{A} + k^2 \mathbf{A} = 0.$$

We have used this result before in section 2.3 (see equations (2.26)–(2.28)). The constant β will be adjusted later. If we now put

$$\mathbf{H}^i = \text{rot}\,\mathbf{A}$$

the equations $\text{div}\,\mathbf{H}^i = 0$ and $\Delta \mathbf{H}^i + k^2 \mathbf{H}^i = 0$ are fulfilled, as required. To evaluate the somewhat complicated expression $\mathbf{H}^i = \beta\,\text{rot}\,\text{rot}(f\mathbf{H})$, we use formula (A.6) and the easy-to-prove relations

$$\text{div}(f\mathbf{H}) = \mathbf{H}\cdot\nabla f \qquad \nabla f = \mathbf{e}\,\frac{k\cos kr - f}{r} \qquad \Delta(f\mathbf{H}) = -k^2 f\mathbf{H}.$$

Here $\mathbf{e}$ is a unit vector in the direction of $\mathbf{r}$. When one works it out, one obtains for the magnetic field inside the sphere

$$\mathbf{H}^i = \beta\left(\frac{f'}{r} + k^2 f\right)\mathbf{H} - \beta\left(\frac{3f'}{r} + k^2 f\right)\mathbf{e}(\mathbf{e}\cdot\mathbf{H}).$$

3.2.2 The magnetic polarizability

In analogy to (3.11), the (perturbed) field $\mathbf{H}^e$ *outside the sphere*, for which $\mathrm{rot}\,\mathbf{H}^e = 0$ and $\mathrm{div}\,\mathbf{H}^e = 0$, can be written as

$$\mathbf{H}^e = \mathbf{H} + V\alpha_m \frac{3\mathbf{e}(\mathbf{e}\cdot\mathbf{H}) - \mathbf{H}}{r^3}.$$

α_m denotes the magnetic polarizability and

$$\mathbf{m} = V\alpha_m\mathbf{H}$$

is the magnetic dipole moment of the sphere of radius a and volume V. The factors α_m and β follow from the boundary conditions on its surface by equating $\mathbf{H}^i = \mathbf{H}^e$ for $\mathbf{e}\cdot\mathbf{H} = 0$ and $\mathbf{e}\cdot\mathbf{H} = \mathbf{H}$. We find $\beta = 3/[2k^2 f(a)]$ and

$$\alpha_m = -\frac{3}{8\pi}\left(1 - \frac{3}{a^2 k^2} + \frac{3}{ak}\cot ak\right). \tag{3.21}$$

With α_m, we can determine the dissipation rate W in a grain after (1.57), we do not have to integrate $\sigma\mathbf{E}^2$ over the particle volume. When we divide W by the Poynting vector, we get the absorption cross section of a magnetic dipole.

3.2.3 The penetration depth

Even when the particle is small compared to the scale on which the outer magnetic field changes, as we assume in this chapter, the magnetic field may not fully pervade it. To see how far it can penetrate, we apply the wave equation (3.20) to a simple one-dimensional situation, where a plane wave falls on a metal surface. Suppose the field vector $\mathbf{H}$ is parallel to the x-axis and a function of z only,

$$\mathbf{H} = (H_x(z), 0, 0)$$

and the (x, y)-plane marks the surface of the metallic body. Then with $\mu = 1$,

$$\frac{\partial^2 H_x}{\partial z^2} + k^2 H_x = 0 \qquad \text{where } k = \frac{\sqrt{i4\pi\sigma\omega}}{c} = \frac{\sqrt{2\pi\sigma\omega}}{c}(1+i). \tag{3.22}$$

At the boundary of the body, $\mathbf{H} = \mathbf{H}_0 e^{-i\omega t}$ and inside it ($z > 0$), the field has the form

$$\mathbf{H}^i = \mathbf{H}_0 e^{-i\omega t}\, e^{ikz} = \mathbf{H}_0 e^{-z/\delta}\, e^{i(-\omega t + z/\delta)}.$$

It falls off exponentially in the metal and, therefore,

$$\delta = \frac{c}{\sqrt{2\pi\sigma\omega}} = \frac{1+i}{k} \tag{3.23}$$

is a characteristic scale for the penetration depth of the magnetic field into the particle: the lower the frequency, the greater δ becomes. The electric field has the same penetration depth and follows from (1.114),

$$\mathbf{E} = (1 - i)\sqrt{\frac{\omega}{8\pi\sigma}}\mathbf{H}^i \times \mathbf{e}_z \tag{3.24}$$

where $\mathbf{e}_z$ is the unit vector in the direction of the z-axis. $\mathbf{E}$ is parallel to the y-axis.

3.2.4   Limiting values of the magnetic polarizability

The formula (3.21) for the magnetic polarizability α_m of a sphere simplifies further when the penetration depth δ, which we just introduced, is small or large compared to the grain radius. At any rate, the grain is always small compared to the wavelength, $a \ll \lambda = 2\pi c/\omega$.

* If the penetration depth is large ($a \ll \delta$), we develop the cotangent function in (3.21) to fifth order in ka, where

$$ka = (1 + i)\frac{a}{\delta} = (1 + i)\frac{a\sqrt{2\pi\sigma\omega}}{c} \ll 1.$$

  Then the first two terms in the bracket of (3.21) cancel out and we obtain for the magnetic polarizability in the low frequency limit

$$\alpha_m = -\frac{1}{105\pi}\left(\frac{a}{\delta}\right)^4 + i\frac{1}{20\pi}\left(\frac{a}{\delta}\right)^2. \tag{3.25}$$

  A metallic sphere of radius a and conductivity σ has then at frequency ω the cross section for magnetic dipole absorption

$$C^{\text{abs}} = \frac{8\pi^2\sigma}{15c^3}\omega^2 a^5. \tag{3.26}$$

* If the penetration depth is small ($\delta \ll a \ll \lambda$), the magnetic polarizability becomes

$$\alpha_m = -\frac{3}{8\pi}\left[1 - \frac{3\delta}{2a}\right] + i\frac{9}{16\pi}\frac{\delta}{a} \tag{3.27}$$

  and

$$C^{\text{abs}} = 3\sqrt{\frac{\pi}{2\sigma}}a^2\omega^{1/2}. \tag{3.28}$$

3.3   Tiny ellipsoids

The treatment of a sphere in a constant electric field may be extended to ellipsoids. In analogy to the previous discussion, we can determine their scattering and absorption cross section once we have worked out the dipole moment that they acquire and the strength of their internal field. An ellipsoid has three principal axes a, b, c. We use here the convention

$$a \geq b \geq c$$

and that a is aligned along the x-axis of a Cartesian coordinate system.

3.3.1 Elliptical coordinates

Ellipsoids are naturally handled in elliptical coordinates. They are defined as follows: For any three numbers a, b, c with

$$a > b > c > 0$$

the function

$$f(u) = \frac{x^2}{a^2 + u} + \frac{y^2}{b^2 + u} + \frac{z^2}{c^2 + u} - 1 \tag{3.29}$$

is of third order in u. It has poles at $-a^2, -b^2, -c^2$ and for reasons of continuity must vanish somewhere in each of the intervals $(-a^2, -b^2)$, $(-b^2, -c^2)$ and $(-c^2, +\infty)$. Therefore, $f(u)$ has three real roots, named ξ, η, ζ, with

$$\xi > -c^2 \qquad -c^2 > \eta > -b^2 \qquad -b^2 > \zeta > -a^2. \tag{3.30}$$

(ξ, η, ζ) are called elliptical coordinates because for $u = \xi$, the equation $f(\xi) = 0$ describes an ellipsoid that has the same foci as the ellipsoid

$$\frac{x^2}{a^2} + \frac{y^2}{b^2} + \frac{z^2}{c^2} = 1. \tag{3.31}$$

For $u = \eta$ or $u = \zeta$, one obtains confocal hyperboloids. The essential point is that ξ is constant on the surface of an ellipsoid. If $a = b = c$, we are reduced to a sphere and $f(u)$ has only one root.

Equation (3.29) constitutes a set of *three equations* when we put $u = \xi$, $u = \eta$ or $u = \zeta$. To transform the elliptical into Cartesian coordinates, one has to solve this set for x, y and z. This gives, for example,

$$x = \pm\sqrt{\frac{(\xi + a^2)(\eta + a^2)(\zeta + a^2)}{(b^2 - a^2)(c^2 - a^2)}}$$

and corresponding expressions for y and z. We will again need the Laplace equation $\Delta\varphi = 0$, this time in elliptical coordinates. Dividing the Laplace operator $\Delta\varphi$ by $\frac{1}{4}(\xi - \eta)(\zeta - \xi)(\eta - \zeta)$, one gets

$$(\eta - \zeta)R_\xi \frac{\partial}{\partial\xi}\left(R_\xi \frac{\partial\varphi}{\partial\xi}\right) + (\zeta - \xi)R_\eta \frac{\partial}{\partial\eta}\left(R_\eta \frac{\partial\varphi}{\partial\eta}\right) + (\xi - \eta)R_\zeta \frac{\partial}{\partial\zeta}\left(R_\zeta \frac{\partial\varphi}{\partial\zeta}\right) = 0. \tag{3.32}$$

The expressions

$$R_u = \sqrt{(u + a^2)(u + b^2)(u + c^2)} \qquad (u = \xi, \eta, \zeta)$$

are related to the length element ds in elliptical coordinates by

$$ds^2 = h_\xi^2 d\xi^2 + h_\eta^2 d\eta^2 + h_\zeta^2 d\zeta^2$$

with

$$h_\xi = \frac{\sqrt{(\xi - \eta)(\xi - \zeta)}}{2R_\xi} \qquad h_\eta = \frac{\sqrt{(\eta - \zeta)(\eta - \xi)}}{2R_\eta} \qquad h_\zeta = \frac{\sqrt{(\zeta - \xi)(\zeta - \eta)}}{2R_\zeta}.$$

3.3.2   An ellipsoid in a constant electric field

3.3.2.1   The dipole potential

Imagine an ellipsoidal grain with principal axes $a > b > c$ at the center of a Cartesian coordinate system. Its boundary is given by (3.31) or, equivalently, by $\xi = 0$. Let the outer electric field **E** be directed parallel to the x-axis and thus to the axis a of the ellipsoid. In analogy to (3.6), we write

$$\varphi = \begin{cases} \varphi_0[1 + F(\xi)] & \text{outside ellipsoid} \\ c_2\varphi_0 & \text{inside ellipsoid} \end{cases} \qquad (3.33)$$

where $\varphi_0 = -\mathbf{E} \cdot \mathbf{r}$ is the potential of the unperturbed field. The function F depends only on ξ, and $\varphi_0 F$ is the perturbation evoked by the grain; c_2 is a constant. When we insert $\varphi = \varphi_0(1 + F)$ into (3.32) and take into account that also for the perturbation $\Delta(\varphi_0 F) = 0$, we arrive at the differential equation

$$0 = F'' + F'\left(\frac{1}{\varphi_0^2} + \frac{R_\xi}{R_\xi'}\right) = F'' + F'\frac{d}{d\xi}\ln[R_\xi(\xi + a^2)]. \qquad (3.34)$$

A prime here and later denotes the derivative with respect to ξ. There are two solutions to (3.34): $F = $ constant, which applies to the interior of the grain, and

$$F(\xi) = c_1 \int_\xi^\infty \frac{dx}{(x + a^2)R_x}. \qquad (3.35)$$

Far away from the grain, the perturbation $\varphi_0 F$ has the form characteristic of a dipole. At great distance $r = \sqrt{x^2 + y^2 + z^2}$, at least one of the Cartesian coordinates is large, so the terms a^2, b^2, c^2 in (3.29) are negligible and $r \simeq \xi^{1/2}$. This allows us to evaluate the integral in (3.35) and one obtains the dipole potential

$$F(\xi) \simeq \frac{2c_1}{3}\xi^{-3/2} \propto r^{-3}.$$

3.3.2.2   The shape factor

As in the case of a sphere, we derive the constants c_1, c_2 from the continuity conditions on the surface where $\xi = 0$. Continuity of the *tangential* electric field, or of φ, gives $c_2 = 1 + F(0)$. The *normal* component of the displacement yields

$$\varepsilon c_2 \varphi_0'(0) = \varphi_0'(0) \cdot [1 + F(0)] + \varphi_0(0) \cdot F'(0)$$

and from there follows

$$c_2 = 2a^2 \frac{F'(0)}{\varepsilon - 1}.$$

Defining the *shape factor*

$$L_a = \frac{abc}{2} \int_0^\infty \frac{dx}{(x + a^2) R_x} \tag{3.36}$$

we find

$$c_2 = \frac{1}{1 + L_a(\varepsilon - 1)} \qquad c_1 = -\frac{abc(\varepsilon - 1)}{2[1 + L_a(\varepsilon - 1)]}.$$

3.3.2.3   The polarizability of an ellipsoid

With c_1 and c_2 being determined, we know the electric field in and around an ellipsoid. The internal field is constant and parallel to $\mathbf{E}$,

$$\mathbf{E}^{\mathrm{i}} = \frac{\mathbf{E}}{1 + L_a(\varepsilon - 1)}. \tag{3.37}$$

With the same arguments as in (3.8) for a sphere, the dipole moment of the ellipsoid becomes

$$\mathbf{p} = \frac{\varepsilon - 1}{3[1 + L_a(\varepsilon - 1)]} \mathbf{E} \tag{3.38}$$

and its polarizability, when the electric field is parallel to axis a,

$$\alpha_a = \frac{\varepsilon - 1}{4\pi [1 + L_a(\varepsilon - 1)]}. \tag{3.39}$$

3.3.3   Cross section and shape factor

Because of our experience with spheres, we can immediately write down the formulae of the scattering and absorption cross section, C^{sca} and C^{abs}, for ellipsoids. In the case of scattering, equation (3.12) for dipole radiation is applicable with the dipole moment $\mathbf{p}$ from (3.38). Absorption is still proportional to grain volume V, so we use (3.16) but now with $E_0 = |\mathbf{E}^{\mathrm{i}}|$ from (3.37). This gives

$$C^{\mathrm{abs}} = \frac{2\pi V}{\lambda} \mathrm{Im} \left\{ \frac{\varepsilon - 1}{1 + L_a(\varepsilon - 1)} \right\} \tag{3.40}$$

$$C^{\mathrm{sca}} = \frac{8\pi^3 V^2}{3\lambda^4} \left| \frac{\varepsilon - 1}{1 + L_a(\varepsilon - 1)} \right|^2. \tag{3.41}$$

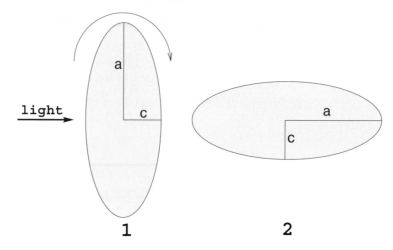

Figure 3.1. A cigar rotating about axis b, which is perpendicular to the page in this book. Light is traveling in the indicated direction. After a quarter of the rotation cycle, the cigar has changed from position 1 to position 2.

When the electric field is parallel to axis b or c, and not to axis a as we have assumed so far, the only thing that changes for the cross section is the shape factor of (3.36). We then have to replace L_a by

$$L_b = \frac{abc}{2} \int_0^\infty \frac{dx}{(x+b^2)R_x} \quad \text{or} \quad L_c = \frac{abc}{2} \int_0^\infty \frac{dx}{(x+c^2)R_x}$$

respectively. We can easily check that the sum over all shape factors is one:

$$L_a + L_b + L_c = 1$$

so only two of the three L values are independent. Obviously, with $a = b = c$, all Ls are equal to $\frac{1}{3}$ and we recover the formulae of the polarizability and cross sections for spheres.

Only in the Rayleigh limit does the cross section C depend solely on the direction of the electric field and not on the direction of wave propagation. Consider, for example, the cigar in figure 3.1 and let the electric vector **E** swing parallel to axis c. The cross section is then the same for light that falls in parallel to axis a or parallel to axis b (which is perpendicular to a and c). In the first case, the projected surface is a small circle, in the second it is a broad ellipse and much bigger.

When the small ellipsoid is very transparent ($|\varepsilon| \simeq 1$), the shape factors loose their importance, the electric field inside and outside are more or less equal, $\mathbf{E}^i = \mathbf{E}$, and the polarizability $\alpha = (\varepsilon - 1)/4\pi = \chi$. The absorption and scattering coefficient are then independent of the axial ratios and of orientation,

$$C^{abs}(\lambda) = \frac{2\pi V}{\lambda} \varepsilon_2(\lambda) \tag{3.42}$$

$$C^{sca}(\lambda) = \frac{8\pi^3 V^2}{3\lambda^4} |\varepsilon - 1|^2. \tag{3.43}$$

3.3.4 Randomly oriented ellipsoids

When an ellipsoid of fixed orientation in space is illuminated by an electromagnetic wave, the grand principle of superposition allows us to split the electric field vector $\mathbf{E}$ of the wave into components along the orthogonal ellipsoidal axes a, b, c:

$$\mathbf{E} = (E \cos\alpha, E \cos\beta, E \cos\gamma).$$

Here $E = |\mathbf{E}|$ and $\cos^2\alpha + \cos^2\beta + \cos^2\gamma = 1$. Interestingly, for arbitrary grain orientation the internal field $\mathbf{E}^i$ is not parallel to the outer field $\mathbf{E}$, even if the grain material is isotropic; $\mathbf{E}^i$ and $\mathbf{E}$ are parallel only when $\mathbf{E}$ is directed along one of the principal axes.

If C_a, C_b, C_c denote the cross sections of the ellipsoid when the principal axes a, b, c are parallel to the electric vector $\mathbf{E}$ of the incoming wave, the total cross section of the grain can be written as

$$C = C_a \cos^2\alpha + C_b \cos^2\beta + C_c \cos^2\gamma. \tag{3.44}$$

Without alignment, we expect, in interstellar space, random rotation and thus random orientation of the grains. For an ensemble of particles, all directions are equally likely and the mean of $\cos^2 x$ over 4π is (see (2.2))

$$\langle \cos^2 x \rangle = \frac{1}{4\pi} \int_0^{2\pi} dy \int_0^{\pi} dx \, \cos^2 x \sin x = \frac{1}{3}.$$

As the terms on the right-hand side of (3.44) are independent of each other, the average cross section for identical ellipsoids under random orientation is equal to the arithmetic mean:

$$\langle C \rangle = \tfrac{1}{3} C_a + \tfrac{1}{3} C_b + \tfrac{1}{3} C_c. \tag{3.45}$$

Equations (3.44) and (3.45) are true only in the electrostatic approximation.

3.3.5 Pancakes and cigars

When two of the principal axes are equal, the ellipsoid is called a spheroid.

- If then the two equally long axes are larger than the third one, $a = b > c$, the body has the shape of a *pancake*.

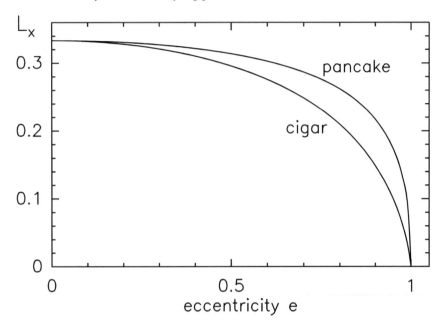

Figure 3.2. The shape factor for oblate (pancake) and prolate (cigar) spheroids. It can be found by numerically evaluating the integral in (3.36) or by using the analytical expressions (3.47) and (3.48).

- Otherwise, if $a > b = c$, it resembles a *cigar*.

More educated terms are *oblate* and *prolate* spheroids. We will see in chapter 10 that pancakes and cigars, besides being nourishing or fragrant, can explain why and how stellar light is polarized by dust clouds. The shape factor of (3.36) now permits an analytical solution (figure 3.2). Defining the eccentricity e through

$$e^2 = 1 - \frac{c^2}{a^2},$$
(3.46)

- cigars have $a > b = c$, $L_b = L_c$ and

$$L_a = \frac{1 - e^2}{e^2} \left[-1 + \frac{1}{2e} \ln\left(\frac{1+e}{1-e}\right) \right]$$
(3.47)

- pancakes have $a = b > c$, $L_a = L_b$, and

$$L_a = \frac{g(e)}{2e^2} \left[\frac{\pi}{2} - \arctan g(e) \right] - \frac{g^2(e)}{2} \quad \text{with } g(e) = \sqrt{\frac{1 - e^2}{e^2}}.$$
(3.48)

For very long cigars (needles) and very flat pancakes $e = 1$ and $L_a = 0$, so $L_b = L_c = \frac{1}{2}$. A sphere is the subcase when $e = 0$ and all shape factors are equal to $\frac{1}{3}$.

Figure 3.3 displays, over a broad wavelength band, what happens to the optical depth when we replace spheres of a fixed radius by randomly oriented spheroids, identical in shape and *of the same volume as the spheres*. It turns out, although not as a strict rule, that ellipsoids have a higher *average* cross section, C_{av}, than spheres and the effect increases with eccentricity. We can understand this because C_{av} is, from (3.45), the arithmetic mean over C_a, C_b, C_c. When we evaluate the Cs assuming $a \geq b \geq c$ we find that (mostly) $C_a \geq C_b \geq C_c$ in such a way that the increase in C_a more than offsets the decline in C_b or C_c.

With fixed eccentricity, the cross section is still a function of the optical constant m and thus of wavelength λ. Any wiggle in m due to a resonance will be reflected in the curves. We show results for silicate and carbon particles and we can readily identify the resonances in silicate at 10 and 18 μm by comparison with figure 7.19. The difference between cigars and pancakes is mild. Overall, the effect of particle elongation on the optical depth is more pronounced in the far than in the near infrared. For example, grains with an axial ratio of two, which implies an eccentricity of 0.866, are at $\lambda \geq 100 \ \mu$m by some 30% better emitters or absorbers than spheres.

As a further sophistication, one can treat optically anisotropic ellipsoids. Graphite particles are an example. When the carbon sheets are stacked parallel to the (x, y)-plane, one has $\varepsilon_x = \varepsilon_y \neq \varepsilon_z$. It is also not difficult to extend the computations of cross sections to coated ellipsoids but for particulars.

3.3.6   Rotation about the axis of greatest moment of inertia

The cross section of spinning particles changes periodically and one has to take time averages. Suppose a spheroid rotates about the major axis where the moment of inertia is greatest. For cigars, this is axis b or c, for pancakes it is axis c. The rotation axis stays fixed in space, while the other two are spinning. Let the light propagate in a direction perpendicular to the rotation axis and let

$$C_{E\|rot} \qquad C_{E\perp rot}$$

denote the *time-averaged cross sections* in the case when the electric vector of the incident wave is parallel and perpendicular to the rotation axis, respectively.

Figure 3.1 shows such a situation for a spinning cigar with rotation axis b. Because the mean of $\cos^2 x$ over a half-cycle is

$$\langle \cos^2 x \rangle = \frac{1}{\pi} \int_0^{\pi} \cos^2 x \, dx = \frac{1}{2}$$

we find that the average cross sections, $C_{E\|rot}$ and $C_{E\perp rot}$, depend on the direction of linear polarization of the incident wave,

$$C_{E\perp rot} = \tfrac{1}{2}\big[C_a + C_c\big]$$

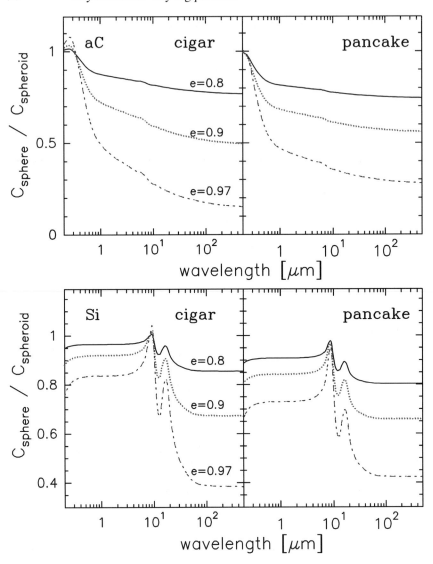

Figure 3.3. The cross section of spheres over that of randomly oriented spheroids of the same volume. Calculations are done in the electrostatic approximation implying grains much smaller than the wavelength. The eccentricity of the spheroids defined in (3.46) is indicated. The grains consist of amorphous carbon (aC, top) or silicate (Si, bottom); optical constants for the two materials from figure 7.19.

$$C_{E\|rot} = C_b < C_{E\perp rot}.$$

If a pancake rotates about its axis of maximum moment of inertia, which is

c, and light comes in perpendicular or parallel to c, no averaging is necessary and

$$C_{E\perp rot} = C_a$$
$$C_{E\|rot} = C_c < C_{E\perp rot}.$$

The *effective cross section* for incident unpolarized light is

$$C = \tfrac{1}{2}\left[C_{E\|rot} + C_{E\perp rot}\right].$$

To calculate in chapter 10 the degree of polarization produced by grains, we need the difference

$$\Delta C = C_{E\perp rot} - C_{E\|rot}.$$

- For cigars ($a > b = c$ and $C_b = C_c$):

$$C = \tfrac{1}{4}C_a + \tfrac{3}{4}C_c \qquad \Delta C = \tfrac{1}{2}\left[C_a - C_c\right] \tag{3.49}$$

- for pancakes ($a = b > c$ and $C_a = C_b$):

$$C = \tfrac{1}{2}\left[C_a + C_c\right] \qquad \Delta C = C_a - C_c. \tag{3.50}$$

If the light travels parallel to the rotation axis of the spheroid (axis b for cigars, axis c for pancakes), there are, on average, no polarization effects.

3.4 The fields inside a dielectric particle

3.4.1 Internal field and depolarization field

We determined in equation (3.7) the field $\mathbf{E}^i$ inside a dielectric sphere that sits in a time-constant homogeneous outer field $\mathbf{E}$. In section 3.3 we generalized to ellipsoids. In both cases, the polarization $\mathbf{P}$ of the medium is constant and the field inside smaller than outside. Writing the *internal field* $\mathbf{E}^i$ in the form

$$\mathbf{E}^i = \mathbf{E} + \mathbf{E}_1$$

defines a new field $\mathbf{E}_1$. It arises from all atomic dipoles and is directed opposite to the polarization $\mathbf{P}$. Because $E^i < E$, one calls $\mathbf{E}_1$ the *depolarization field* (see figure 3.4). For example, from (3.7) we find for a sphere

$$\mathbf{E}_1 = -\frac{4\pi}{3}\mathbf{P}. \tag{3.51}$$

It is important to make a distinction between the local field at exactly one point and macroscopic averages. $\mathbf{E}_1$ and $\mathbf{E}^i$ are such averages over many atoms, a hundred or so but at least over one unit cell in a crystalline structure. On a microscopic level, the field has tremendous gradients. Atoms are not at random positions but at privileged sites (lattice grid points are loci of minimum potential energy) and the local field $\mathbf{E}^{loc}$ acting on an atom is usually different from the average field $\mathbf{E}^i$, so generally

$$\mathbf{E}^{loc} \neq \mathbf{E}^i.$$

To find $\mathbf{E}^{loc}$ at a specific locus in the grain, say at $\mathbf{x}_0$, we have to add to $\mathbf{E}$ the fields from all dipoles. As the dipoles are at *discrete* lattice points, their distribution is not smooth, at least, it does not appear to be so close to $\mathbf{x}_0$. We, therefore, imagine a spherical cavity around $\mathbf{x}_0$ of such a size that beyond the cavity the dipole distribution may be regarded as smooth, whereas inside it, it is discontinuous, so

$$\mathbf{E}^{loc} = \mathbf{E} + \sum_{\text{outside}} \mathbf{E}_{dip} + \sum_{\text{cavity}} \mathbf{E}_{dip}. \tag{3.52}$$

The field arising from the smooth distribution outside the cavity can be expressed as a volume integral, the field from the dipoles within the cavity has to be explicitly written as a sum.

3.4.2 Depolarization field and the distribution of surface charges

A body of constant polarization $\mathbf{P}$ has, on its outside, a surface charge σ of strength

$$\sigma = \mathbf{e} \cdot \mathbf{P} \tag{3.53}$$

where $\mathbf{e}$ is the outward surface normal. This expression for σ follows when we recall that according to (1.20) a non-uniform polarization creates a charge $\rho_{pol} = -\operatorname{div}\mathbf{P}$. Inside the body, the divergence of the polarization vector is zero and $\rho_{pol} = 0$ but on its surface, $\mathbf{P}$ is discontinuous and a charge appears.

There is a theorem which states that for any body of *constant polarization* $\mathbf{P}$, the depolarization field $\mathbf{E}_1$ is identical to the field that arises in vacuum from the distribution of surface charges as given by (3.53). Let $\varphi(\mathbf{r})$ be the electrostatic potential from all dipoles in the body. We prove the theorem by writing $\varphi(\mathbf{r})$ as a volume integral,

$$\varphi(\mathbf{r}) = -\int (\mathbf{P} \cdot \nabla r^{-1})\, dV.$$

This is correct because a single dipole $\mathbf{p}$ has a potential

$$\varphi_{dip} = -\mathbf{p} \cdot \nabla \left(\frac{1}{r}\right).$$

Now we transform the volume integral into a surface integral employing the relation $\operatorname{div}(f\mathbf{P}) = f \operatorname{div}\mathbf{P} + \mathbf{P} \cdot \nabla f$. With $f(\mathbf{r}) = 1/r$ and $\operatorname{div}\mathbf{P} = 0$, we get

$$\varphi(\mathbf{r}) = -\oint \frac{1}{r}\mathbf{P} \cdot d\mathbf{S} = \oint \frac{\sigma}{r}\, dS.$$

The second integral sums up the potentials from all surface charges and their total field is thus equivalent to the field of all atomic dipoles.

3.4.3 The local field at an atom

The 'outside' sum in (3.52) may be evaluated by an integral. According to our theorem, it is equivalent to the field from the surface charges. However, there are now two boundaries. The outer, S_1, gives the depolarization field $\mathbf{E}_1$. The inner, S_2, has the same surface charge as a sphere in vacuum of constant polarization $\mathbf{P}$, only of inverted sign. Comparing with (3.51), we find that the surface charge on S_2 produces the field $(4\pi/3)\mathbf{P}$.

It remains to include the atoms in the spherical cavity S_2. Suppose we have a *cubic lattice*, the dipoles are at positions (x_i, y_i, z_i), their moments are all of strength p and aligned in the z-direction, the point $\mathbf{x}_0$ is at the origin. Then the z-component of the total field from all dipoles at $\mathbf{x}_0$ is (see (3.10))

$$\sum_{\text{cavity}} (\mathbf{E}_{\text{dip}})_z = p \sum_i \frac{3z_i^2 - r_i^2}{r_i^5} = p \sum_i \frac{2z_i^2 - x_i^2 - y_i^2}{r_i^5} = 0$$

where $r_i = \sqrt{x_i^2 + y_i^2 + z_i^2}$ is the distance of dipole i to the origin. The sum is zero because of the symmetry of the grid. Likewise the x- and y-components of the total field vanish. So we can neglect the influence of the nearest atoms altogether. We expect that we may also neglect it if there is no grid order at all, i.e. in an amorphous substance. For such a situation, therefore,

$$\mathbf{E}^{\text{loc}} = \mathbf{E}^{\text{i}} + \frac{4\pi}{3}\mathbf{P}. \tag{3.54}$$

Given the outer field $\mathbf{E}$, the local field $\mathbf{E}^{\text{loc}}$ at an atom depends on the shape of the particle. For example, for a spherical grain we derive from (3.54) with the help of (3.7) and (3.8):

$$\mathbf{E}^{\text{loc}} = \mathbf{E}.$$

For a body in the shape of a thin slab with parallel surfaces perpendicular to $\mathbf{E}$, which is the configuration of a parallel–plate condenser, one finds

$$\mathbf{E}^{\text{loc}} = \mathbf{E} - \frac{8\pi}{3}\mathbf{P}.$$

3.4.4 The Clausius–Mossotti relation

The local field $\mathbf{E}^{\text{loc}}$ produces in each atom of volume V a dipole moment

$$p = \alpha V E^{\text{loc}}.$$

Hence α is called the atomic polarizability. The formula is analogous to (1.8) which we applied to a grain as a whole. If there are N atoms per unit volume, the polarization of the matter is, in view of (3.54),

$$P = N V \alpha E^{\text{loc}} = N V \alpha \left(E^{\text{i}} + \frac{4\pi}{3} P \right).$$

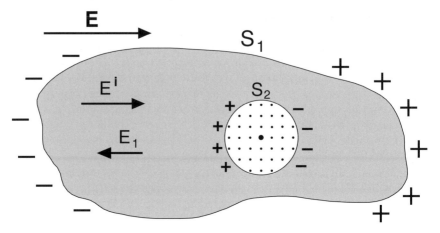

Figure 3.4. A field **E** produces on a dielectric grain a surface charge σ. According to a theorem of electrostatics, σ gives rise to a field $\mathbf{E}_1$ that combines with **E** to form the average internal field $\mathbf{E}^i$. The latter is not necessarily equal to the local field $\mathbf{E}^{loc}$ at a particular atom or unit cell (central larger dot in inner circle) in a crystal. To find $\mathbf{E}^{loc}$, one has to take into account the regularly but discontinuously arranged dipoles (small dots) in the vicinity as well as those farther away which can be considered as being distributed smoothly (shaded area). S_1 and S_2 denote surfaces.

Because $P = \chi E^i$ (see (1.6)), one can relate the dielectric susceptibility χ of the medium to the polarizability α of the atoms. This is done in the Clausius–Mossotti formula:

$$\chi = \frac{NV\alpha}{1 - \frac{4\pi}{3}NV\alpha}.\qquad(3.55)$$

The field exerts forces on an atom and distorts the cloud of electrons; it attracts them one way and pushes the heavy nuclei the opposite way. This can be modeled with the harmonic oscillator of section 1.3. When we take for the amplitude of the electron, x_0, the value from (1.67) but without damping ($\gamma = 0$), and equate the dipole moment $p = ex_0$ to αE^{loc}, we get the electronic polarizability

$$\alpha(\omega) = \frac{e^2}{m_e(\omega_0^2 - \omega^2)}.\qquad(3.56)$$

It is relevant at optical frequencies because the electrons have little inertia and can swiftly follow the field. The resonance frequency ω_0 of the atom lies typically in the ultraviolet. The static value of the atomic polarizability in a constant field is $\alpha_0 = e^2/m_e\omega_0^2$.

3.5 Very large particles

A particle is defined to be very large when its size is much bigger than the wavelength. As in the case of tiny grains, the definition is relative and the same particle may be both small and large, depending on the wavelength. To quantify the diffraction phenomena that occur around very big grains, we have to study basic optical principles.

3.5.1 Babinet's theorem

For any very large particle the extinction efficiency,

$$Q^{ext} = Q^{abs} + Q^{sca}$$

approaches two, independently of the chemical composition or shape of the particle. This important result is called Babinet's theorem or extinction paradox. For spheres,

$$Q^{ext} \to 2 \qquad \text{when } x = \frac{2\pi a}{\lambda} \to \infty. \qquad (3.57)$$

We illustrate Babinet's theorem with an experiment carried out in three steps as sketched in figure 3.5:

(1) When a parallel wavefront falls on an orifice, which is much bigger than the wavelength, it produces, on a far-away screen, a bright spot with a blurred rim. Outside the bright spot and the rim, the screen is dark. We restrict the discussion to this dark area.

(2) If one places a small, but still much bigger than λ, obstacle into the orifice, a diffraction pattern appears and light is scattered beyond the blurred rim into the former dark area.

(3) If we cover the orifice with black paper leaving just a hole of the same size and shape as the obstacle, the diffraction patterns of the hole and the obstacle are identical. According to Huygens' principle for wave propagation (this result will be shown later), the diffraction pattern of the obstacle arises because each point in the plane of the orifice, except for the obstacle itself, is the origin of a spherical wave; the diffraction pattern of the hole arises because each point in the plane of the hole is the origin of a spherical wave.

In the case of the completely uncovered orifice (1), both diffraction patterns are present simultaneously. Because the region beyond the blurred rim is then dark, the patterns from the obstacle and the hole must cancel each other exactly, i.e. they must have the same intensity but be phase-shifted by 180°. As the hole scatters all the light falling onto it, the obstacle must scatter exactly the same amount. Altogether the obstacle thus removes twice as much light than that which corresponds to its projected geometrical surface: half of it through scattering, the other half by absorption and reflection,

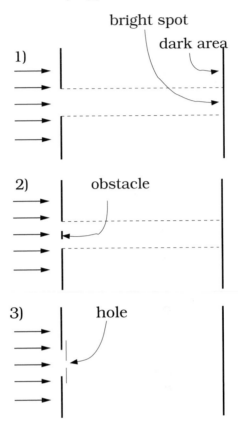

Figure 3.5. A diffraction experiment to explain Babinet's theorem.

Scattering at the edge of a large obstacle is, however, predominantly forward. Therefore $Q^{\text{ext}} = 2$ can only be verified at far distances; it is always valid for interstellar grains. At short distances, we know from everyday experience that a brick removes only as much sunlight as falls onto its projected surface and not twice as much.

3.5.2 Reflection and transmission at a plane surface

3.5.2.1 *Normal incidence*

A ray of light travels in the positive z-direction and hits, under normal incidence, a large particle as shown in figure 3.6. We wish to evaluate which fraction of the incident flux is reflected. This quantity is denoted by r and called the reflectance. If $k_1 = \sqrt{\varepsilon_1 \mu_1}\,\omega/c$ is the wavenumber from (1.41) in the medium on the left ($z < 0$), and $k_2 = \sqrt{\varepsilon_2 \mu_2}\,\omega/c$ in the medium on the right ($z > 0$), we obtain for

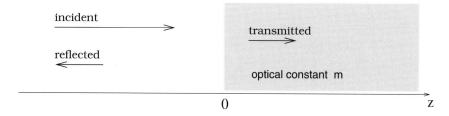

Figure 3.6. Light falls perpendicular on a plane surface, some is reflected and some transmitted. The medium to the right ($z > 0$) has an optical constant $m = n + ik$.

the electric field

$$E = E_i e^{i(k_1 \cdot z - \omega t)} + E_r e^{i(-k_1 \cdot z - \omega t)} \qquad \text{for} \quad z < 0 \qquad (3.58)$$

$$E = E_t e^{i(k_2 \cdot z - \omega t)} \qquad \text{for} \quad z > 0. \qquad (3.59)$$

To the left-hand side, we have the *incident* and the *reflected* wave (indices i and r), to the right only the *transmitted* wave (index t). At the surface of the particle ($z = 0$), the tangential components of the electric and magnetic field are continuous. When we express the magnetic field via the rotation of the electric field (see (1.32)), we find, at $z = 0$,

$$E_i + E_r = E_t$$

$$E_i - E_r = E_t \sqrt{\frac{\varepsilon_2 \mu_1}{\varepsilon_1 \mu_2}}.$$

This immediately allows one to calculate the reflectance $r = |E_r / E_i|^2$. The most common case is the one where there is a vacuum on the left and the grain is non-magnetic ($\mu_2 = 1$). The optical constant of the particle then equals $m = \sqrt{\varepsilon_2}$ and the reflectance is given by

$$r = \left| \frac{1 - m}{1 + m} \right|^2 = \frac{(n - 1)^2 + k^2}{(n + 1)^2 + k^2}. \qquad (3.60)$$

For a large sphere ($x = 2\pi a / \lambda \gg 1$), the reflectance equals the backscatter efficiency Q^{back} of (2.67).

Should n or k be large, the reflectivity is high and absorption low. Metal surfaces make good mirrors because they have a large optical constant m and $n \simeq k$ (see (1.117)). However, if $k \simeq 0$, the reflectivity grows with n. A pure diamond sparkles at visual wavelengths because k is small and $n = 2.4$, so the reflectivity is high ($r = 0.17$). If the stone in a ring is a fake, of standard glass which has $n = 1.5$, it will catch less attention because it reflects only 4% of the light, not 17%.

We see from (3.59) how the light that enters into a large and absorbing particle is attenuated. The amplitude of its electric field is weakened proportionally to $\exp(-2\pi kz/\lambda)$, so the intensity I_t of the transmitted light diminishes like $\exp(-4\pi k/\lambda z)$. Per wavelength of penetration, the intensity decreases by a factor $\exp(-4\pi k)$. Unless k is very small, the transmitted light is removed very quickly and the penetration depth is only a few wavelengths.

One may wonder about the implications of formula (3.60) for a blackbody. By definition, it has zero reflectance and would, therefore, require the optical constant of vacuum ($m = 1$). But a vacuum is translucent. So no real substance is a perfect absorber. A blackbody can be approximated by a particle with $n \rightarrow 1$ and $k \rightarrow 0$; the particle must also have a very large size d such that kd/λ is much greater than one, despite the smallness of k.

3.5.2.2 Oblique incidence

We generalize the reflectance r of (3.60) to the case of oblique incidence. If the incident beam is inclined to the normal of the surface element of the particle by some angle $\theta_i > 0$, the reflected beam forms an angle $\theta_r = \theta_i$ with the normal. The angle of the transmitted beam is given by *Snell's law*:

$$\sin\theta_t = \frac{\sin\theta_i}{m}. \tag{3.61}$$

Whereas for a non-absorbing medium, m and θ_t are real, in the generalized form of (3.61) m and also θ_t are complex. One now has to specify in which direction the incident light is polarized which is not necessary under normal incidence. For incident *unpolarized light* the reflectance is

$$r = \frac{1}{2}\left|\frac{\cos\theta_t - m\cos\theta_i}{\cos\theta_t + m\cos\theta_i}\right|^2 + \frac{1}{2}\left|\frac{\cos\theta_i - m\cos\theta_t}{\cos\theta_i + m\cos\theta_t}\right|^2. \tag{3.62}$$

Using this formula, we can determine the limiting value of Q^{abs} and Q^{sca} for very large spheres. If the particle is translucent ($k = 0$), it only scatters the light; then $Q^{abs} = 0$ and $Q^{sca} = Q^{ext} = 2$ according to Babinet's theorem. If $k > 0$, any light that enters the grain will, under the assumption of ray optics, eventually peter out within it. One can define a reflection efficiency Q^{ref} for spheres by

$$\pi a^2 Q^{ref} = \int_0^a 2\pi r(x)\, dx \tag{3.63}$$

where $r(x)$ is the reflectance after (3.62) for an incident angle $\theta_i = \arcsin(x/a)$ and a is the grain radius. Q^{ref} lies between 0 and 1 and is related to Q^{abs} through $Q^{abs} = 1 - Q^{ref}$. So the absorption efficiency for large particles can never be greater than one, only smaller if some fraction of light is reflected off the surface.

3.5.3 Huygens' principle

In figure 3.7, a point source Q emits isotropically radiation. On any spherical surface S around Q, the oscillations of the electromagnetic field are in phase

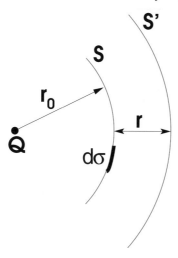

Figure 3.7. A spherical wavefront traveling from a point source Q at velocity c.

forming a wavefront. According to Huygens' principle, each surface element $d\sigma$ is the source of an elementary (or secondary) spherical wave. The strength of the elementary waves is proportional to the amplitude of the primary field and to the area $d\sigma$. It also depends on direction, being greatest radially away from Q and zero towards the rear; a more detailed description is not given. The superposition (interference) of all elementary waves of equal radius originating from the surface S fixes the position of the wavefront in the future and thus describes the propagation of the primary wave. For instance, if the wavefront is, at time t, on the surface S of figure 3.7, a time $\Delta t = r/c$ later it will be again on a spherical surface but of radius $r_0 + r$. The new surface S' is the envelope of all elementary waves of radius r. Outside S', the field of elementary waves is extinguished by interference.

Applied to plane waves, which are spherical waves of very large radius, Huygens' principle makes understandable why light propagates along straight lines. It also explains the laws of refraction and reflection in geometrical optics. To derive them from Huygens' principle, one just has to assume that all surface elements on the plane separating two media of different optical constants, n_1 and n_2, oscillate in phase and emit elementary waves which propagate with the phase velocity $v = c/n$ of the respective medium.

If there is an obstacle in the way of the primary wave, some region on the surface of its wavefront, corresponding in size and shape to the projected area of the obstacle, does not create secondary waves. Then the superposition of the remaining elementary waves leads to diffraction patterns. Interesting examples are the intensity distribution of light behind a straight wire or a slit, or the fuzzy edges of shadows. Light no longer travels only on straight lines but can bend around corners.

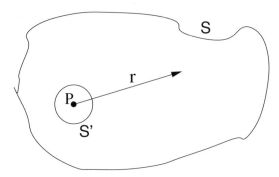

Figure 3.8. To evaluate the field u at point P inside a region G when u is known on its surface S.

3.5.3.1 Kirchhoff's strict formulation of Huygens' principle

Kirchhoff put Huygens' principle in a mathematically rigorous form. He was able to determine the electromagnetic field u at any point P inside a region G when u is known on the surface S of the region G (see figure 3.8). The field u must obey the wave equation:

$$\Delta u + k^2 u = 0. \tag{3.64}$$

We have already discussed plane waves, $u = e^{i(\mathbf{k}\cdot\mathbf{x} - \omega t)}$, as a solution to (3.64) in section 1.2. Spherical waves where u depends only on the distance $r = \sqrt{x^2 + y^2 + z^2}$ so that

$$\frac{1}{r}\frac{d^2}{dr^2}(ru) + k^2 u = 0$$

are another type of solution. By introducing the auxiliary function $v = ru$, one finds

$$u = \frac{e^{i(kr - \omega t)}}{r}. \tag{3.65}$$

Equation (3.65) describes an outward-going spherical wave; the time factor $e^{-i\omega t}$ is again included.

As formula (3.64) is akin to the Laplace equation, $\Delta\phi = 0$, which is encountered in electrostatics and where ϕ is the electric potential, one can use the familiar formalism of electrostatics to solve it. To determine u at a point P inside a bounded region G when u is given on its surface S, we use Green's identity (see (A.14)),

$$\int_G (u\Delta v - v\Delta u)\, dV = \oint_S \left(u\frac{\partial v}{\partial n} - v\frac{\partial u}{\partial n} \right) d\sigma$$

where the function v is defined as

$$v = \frac{e^{ikr}}{r}.$$

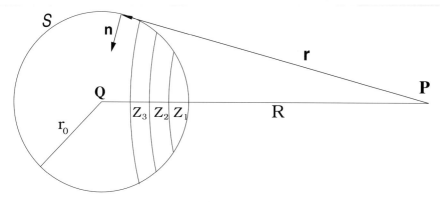

Figure 3.9. A point source Q isotropically emits light. The field u in P at a distance $R = \overline{QP}$ is calculated from Huygens' principle in the form (3.66). It is assumed that u is known on the spherical surface S around Q with radius r_0. To carry out the integration, one constructs Fresnel zones $Z_1, Z_2, Z_3, \ldots$. They are the intersection of S with shells around P of inner and outer radius $R - r_0 + (j - 1)\lambda/2$ and $R - r_0 + j\lambda/2$, respectively, $j = 1, 2, 3, \ldots$.

The radius r counts the distance from the point P where we want to evaluate the field u. At P, there is a singularity because $r = 0$. To circumvent it, one cuts out from G a small sphere around P of surface S'. The volume integral in Green's identity is then taken only over the remaining region without the small sphere; this remaining region is called $\overline{G}$. The surface integral in Green's identity now extends over S and S'. Because in $\overline{G}$, the wave equation is fulfilled for u and v, i.e. $\Delta u + k^2 u = 0$ and $\Delta v + k^2 v = 0$, the volume integral vanishes and only the surface integral is left. Letting the radius of the small sphere around P go to zero, one eventually arrives at an equation that incorporates Huygens' principle,

$$4\pi u_p = -\oint_S \left(u \frac{\partial}{\partial n} \frac{e^{ikr}}{r} - \frac{e^{ikr}}{r} \frac{\partial u}{\partial n} \right) d\sigma. \tag{3.66}$$

3.5.4 Fresnel zones and a check on Huygens' principle

To convince ourselves that (3.66) is indeed a strict form of Huygens' principle, we show, as had been asserted before (figure 3.7), that the field from a point source emitting spherical waves can be considered to result from interference of secondary spherical waves. The new configuration is depicted in figure 3.9. At P, a distance R from the point source Q, the field is, of course,

$$u_p = \frac{e^{ikR}}{R}$$

We claim to obtain the same result from superposition of all secondary waves emitted from the spherical surface S. So we calculate u_p from (3.66) supposing that we know the field and its normal derivative on S. On the surface S,

$$u = u_0 = \frac{e^{ikr_0}}{r_0} \implies \frac{\partial u}{\partial n}\bigg|_0 = -\frac{e^{ikr_0}}{r_0}\left(ik - \frac{1}{r_0}\right).$$

The region G surrounding P, over whose surface one has to integrate, comprises now all space *outside* the sphere of radius r_0. The contribution to the integral (3.66) of the surface which is at infinity may be neglected, we only have to consider the surface S. Because

$$\frac{\partial}{\partial n}\frac{e^{ikr}}{r} = \frac{e^{ikr}}{r}\left(ik - \frac{1}{r}\right)\cos(\mathbf{n}, \mathbf{r})$$

where $\mathbf{r}$ is the radius vector from P to S and $r = |\mathbf{r}|$, we get

$$4\pi u_p = -\oint_S \frac{e^{ik(r+r_0)}}{rr_0}\left\{(ik - r^{-1})\cos(\mathbf{n}, \mathbf{r}) + (ik - r_0^{-1})\right\}d\sigma. \qquad (3.67)$$

If the distances are much greater than the wavelength $(r, r_0 \gg k^{-1} = \lambda/2\pi)$,

$$4\pi u_p = -ik\frac{e^{ikr_0}}{r_0}\oint_S \frac{e^{ikr}}{r}[1 + \cos(\mathbf{n}, \mathbf{r})]\,d\sigma.$$

We evaluate this integral as a sum and, for this purpose, divide the surface S into so called Fresnel zones. These are segments on S lying between constant radii r and $r + \lambda/2$. The surface normal $\mathbf{n}$ of a given segment forms an approximately constant angle $\cos(\mathbf{n}, \mathbf{r})$ with the radius vector $\mathbf{r}$. The area of a segment is

$$d\sigma = \frac{2\pi rr_0}{R}\,dr.$$

The contribution u_j of the jth zone Z_j to the field at P follows from a simple integration over r, in the limits $R - r_0 + j\lambda/2$ and $R - r_0 + (j-1)\lambda/2$, at fixed $[1 + \cos(\mathbf{n}, \mathbf{r})]$,

$$u_j = \frac{e^{ikR}}{R}(-1)^{j+1}\left[1 + \cos(\mathbf{n}, \mathbf{r})\right].$$

The first zone alone, if all others were covered by some dark material, yields

$$u_1 = \frac{2e^{ikR}}{R}.$$

The contributions from the following zones alternate in sign. The first two zones *together* produce almost nothing but darkness. The last zone is the one on the far side of the sphere which, like the first, is intersected by the straight line through P and Q. It is not difficult to work out the sum $\sum u_j$ over all Fresnel zones. One really obtains e^{ikR}/R, and this strengthens our confidence in Huygen's principle.

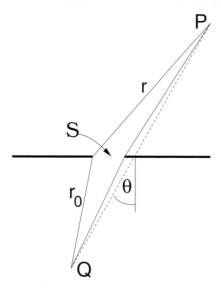

Figure 3.10. Light falls from Q though an orifice S and is observed at P.

3.5.5 The reciprocity theorem

As very big particles block practically all light that falls onto them, interesting phenomena occur only through diffraction at their edges. Suppose we have a source Q behind an opaque plane with an orifice S of arbitrary shape and we want to determine the intensity u_p at a point P in front of the plane (see figure 3.10). We compute it from superposition of secondary waves from the orifice S.

To first order, geometrical optics is valid and the hole deflects light only by small angles. Therefore we approximate equation (3.67) under the assumption $r, r_0 \gg k^{-1}$ by

$$u_p = -\frac{i}{\lambda}\frac{\cos\theta}{rr_0} \oint_S e^{ik(r+r_0)}\, d\sigma. \tag{3.68}$$

Here $\cos\theta \simeq \frac{1}{2}[\cos(\mathbf{n}, \mathbf{r}) - \cos(\mathbf{n}, \mathbf{r_0})]$, and θ is the angle between $\overline{QP}$ and the normal of the plane. The distances r and r_0 are also almost constant and may therefore stand before the integral; however, $e^{ik(r+r_0)}$ varies of course over the hole because it is many wavelengths across. The symmetry of formula (3.68) with respect to r and r_0 implies that when a source at Q produces a field u_p at P, the same source brought to the point P produces the same field at Q. This is the reciprocity theorem of diffraction theory.

3.5.6 Diffraction by a circular hole or a sphere

Equation (3.68) gives the field of an electromagnetic wave that has passed through an orifice. We now apply it to a circular hole of radius a and remember from the

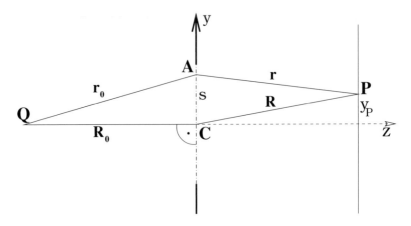

Figure 3.11. The source Q is behind a circular hole (dash-dot line) and illuminates the point P on a screen. The hole is in the (x, y)-plane of the Cartesian coordinate system with origin C. The x-axis is perpendicular to the page of the book.

subsection on Babinet's theorem that a spherical obstacle will produce the same diffraction pattern.

The configuration is shown in figure 3.11. It is assumed that the hole is small ($s \ll R, R_0$) and that Q, C and P lie roughly on a straight line so that $\cos\theta$ (see figure 3.10) is close to one and may be omitted. R_0 and R are the distances from the center C of the Cartesian coordinate system (x, y, z) to the source Q and the observing point P, respectively; r is the distance between P and an arbitrary surface element $d\sigma$ located at A; r_0 is the distance between Q and $d\sigma$. The coordinates of the points marked in figure 3.11 are

$$A = (x, y, 0) \qquad Q = (0, 0, -R_0) \qquad P = (0, y_p, z_p) \qquad C = (0, 0, 0).$$

As we have to integrate over the hole, it is advantageous to use cylindrical coordinates (s, φ, z), where

$$s^2 = x^2 + y^2 \qquad x = s \sin\varphi \qquad y = s \cos\varphi.$$

Because

$$r_0^2 = R_0^2 + s^2 \qquad r^2 = x^2 + (y_p - y)^2 + z_p^2 \qquad R^2 = y_p^2 + z_p^2$$

it follows that $r_0 \simeq R_0 + s^2/2R_0$ and $r \simeq R + (s^2 - 2yy_p)/2R$. Neglecting terms of order s^2 or higher, which implies $s \ll y_p$, one obtains

$$r + r_0 \simeq R + R_0 - \frac{y_p}{R}s\cos\varphi.$$

Because $d\sigma = s \cdot ds\, d\varphi$, the integral in (3.68) becomes,

$$\oint_S e^{ik(r+r_0)}\, d\sigma = e^{ik(R+R_0)} \int_0^a ds\, s \int_0^{2\pi} d\varphi\, e^{-ik\alpha s \cos\varphi}.$$

We have introduced the deflection angle $\alpha = y_p/R$. To evaluate the double integral, one uses complex integer Bessel functions $J_n(z)$ defined in (A.21). The formulae (A.22)–(A.24) yield for the double integral

$$2\pi \int_0^a ds\, s\, J_0(k\alpha s) = 2\pi a^2 \frac{J_1(ak\alpha)}{ak\alpha}$$

and thus

$$|u_p|^2 = \frac{4\pi^2 a^4}{\lambda^2 r^2 r_0^2} \cdot \left(\frac{J_1(ak\alpha)}{ak\alpha} \right)^2. \tag{3.69}$$

The Bessel function $J_1(x)$ is tabulated in mathematical encyclopediae. For small x, $J_1(x) \to \frac{1}{2}x$; at $x = 1$, this approximation has an error of roughly 10%. The intensity $|u_p|^2$ drops to half its maximum value at $\alpha = 0$ at a deflection angle

$$\alpha_{\text{half}} \simeq 1.617 \frac{\lambda}{2\pi a}. \tag{3.70}$$

For example, in the visual bound, for a grain with 1 mm diameter, α_{half} is about 1 arcmin.

It may seem puzzling that the intensity of the light which one observes at P and which is equal to $|u_p|^2$ rises with the fourth power of the hole radius a, although the flux passing through the hole increases only with a^2. But as the hole becomes bigger, the diffraction pattern gets smaller and, in the end, as much energy as falls onto the hole from the source emerges on the other side; energy is conserved.

3.5.7 Diffraction behind a half-plane

As a final example, we derive the intensity pattern on a screen behind a half-plane (wall). This problem affords a nice graphical solution. The situation is depicted in figure 3.12. To determine the field at P, opposite to S, one has to sum up all elementary waves from the plane in which the wall lies. Because we assume $y_p - y \ll R$, the path $\overline{TP}$ is approximately equal to $R + (y_p - y)^2/2R$. Therefore, the wave from T has, relative to the wave emitted from S, a phase lag

$$\delta = \frac{2\pi}{\lambda} \frac{(y_p - y)^2}{2R}.$$

Note that the phase lag δ increases quadratically with $(y_p - y)$. At P, the field vectors $\mathbf{E}_j$ of all elementary waves have to be added up. They have practically all the same length but different directions. This leads to the so called cornu-spiral

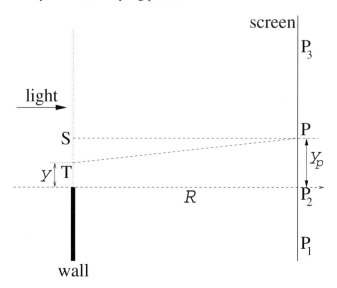

Figure 3.12. Light comes in from the left, is blocked by a wall and observed at various points on a screen. A blurred rim appears near P_2.

of figure 3.13. The asymptotic points of the spiral are A and B. If one connects any two neighboring dots in figure 3.13 by a small arrow, this arrow corresponds to a particular vector $\mathbf{E}_j$. The following vector $\mathbf{E}_{j+1}$ is rotated by an angle that increases as one approaches the asymptotic points, thus the curvature of the spiral increases too. The sum of *all* field vectors $\mathbf{E}_j$ at the point P has two parts:

- the first includes all $\mathbf{E}_j$ of elementary waves emitted from points above S, up to infinity; and
- the second consists of all $\mathbf{E}_j$ from points below S, down to the edge of the wall.

Their sum, which is the resulting field at P, is represented in figure 3.13 by a large arrow. It starts at A and ends at the dot which corresponds to the elementary wave from the rim of the wall.

If we observe at position P_2, opposite to the edge of the wall, the large arrow in figure 3.13 would end in the middle of the spiral at the fat point. Observing at P_1, a position below the edge of the wall, it ends in the lower half of the spiral. At point P_3, which is far up on the screen, the arrow goes from A almost to B. Obviously, the intensity at P_2 is four times smaller than at P_3.

Figure 3.14 also gives roughly the variation of light at the 'blurred rim' around the bright spot mentioned in the discussion of Babinet's theorem (see the top of figure 3.5); the hole was assumed to be much bigger than the wavelength so that sections of its rim can be approximated by a straight line. But exactly behind a sphere at large distances there is, from (3.69), always a maximum intensity

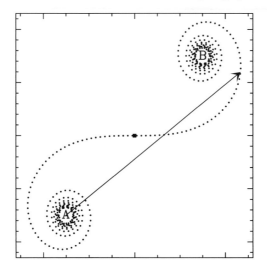

Figure 3.13. The cornu-spiral. The squared length of the plotted vector gives the intensity at a point somewhere above P_2 in figure 3.12, see text.

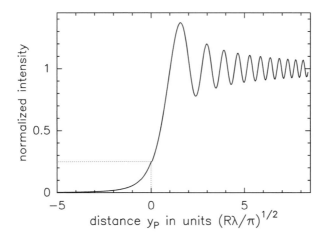

Figure 3.14. The normalized intensity around position P_2 in figure 3.12. On the bright side of the screen, the intensity oscillates and approaches unity. On the dark side, it falls smoothly to zero which is geometrically clear from figure 3.13. At exactly P_2 ($y_p = 0$), the intensity equals 0.25.

because the waves from the circular rim are all in phase on the center line and interfere there constructively.

The separation between the first two maxima in figure 3.14 is 1.42 in units $\sqrt{\lambda R/\pi}$. Here is an astronomical application of this result which can, and has been, observationally checked. The silvery moon on its heavenly orbit ($R = 384\,000$ km) moves relative to the stars with an angular velocity of 0.54 arcsec s^{-1}. If it occults a quasar, which happens to lie on its path, its intensity will oscillate as shown in figure 3.14. If the quasar is observed at the radio wavelength $\lambda = 11$ cm and if it is much smaller than 1 arcsec and crosses the rim of the moon at a right angle, the second intensity maximum follows the first after 5.2 s.

3.5.8 Particles of small refractive index

A grain that fulfils the conditions that

- its optical constant is close to unity,

$$|m - 1| \ll 1 \qquad \text{or} \qquad n \simeq 1,\ k \simeq 0 \qquad (3.71)$$

- and that it is more or less transparent,

$$\frac{d}{\lambda}|m - 1| \ll 1, \qquad (3.72)$$

is called a *Rayleigh–Gans* particle. Otherwise it may be of *arbitrary shape and size*; it may, therefore, be big compared to the wavelength. Inside such a particle, the electromagnetic field is only weakly deformed and practically the same as in the incoming wave. If we imagine the total grain volume V to be divided into many tiny subvolumes v_i with $\sum v_i = V$, each of them absorbs and scatters independently of the others. For example, if the subvolumes are ellipsoids, their absorption and scattering cross sections are given by (3.42) and (3.43), which we derived in the Rayleigh limit. Consequently, the absorption cross section C^{abs} of the total grain is also given by formula (3.42). Therefore, the absorption efficiency of a Rayleigh–Gans sphere for any x is

$$Q^{abs} = \frac{8x}{3} \operatorname{Im}\{m - 1\}.$$

To find the *scattering* cross section or the scattered intensity in a certain direction, one cannot simply sum the contributions of the subvolumes because there is a phase difference between them which leads to interference. It is this effect which, for large bodies, tends to reduce $C^{sca}(\lambda)$ with respect to (3.43). The formalism for how waves originating from different points (subvolumes) are added up is basically the same as that discussed in section 2.2 on the optical theorem. The analytical computation of the scattering cross section or the phase function is, however, complicated, even for simple grain geometries (see examples in [Hul57]). Of course, everything can be extracted from the full Mie theory but not in analytical form.

Here are some results for spheres: If x is the size parameter and $\gamma = 0.577\,215\,66$, Euler's constant,

$$Q^{\text{sca}} = |m - 1|^2 \varphi(x)$$

with

$$\varphi(x) = \frac{5}{2} + 2x^2 - \frac{\sin 4x}{4x} - \frac{7}{16x^2}(1 - \cos 4x)$$
$$+ \left(\frac{1}{2x^2} - 2\right)\left[\gamma + \log 4x + \int_{4x}^{\infty} \frac{\cos u}{u}\, du\right].$$

For $x \ll 1$, we obtain the Rayleigh limit

$$Q^{\text{sca}} = \tfrac{32}{27}|m - 1|^2 x^4$$

and for $x \gg 1$, we get

$$Q^{\text{sca}} = 2|m - 1|^2 x^2.$$

The latter is much smaller than for large, non-transparent spheres which always have $Q^{\text{sca}} \sim 1$. The scattered intensity $I(\theta)$ is proportional to

$$I(\theta) \propto G^2(2x \sin \tfrac{1}{2}\theta)(1 + \cos^2 \theta)$$

with

$$G(u) = \left(\frac{9\pi}{2u^3}\right)^{1/2} J_{3/2}(u)$$

where $J_{3/2}(u)$ is a half-integer Bessel function.

3.5.9   X-ray scattering

We now apply the results obtained for particles of small refractive index to X-rays. At a wavelength of, say 10 Å, interstellar dust particles are always big because $x = 2\pi a/\lambda \gg 1$. The value of the dielectric permeability in this range may be estimated from the high-frequency approximation (1.79), which gives

$$\varepsilon_1 = 1 - \frac{\omega_{\text{p}}^2}{\omega^2} \simeq 1 \qquad \varepsilon_2 = \frac{\omega_{\text{p}}^2 \gamma}{\omega^3} \ll 1$$

where ω_{p} is the plasma frequency and γ a damping (not Euler's) constant. As the frequency ω is large, we expect ε_1 to be close to one and ε_2 to be small, so $|\varepsilon - 1| \ll 1$; likewise for the optical constant, $|m - 1| \ll 1$. The grains are, therefore, very transparent and even satisfy condition (3.72):

$$\frac{2\pi a}{\lambda}|m - 1| \ll 1.$$

For big spheres, scattering is very much in the forward direction defined by $\theta = 0$. Therefore, the angular dependence of the scattered intensity is given by

$$I(\theta) \propto G^2(u) \qquad \text{with } u = 2x \sin \tfrac{1}{2}\theta.$$

When one numerically evaluates the function $G^2(u)$, one finds that it has its maximum at $u = 0$, drops monotonically, like a bell-curve, reaches zero at about 4.5 and then stays very small. Consequently, there is forward scattering over an angle such that $\theta x \sim 2$. For grains of 1000 Å radius, this corresponds to $\theta \sim 10$ arcmin. One can, therefore, observe towards strong X-ray sources located behind a dust cloud an X-ray halo of this size. The intensity towards it follows from the fact that the total scattered X-ray flux is smeared out over the halo.

Chapter 4

Case studies of Mie calculus

We have collected the tools to calculate how an electromagnetic wave interacts with a spherical grain. Let us apply them. With the help of a small computer, the formulae presented so far allow us to derive numbers for the cross sections, scattering matrix, phase function, and so forth. In this chapter, we present examples to deepen our understanding. To the student who wishes to perform similar calculations, the figures also offer the possibility of checking his computer program against the one used in this text.

4.1 Efficiencies of bare spheres

4.1.1 Pure scattering

As a first illustration, we consider the scattering efficiency of a non-dissipative sphere, i.e. one that does not absorb the incident light. The imaginary part k of the optical constant $m = n + ik$ is, therefore, zero, the absorption coefficient vanishes[1] and

$$Q^{\text{sca}} = Q^{\text{ext}}.$$

The efficiency depends only on m and on the size parameter $x = 2\pi a/\lambda$. To better interpret the dependence of Q^{sca} on x, we may envisage the wavelength λ to be fixed so that x is proportional to grain radius a. Figure 4.1 demonstrates two features valid for any particle:

- When a particle is small, we naturally expect the cross section C to be small, too. But even the extinction *efficiency*, which is the cross section over projected area, goes to zero,

$$Q^{\text{ext}} \to 0 \qquad \text{for } x \to 0.$$

---

[1] Astronomers sometimes call a non-dissipative medium, i.e. one with a *purely real* optical constant, a dielectric. However, we understand a dielectric to be the opposite of a metal. So a dielectric may well have a truly complex m, where the imaginary part is not zero.

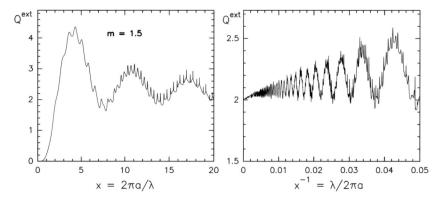

Figure 4.1. The extinction efficiency Q^{ext} as a function of size parameter x and its inverse x^{-1} for an optical constant $m = 1.5$; this value is appropriate for glass in the visible part of the spectrum. The left box shows the range $0 \le x \le 20$, the right the range $0.002 \le x^{-1} \le 0.05$ corresponding to $20 \le x \le 500$. Because in any real material, m depends on wavelength, whereas we have kept it fixed in the figure, it is best to envisage λ as constant. Then the abscissa of the left frame is proportional to the grain radius.

- In accordance with Babinet's theorem, when the grain is very large, the extinction efficiency approaches two (see the right-hand frame of figure 4.1),

$$Q^{\mathrm{ext}} \to 2 \qquad \text{for } x^{-1} \to 0.$$

There are two interesting aspects pertaining to a pure scatterer or weak absorber:

- Q^{ext} displays an overall undulating character with broad waves on which semi-regular ripples are superimposed. The phenomenon of ripples and waves can be traced to the behavior of the coefficients a_n, b_n of (2.50), (2.51) and is due to interference in the scattered wave (see(2.46)). The ripples correspond to small denominators in a_n, b_n, the broad crests of the waves to maxima in the sum $a_n + b_n$. The first maximum of Q^{ext} is achieved roughly at $8(n-1)$, in our particular example this occurs at $x \simeq 4$; it is followed by gradually declining maxima around $x = 11, 17, \ldots$. The minima in between stay more or less all around two.
- The extinction efficiency can be quite large. In figure 4.1, its maximum value is 4.4 times (!) bigger than the projected surface of the particle.

4.1.2 A weak absorber

Figure 4.2 exemplifies the effects of adding impurities to the material. The top frame is identical to the left box of figure 4.1 and displays a medium with $m = 1.5$. In the middle, the optical constant is complex, $m = 1.5 + i0.02$. Now

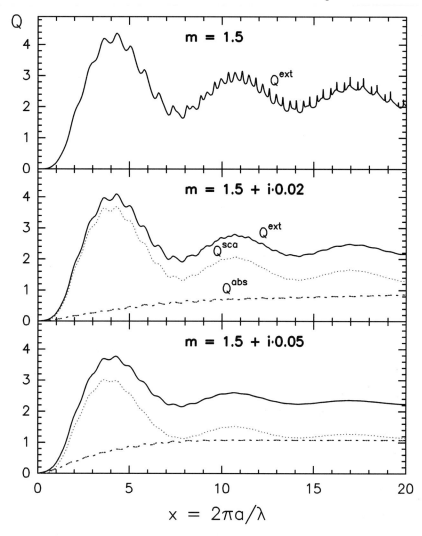

Figure 4.2. The efficiencies for absorption (dashes), scattering (dots) and extinction (full line) for three different substances: one is not dissipative, the others are weakly absorbing. An optical constant $m = 1.5 + 0.05i$ may be appropriate for astronomical silicate around $\lambda = 6\ \mu$m.

the particle not only scatters the light but also absorbs it. Impurities effectively suppress the ripples and they also lead to a decline in the amplitudes of the waves. The damping is sensitive to k, as can be seen by comparing the two bottom panels where k changes from 0.02 to 0.05.

For interstellar grains, the ripples and the waves are irrelevant. For example,

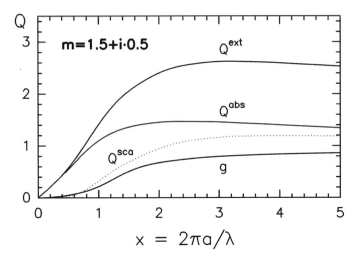

Figure 4.3. The efficiency for absorption, scattering and extinction, as well as the asymmetry factor g for a strongly absorbing material with $m = 1.5 + 0.5i$.

when we measure the intensity of a star behind a dust cloud, the observation is not monochromatic but comprises a certain bandwidth $\Delta\lambda$. The optical constant $m(\lambda)$ will change somewhat over this bandwidth and, more importantly, interstellar dust displays a range of sizes, so the peaks produced by particles of a certain diameter are compensated by minima of slightly smaller or bigger ones.

4.1.3   A strong absorber

In figure 4.3 we retain the value of $n = 1.5$ of the previous two plots but choose for the imaginary part of the optical constant $k = 0.5$. This implies strong absorption. The efficiencies change now very smoothly and there are no signs of ripples or waves left. Because of the strong absorption, Q^{abs} increases quickly with x at $x < 1$. We discern, in this particular example, the general features of small-sized grains. When $x \to 0$:

- Q^{abs} and x are proportional, as predicted by (3.3).
- The scattering efficiency changes after (3.4) as

$$Q^{sca} \propto x^4$$

  so Q^{sca} falls off more steeply than Q^{abs} (see logarithmic plot of figure 4.5).
- Absorption dominates over scattering,

$$Q^{abs} \gg Q^{sca}.$$

- The asymmetry factor g of (2.6), which describes the mean direction of the scattered light, vanishes. Although isotropic scattering implies $g = 0$, the reverse is not true.

Figure 4.3 plots only the range up to a size parameter of five but the behavior for larger x is smooth and can be qualitatively extrapolated. In this particular example of $m = 1.5 + i0.5$, a very large particle ($x \gg 1$)

- scatters mostly in the forward direction, so g is not far from one; and
- it has a large absorption efficiency and, therefore, a small reflectance $Q^{ref} = 1 - Q^{abs}$, as defined in equation (3.63). The asymptotic values for $x \to \infty$ are: $g = 0.918$, $Q^{abs} = 0.863$ and $Q^{ref} = 0.137$.

Surprisingly, the absorption efficiency can exceed unity, which happens here when $x > 0.9$. Unity is the value of Q^{abs} of a blackbody and may thus appear to be an upper limit for any object. Nevertheless, the calculations are correct, only the concept of a blackbody is reserved to sizes much larger than the wavelength.

4.1.4   A metal sphere

- Ideally, a metal has (at low frequencies) a purely imaginary dielectric permeability so that $n \simeq k$ goes to infinity (see (1.117)).
- A metallic sphere is, therefore, very reflective and Q^{abs} tends to zero.
- When the sphere is big, Q^{ext} is, as always, close to the Babinet limit of two.
- When the sphere is big, of all the light removed, half is scattered isotropically and it alone would yield an asymmetry factor $g = 0$. The other half is scattered at the particle's rim entirely in the forward direction, which alone would produce $g = 1$. The combination of both effects gives $g = \frac{1}{2}$.
- When the sphere has a size comparable or smaller than the wavelength, backward scattering dominates ($g < 0$).
- For small values of $|mx|$, the metallic sphere also obeys the relation $Q^{abs} \gg Q^{sca}$. But as x grows, the scattering increases and soon takes over, whereas the absorption stays at some low level.

Figure 4.4 illustrates these items in case of large but not extreme metallicity ($m = 20 + i\, 20$).

4.1.5   Efficiency versus cross section and volume coefficient

In figure 4.5, we compare efficiencies, cross sections and volume coefficients using an optical constant $m = 2.7 + i\, 1.0$ which applies to amorphous carbon at a wavelength $\lambda = 2.2\mu m$. The presentation is different from the preceding figures as we choose for the abscissa the grain radius a, and not the size parameter x; the wavelength is kept constant at 2.2 μm. The logarithmic plots reveal the limiting behavior of small and large particles. These features, compiled in table 4.1, are independent of the choice of m or λ.

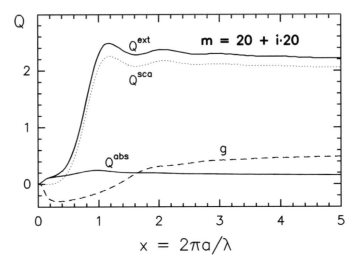

Figure 4.4. Same as figure 4.3 but now for a metal for which n and k are large and equal. The situation applies approximately to amorphous carbon at millimeter wavelengths (see figure 7.19). Note that g becomes negative.

Table 4.1. The asymptotic dependence of efficiency, cross section and mass or volume coefficient on the radius, a, of a spherical grain.

	Symbol	Large radii $a \gg \lambda$	Small radii $a \ll \lambda$
Efficiency	Q	$Q^{abs} = $ constant $Q^{sca} = $ constant	$Q^{abs} \propto a$ $Q^{abs} \propto a^4$
Cross section per grain	C	$C^{abs} \propto a^2$ $C^{sca} \propto a^2$	$C^{abs} \propto a^3$ $C^{sca} \propto a^6$
Mass or volume coefficient	K	$K^{abs} \propto a^{-1}$ $K^{sca} \propto a^{-1}$	$K^{abs} = $ constant $K^{sca} \propto a^3$

4.1.5.1 Spheres of small radii

- The efficiencies depend on wavelength as given by (3.3) and (3.4). If m is constant, $Q^{abs} \propto \lambda^{-1}$ and $Q^{sca} \propto \lambda^{-4}$. Scattering is, therefore, negligible relative to absorption.
- As interstellar dust has typical sizes of $0.1\ \mu$m, one may assume the Rayleigh limit to be valid at all wavelengths greater than $10\ \mu$m.
- The mass absorption coefficient K_λ^{abs} is proportional to the grain volume V, irrespective of the size distribution of the particles.

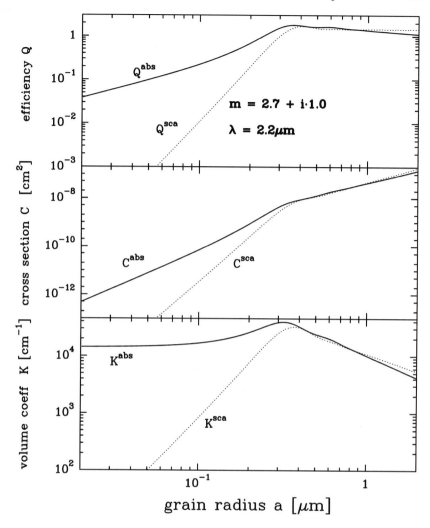

Figure 4.5. The top frame gives the absorption and scattering efficiency Q; the middle one the cross section per grain $C = \pi a^2 Q$; and the bottom one the volume coefficient K referring to 1 cm^3 of grain volume. The optical constant $m = 2.7 + i1.0$, wavelength $\lambda = 2.2\ \mu$m.

- The mass scattering coefficient K_λ^{sca} follows V^2. Although the scattering process is identical in all subvolumes of a grain, the net effect is not linear but proportional to the square of the grain volume because of interference of the scattered waves.

- The emission coefficient

$$\epsilon_\lambda = K_\lambda^{\mathrm{abs}} B_\lambda(T_{\mathrm{dust}})$$

also depends linearly on the grain volume; $B_\lambda(T_{\mathrm{dust}})$ is the Planck function at the dust temperature (see (8.1)). Provided the infrared photons are not trapped in the cloud, a condition that is usually fulfilled, the observed flux is directly proportional to the total dust volume.

4.1.5.2   Spheres of large radii

The efficiencies, Q, reach a finite value with $Q^{\mathrm{ext}} = Q^{\mathrm{abs}} + Q^{\mathrm{sca}} = 2$. The cross sections per grain, C, increase without bounds proportional to the projected area of the grain. The mass or volume coefficients, K, decrease inversely proportional to the radius.

4.1.6   The atmosphere of the Earth

The atmosphere of the Earth presents another instructive example. We can now understand why the daylight sky is blue. If there were no atmosphere, it would appear black, except for the stars, Sun and planets. However, the tiny air molecules scatter the sunlight towards us. Because $Q^{\mathrm{sca}} \propto \lambda^{-4}$ and $\lambda_{\mathrm{blue}} < \lambda_{\mathrm{red}}$, the sky is painted blue. An additional requirement is that the air molecules are randomly moving, as otherwise interference would cancel out the intensity except in the forward direction. The scattered light is polarized and maximum polarization occurs 90° away from the Sun (see (3.19)), or at night time 90° away from the moon.

As preferentially blue light is removed in scattering, the transmitted light becomes redder. But this does not explain the occasional romantic sunsets. More relevant for the reddening towards the horizon is the removal of blue light through absorption by tiny solid particles. These may arise from industrial exhausts, a sand storm or a volcano eruption. The dependence of absorption on λ is less steep than for scattering but qualitatively similar ($Q^{\mathrm{abs}} \propto \lambda^{-1}$) and also results in reddening.

We can also explain why we see the water in the air only when it has liquefied into clouds, and not as vapor. When on a summer day the solar heating induces convection and a parcel of air rises we see the parcel only at the moment the water condenses, although the H_2O molecules were there before. This seems puzzling because upon condensation the water molecules do not drastically change their properties as radiating oscillators. The reason that they become visible when they cluster into droplets is the following: The mass scattering coefficient K^{sca} is, according to table 4.1, proportional to the volume V of a *single* scatterer (one H_2O molecule or one water droplet). During the phase transition from gas to liquid, V grows by a factor of $\sim 10^{10}$. The size of the water droplets is then still

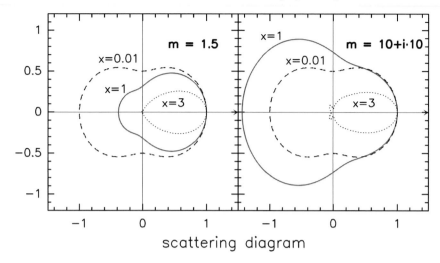

Figure 4.6. Diagrams of the intensity distribution of scattered light for three sizes for parameter x and two optical constants m. Light enters from the left and the grain is at the position $(0, 0)$. Each diagram is normalized in such a way that the intensity equals one in the forward direction $(\theta = 0)$; its value in any other direction is given by the distance between the frame center and the contour line. All contours are mirror-symmetric with respect to the horizontal arrows, see also table 4.2.

smaller than the wavelength of visible light so that the dipoles in the droplet swing in phase.

4.2 Scattering by bare spheres

4.2.1 The scattering diagram

The pattern of how the intensity of scattered light changes with angle is described by the phase function of (2.63) and visualized in figure 4.6. In these examples, the left frame refers to a non-dissipative sphere with $m = 1.5$. The curve labeled $x = 0.01$ represents a dipole pattern which is characteristic for a particle small compared with the wavelength *irrespective* of the choice of the optical constant m. It is symmetric relative to the scattering angle of $90°$ (vertical line in the figure) and therefore has equal maxima in the forward $(\theta = 0)$ and backward $(\theta = 180°)$ directions. Note that the scattering is not isotropic, although $g = 0$. The pattern is sensitive to the size parameter $x = 2\pi a/\lambda$. When $x \geq 1$, scattering becomes very forward and for a grain diameter just equal to the wavelength $(x \simeq 3)$, almost all light goes into a forward cone.

In the right frame where $m = 10 + i10$, we are dealing with a reflective material. The change in m relative to the non-dissipative sphere shown on the

Table 4.2. Scattering efficiency, asymmetry factor and back scatter for spheres whose scattering diagram is displayed in figure 4.6.

$x = 2\pi a/\lambda$	m	Q^{sca}	Asymmetry factor g	Q^{back}
0.01	1.5	2.31×10^{-9}	1.98×10^{-5}	3.5×10^{-9}
1	1.5	2.15	0.20	0.19
3	1.5	3.42	0.73	0.53
0.01	$10 + i\,10$	2.67×10^{-8}	1.04×10^{-5}	4.0×10^{-8}
1	$10 + i\,10$	2.05	0.111	3.31
3	$10 + i\,10$	2.08	0.451	0.42

left does not affect the curve $x = 0.01$. There is also not much difference for $x = 3$, although curious tiny rear lobes appear. However, at the intermediate value $x = 1$, we now have preferential backscatter and the asymmetry factor is negative ($g = -0.11$), whereas in the left-hand panel, at $x = 1$, the scattering is mainly forward.

4.2.2 The polarization of scattered light

For the same optical constants and size parameters as previously, figures 4.7 and 4.8 depict the degree of linear polarization p of light scattered by a sphere (see (2.77)). They display the basic features, although the details are sensitive to size, shape and optical constant of the particle.

- The polarization curve $p(\theta)$ of tiny and weakly absorbing spheres ($x = 0.01$, $k \ll 1$; dashes in figure 4.7) is symmetric around 90° and has the shape of a bell. The polarization is 100% at a scattering angle of 90°. For weak absorbers, $p(\theta)$ hardly changes as x grows from 0.01 up to 1.
- Tiny metallic particles ($x = 0, 01, m = 10 + i\,10$; figure 4.8) have the same pattern as tiny dielectrics but at $x = 1$, the polarization $p(\theta)$ is no longer symmetric and attains a maximum value of only 78%.
- The behavior of $p(\theta)$ at intermediate sizes ($x = 3$) is more complicated in both materials. The polarization changes sign and there are several maxima and minima of varying height. In our examples, p vanishes at three intermediate angles. Here the matrix element S_{12} switches sign (see (2.75)) and the direction of polarization with respect to the scattering plane changes from perpendicular to parallel or back. The scattering plane is defined by the unit vectors in the direction from where the light is coming and where it is going.

  When one measures the linear polarization of scattered light at various locations around the exciting star of a reflection nebula. the observed values

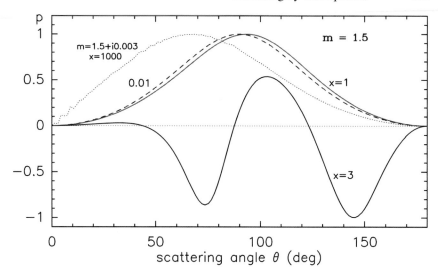

Figure 4.7. The degree of linear polarization p of scattered light as a function of scattering angle for various size parameters x and an optical constant $m = 1.5$. For intermediate sizes ($x = 3$), the polarization at certain scattering angles goes through zero and becomes negative. At the points where $p = 0$, the polarization vector flips. The dotted line depicts a very large sphere ($x = 1000$) with some slight contamination ($m = 1.5 + i0.003$).

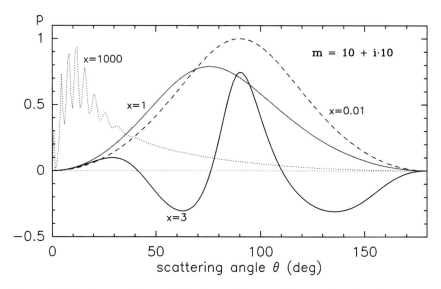

Figure 4.8. The degree of linear polarization p of scattered light as a function of scattering angle for various size parameters x and an optical constant $m = 10 + i10$.

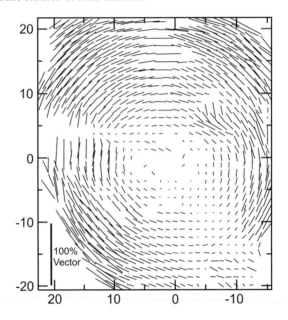

Figure 4.9. The infrared source IRS4 (at figure center) in the bipolar nebula S106 illuminates its vicinity. Some light is scattered by dust particles producing in the K-band (2.2 μm) a polarization pattern of wonderful circular symmetry (adapted from [Asp90]). Reproduced from [Asp90] with permission of Blackwell Science.

are always an average along the line of sight. If the particles are small, $S_{12} < 0$ and the polarization vector is perpendicular to the radial vector pointing towards the star. However, as the figure suggests, it can happen, although it is an unfamiliar scenario to astronomers, that the radial and polarization vectors are parallel.

- Very big dielectric spheres ($x \simeq 1000$) again show a simple pattern with no sign reversals. There is 100% polarization in a slightly forward direction ($\theta \simeq 65°$). For metals of this size, $p(\theta)$ has more structure.
- Polarization is always zero in the forward and backward directions ($\theta = 0$ or 180°).

An impressive example of a very regular polarization pattern is presented by the bipolar nebula S106 (figure 4.9). This is a well-known star-forming region. At optical wavelengths, one observes light from two nebulous lobes separated by a dark lane that is interpreted as a disk inside which sits the central and exciting source named IRS4. We are viewing the disk from the side. There is plenty of visual obscuration towards IRS4 ($\sim$20 mag) but in the near infrared one begins to penetrate most of the foreground and outer disk material. IRS4 is an infrared source and its radiation is reflected by the dust particles that surround it. When looking directly towards IRS4, the angle of deflection is close to 0° or 180°

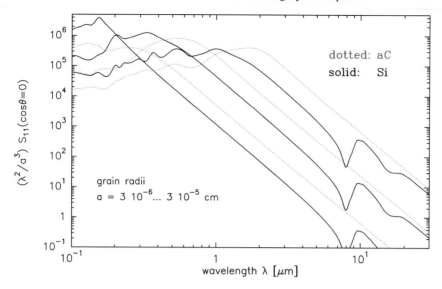

Figure 4.10. The term $(\lambda^2/a^3)S_{11}(\cos\theta)$ in equation (4.1) for right angle scattering ($\cos\theta = 0$). It is computed as a function of wavelength for silicate and carbon grains with radii $a = 3 \times 10^{-6}, 10^{-5}$ and 3×10^{-5} cm (i.e. $a = 0.03, 0.1, 0.3\ \mu\mathrm{m}$). Optical constants from figure 7.19. The 10 μm feature of silicates is discernible.

and the ensuing degree of linear polarization is small. But at positions a few arcseconds off, scattering occurs at right angles and the light becomes strongly polarized. Note that S106 must be optically thin to scattering in the K-band (see table 7.2) as otherwise multiple scattering would lead to depolarization.

4.2.3 The intensity of scattered light in a reflection nebula

To evaluate the intensity of scattered light, we consider 1 g of dust in interstellar space, consisting of $N_d = 3/4\pi a^3\rho_d$ identical spherical grains of radius a and density $\rho_d = 2.5$ g cm^{-3}, which is illuminated by a flux $F_{0\lambda}$ from a nearby star. Some of the radiation is scattered under an angle $\cos\theta$ towards the observer who is at a distance D. The flux F_λ that he receives is found from formula (2.1), where we replace $\mathcal{L}(\theta, \phi)$ by the element S_{11} of equation (2.63),

$$F_\lambda = \frac{3}{16\pi^3\rho_d} \frac{F_{0\lambda}}{D^2} \cdot \frac{\lambda^2}{a^3} S_{11}(\cos\theta). \tag{4.1}$$

The expression $(\lambda^2/a^3)S_{11}(\cos\theta)$ is delineated in figure 4.10 for a scattering angle $\theta = 90°$.

In the infrared where $a \ll \lambda$, the matrix element $S_{11}(\cos\theta)$ is very sensitive to both grain size and wavelength, roughly $S_{11} \propto (a/\lambda)^6$. The size of the largest grains, therefore, determines the intensity of the scattered infrared light. Of course, the shape of the particle also plays some role. Elongated grains, even of arbitrary orientation, scatter more efficiently than spheres of the same volume. For spheres, isotropic scattering at infrared wavelengths is not a bad assumption.

In the ultraviolet, the dependence of $S_{11}(\cos\theta)$ on the scattering angle is complicated and the curves in figure 4.10 may only be used if θ is not too far from 90°.

As an exercise, we apply figure 4.10 to answer the question: What is the approximate V-band surface brightness of scattered light in a reflection nebula? The nebula is $D = 1$ kpc away and excited by a main sequence star of spectral type B1 with $L = 10^4 L_\odot$ and $T_* = 2 \times 10^4$ K. We calculate this quantity at a projected distance from the star of $r = 0.1$ pc assuming that the extinction along the line of sight through the nebula equals $A_V = 0.1$ mag.

An extinction of 0.1 mag corresponds after (7.23) to a hydrogen column density $N_H = 2 \times 10^{20}$ cm^{-2}. This gives, with a gas-to-dust mass ratio of 150 and accounting for 10% of helium, a dust mass of 7.0×10^{26} g in a column with a cross section of 1 arcsec^2. If the star is approximated by a blackbody, its visual flux at a distance $r = 0.1$ pc is $F_{0,V} = 9.80 \times 10^{-15}$ erg s^{-1} cm^{-2}. For a rough estimate, we neglect that the distance of the scattering dust to the star changes along the line of sight. Let the particles have radii $a = 10^{-5}$ cm. Then we read off from figure 4.10 at a wavelength of 0.55 μm for right angle scattering that $(\lambda^2/a^3) \times S_{11} \simeq 8 \times 10^5$. With the factor $w_\lambda = 3640$ to convert Jy into magnitudes from table 7.2, we find a surface brightness in the nebula at V of 18.5 mag arcsec^{-2}.

4.3   Coated spheres

The mathematical formalism of Mie theory presented in section 2.3 can readily be extended to coated spheres. Only the appearance of two more boundary conditions at the interface between the two dust materials is new. The relevance of calculating coated particles stems from the observational fact that grains in cold clouds acquire ice mantles. Such particles may be approximated by two concentric spheres. There is a refractory core of radius a_{core} and optical constant m_{core} surrounded by a volatile shell of thickness d and optical constant m_{shell}. So the total grain radius is

$$a_{tot} = a_{core} + d.$$

The left box in figure 4.11 shows the absorption efficiency Q^{abs} as a function of size parameter x for *homogeneous* particles with optical constants $m = 2.7 + i$ and $m = 1.33 + 0.03i$. In the right box, we plot Q^{abs} for *coated* spheres as a function of

$$r = \frac{a_{core}}{a_{tot}}.$$

For $r = 0$, the grain consists only of core matter with $m = 2.7 + i$; for $r = 1$, only of mantle material ($m = 1.33 + 0.03i$). The four curves represent

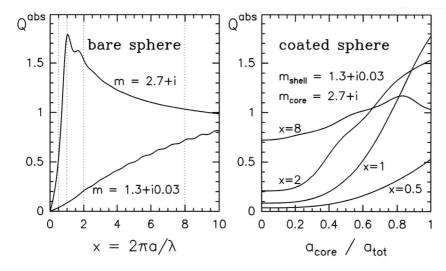

Figure 4.11. The left frame gives the absorption efficiency as a function of size parameter for homogeneous spheres. The optical constant $m = 2.7+i$ is representative of amorphous carbon at $\lambda = 2.2$ μm and $m = 1.33 + 0.03i$ is applicable to dirty ice at the same wavelength. Dotted vertical lines for reference to right box. The right frame shows the absorption efficiency of a coated sphere, x denotes the size parameter of the total grain.

size parameters $x = 2\pi a_{\text{tot}}/\lambda$ of the total sphere between 0.5 and 8. For easy comparison, these values are also indicated by dotted lines in the left frame. The right figure demonstrates how the absorption efficiency of a coated grain changes when the core grows at the expense of the mantle while the outer radius a_{tot} is fixed.

4.4   Surface modes in small grains

4.4.0.1   Small graphite spheres

When grains display, at some wavelength, an extinction peak, the reason is generally a resonance in the grain material. However, in the case of grains which are small compared with the wavelength, the extinction cross section may be large without an accompanying feature in the bulk matter. As the extinction efficiency is in the Rayleigh limit after (3.3) proportional to $\varepsilon_2/|\varepsilon + 2|^2$, we expect a large value for Q^{ext} at a wavelength where

$$\varepsilon + 2 \simeq 0.$$

Because $\varepsilon = m^2$, this happens when the real part n of the optical constant is close to zero and $k^2 \approx 2$. The phenomenon bears the name *surface mode*.

For the major components of interstellar dust, amorphous carbon and silicate, there is no wavelength where the condition of small $\varepsilon+2$ is approximately fulfilled (see figure 7.19) but graphite around 2200 Å is a candidate. Graphite is optically anisotropic and the effect appears when the electric vector of the incoming wave lies in the basal plane of the stacked carbon sheets; the dielectric constant is then denoted by $\varepsilon_\perp$. The left-hand side of figure 4.12 shows $\varepsilon_\perp = \varepsilon_1 + i\varepsilon_2$ and the resulting extinction efficiency for spheres of different sizes. The wavelength interval stretches from 1670 to 3330 Å or from 6 to 3 μm^{-1}. To underline the effect of the surface mode, Q^{ext} is normalized to one at $\lambda^{-1} = 3\ \mu m^{-1}$. Small particles ($a \leq 300$ Å) clearly display a resonance around 2200 Å, which is absent in the bulk material because particles with a radius greater than 1000 Å do not show it. The graphite surface mode is invoked to explain the strong feature around 2200 Å in the interstellar extinction curve (figure 7.8).

For the Rayleigh approximation to be valid, the particles must have diameters of 100 Å or less. But the position of the resonance contains further information on grain size. The upper left box of figure 4.12 illustrates how the peak in Q^{ext} drifts towards smaller wavelengths (greater λ^{-1}) with decreasing radius a; it moves from 2280 to 2060 Å as the radius shrinks from 300 to 30 Å. There is no further shift for still smaller particles.

In astronomical observations of the extinction curve, the relevant quantities are, however, not the extinction efficiencies Q^{ext} but the extinction coefficients per cm^3 of dust volume, K^{ext}. They are shown in the upper right frame of figure 4.12. The dependence of K^{ext} on λ^{-1} is qualitatively similar to that of Q^{ext} but the discrimination against size is more difficult and not possible below 100 Å.

4.4.0.2 Ellipsoids

From comparison of the electric polarizability for spherical and ellipsoidal particles (see (3.9) and (3.39)), it follows that the condition $\varepsilon + 2 \simeq 0$, necessary for the appearance of a surface mode in spheres, must for ellipsoids be replaced by

$$1 + L_x(\varepsilon - 1) \simeq 0$$

or similar expressions for the y- or z-axis. The form of the particle therefore influences, via the shape factor L_x, the strength of the surface mode as well as its center wavelength.

4.4.0.3 Metals versus dielectrics

Figure 4.13 illustrates the interplay between ε and Q^{ext} in a surface mode for a dielectric consisting of identical harmonic oscillators. The variation of ε is, of course, similar to figure 1.2, only the resonance is here sharper ($\gamma = 0.05$). We see that a surface mode cannot happen far from a resonance in the bulk material and if it does, only at a higher frequency. We wish to make two points:

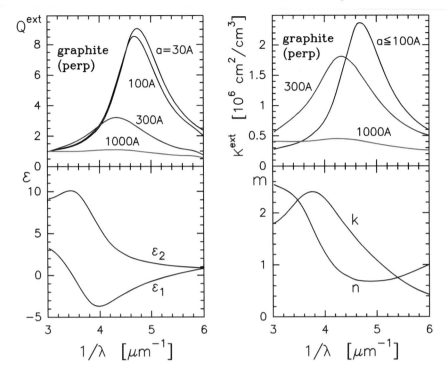

Figure 4.12. The upper left frame gives the extinction efficiency for graphite spheres of various radii, a, in the 2200 Å wavelength region when the electric vector lies in the basal plane. The efficiencies are normalized to one at $3\ \mu m^{-1}$. The increase for small sizes is due to a surface mode and not to an intrinsic quality of the bulk material. The lower left frame gives the dielectric constant $\varepsilon = \varepsilon_1 + \varepsilon_2$; the upper right frame the volume coefficients, not normalized; and the lower right frame the optical constant $m = n + ik$.

- There is a clear offset between the peaks in ε_2 and Q^{ext}.
- Although the grain is in the Rayleigh limit ($|mx| \leq 0.1$), the extinction efficiency is around one and, surprisingly, not small. Nevertheless, $Q \to 0$ for $|mx| \to 0$ is always correct.

Not every dielectric material displays a surface mode but for metallic grains the phenomenon is inevitable when damping is weak. With the dielectric permeability for metals given in equation (1.123), the denominator in (3.3) vanishes when

$$\left[3(\omega^2 + \gamma^2) - 1\right]^2 \omega^2 + \gamma^2 = 0.$$

Here the frequency ω and the damping constant γ are in units of the plasma frequency ω_p. As γ is usually much smaller than ω_p, a surface mode, i.e. a strong enhancement of Q^{abs}, appears at $\omega = \omega_p/\sqrt{3}$ (see figure 4.14).

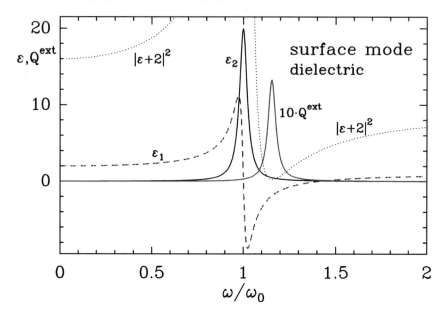

Figure 4.13. For small spheres, the absorption efficieny Q^{ext} can be calculated from the Rayleigh limit (3.3). The dielectric permeability ε of the grain material in the figure is that of an harmonic oscillator (see (1.77)). When the resonance frequency ω_0 and the damping constant γ are measured in units of the plasma frequency, $\omega_0 = 1$ and $\gamma = 0.05$. The grain radius a is determined by the condition that the size parameter equals $x = \omega a/c = 0.05$ at the frequency ω_0. The peak in the absorption efficiency is due to a surface mode, because Q^{ext} attains its maximum, where the denominator $|\varepsilon + 2|$ in (3.3) has its minimum. The dotted curve of $|\varepsilon+2|^2$ is not always within the box. For better visibility, Q^{ext} is magnified by a factor of 10.

4.5   Efficiencies of idealized dielectrics and metals

4.5.1   Dielectric sphere consisting of identical harmonic oscillators

The functional form of the dielectric permeability ε of an harmonic oscillator with resonance at frequency ω_0 is given by equation (1.77). We plotted ε in figure 1.2 for a damping constant $\gamma = 0.2\omega_0$ and a plasma frequency $\omega_p = \omega_0$. It is instructive to compute in Mie theory the efficiencies of a particle with such an idealized dielectric constant to see how the complex ε translates into Q^{abs} or Q^{ext}.

Figure 4.15 shows the extinction efficiency Q^{ext} when there is a broad resonance in the bulk material ($\gamma = 0.2\omega_0$, as in figure 1.2). When the grains are large, Q^{ext} is basically constant over the resonance and does not seem to depend on ε at all. At intermediate sizes, one can discern some correspondence between Q^{ext} and the imaginary part ε_2. When the grain is much smaller than

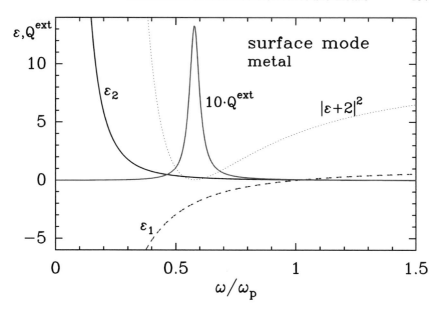

Figure 4.14. As figure 4.13 but for a metal with a dielectric permeability ε from the Drude profile (1.123): $\gamma = 0.05\omega_p$, size parameter $x = \omega_p a/c = 2\pi a/\lambda = 0.05$.

the wavelength, all charges move synchronously in the electromagnetic field. Scattering is then negligible and one would expect Q^{abs} to mimic ε_2 rather closely. However, as already discussed in figure 4.13, this is not so.

Figure 4.16 shows the same Q^{ext} as in figure 4.15 but over a wider frequency range and on a logarithmic scale. One may wonder why Q^{ext} falls and does not approach two when the particle is much bigger than the wavelength. However, there is no conflict with Babinet's theorem. The latter states that $Q^{ext} \rightarrow 2$ for $a/\lambda \rightarrow \infty$ if the dielectric permeability is kept constant, whereas here ε becomes smaller and smaller as the frequency rises.

The curve in figure 4.16 labeled *big grains* (radius $a = 10c/\omega_0$) demonstrates that when there is a resonance in the bulk material at some frequency ω_0, the corresponding signature in the emission or absorption of the grain gets lost if the particle is very large. For example, silicate grains of a size much exceeding 10 μm will not display the famous 10 μm feature. Another way to swamp a resonance is by high optical depth; this will be discussed in section 8.3.

When we insert the low-frequency limit of the dielectric permeability ε of an harmonic oscillator after (1.78) into the Rayleigh approximation (3.3) for the

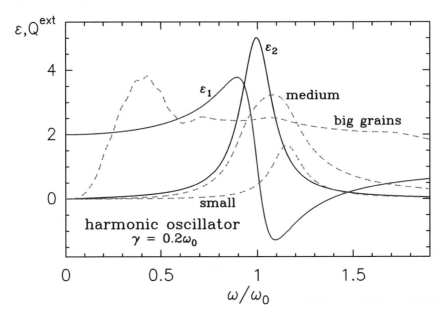

Figure 4.15. Broken lines represent the extinction efficiency of spheres with the dielectric constant ε of a harmonic oscillator given in equation (1.77); here with $\gamma/\omega_0 = 0.2$ and $\omega_0/\omega_p = 1$, where ω_0 is the resonance frequency. Q^{ext} is plotted for three grain radii, a. When $a = 10c/\omega_0$, the particle is big (size parameter $x = 10$ at frequency ω_0); the case $a = c/\omega_0$ represents medium-sized grains ($x = 1$ at ω_0) and $a = 0.1c/\omega_0$ small ones ($x = 0.05$ at ω_0); this line is enhanced for better visibility by a factor of 5. The full curves represent $\varepsilon = \varepsilon_1 + i\varepsilon_2$.

absorption efficiency, we get

$$Q^{\text{abs}} = 48c\pi^2 \frac{\omega_p^2 \gamma}{\omega_0^4 \left(3 + \frac{\omega_p^2}{\omega_0^2}\right)^2} \frac{a}{\lambda^2}. \tag{4.2}$$

The behavior at long wavelengths far away from the resonance is, therefore,

$$Q^{\text{abs}} \propto \nu^2 \qquad \text{for } \nu \to 0.$$

This is the reason why one often assumes that optically thin dust emission follows a modified Planck curve $\nu^2 B_\nu(T)$ in the long wavelength limit..

4.5.2 Dielectric sphere with Debye relaxation

The dielectric permeability for Debye relaxation is given by formula (1.146). The absorption efficiency Q^{abs} is now temperature-dependent and varies with $1/T$.

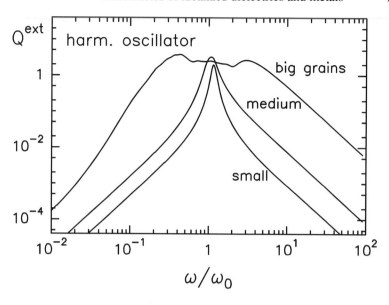

Figure 4.16. The efficiencies of figure 4.15 but now on a logarithmic plot and over a broader frequency range.

For small ω,

$$Q^{\text{abs}} = 48c\pi^2 \frac{\alpha t_{\text{rel}}}{(3+\alpha)^2} \frac{a}{\lambda^2} \quad \text{with } \alpha = \frac{4\pi Np^2}{3kT} \tag{4.3}$$

so again

$$Q^{\text{abs}} \propto \nu^2.$$

The inverse of the relaxation time t_{rel} is a characteristic frequency denoted by $\omega_{\text{rel}} = t_{\text{rel}}^{-1}$; it corresponds, in some sense, to the resonance frequency of an harmonic oscillator. The efficiency Q depends on $\omega/\omega_{\text{rel}}$ and on the grain radius in a way that is qualitatively similar to what is plotted in figure 4.15.

4.5.3 Magnetic and electric dipole absorption of small metal spheres

We derived in equation (1.57) the average rate W at which a small grain of volume V and polarizability α_{m} oscillating at frequency ω in a time-variable magnetic field of amplitude H_0 dissipates energy:

$$W = \tfrac{1}{2} V \omega \, \text{Im}\{\alpha_{\text{m}}\} H_0^2.$$

In section 3.2 we learnt that a piece of metal in an electromagnetic wave becomes such an oscillating magnetic dipole because circular currents are induced in its interior which are surrounded by a magnetic dipole field. We also evaluated

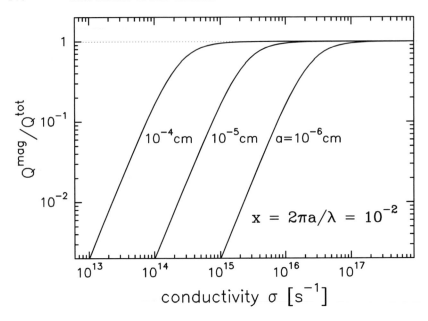

Figure 4.17. The ratio of magnetic dipole absorption over total absorption for metallic spheres with dielectric permeability $\varepsilon = i4\pi\sigma/\omega$ as a function of the conductivity σ. The size parameter $x = 10^{-2}$ is small and kept constant. Therefore, given the particle radius a, the wavelength follows from x. For example, $\lambda = 62.8\ \mu\text{m}$ for $a = 10^{-5}$ cm.

in section 3.2 the magnetic polarizability α_m of small metal spheres analytically. We thus have all the tools to calculate the cross section due to *magnetic dipole absorption*.

But the metal sphere in the electromagnetic wave is also a variable *electric dipole* because its mobile electrons are shifted about by the electric field. The dissipation rate of the latter is given by (1.55), an expression formally identical to (1.57). There is also magnetic dipole scattering but it is much weaker in the Rayleigh limit and we neglect it here.

In figure 4.17, we assess the importance of magnetic dipole absorption for pure metal spheres by comparing the corresponding efficiency Q^{mag} with the total Q^{tot}, which also contains electric dipole absorption. Barring quadrupole terms, we have

$$Q^{\text{tot}} = Q^{\text{mag}} + Q^{\text{el}}.$$

We assume that ε has no dielectric contribution and put (see (1.115))

$$\varepsilon = i\frac{4\pi\sigma}{\omega}$$

where σ is the direct-current conductivity of (1.120). We then calculate the *total* absorption efficiency Q^{tot} correctly from formula (2.55) of Mie theory. The size parameter in figure 4.17 is small and kept fixed, $x = 2\pi a/\lambda = 10^{-2}$. Therefore we can write

$$Q^{\text{mag}} = \frac{32\pi^2}{3} \frac{a}{\lambda} \text{Im}\{\alpha_{\text{m}}\}; \tag{4.4}$$

the magnetic polarizability α_{m} is taken from (3.21). An equivalent formula for the Rayleigh limit of the electric dipole absorption efficiency Q^{el} is not valid everywhere in figure 4.17; although the particle itself is small, being a metal, the relevant quantity $|mx|$ is not necessarily so.

Obviously, magnetic dipole absorption increases with the conductivity and it can be the dominant absorption process. We consider in figure 4.17 the Q's for three radii in the likely range of interstellar grains. Because the size parameter is constant, the implied frequencies ω lie between 3×10^{12} and 3×10^{14} s^{-1}.

- A large size favours magnetic dipole dissipation.
- At low conductivities, where the curves in figure 4.17 still increase, the penetration depth of the magnetic field, as defined in equation (3.23), is large compared with the grain size and (see (3.26))

$$Q^{\text{mag}} \propto a^3/\lambda^2 \qquad \delta \gg a \qquad \text{(for small } \sigma\text{)}.$$

- At high conductivities, where the curves are flat, the penetration depth is small. Q^{mag} does not depend on radius and is proportional to the wavelength (see (3.28)):

$$Q^{\text{mag}} \propto \lambda^{-1/2} \qquad \delta \ll a \qquad \text{(for large } \sigma\text{)}.$$

4.5.4 Efficiencies for Drude profiles

We have already computed the efficiency Q of a metal-like sphere in figure 4.4. It applies to the low frequency limit where $\varepsilon \simeq i4\pi\sigma/\omega$ and σ is the direct current conductivity. The optical constants n and k are then large and roughly equal. As $m = n + ik$ is constant in figure 4.4, the plotted values of Q refer to a fixed frequency, and in the size parameter x only the grain radius changes. In figure 4.18, however, we use, for ε, the Drude profile from (1.123) and keep the grain diameter constant. We now plot Q as a function of frequency (in units of ω_{p}) in the same manner as in figure 4.17 for a dielectric. The damping constant is the only free parameter (here $\gamma = 0.1\omega_{\text{p}}$).

- At long wavelengths, we get for the electric dipole absorption (see (3.3)),

$$Q^{\text{el}} = \frac{6}{x^2} \text{Re}\{a_1\} = \frac{48\pi^2 c\gamma}{\omega_{\text{p}}^2} \frac{a}{\lambda^2} \tag{4.5}$$

- and for the magnetic dipole absorption

$$Q^{\text{mag}} = \frac{6}{x^2} \text{Re}\{b_1\} = \frac{8\pi^2 \omega_{\text{p}}^2}{15\gamma c} \frac{a^3}{\lambda^2}. \tag{4.6}$$

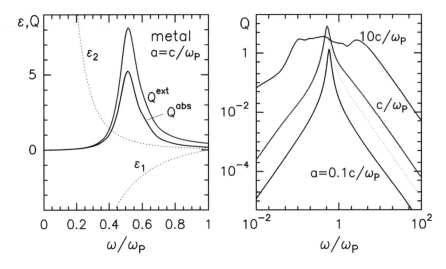

Figure 4.18. The left panel shows the extinction and absorption efficiency of a metal sphere whose dielectric permeability ε follows a Drude profile with a ratio of damping constant over plasma frequency $\gamma/\omega_p = 0.1$. The grain radius equals $a = c/\omega_p$. The right panel shows the extinction efficiencies for three particle radii, a, for the same ε, as indicated. They are plotted here on a logarithmic ordinate to demonstrate their low- and high-frequency behavior. The dotted line gives the scattering efficiency for $a = c/\omega_p$; note that $Q^{sca} \simeq Q^{ext}$ holds only at low frequencies ($\omega/\omega_p \ll 1$).

- Both decline at low frequencies like ν^2; however, their dependence on radius is different:

$$Q^{el} \propto a \qquad Q^{mag} \propto a^3.$$

Figure 4.17 suggests that for graphite grains ($\sigma = 7 \times 10^{15}$ s^{-1}) of sizes ~ 0.1 μm, magnetic dipole absorption is important in the far infrared around 100 μm. We, therefore, repeat in figure 4.19 the calculations of Q^{mag} and Q^{el} for a metal sphere with damping constant and plasma frequency like graphite but without a dielectric contribution. The permeability $\varepsilon(\omega)$ is computed from the Drude profile and the particle size ($a = 0.1$ μm) is small enough for the Rayleigh limit to be valid for both Q^{mag} and Q^{el}. We learn from figure 4.19 that in graphitic particles magnetic dipole absorption dominates at wavelengths greater than 100 μm.

4.5.5 Elongated metallic particles

The absorption by infinitely long cylinders, which are an extreme version of elongated particles and are treated in section 10.1, depends on the polarization of the wave. An example is plotted in figure 4.20 for idealized graphite material (see previous subsection) and a cylinder radius $a = 10^{-5}$ cm. When the direction

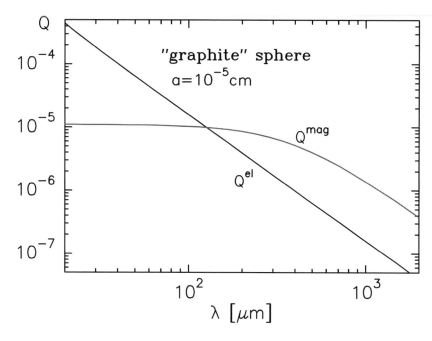

Figure 4.19. The magnetic and electric dipole absorption efficiency of a metal sphere of radius $a = 10^{-5}$ cm that has the damping constant and plasma frequency of graphite in the basal plane ($\gamma = 5 \times 10^{12}$ s^{-1}, $\omega_p = 6.7 \times 10^{14}$ s^{-1}). The permeability ε is from (1.123).

of the electric vector is perpendicular to the cylinder axis, the corresponding efficiency $Q_\perp$ is similar to that of a sphere of the same radius, at least for $\omega < \omega_p$. At very low frequencies, $Q_\perp$ is due to magnetic dipole absorption and proportional to a^3.

The behavior of $Q_\parallel$, when the electric vector is parallel to the cylinder axis, is radically different. The efficiency does not fall with wavelength but levels off at some high value. To understand this behavior, we consider a spheroid with an eccentricity e close to unity so that it resembles an infinitely long cylinder. In the long wavelength limit, $Q_\parallel$ is due to *electric* dipole absorption and can be computed from equation (3.40). As we keep stretching the cigar, e approaches one and the shape factor L_x goes to zero, as displayed in figure 3.2. For $L_x = 0$ and $\omega \to 0$, we indeed find that Q^{abs} is constant:

$$Q^{abs} = \frac{16\pi\sigma a}{3c}$$

where a is the short major axis. All real cigars are finite, and their efficiency is, in the long-wavelength limit, proportional to $a\lambda^{-2}$. But if the cigars are very elongated, the steep λ^{-2}-decline sets in only when λ is much larger than the big

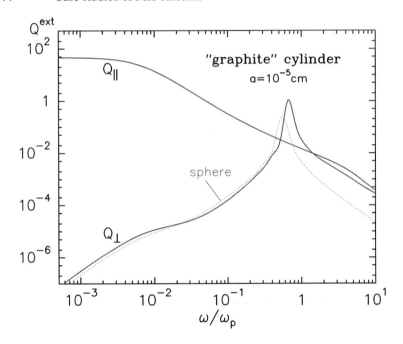

Figure 4.20. The absorption efficiency of an infinitely long cylinder under normal incidence. The cylinder is purely metallic with damping constant and plasma frequency like graphite ($\gamma = 5 \times 10^{12}$ s^{-1}, $\omega_p = 6.7 \times 10^{14}$ s^{-1} corresponding to a wavelength $\lambda_p = 2.8$ μm). For comparison, the dotted curve presents Q^{abs} for a sphere composed of the same metal and with the same radius. The permeability ε is from (1.123).

major axis of the cigar, i.e. at extremely long wavelengths, outside the observable range.

The cylinder with $Q^{abs} = 16\pi\sigma a/3c$ absorbs per unit length from an electromagnetic wave with Poynting vector $S = (c/8\pi)E^2$ the power $W = (4/3)a^2\sigma E^2$. A constant field E in the cylinder, however, leads to a current and the ohmic losses per unit length are $\pi a^2\sigma E^2$. So the infinite cylinder is similar to a wire in which a current flows.

Chapter 5

Particle statistics

This chapter presents purely physical topics selected for our study of the interstellar medium. They concern the statistical distribution of atoms and photons, basic thermodynamic relations and the radiation of blackbodies.

5.1 Boltzmann statistics

For a large number of atoms in thermodynamic equilibrium, their distribution in energy space is just a function of temperature and described by the Boltzmann equation. We outline the derivation of this fundamental statistical relation and append a few loosely connected items.

5.1.1 The probability of an arbitrary energy distribution

We first compute the number of ways in which the energy of N identical particles, for example atoms, can be distributed and then find among them the most likely distribution. To this end, we divide the energy space, which ranges from zero to some maximum E_{max}, into n ordered cells with mean energy E_i so that

$$E_1 < E_2 < \cdots < E_n.$$

Each cell has a size ΔE_i which determines its statistical weight $g_i = \Delta E_i / E_{max}$. The statistical weight gives the fractional size of a cell, so

$$\sum_i g_i = 1$$

Let each level E_i be populated by N_i atoms. If the atoms are labeled from a_1 to a_n, this particular configuration is represented, for example, by the following sequence:

$$\underbrace{a_1 a_2 \ldots a_{N_1}}_{N_1} \underbrace{a_{N_1+1} \ldots a_{N_1+N_2}}_{N_2} \cdots \underbrace{a_{N-N_n+1} \ldots a_{N_n}}_{N_n}.$$

The N_i under the horizontal brackets give the number of atoms in the energy bins E_i. But there are many others ways in which the same energy distribution can be realized. According to elementary probability theory, a configuration where N particles are grouped as shown here can be achieved in

$$\Omega = \frac{N!}{N_1! N_2! \ldots N_n!} \tag{5.1}$$

ways. Here it is assumed that the particles are distinguishable. Ω is called the thermodynamic probability and is a very large number. When the N atoms are arbitrarily distributed over the available cells, the probability that the first N_1 atoms fall into cell E_1, the following N_2 atoms into cell E_2, and so forth equals

$$g_1^{N_1} \cdot g_2^{N_2} \cdot \ldots \cdot g_n^{N_n}.$$

The probability ω for the particular configuration of the N atoms is, therefore,

$$\omega = \frac{N!}{N_1! N_2! \ldots N_n!} \cdot g_1^{N_1} \cdot g_2^{N_2} \cdot \ldots \cdot g_n^{N_n}. \tag{5.2}$$

Equation (5.2) is at the heart of classical statistics. ω is, of course, smaller than one. Summing the probabilities ω of all possible distributions gives the total probability. In view of the binomial theorem, this sum is $(g_1 + g_2 + \cdots + g_n)^N$, which equals one, as it should. When all cells are of the same size, each cell has the same chance that a particular atom will fall into it. Then $g_i = 1/n$ for all i and the probability ω in equation (5.2) becomes

$$\omega = \frac{N!}{n^N \prod N_i!}.$$

5.1.2 The distribution of maximum probability

As the numbers N_i in the expression for the probability ω are large, their faculties are much larger still. To handle them, one uses Stirling's formula:

$$\ln N! \simeq N \ln N - N + \ln \sqrt{2\pi N} + \frac{\vartheta_N}{N} \qquad (0 \le \vartheta_N \le 1). \tag{5.3}$$

The accuracy of this formula is quite remarkable (see the error estimate ϑ_N). Retaining only the first term, this yields for (5.2)

$$\ln \omega \simeq N \ln N + \sum_{i=1}^{n} N_i \ln \frac{g_i}{N_i}.$$

In the equilibrium distribution achieved by nature, the probability has the maximum possible value. Because $N \ln N$ is constant, we are looking for an extremum of the function

$$f(N_1, N_2, \ldots, N_n) = \sum_{i=1}^{n} N_i \cdot \ln \frac{g_i}{N_i}.$$

There are two additional and obvious requirements:

- The total number of particles N should be constant and equal to $\sum N_i$,

$$\phi = \sum_{i=1}^{n} N_i - N = 0.$$

- The mean particle energy $\langle E_i \rangle$ is related to the total energy E of all particles through

$$E = N \cdot \langle E_i \rangle$$

therefore

$$\psi = N\langle E_i \rangle - \sum_{i=1}^{n} N_i E_i = 0. \tag{5.4}$$

The maximum of f, subject to the conditions expressed by the auxiliary functions ϕ and ψ, can be found using the method of Lagrangian multiplicators. If we denote them by α and β, we get the equation for the differentials

$$df + \alpha\, d\phi + \beta\, d\psi = 0$$

leading to

$$\sum dN_i \left(\ln \frac{g_i}{N_i} - 1 + \alpha - \beta E_i \right) = 0.$$

In this sum, all brackets must vanish. So for every i,

$$N_i = g_i e^{\alpha-1} e^{-\beta E_i}.$$

Exploiting $\sum N_i = N$ eliminates α and gives the fractional population

$$\frac{N_i}{N} = \frac{g_i e^{-\beta E_i}}{\sum_{j} g_j e^{-\beta E_j}}. \tag{5.5}$$

5.1.3 Partition function and population of energy cells

We have found N_i/N but we still have to determine β. When we define, in (5.5),

$$Z = \sum_{j} g_j e^{-\beta E_j}$$

equations (5.1), (5.3) and (5.5) yield

$$\ln \Omega = -\sum N_i \ln \frac{N_i}{N} = -\sum N_i \left(\ln g_i - \beta E_i - \ln Z \right)$$
$$= \beta E + N \ln Z - \sum N_i \ln g_i.$$

We now plug in some thermodynamics from section 5.3. The entropy S of the system equals the Boltzmann constant multiplied by the logarithm of the thermodynamic probability Ω of (5.1),

$$S = k \ln \Omega,$$

and the differential of the entropy is (see (5.51)),

$$dS = \frac{dE + P\,dV}{T}.$$

This allows us to express the multiplicator β in terms of the temperature. Because at constant volume V,

$$\frac{1}{T} = \left(\frac{\partial S}{\partial E}\right)_V = k\frac{\partial \ln \Omega}{\partial E}$$

it follows that

$$\frac{1}{T} = k\beta + kE\frac{\partial \beta}{\partial E} - kN \cdot \frac{\sum g_i E_i e^{-\beta E_i}}{\sum g_j e^{-\beta E_j}} \cdot \frac{\partial \beta}{\partial E}.$$

The last two terms on the right-hand side cancel and thus

$$\beta = \frac{1}{kT}. \tag{5.6}$$

The expression Z introduced above now becomes

$$Z(T) = \sum_j g_j e^{-E_j/kT}. \tag{5.7}$$

This is called the partition function, which depends only on temperature. The fractional abundance of atoms with energy E_i is, therefore,

$$\frac{N_i}{N} = \frac{g_i e^{-E_i/kT}}{\sum g_j e^{-E_j/kT}} = \frac{g_i e^{-E_i/kT}}{Z} \tag{5.8}$$

and the abundance ratio of two levels i and j,

$$\frac{N_j}{N_i} = \frac{g_j}{g_i} e^{-(E_j - E_i)/kT}. \tag{5.9}$$

Although the population numbers are statistical averages, they are extremely precise and, practically, important deviations from the mean never occur.

5.1.4 The mean energy of harmonic oscillators

With some reinterpretation, equations (5.7)–(5.9) are also valid in quantum mechanics (more general forms are given by (5.26) and (5.36)). For atoms with internal energy levels E_i, the statistical weight g_i denotes then the degeneracy of the level.

We compute the partition function of a system of N quantized identical harmonic oscillators in equilibrium at temperature T. The energy levels are non-degenerate ($g_i = 1$) and equidistant, $E_i = (i + \frac{1}{2})\hbar\omega$ for $i = 0, 1, 2, \ldots$ (see (6.25)). With $\beta^{-1} = kT$,

$$Z(T) = \sum_{i=0}^{\infty} e^{-\beta E_i} = e^{-\frac{1}{2}\beta\hbar\omega} \sum_{i=0}^{\infty} e^{-i\beta\hbar\omega}.$$

The last sum is an infinite series where each term is a factor $e^{-\beta\hbar\omega}$ smaller than the preceding one, therefore,

$$Z(T) = \frac{e^{-\frac{1}{2}\beta\hbar\omega}}{1 - e^{-\beta\hbar\omega}}. \tag{5.10}$$

We can also easily compute the mean oscillator energy $\langle E \rangle$. If N_i oscillators have an energy E_i, the average is

$$\langle E \rangle = \frac{\sum N_i E_i}{N} = \sum E_i \frac{N_i}{N} = \sum E_i \frac{e^{-\beta E_i}}{Z} = -\frac{1}{Z}\frac{\partial Z}{\partial \beta} = -\frac{\partial \ln Z}{\partial \beta}.$$

Because $\ln Z = -\frac{1}{2}\beta\hbar\omega - \ln(1 - e^{-\beta\hbar\omega})$, it follows that

$$\langle E \rangle = \frac{1}{2}\hbar\omega + \frac{\hbar\omega}{e^{\hbar\omega/kT} - 1}. \tag{5.11}$$

At high temperatures ($\hbar\omega \ll kT$), we get the classical result that the sum of potential plus kinetic energy of an oscillator equals kT,

$$\langle E \rangle = kT.$$

At low temperatures ($\hbar\omega \gg kT$), the average $\langle E \rangle$ is close to the zero point energy,

$$\langle E \rangle = (\frac{1}{2} + e^{-\hbar\omega/kT})\hbar\omega.$$

5.1.5 The Maxwellian velocity distribution

The probability that an arbitrary gas atom has a velocity in a certain range follows as a corollary from the equilibrium energy distribution (5.8). We substitute in this formula for E_i the kinetic energy of an atom, $p^2/2m$. All directions of motion are equally likely and the square of the length of the momentum vector is

$$p^2 = p_x^2 + p_y^2 + p_z^2.$$

For the size of an energy cell, i.e. for the statistical weight, we take

$$g = 4\pi p^2 \, dp. \tag{5.12}$$

In the ratio N_i/N of (5.8), it is the relative not the absolute value of the statistical weight g that matters;, therefore, the sum over all g need not be one. When we replace the sum in the partition function $Z(T)$ of (5.7) by an integral, we obtain the momentum distribution. Changing from momenta $p = mv$ to velocities gives the familiar Maxwellian velocity distribution (figure 5.1) which states that if there are N gas atoms altogether, the number of those whose absolute velocities are in the range $v \ldots v + dv$ equals

$$N(v) \, dv = N 4\pi v^2 \left(\frac{m}{2\pi kT}\right)^{3/2} e^{-mv^2/2kT} \, dv \tag{5.13}$$

with

$$\int_0^\infty N(v) \, dv = N.$$

In any arbitrary direction x, the number of atoms with velocities in the interval $[v_x, v_x + dv_x]$ is given, except for a normalizing constant, by the Boltzmann factor $e^{-mv_x^2/2kT}$,

$$N(v_x) \, dv_x = N \left(\frac{m}{2\pi kT}\right)^{1/2} e^{-mv_x^2/2kT} \, dv_x. \tag{5.14}$$

$N(v_x)$ is symmetric about $v_x = 0$ and has its maximum there (contrary to $N(v)$ in (5.13)) because near zero velocity, $N(v_x)$ includes all particles with any velocity in the y- or z-direction. The total number of atoms, N, follows from (5.14) by integrating from $-\infty$ to ∞.

There are three kinds of averages of the absolute velocity v. To compute them, we exploit the integrals (A.27)–(A.29) in appendix A associated with the bell-curve.

● The most *probable* velocity, v_p, is the one at the maximum of the curve $N(v)$,

$$v_p = \sqrt{\frac{2kT}{m}}. \tag{5.15}$$

● The mean velocity $\langle v \rangle$ is relevant for calculating collision rates,

$$\langle v \rangle = \sqrt{\frac{8kT}{\pi m}}. \tag{5.16}$$

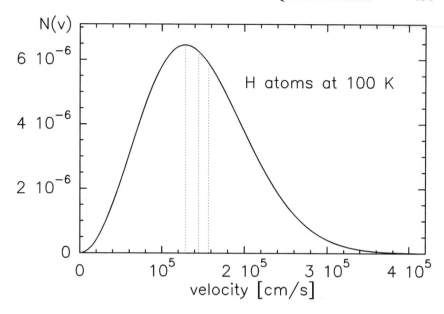

Figure 5.1. The Maxwellian velocity distribution for hydrogen atoms at 100 K. The ordinate gives the probability density $N(v)$ after (5.13) with respect to one atom ($N = 1$). The vertical dotted lines show from left to right the means v_p, $\langle v \rangle$ and $\sqrt{\langle v^2 \rangle}$ according to (5.15)–(5.17).

• The mean energy of the atoms is given by the average of v^2 such that $\frac{1}{2} m \langle v^2 \rangle = \frac{3}{2} kT$ or

$$\langle v^2 \rangle = \frac{3kT}{m}. \tag{5.17}$$

The averages are ordered (see also figure 5.1),

$$v_p < \langle v \rangle < \sqrt{\langle v^2 \rangle}. \tag{5.18}$$

5.2 Quantum statistics

5.2.1 The unit cell h^3 of the phase space

In deriving the energy distribution for an ensemble of particles after Boltzmann, we divided the energy or momentum space into arbitrarily small cells. The real coordinate space was not considered. In quantum statistics, one divides the *phase space*, which consists of coordinates and momenta, into cells. If V is the volume and p the momentum, we form cells of size

$$d\Phi = V \cdot 4\pi p^2 \, dp$$

where $4\pi p^2 dp$ is the volume of a shell in momentum space with radius p and thickness dp. According to the uncertainty principle, position q and momentum p of a particle can only be determined to an accuracy Δp and Δq such that

$$\Delta p \cdot \Delta q \gtrsim \hbar.$$

This relation results from the wave character possessed by all particles, not just photons. The wavelength of a particle is related to its momentum (after *de Broglie*) by $p = h/\lambda$. Because of the wave character, particles in a box of volume V can only have discrete momenta, otherwise they would be destroyed by interference. The situation is formally almost identical to standing waves in a crystal, which we treat in section 8.4. We may, therefore, exploit equation (8.31) of that section and find that the possible number of eigenfrequencies within the momentum interval p to $p + dp$ in a box of volume V is

$$g = \frac{V \cdot 4\pi p^2 \, dp}{h^3}. \tag{5.19}$$

The eigenfrequencies are the only allowed states in which a particle can exist and correspond to standing waves. The quantity g (named dZ in (8.31)) is called the statistical weight (see also (5.12)). It states how many *unit cells*, each of size h^3, fit into the phase space element $d\Phi$. Note that classically the number of possible (physically different) states is unlimited, which would correspond to a unit cell of size zero.

To determine the state of a particle with non-zero spin, one needs, besides place and momentum, the spin direction. Therefore, the number of quantum states of an electron (spin $\frac{1}{2}$) is twice the value given in (5.19). The same is true for photons (spin 1) which, traveling at the speed of light, also have two (not three) spin directions.

5.2.2 Bosons and fermions

Particles are described by their position in phase space. Any two particles of the same kind, for example, two photons or two electrons, are otherwise identical. This means that when we swap their positions in phase space, nothing changes physically, the particles are indistinguishable.

In quantum mechanics, particles are represented by their wavefunctions. Let $\Psi(x_1, x_2, \ldots, x_n)$ denote the wavefunction of a system of n particles, all of one kind and not interacting. Their identity then implies that $|\Psi|^2$ stays the same when any two of the particles are interchanged. So after they have been swapped, either Ψ itself is not altered, then one speaks of a symmetric wavefunction, or Ψ is transferred into $-\Psi$, then the wavefunction is called anti-symmetric.

Let particle i have the coordinate x_i, the wavefunction ψ_i and the energy E_i. The total energy of the system is then

$$E = \sum_{i=1}^{n} E_i.$$

Because the particles are non-interacting, the wavefunction of the whole system can be factorized into the N wavefunctions ψ_i of the individual particles. We, therefore, write $\Psi(x_1, x_2, \ldots, x_n)$ as a sum of such terms but with interchanged arguments; this corresponds to swapping particles. For example, the first term of the sum is

$$\psi_1(x_1)\psi_2(x_2) \cdot \ldots \cdot \psi_n(x_n) \tag{5.20}$$

and the next, where particles 1 and 2 have been exchanged,

$$\psi_1(x_2)\psi_2(x_1) \cdot \ldots \cdot \psi_n(x_n)$$

and so forth. Each such product is a wavefunction that gives the same total energy E but fails to fulfil the symmetry condition. Note that in a multi-dimensional function it does not matter by which symbols the arguments are presented, what counts is the sequence in which they appear. So $f(x_1, x_2) = x_1 x_2^2$ and $g(x_2, x_1) = x_1^2 x_2$ are identical functions but $f(x_1, x_2)$ and $h(x_1, x_2) = x_1^2 x_2$ are different.

- It can readily be checked that the linear combination

$$\Psi_{\text{Bose}}(x_1, x_2, \ldots, x_n) = \sum_P \hat{\mathbf{P}}\psi_1(x_1)\psi_2(x_2) \cdot \ldots \cdot \psi_n(x_n) \tag{5.21}$$

  and no other, is symmetric and yields the correct energy E. Here $\hat{\mathbf{P}}$ is an operator that performs a permutation of the coordinates $x_1, x_2, \ldots, x_n$; the sum extends over all $N!$ possible permutations. All coefficients before the products $\psi_1\psi_1 \ldots \psi_n$ equal one (we neglect normalization).
- Likewise, the only anti-symmetric wavefunction of the whole system is the linear combination formed by the determinant

$$\Psi_{\text{Fermi}}(x_1, x_2, \ldots, x_n) = \det \begin{pmatrix} \psi_1(x_1) & \psi_2(x_1) & \ldots & \psi_n(x_1) \\ \psi_1(x_2) & \psi_2(x_2) & \ldots & \psi_n(x_2) \\ \ldots & \ldots & \ldots & \ldots \\ \psi_1(x_n) & \psi_2(x_n) & \ldots & \psi_n(x_n) \end{pmatrix}. \tag{5.22}$$

As indicated by the subscripts on Ψ, if the wavefunction is symmetric, the particles are called bosons, if it is anti-symmetric, fermions. When two rows or columns of the matrix (5.22) are equal, $\psi_i = \psi_j$, they are linearly dependent and the determinant in (5.22) vanishes. This has the fundamental consequence, known as *Pauli's exclusion principle*, that two fermions cannot have the same wavefunction, i.e. they cannot occupy the same quantum state or unit cell. Bosons, however, are not subject to Pauli's exclusion principle. It is an experimental fact and is also claimed theoretically that particles with half-integer spin, like electrons, protons or neutrons are fermions, while particles with integer spin, like photons, are bosons.

5.2.3   Bose statistics

5.2.3.1   Counting states

Let us assume N identical particles of mass m in a box of volume V and with a total energy E. We divide the phase into cells of size $d\Phi = V4\pi p_i^2\, dp$. Each cell contains N_i particles of energy

$$E_i = \frac{p_i^2}{2m}$$

and comprises

$$g_i = \frac{d\Phi}{h^3} = \frac{V4\pi p_i^2\, dp}{h^3} \tag{5.23}$$

unit cells. A particular distribution of the system is completely determined by specifying for each cell $d\Phi$ its population N_i, energy E_i and statistical weight g_i (size), i.e. through the three sets of values

$$
\begin{array}{ccccc}
N_1, & N_2, & \ldots & N_i, & \ldots \\
E_1, & E_2, & \ldots & E_i, & \ldots \\
g_1, & g_2, & \ldots & g_i, & \ldots
\end{array}
\tag{5.24}
$$

The probability of finding this particular distribution is proportional to the number of ways W in which it can be realized.

First, we consider only one cell, say cell i of size $d\Phi$. Its N_i atoms constitute a subsystem. We ask in how many ways can one distribute its N_i particles over its g_i unit cells (quantum states). Let us label the unit cells within $d\Phi$ by $Z_1, Z_2, \ldots, Z_{g_i}$ and the atoms by $a_1, a_2, \ldots, a_{N_i}$. If we write behind each unit cell the atoms which it contains, one particular distribution is given, for instance, by the sequence

$$\underbrace{Z_1 a_1 a_2 a_3}\ \underbrace{Z_2}\ \underbrace{Z_3 a_4 a_5}\ \underbrace{Z_4 a_6}\ \underbrace{Z_5 a_7} \ldots.$$

In this sequence, the atoms a_1, a_2, a_3 are in Z_1, cell Z_2 is empty, the atoms a_4, a_5 are in Z_3, and so forth. In boson statistics, there is no limit as to how many atoms can be squeezed into one unit cell. All such sequences of symbols Z_l and a_k, like the one here, start with some Z, because we always first specify the cell and then the atoms it contains. Therefore the total number of mathematically different sequences is $g_i(g_i + N_i - 1)!$. As permuting the atoms or unit cells does not result in a new quantum state of the subsystem $d\Phi$, only

$$\frac{g_i(g_i + N_i - 1)!}{N_i! g_i!} = \frac{(g_i + N_i - 1)!}{N_i!(g_i - 1)!}$$

of the sequences are physically different. Therefore, when considering the *total* phase space, there are

$$\Omega = \prod_i \frac{(g_i + N_i - 1)!}{N_i!(g_i - 1)!} \tag{5.25}$$

ways in which the configuration of (5.24) can be achieved. The maximum of Ω, which is what nature chooses, is found in the same manner as the maximum of the probability ω in (5.2) for Boltzmann statistics. One forms the logarithm using the first two terms of the Stirling formula (5.3),

$$\ln \Omega = \sum_i \left[(g_i + N_i - 1) \ln(g_i + N_i - 1) - (g_i - 1) \ln(g_i - 1) - N_i \ln N_i \right]$$

and exploits the conditions for the conservation of energy and number of particles,

$$E - \sum_i N_i E_i = 0 \qquad N - \sum_i N_i = 0.$$

Following the procedure outlined in section 5.1, this leads to

$$\ln \frac{g_i + N_i}{N_i} = \alpha + \beta E_i.$$

5.2.3.2 The occupation number in Bose statistics

When, in the last formula, we drop the subscript i, write dn instead of N_i (dn is the number of particles in the phase space element $V 4\pi p^2 \, dp$), and insert (5.23) for the statistical weight, we get

$$dn = \frac{V 4\pi p^2}{h^3} \frac{dp}{e^{\alpha + \beta E} - 1} = \frac{g}{e^{\alpha + \beta E} - 1}. \tag{5.26}$$

As the g unit cells are, according to this formula, populated by dn particles, one calls

$$\mathcal{N} = \frac{dn}{g} = \frac{1}{e^{\alpha + \beta E} - 1} \tag{5.27}$$

the occupation number, which is the number of particles per unit cell. We still have to determine the coefficients α and β. Using the same thermodynamic arguments as for Boltzmann statistics gives (see (5.6))

$$\beta = \frac{1}{kT}.$$

α follows from the requirement that the number of particles is conserved, i.e. from

$$\int dn = \frac{4\pi V}{h^3} \int \frac{p^2 \, dp}{e^{\alpha + \beta E} - 1} = NV. \tag{5.28}$$

The coefficient α is always positive, as otherwise the denominator under the integral might vanish, so $e^{-\alpha}$ lies between zero and one. We only mention that α is related to the chemical potential μ by $\alpha = -\mu/kT$. When $e^{\alpha + \beta E} \gg 1$, the integrand in (5.28) becomes $p^2 e^{-\alpha} e^{-E/kT}$, which yields

$$e^{-\alpha} = \frac{N}{V} \frac{h^3}{(2\pi m k T)^{3/2}} \tag{5.29}$$

and one recovers the Maxwell distribution (5.13),

$$dn = N \frac{4\pi p^2 \, dp}{(2\pi mkT)^{3/2}} e^{-p^2/2mkT}.$$

5.2.4 Bose statistics for photons

For photons, which have a momentum $p = h\nu/c$, the statistical weight is

$$g = 2\frac{V4\pi \nu^2 \, d\nu}{c^3}. \tag{5.30}$$

The factor 2 accounts for the two possible modes of circular polarization. As radiative equilibrium is established through emission and absorption processes, in which photons are created and destroyed, their number is not conserved. Consequently, there is no such constraint as

$$N - \sum_i N_i = 0$$

the parameter α in (5.26) is zero and

$$dn = \frac{g}{e^{h\nu/kT} - 1} = \frac{V8\pi \nu^2}{c^3} \frac{d\nu}{e^{h\nu/kT} - 1}.$$

The radiative energy density u_ν now follows from $u_\nu \, d\nu = h\nu \, dn/V$,

$$u_\nu = \frac{8\pi h}{c^3} \frac{\nu^3}{e^{h\nu/kT} - 1}. \tag{5.31}$$

Formula (5.31) describes the distribution of photons in equilibrium at temperature T—the so called blackbody radiation. Although we did not explicitly prescribe the total number of photons within a unit volume, n_{phot}, the average is fixed and follows from integrating $u_\nu/h\nu$ over frequency,

$$n_{\text{phot}} = \int_0^\infty \frac{u_\nu}{h\nu} \, d\nu = 8\pi \left(\frac{kT}{hc}\right)^3 \int_0^\infty \frac{x^2}{e^x - 1} \, dx$$

(see (A.17) for an evaluation of the integral). A unit volume of gas in thermodynamic equilibrium, however, may be filled by an arbitrary number of atoms.

The occupation number of photons in equilibrium is, from (5.27),

$$\mathcal{N}_{\text{equil}} = \frac{1}{e^{h\nu/kT} - 1}. \tag{5.32}$$

The occupation number equals dn/g and, therefore, also has a value when the photons are out of equilibrium and no temperature is defined. Because there are $dn = g\mathcal{N}$ photons in the phase cell $d\Phi$, which have a total energy $g\mathcal{N}h\nu$, the occupation number may be used to specify the radiative energy density at a particular frequency,

$$u_\nu = \frac{8\pi h\nu^3}{c^3}\mathcal{N}. \qquad (5.33)$$

Likewise, $\mathcal{N}$ may define the intensity of radiation. We will return to blackbody radiation in section 5.4 and rediscover the basic equation (5.31) from less abstract and less strict arguments.

5.2.5   Fermi statistics

Fermi statistics deals with particles that have an anti-symmetric wavefunction and which, therefore, obey Pauli's exclusion principle, like electrons. We again divide the phase space into cells of size $d\Phi = V4\pi p^2\,dp$. Now there can only be one particle per unit cell h^3 with a given spin direction, so the statistical weight is

$$g_i = \frac{V\cdot 4\pi p_i^2\,dp}{h^3}. \qquad (5.34)$$

However, there may be two particles of opposite spin per unit cell because their quantum states differ. In counting states, we repeat the steps which we carried out for Bose statistics. We consider a cell $d\Phi$ consisting of g_i unit cells, each harbouring N_i electrons. A unit cell is either empty or filled with one electron. The essential difference from Bose statistics is that there are now

$$\frac{g_i!}{(g_i - N_i)!N_i!}$$

ways to distribute N_i electrons over the g_i unit cells, thus in all

$$\Omega = \prod_i \frac{g_i!}{(g_i - N_i)!N_i!} \qquad (5.35)$$

different possibilities. This gives, with the same procedure as before, the counterpart to (5.26) for the number of particles dn in the phase space element $V4\pi p^2\,dp$ which contains g unit cells,

$$dn = \frac{g}{e^{\alpha+\beta E} + 1} \qquad (5.36)$$

and for the occupation number (cf. (5.27))

$$\mathcal{N} = \frac{dn}{g} = \frac{1}{e^{\alpha+\beta E} + 1}. \qquad (5.37)$$

In the denominator, it is now $+1$, and not -1. Of course, as before $\beta = 1/kT$ and the parameter α follows from the condition that the total number of particles comes out correctly (cf. (5.28)),

$$\int dn = NV.$$

But now α, which is also called the degeneracy parameter, may be positive as well as negative (from $-\infty$ to $+\infty$). Degeneracy is absent for large positive α, when the gas is hot and rarefied (see (5.29)). As in Bose statistics: $\alpha = -\mu/kT$ where μ is the chemical potential, and if $e^{\alpha+\beta E} \gg 1$, we are in the realm of classical physics.

When the temperature T goes to zero, the occupation number $\mathcal{N}$ changes discontinuously from one for energies below the Fermi limit E_F to zero for $E > E_F$ (subsection 6.4.2); in Bose statistics, *all* particles are in their lowest (zero momentum) state at $T = 0$.

In quantum statistics, particles are indistinguishable and states are counted either after Bose as in (5.25) or after Fermi as in (5.35). The quantum mechanical way of counting is the right way and takes into account the fact that the phase space is partitioned into unit cells h^3. In Fermi statistics, there is not more than one particle (of the same spin) per unit cell. In classical statistics, however, particles are distiguishable and the probability is, from (5.2),

$$\Omega_{\text{class}} = N! \prod_i \frac{g_i^{N_i}}{N_i!}.$$

In accordance with the correspondence principle, the quantum mechanical probabilities (5.25) and (5.35) approach Ω_{class} when the Planck constant h goes to zero and the g_i become very large ($g_i \gg N_i$).

5.2.6   Ionization equilibrium and the Saha equation

In a hot gas, the atoms of any element X exist in different stages r of ionization. Denoting the corresponding ion by X_r, there are ionizing and recombining processes symbolized by

$$X_r \quad \rightleftharpoons \quad X_{r+1} + e.$$

Suppose that in 1 cm^3 there are N_e electrons and N_r particles of species X_r, of which $N_{1,r}$ are in their ground state. The Ns of two neighboring ionization stages are related in thermodynamic equilibrium by the Saha formula

$$\frac{N_{1,r+1} N_e}{N_{1,r}} = 2 \frac{(2\pi m_e kT)^{3/2}}{h^3} \frac{g_{1,r+1}}{g_{1,r}} e^{-\chi/kT}. \tag{5.38}$$

$g_{1,r}$ is the statistical weight of the ground state of ionization stage r, and $\chi > 0$ is the energy difference between the ground state of ion X_r and X_{r+1}. The very form of (5.38) suggests that the Saha equation can be derived from the

Boltzmann equation. For simplicity, we restrict the discussion to pure hydrogen gas consisting of electrons, protons and neutral hydrogen atoms. Let us consider

- the atom with the electron bound in level $n=1$ as the ground state
- the proton *together* with a free electron of momentum p as the excited state.

The energies of the excited states form a continuum because the electrons may have any velocity. Energies are counted in the following way: for the hydrogen atom in level n, E_n is negative. Thus in the ground state, $-E_1 = \chi > 0$, where χ is the ionization potential of hydrogen. At the ionization threshold, the energy is zero, and for the proton–free-electron system it is positive and equals $p^2/2m_e$.

Let $N(p) \, dp$ be the number density of excited particles (one such particle consists of a proton and an electron) where the electron has momentum in the range $[p, p + dp]$. Let $g(p)$ be the statistical weight of the excited state. If $N_{1,H}$, $g_{1,H}$ are the corresponding numbers for the hydrogen atoms in the ground state, then according to (5.9)

$$\frac{N(p) \, dp}{N_{1,H}} = \frac{g(p)}{g_{1,H}} \exp\left(-\frac{\chi + p^2/2m_e}{kT}\right).$$

The statistical weight $g(p)$ is the number of quantum states of the proton–free-electron particle and thus the product of the statistical weight of the proton and the electron,

$$g(p) = g_p \cdot g_e.$$

After (5.34), $g_e = 2 \cdot 4\pi p^2 \, dp \, V/h^3$. The factor 2 comes from the two electron spin directions. Because

$$\int_0^\infty N(p) \, dp = N_e = N_p,$$

one gets

$$\frac{N_p}{N_{1,H}} = 2 \frac{(2\pi m_e kT)^{3/2}}{h^3} e^{-\chi/kT} V \frac{g_p}{g_{1,H}}.$$

To determine the volume V, we note that the phase space cell of a free electron, which always pertains to one proton, must contain exactly one electron, so

$$V = 1/N_e.$$

The statistical weight of the neutral atom with the electron in level n is equal to the maximum number of electrons in that level, so $g_{n,H} = 2n^2$ and $g_{1,H}/g_p = 2$, altogether

$$\frac{N_p N_e}{N_{1,H}} = \frac{(2\pi m_e kT)^{3/2}}{h^3} e^{-\chi/kT}. \tag{5.39}$$

When the partition function of the neutral hydrogen atom $Z = \sum g_i e^{-E_i/kT}$ is approximated by its first term $g_1 e^{-E_1/kT}$, which is often possible and implies that almost all atoms reside in the ground state, then $N_{1,H}$ is equal to the number N_H of all hydrogen atoms and

$$\frac{N_p N_e}{N_H} = \frac{(2\pi m_e kT)^{3/2}}{h^3} e^{-\chi/kT}.$$

We can also immediately calculate the number N_n of atoms in level n under conditions of thermodynamic equilibrium,

$$\frac{N_n}{N_e N_p} = \frac{h^3 n^2}{(2\pi m_e kT)^{3/2}} e^{-E_n/kT}. \tag{5.40}$$

Note that $-E_n$ is positive and for high levels very small.

The Saha equation goes wrong when the mean distance d between ions becomes comparable to the size $2r$ of an atom. This happens in stellar interiors where the density is high, although LTE (local thermodynamic equilibrium) certainly prevails. An atom is then closely surrounded by ions and the overlap of their electric fields disturbs the higher quantum states and lowers the ionization energy. Atoms in high quantum number n are big. For hydrogen, their radius is $r = a_0 n^2$ where a_0 is the Bohr radius. In the central region of the Sun ($\rho \sim 10^2$ g cm^{-3}, $T \sim 15 \times 10^6$ K), d is less than the Bohr radius and even the ground state of hydrogen cannot exist, although the Saha equation predicts that about half of all atoms are neutral.

5.3 Thermodynamics

This section summarizes the results of thermal and statistical physics. For the computation of the population of energy levels in the smallest grains (see section 12.1) we have to recall the definition of temperature, entropy and number of accessible states. The formulae around the specific heat find their application mainly in the theory of magnetic dissipation in section 11.2. Some of the thermodynamic elements presented here are also indispensable in the discussion of phase transition during the formation and evaporation of grains (see section 9.5).

5.3.1 The ergodic hypothesis

Boltzmann's equilibrium population of the energy levels of a system was derived by determining the energetically most likely distribution. More generally, instead of energy space one considers the phase space of the system; the path of the atoms in phase space contains the full mechanical information. A system of f degrees of freedom is described by the spatial coordinates $q_1, \ldots, q_f$ and the conjugate momenta $p_1, \ldots, p_f$. Usually one needs three coordinates (x, y, z) for one particle, so if there are N particles,

$$f = 3N.$$

Let the system be isolated so that it cannot exchange energy with its surroundings. The total energy E is, therefore, constant or, rather, it is in a narrow range from E to $E + \delta E$, because in quantum mechanics, E cannot be defined with absolute precision in a finite time.

At any instant, the system may be represented by a point in the $2f$-dimensional phase space with the coordinates $q_1, \ldots, q_f, p_1, \ldots, p_f$. As the system evolves in time t, it describes a trajectory parametrized by $(q_1(t), \ldots, p_f(t))$. If one divides the phase space into small cells Z of equal size, the *microstate* of the system at a particular moment is determined by the cell in which it is found. The cells may be enumerated. All cells (states) Z_i that correspond to an energy between E and $E + \delta E$ are called *accessible*. In equilibrium, the probability to find the system in a certain accessible cell Z_r is, by definition, independent of time.

- The fundamental (*ergodic*) postulate asserts that when the system is in equilibrium, all accessible states are equally likely, none has any preference.

Let $\Omega(E)$ be the total number of accessible states (cells), i.e. those with energy in the interval $[E, E + \delta E]$. If there are many degrees of freedom ($f \gg 1$), the function $\Omega(E)$ increases *extremely* rapidly with the total energy E of the system. To see how fast $\Omega(E)$ rises, we make a rough estimate. Let our system consist of identical quantum oscillators, each corresponding to one degree of freedom. Their energy levels are equally spaced in multiples of $\hbar\omega$. Each oscillator has on average an energy E/f. After equation (6.26), it is spread out in one-dimensional phase space over a region $\Delta x \Delta p = E/f\omega$ where it thus occupies $E/f\hbar\omega$ cells of size $\hbar$. It does not make sense to consider cells smaller than $\hbar$ because one cannot locate a particle more accurately. The number of states, N_s, of the total system whose energy does not exceed E is, therefore,

$$N_s = \left(\frac{E}{f\hbar\omega} \right)^f. \tag{5.41}$$

Hence the number of cells, $\Omega(E)$, with energies from $[E, \ E + \delta E]$ becomes

$$\Omega(E) = N_s(E + \delta E) - N_s(E) \simeq \frac{dN_s}{dE} \delta E.$$

We denote by

$$\rho(E) = \frac{dN_s}{dE}$$

the density of states around energy E. It is independent of the width of the chosen energy interval δE, contrary to $\Omega(E)$ which obviously increases with δE. As a first approximation, one gets $\rho(E) \simeq f E^{f-1}/(f\hbar\omega)^f$. If one counts the states more properly but neglects powers of E with exponents $f - 2$ or smaller, one finds

$$\rho(E) \simeq \frac{E^{f-1}}{(f-1)!(\hbar\omega)^f}. \tag{5.42}$$

If the total energy E of the system is n times greater than the natural energy unit $\hbar\omega$, the density of states becomes proportional to $n^{f-1}/(f-1)!$. Note that ρ and Ω rise with E^{f-1}. When we evaluate Ω for something from everyday life, like the gas content in an empty bottle, the number of degrees of freedom of the atoms is $\sim 10^{24}$. This, by itself, is a huge number but because f appears as an exponent, Ω is so large that even its logarithm $\ln \Omega$ is of order f, or 10^{24}.

The probability that all gas atoms in a box are huddled in the upper left-hand corner and the rest of the box is void is extremely low, one will never encounter such a configuration. After the ergodic principle, all accessible cells in a phase space have an equal chance of being populated. The probability of a particular microstate which actually is encountered and where the atoms are very evenly distributed must, therefore, be equally low. There is no contradiction because there is just one cell corresponding to all atoms being clustered in one corner but a multitude corresponding to a very smooth distribution of atoms over the available box volume.

5.3.2 Definition of entropy and temperature

In statistical physics, the starting point is the number of accessible states $\Omega(E)$ with energies from $[E, E + \delta E]$; it determines the entropy,

$$S = k \ln \Omega. \tag{5.43}$$

Of course, $\Omega(E)$ depends on the width of the energy interval δE but because Ω is so large, one can easily show that δE is irrelevant in (5.43). The absolute temperature T is then defined by

$$\frac{1}{k}\frac{\partial S}{\partial E} = \frac{\partial \ln \rho}{\partial E} = \frac{1}{kT}. \tag{5.44}$$

Let us see what these two definitions mean for a bottle of warm lemonade (system A) in an icebox (system A'). The two systems are in thermal contact, i.e. heat can flow between them but the external parameters (like volume) are fixed. The total system, $A + A'$, is isolated (we do not have an electric icebox), so the joint energy $E_{\text{tot}} = E + E'$ is constant; otherwise E and E' are arbitrary. The number of accessible states of the whole system, subject to the condition that subsystem A has an energy E, is given by the product $\Omega(E)\Omega'(E_{\text{tot}} - E)$, where Ω and Ω' refer to A and A', respectively. The probability $P(E)$ of finding the whole system in a state where A has the energy E is evidently proportional to this product,

$$P(E) \propto \Omega(E) \cdot \Omega'(E_{\text{tot}} - E).$$

When $\Omega(E)$ rises, $\Omega'(E_{tot} - E)$ must fall. Because of the extreme dependence of the number of accessible states Ω on E, or of Ω' on E', the probability $P(E)$ must have a *very sharp* maximum at some value E_{eq}. To find system A at an energy $E \neq E_{eq}$ is totally unlikely. The probability $P(E)$ is almost a δ-function, the value where it is not zero determines equilibrium. The energy E where P, or the entropy $S + S'$, of the whole system has its maximum follows from

$$\frac{\partial \ln P}{\partial E} = 0.$$

This equation immediately gives

$$\left.\frac{\partial S}{\partial E}\right|_{E=E_{eq}} = \left.\frac{\partial S'}{\partial E'}\right|_{E'=E'_{eq}}$$

which means that in equilibrium the temperature of the icebox and lemonade are equal, the lemonade is cool. Note that nothing has been said about how long it takes, starting from some arbitrary microstate, to arrive at equilibrium.

5.3.3   The canonical distribution

Let a system A be in thermal contact with a much larger heat bath A' of temperature T. The energy of A is not fixed, only the joint energy E_{tot} of the combined system A + A'. Suppose now that A is in a *definite* state r of energy E_r. The heat bath has then the energy $E' = E_{tot} - E_r$ and $\Omega'(E_{tot} - E_r)$ accessible states. The probabity P_r for A being in state r is proportional to the number of states of the total system A + A' under that condition, therefore, as A is fixed to state r,

$$P_r = C'\Omega'(E_{tot} - E_r).$$

The constant C' follows from the condition $\sum_s P_s = 1$ in which the sum includes all possible states s of A, irrespective of their energy. Because $E_{tot} \gg E_r$, one can develop $\ln \Omega'$ around E_{tot},

$$\ln \Omega'(E_{tot} - E_r) = \ln \Omega'(E_{tot}) - \beta E_r$$

where $\beta = \partial \ln \Omega'/\partial E' = (kT)^{-1}$ (see (5.44)), and obtains the canonical distribution

$$P_r = \frac{e^{-\beta E_r}}{\sum_s e^{-\beta E_s}}. \tag{5.45}$$

The sum in the denominator represents the partition function $Z(T)$. The probability $P(E)$ to find system A in the energy range δE around E becomes

$$P(E) = \frac{\Omega(E)\,e^{-\beta E_r}}{\sum_s e^{-\beta E_s}} \tag{5.46}$$

where $\Omega(E)$ is the corresponding number of states accessible to A. Formula (5.46) is in agreement with the Boltzmann distribution (5.9): As the size of system A has not entered the derivation of $P(E)$, the expression is also correct when A is a microscopic system. Thermal contact then just means that A can exchange energy with A'. If A consists of just one atom with discrete energy levels E_j, there is only one state for which $E = E_j$, so $\Omega(E) = 1$ and one recovers the Boltzmann distribution (5.9).

5.3.3.1 *Constraints on the system*

Sometimes a macroscopic system A may have to fulfil additional constraints, besides having an energy between E and $E + \delta E$. For example, some parameter Y may have to take up a certain precise value y, or lie in the range $[y, y + \delta y]$; we then symbolically write $Y = y$. The parameter Y may be anything, for example, the energy of ten selected atoms or the integrated magnetic moment of all particles. The states for which $Y = y$ form a subclass of all accessible states, and we denote their total number by $\Omega(E; y)$. In view of the ergodic postulate, the probability P of finding the system in the desired configuration $Y = y$ is

$$P(y) = \frac{\Omega(E; y)}{\Omega(E)}. \tag{5.47}$$

Likewise the probability P can be expressed by the ratio of the densities of states,

$$P(y) = \frac{\rho(E; y)}{\rho(E)}. \tag{5.48}$$

5.3.4 Thermodynamic relations for a gas

When one transfers to a system a small amount of heat δQ and does the work δA on the system, then, according to the first law of thermodynamics, its internal energy U is altered by the amount

$$dU = \delta Q + \delta A. \tag{5.49}$$

If the infinitesimal work is due to a change of volume dV,

$$\delta A = -p\, dV.$$

So when V is decreased, δA is positive. If U depends on temperature and volume, $U = U(T, V)$ and

$$dU = \left(\frac{\partial U}{\partial T}\right)_V dT + \left(\frac{\partial U}{\partial V}\right)_T dV$$

and because of equation (5.49),

$$\delta Q = \left(\frac{\partial U}{\partial T}\right)_V dT + \left[\left(\frac{\partial U}{\partial V}\right)_T + p\right] dV. \tag{5.50}$$

The second law of thermodynamics brings in the entropy S. When the heat δQ is added reversibly to a system at temperature T, its entropy increases by

$$dS = \frac{\delta Q}{T} = \frac{dU - \delta A}{T}.$$ (5.51)

The entropy is a state function and the integral $\int \delta Q/T$ in the (p, V)-plane along any closed path is, therefore, zero. Because dS is a full differential, i.e.

$$\frac{\partial^2 S}{\partial T \partial V} = \frac{\partial^2 S}{\partial V \partial T}$$

we get from (5.51) with $\delta A = -p\, dV$,

$$\left(\frac{\partial U}{\partial V}\right)_T = T \left(\frac{\partial p}{\partial T}\right)_V - p.$$ (5.52)

Other common thermodynamic state functions besides U and S are as follows.

- The free energy or Helmholtz potential

$$F = U - TS.$$ (5.53)

  To connect a physical meaning to F, we note that if, in an isothermal reversible process, one applies the work δA on a system, for example, by compressing a gas, the free energy of the system increases by $\Delta F = \delta A$.
- The enthalpy

$$H = U + pV.$$ (5.54)

  When one presses gas through a small hole from one vessel to another, as in the famous Joule–Thomson experiment, it is the enthalpy that stays constant.
- The free enthalpy or Gibbs potential

$$G = U + pV - TS.$$ (5.55)

As convenient variables one usually uses for G temperature, pressure and number N_j of atoms of the jth gas component, $G = G(T, V, N_j)$. With some elementary calculus manipulations one obtains the full differential

$$dG = \frac{\partial G}{\partial T} dT + \frac{\partial G}{\partial p} dp + \frac{\partial G}{\partial N_j} dN_j = -S\, dT + V\, dp + \sum_j \mu_j\, dN_j$$

where

$$\mu_j = \frac{\partial G}{\partial N_j}$$ (5.56)

defines the chemical potential. In a homogeneous phase, in which for any α

$$S(\alpha U, \alpha V, \alpha N_j) = \alpha S(U, V, N_j)$$

one finds (without proof but it is not difficult)

$$G = \sum_j \mu_j N_j.$$

Therefore, the free enthalpy of a system consisting of a one-component fluid and its vapor (only one kind of molecules, i.e. $j = 1$) is

$$G = G_{\text{fl}} + G_{\text{gas}} = \mu_{\text{fl}} N_{\text{fl}} + \mu_{\text{gas}} N_{\text{gas}}. \tag{5.57}$$

5.3.5 Equilibrium conditions of the state functions

Let us recall the conditions under which a system is in equilibrium. For a *mechanical* system, all forces F_i acting on the particles must vanish. The forces derive from a potential ϕ, so $F_i = -\partial\phi/\partial x_i$. In equilibrium, the potential ϕ is at its minimum and the differential $\Delta\phi = 0$.

For an isolated *thermodynamic system* A of constant energy U and volume V, the entropy S_A attains its maximum. Therefore, the differential ΔS_A is zero under small variations of the variables (U, V, N_j). If A is not isolated but in contact with a heat bath A$'$, the equilibrium condition for the total system A + A$'$ reads:

$$\Delta S_{\text{tot}} = \Delta(S_A + S_{A'}) = 0.$$

System A receives from the environment A$'$ the heat $\delta Q = -T\Delta S_{A'}$ which increases its internal energy by ΔU and does the work δA on the system (see (5.49)),

$$\delta Q = T\Delta S_A = \Delta U - \delta A.$$

As all variables here refer to system A, we may drop the index A in the entropy and get

$$\Delta U - T\Delta S - \delta A = 0.$$

In the case of pure compressional work, $\delta A = -p\Delta V$ and

$$\Delta U - T\Delta S + p\Delta V = 0. \tag{5.58}$$

Equation (5.58) leads in a straightforward way to the following equilibrium conditions:

(1) In a system A of constant temperature and volume ($\Delta T = \Delta V = 0$) and in contact with a heat reservoir A$'$, the differential of the free energy vanishes under small changes of the variables,

$$\Delta F = \Delta(U - TS) = 0.$$

F has then its minimum because S_{tot} of the total system A + A$'$ is at its maximum. Note that all variables without an index refer to system A.

It is reassuring that the condition $\Delta F = 0$ is also in line with the Boltzmann equation (5.9): Suppose the N particles of the system possess only two levels, 1 and 2. In an equilibrium state a, the lower level 1 is populated by N_1 and the upper level 2 by N_2 atoms with $N_1 + N_2 = N$. In a nearby, almost-equilibrium state b, the corresponding populations are $N_1 + 1$ and $N_2 - 1$. The thermodynamic probability Ω for the two states states is given by (5.1) and the entropy changes between state a to b by

$$\Delta S = k[\ln \Omega_b - \ln \Omega_a] = k \ln \frac{\Omega_b}{\Omega_a} = k \ln \frac{N_1}{N_2 + 1} \simeq k \ln \frac{N_1}{N_2}.$$

At constant temperature and volume, the condition $\Delta F = \Delta U - T\Delta S = 0$, where ΔU is the excitation energy of level 2, implies $N_2 / N_1 = e^{-\Delta U / kT}$, as in (5.9).

Let system A depend on some parameter Y. The number of accessible states where Y takes up values between y and $y + \delta y$ is (see (5.43))

$$\Omega_{tot}(y) = e^{S_{tot}(y)/k}$$

and the probability $P(y)$ of such states is proportional to $\Omega_{tot}(y)$. Likewise, $P(y') \propto \Omega_{tot}(y') = e^{S_{tot}(y')/k}$ for another value y' of the parameter Y. As Y varies from y to y' (this need not be a small step), the entropy of the total system changes by

$$\Delta S_{tot} = \frac{T\Delta S - \Delta U + \delta A}{T}$$

where δA is the (not necessarily infinitesimally small) work done on system A and $\Delta S = S(y) - S(y')$ and $\Delta U = U(y) - U(y')$. As the volume is constant, there is no compressional work and $\delta A = 0$. With $\Delta F = F(y) - F(y')$, we get, for the population ratio of states where $Y = y$ and $Y = y'$,

$$\frac{P(y)}{P(y')} = e^{-\Delta F / kT}.$$

This result follows also from the canonical distribution (5.45).

(2) At constant temperature and pressure ($\Delta T = \Delta p = 0$), the differential of the free enthalpy

$$\Delta G = \Delta(U + pV - TS) = 0$$

and G has its minimum. In an analagous manner as for the free energy, one derives $\Delta S_{tot} = -\Delta G / T$ and, putting $\Delta G = G(y) - G(y')$, a population ratio of states where the parameter Y has values y and y'

$$\frac{P(y)}{P(y')} = e^{-\Delta G / kT}. \tag{5.59}$$

(3) At constant pressure ($\Delta p = 0$) and constant entropy ($\Delta S = 0$, adiabatic process), $\Delta H = \Delta(U + pV) = 0$ and the enthalpy has its minimum.

5.3.6   Specific heat of a gas

The specific heat at constant pressure, C_P, and at constant volume, C_V, are defined by

$$C_p = \left(\frac{\delta Q}{dT}\right)_P \qquad C_v = \left(\frac{\partial U}{\partial T}\right)_V. \tag{5.60}$$

Let us determine $C_p - C_v$. Because at constant pressure $(dp = 0)$

$$dp = \left(\frac{\partial p}{\partial T}\right)_V dT + \left(\frac{\partial p}{\partial V}\right)_T dV = \left(\frac{\partial p}{\partial T}\right)_V dT + \left(\frac{\partial p}{\partial V}\right)_T \left(\frac{\partial V}{\partial T}\right)_P dT$$

we obtain, using (5.50), (5.52) and (5.60),

$$C_p - C_v = T\left(\frac{\partial p}{\partial T}\right)_V \left(\frac{\partial V}{\partial T}\right)_P = -T\left(\frac{\partial p}{\partial T}\right)_V^2 \left(\frac{\partial p}{\partial V}\right)_T^{-1}. \tag{5.61}$$

It follows immediately that

$$\delta Q = C_v \left(\frac{\partial T}{\partial p}\right)_V dp + C_p \left(\frac{\partial T}{\partial V}\right)_P dV. \tag{5.62}$$

5.3.7   The work done by magnetization

Consider a magnetic body of unit volume in a field H and with magnetization M. Its internal energy depends on the temperature and on the magnetic field, $U = U(T, H)$. If a field H is needed to produce in the body the magnetization M, to change the magnetization by dM requires the infinitesimal work

$$\delta A = H\, dM. \tag{5.63}$$

This follows from the following thought experiment:

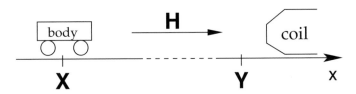

We move an unmagnetized body of unit volume along the x-axis from a position X, where there is no magnetic field, to another place Y where there is some field H_0, say, near a coil through which runs an electric current. The body is attracted by the coil and in the field H_0, it acquires the magnetization M_0. Then we pull the body back to X but while doing so, we fix the magnetization M_0 (that is why it is a thought experiment). The force acting on the body is, at each position, $M(dH/dx)$ as follows from (11.16) for the potential energy of a dipole. The total mechanical work A done on the system over the whole path from X to Y and back becomes

$$A = - \int_0^{H_0} M \, dH + \int_0^{H_0} M_0 \, dH = \int_0^{M_0} H \, dM.$$

The last equality is obvious when one identifies the integrals with the area under the curves $M(H)$ and $H(M)$. So indeed $\delta A = H \, dM$ and, therefore,

$$\delta Q = T \, dS = dU - H \, dM.$$

5.3.8   Susceptibility and specific heat of magnetic substances

The previously derived formulae for C_P and C_V now come in handy. To obtain the relations for the specific heat at constant magnetic field H and at constant magnetization M, defined by

$$C_H = \left(\frac{\partial U}{\partial T} \right)_H \qquad C_M = \left(\frac{\delta Q}{dT} \right)_M \qquad (5.64)$$

we just have to interchange

$$M \leftrightarrow p \qquad \text{and} \qquad H \leftrightarrow V.$$

In complete analogy to (5.61) and (5.62), we write

$$C_M - C_H = T \left(\frac{\partial M}{\partial T} \right)_H \left(\frac{\partial H}{\partial T} \right)_M = -T \left(\frac{\partial M}{\partial T} \right)_H^2 \left(\frac{\partial M}{\partial H} \right)_T^{-1} \qquad (5.65)$$

and

$$\delta Q = C_H \left(\frac{\partial T}{\partial M} \right)_H dM + C_M \left(\frac{\partial T}{\partial H} \right)_M dH. \qquad (5.66)$$

The adiabatic and isothermal (or static) susceptibility are defined as

$$\chi_{ad} = \left(\frac{\partial M}{\partial H} \right)_S \qquad \text{and} \qquad \chi_T = \left(\frac{\partial M}{\partial H} \right)_T. \qquad (5.67)$$

As long as magnetic saturation is excluded, one also has $\chi_T = M/H$. We compute χ_{ad} from (5.66) by putting $\delta Q = 0$ ($S = $ constant),

$$\chi_{ad} = -\frac{C_M}{C_H} \left(\frac{\partial T}{\partial H} \right)_M \left(\frac{\partial M}{\partial T} \right)_H = \frac{C_M}{C_H} \left(\frac{\partial M}{\partial H} \right)_T \qquad (5.68)$$

where we have exploited, after the second equals sign, formula (5.65). Therefore

$$\chi_{ad} = \frac{C_M}{C_H} \chi_T. \qquad (5.69)$$

5.4 Blackbody radiation

5.4.1 The Planck function

The Planck function gives the intensity of radiation in an enclosure in thermal equilibrium at temperature T. It is a universal function depending only on T and frequency ν. Written in the frequency scale,

$$B_\nu(T) = \frac{2h}{c^2} \frac{\nu^3}{e^{\frac{h\nu}{kT}} - 1}. \tag{5.70}$$

The intensity is taken per Hz and has the units erg cm^{-2} s^{-1} ster^{-1} Hz^{-1}. $B_\nu(T)$ is related to the monochromatic radiative energy density u_ν in the enclosure by

$$u_\nu = \frac{4\pi}{c} B_\nu(T) = \frac{8\pi h}{c^3} \cdot \frac{\nu^3}{e^{\frac{h\nu}{kT}} - 1}. \tag{5.71}$$

Alternatively, the Planck function may be referred to wavelength and designated $B_\lambda(T)$; its unit is then erg cm^{-2} s^{-1} ster^{-1} cm^{-1}. The two forms are related through

$$B_\lambda(T)\,d\lambda = -B_\nu(T)\,d\nu. \tag{5.72}$$

Because of $d\nu = -(c/\lambda^2)\,d\lambda$,

$$B_\lambda(T) = \frac{2hc^2}{\lambda^5(e^{\frac{hc}{kT\lambda}} - 1)}. \tag{5.73}$$

In view of the exponential factor, $B_\nu(T)$ is usually very sensitive to both temperature and frequency. The Planck function increases monotonically with temperature, i.e. at any frequency

$$B_\nu(T_2) > B_\nu(T_1) \qquad \text{if } T_2 > T_1.$$

Figure 5.2 depicts the plot of the curve

$$y = \frac{x^3}{e^x - 1}$$

from which one can read off the value of the Planck function for any combination (ν, T) of frequency and temperature. It starts at the origin $(0, 0)$ with a slope of zero, culminates at $x_{max} = 2.822$ and asymptotically approaches zero for large x.

An object that emits at all frequencies with an intensity $B_\nu(T)$ is called a blackbody. The emergent flux F_ν from a unit area of its surface into all directions of the half-sphere is

$$F_\nu = \pi B_\nu(T). \tag{5.74}$$

This expression, for example, is approximately applicable to stellar atmospheres when T is the effective surface temperature.

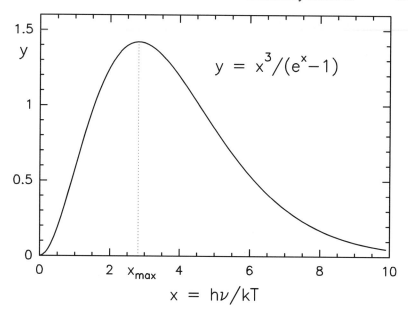

Figure 5.2. The universal shape of the Planck function. As the abscissa x is here in the unit $h\nu/kT$, one has to multiply the ordinate y by the factor $(2k^3/h^2c^2)T^3$ to get $B_\nu(T)$.

5.4.2 Low- and high-frequency limit

There are two asymptotic approximations to the Planck function depending on the ratio x of photon energy $h\nu$ over thermal energy kT,

$$x = \frac{h\nu}{kT}.$$

- In the *Wien limit*, $x \gg 1$ and

$$B_\nu(T) \to \frac{2h\nu^3}{c^2} e^{-\frac{h\nu}{kT}}. \tag{5.75}$$

With respect to dust emission, where wavelengths are typically between 1 μm and 1 mm and temperatures from 10 to 2000 K, Wien's limit is never appropriate.

- In the *Rayleigh–Jeans limit*, $x \ll 1$ and

$$B_\nu(T) \to \frac{2\nu^2}{c^2} kT. \tag{5.76}$$

As the photon energy $h\nu$ is much smaller than the thermal energy kT of an oscillator, one can expand the Planck function into powers of x and enter the realm of classical physics where the Planck constant h

vanishes. The dependence of $B_\nu(T)$ on frequency and temperature is then no longer exponential. The Rayleigh–Jeans approximation is always good for centimeter radio astronomy, and sometimes also applicable to dust emission. However, one must check whether

$$x = \frac{1.44}{\lambda T} \qquad (\lambda \text{ in cm})$$

is really small compared to 1.

5.4.3 Wien's displacement law and the Stefan–Boltzmann law

In the wavelength scale, the Planck function $B_\lambda(T)$ reaches its maximum at λ_{max} given by $\partial B_\lambda / \partial \lambda = 0$. Therefore

$$\lambda_{max} T = 0.289 \text{ cm K}. \tag{5.77}$$

In the frequency scale, maximum emission is determined by $\partial B_\nu / \partial \nu = 0$ and occurs at ν_{max} for which

$$\frac{T}{\nu_{max}} = 1.70 \times 10^{-11} \quad \text{Hz}^{-1} \text{ K}. \tag{5.78}$$

Note that λ_{max} from (5.77), which refers to the wavelength scale, is a factor 1.76 smaller than the corresponding wavelength c/ν_{max} from (5.78). The wavelength where the flux from a blackbody peaks depends thus on whether one measures the flux per Hz (F_ν) or per cm (F_λ). The total energy per s over a certain spectral interval is, of course, the same for F_λ and F_ν. If one wants to detect a blackbody with an instrument that has a sensitivity curve S_ν, one usually tries to maximize $\int S_\nu B_\nu(T) \, d\nu$.

When λ_{max} is known, the displacement law determines the temperature of a blackbody. Interstellar grains are certainly not blackbodies but the shape of the spectral energy distribution from a dusty region may at far infrared wavelengths be approximated by $\nu^m B_\nu(T)$ (see (8.1) for the correct expression) if the emission is optically thin and the absorption coefficient has a power-law dependence $K_\nu \propto \nu^m$. The term

$$\nu^m B_\nu(T)$$

is sometimes called the modified Planck function. Maximum emission follows now from $\partial(\nu^m B_\nu)/\partial \nu = 0$. For $K_\nu \propto \nu^2$, one finds $\lambda_{max} T = 0.206$ cm K, so the λ_{max} of radiating dust is shifted to shorter wavelengths with respect to a blackbody emitter of the same temperature.

Integrating the Planck function over frequency, we obtain (see (A.17))

$$B(T) = \int_0^\infty B_\nu(T) \, d\nu = \frac{\sigma}{\pi} T^4 \tag{5.79}$$

where

$$\sigma = \frac{2\pi^5 k^4}{15c^2 h^3} = 5.67 \times 10^{-5} \, \text{erg cm}^{-2} \, \text{s}^{-1} \, \text{K}^{-4} \qquad (5.80)$$

is the *radiation constant*. The total emergent flux F from a unit area of a blackbody surface into all directions of the half-sphere is given by the *Stefan–Boltzmann law* (see (5.74)):

$$F = \int F_v \, dv = \sigma T^4. \qquad (5.81)$$

Applying this to a star of radius R_* and effective temperature T_*, we find for its bolometric luminosity:

$$L_* = 4\pi \sigma R_*^2 T_*^4. \qquad (5.82)$$

For the total radiative energy density u, we get

$$u = \int u_v \, dv = aT^4 \qquad (5.83)$$

with constant

$$a = \frac{4\sigma}{c} = 7.56 \times 10^{-15} \, \text{erg cm}^{-3} \, \text{K}^{-4}. \qquad (5.84)$$

5.4.4 The Planck function and harmonic oscillators

Because the Planck function is fundamental for the emission processes of dust and molecules, we wish to understand it well. We, therefore, derive it once more by representing the atoms as harmonic oscillators. Consider a harmonic oscillator in an enclosure in thermal equilibrium with its surroundings. According to (1.71), it loses energy through emission at a rate

$$-\dot{E} = \gamma E \qquad (5.85)$$

where γ is the damping constant. In equilibrium, it must pick up the same energy from the radiation in the enclosure. If we denote the radiative intensity by $I(\omega)$,

$$\gamma E = \int_0^\infty 4\pi I(\omega)\sigma(\omega) \, d\omega = 4\pi I(\omega_0) \int_0^\infty \sigma(\omega) \, d\omega = 4\pi I(\omega_0) \frac{2\pi^2 e^2}{m_e c}. \qquad (5.86)$$

The cross section $\sigma(\omega)$ is taken from (1.106). Because the resonance curve is very narrow, we can take $4\pi\omega_0^2 I(\omega_0)$ out from under the integral. The integrated absorption cross section is from (1.108). When we substitute for γ the radiative damping constant from (1.100), remembering that $I(\omega) = I(v)/2\pi$, and drop the subscript 0 in the frequency, we obtain

$$I(v) = \frac{2v^2}{3c^2} E. \qquad (5.87)$$

In (5.86), the integration extends over the solid angle 4π implying that the oscillators absorb radiation from all directions. The oscillators must therefore be able to move freely and have three degrees of freedom, not just one. In the classical picture, $\langle E \rangle = kT$ is the average total energy per degree of freedom, i.e. potential plus kinetic energy. Inserting

$$E = 3\langle E \rangle = 3kT$$

into (5.87), one recovers the Rayleigh–Jeans formula of (5.76) which is valid for blackbody radiation at low frequencies,

$$I_\nu(T) = \frac{2\nu^2}{c^2} kT.$$

This formula, however, fails at photon energies $h\nu$ comparable to or larger than kT because the intensity would rise infinitely with frequency. Something is wrong with the classical ideas. The cross section $\sigma(\omega)$ in (5.86) still holds in quantum mechanics but we have neglected that light comes in packages (photons). We obtain the correct Planck function if we assume equally spaced energy states $E_i = ih\nu$ with $i = 0, 1, 2, \ldots$, for which the average $\langle E \rangle$ per degree of freedom is not kT but, from (5.11),

$$\langle E \rangle = kT \frac{x}{e^x - 1} \qquad x = \frac{h\nu}{kT}.$$

The quantum mechanical mean energy $\langle E \rangle$ is smaller than kT. Only when $x \ll 1$, does one obtain the classical result.

It is also enlightening to apply the first principles of thermodynamics to a photon gas in equilibrium. A gas of volume V and internal energy $U(T) = u(T)V$ has the gas pressure $P = u/3$. Because $dU = u\,dV + V\,du$, the differential $dS = dQ/T$ of the entropy is given by

$$dS = \frac{dU + P\,dV}{T} = \frac{u\,dV + V\,du + (u/3)\,dV}{T}.$$

As dS is a full differential, we get the equation

$$\frac{\partial^2 S}{\partial T \partial V} = \frac{1}{T}\frac{du}{dT} = \frac{4}{3}\frac{d}{dT}\left(\frac{u}{T}\right) = \frac{\partial^2 S}{\partial V \partial T}.$$

Solving for the energy density u, we obtain a function that is proportional to the fourth power of the temperature, $u(T) \propto T^4$, in agreement with the Stefan–Boltzmann law in (5.83).

Chapter 6

The radiative transition probability

Almost all astronomical information eventually comes from light which we detect with our telescopes. Photon emission is, therefore, a fundamental process and a major goal of this chapter is to explain the concept of induced and spontaneous emission. The final section deals with transmission and reflection of a free particle at a potential barrier. It helps to understand quantitatively very diverse topics: how atoms can wander about the grain surface by way of tunneling through a potential barrier; how protons can fuse to deuteron in the interior of the Sun; how electrons are energetically arranged in white dwarfs; or how electronic bands appear in crystallized grains.

6.1 A charged particle in an electromagnetic field

6.1.1 The classical Hamiltonian

Consider a system of particles with generalized coordinates q_i and velocities $\dot{q}_i$. If velocities are small (non-relativistic), one finds the motion of the particles from the *Lagrange function*

$$L = T - V$$

where T is the kinetic energy of the particles and V the potential, by integrating the second-order Lagrange equations

$$\frac{d}{dt}\left(\frac{\partial L}{\partial \dot{q}_i}\right) = \frac{\partial L}{\partial q_i}. \tag{6.1}$$

Alternatively, using the conjugate momenta

$$p_i = \frac{\partial L}{\partial \dot{q}_i} \tag{6.2}$$

one constructs the *Hamiltonian*,

$$H = \sum \dot{q}_i p_i - L \tag{6.3}$$

175

and integrates the first-order equations

$$\dot{q}_i = \frac{\partial H}{\partial p_i} \qquad - \dot{p}_i = \frac{\partial H}{\partial q_i}. \tag{6.4}$$

If H does not explicitly depend on time, it is a constant of motion. Moreover, in our applications, H equals the total energy E of the system,

$$E = T + V = H(q_i, p_i). \tag{6.5}$$

6.1.2 The Hamiltonian of an electron in an electromagnetic field

We are interested in the equation of motion of *one* particle of charge e and mass m moving in an electromagnetic field. We now use Cartesian coordinates (x, y, z). The kinetic and potential energy equal

$$T = \tfrac{1}{2}m\mathbf{v}^2$$

$$V = e\phi - \frac{e}{c}\mathbf{A}\cdot\mathbf{v} \tag{6.6}$$

which leads to the *non-relativistic Lagrangian*

$$L = \frac{1}{2}m\mathbf{v}^2 - e\phi + \frac{e}{c}\mathbf{A}\cdot\mathbf{v}. \tag{6.7}$$

As usual, ϕ denotes the scalar potential and $\mathbf{A}$ the vector potential. The latter is connected to the magnetic field through

$$\mathbf{B} = \text{rot}\,\mathbf{A}$$

and the electric field is given by

$$\mathbf{E} = -\,\text{grad}\,\phi - \frac{1}{c}\dot{\mathbf{A}}. \tag{6.8}$$

Remembering that the full-time derivative of the x-component of the vector potential is

$$\frac{dA_x}{dt} = \dot{A}_x + v_x\frac{\partial A_x}{\partial x} + v_y\frac{\partial A_x}{\partial y} + v_z\frac{\partial A_x}{\partial z}$$

with similar expressions for A_y and A_z, we may readily convince ourselves that equation (6.1) with the Lagrangian of (6.7) yields the Lorentz force

$$\mathbf{F} = e\left(\mathbf{E} + \frac{1}{c}\mathbf{v}\times\mathbf{B}\right).$$

Often the potential V is only a function of coordinates and the force $\mathbf{F} = m\dot{\mathbf{v}}$ on a particle is given by the gradient of the potential, $\mathbf{F} = -\,\text{grad}\,V$. However, in the case of an electromagnetic field, V of (6.6) is a generalized potential that also depends on velocity and the force follows from

$$F_i = -\frac{\partial V}{\partial x_i} + \frac{d}{dt}\left(\frac{\partial V}{\partial \dot{x}_i}\right)$$

where we put $(x_1, x_2, x_3) = (x, y, z)$. Using the conjugate momenta

$$p_i = \frac{\partial L}{\partial \dot{x}_i} = mv_i + \frac{e}{c}A_i \tag{6.9}$$

the Hamiltonian becomes

$$H = \frac{1}{2m}\left(\mathbf{p} - \frac{e}{c}\mathbf{A}\right)^2 + e\phi.$$

It equals the total energy E of the system,

$$E = H = T + e\phi. \tag{6.10}$$

If we think of the motion of an electron in an atom, we may regard ϕ as representing the static electric field of the nucleus and, therefore, separate, from the Hamiltonian, the part

$$H_{\text{stat}} = \frac{\mathbf{p}^2}{2m} + e\phi$$

which describes the electron without an electromagnetic wave. (In fact, for slowly moving charges one can always make a transformation of the given potentials to new ones such that div $\mathbf{A} = 0$ for the new $\mathbf{A}$. The new ϕ is then constant in time and refers only to the static field. This is because of the gauge condition (1.85), div $A + \dot{\phi}/c = 0$.) The remaining part of the Hamiltonian,

$$H_{\text{wave}} = \frac{e}{2mc}(\mathbf{p} \cdot \mathbf{A} + \mathbf{A} \cdot \mathbf{p}) + \frac{e^2}{2mc^2}\mathbf{A} \cdot \mathbf{A} \tag{6.11}$$

describes the perturbation of the electron by an electromagnetic wave.

6.1.3 The Hamilton operator in quantum mechanics

6.1.3.1 The Schrödinger equation

In quantum mechanics, the formulae (6.4) governing the motion of the particles are replaced by the Schrödinger equation

$$i\hbar\frac{\partial \Psi}{\partial t} = \hat{H}\Psi. \tag{6.12}$$

This equation follows from $E = H(x_i, p_i)$ of (6.5) by turning the energy E and the Hamiltonian H into operators using the standard prescription for the conversion of energy, conjugate momentum and coordinates,

$$E \to i\hbar\frac{\partial}{\partial t} \qquad p_i \to \hat{p}_i = -i\hbar\frac{\partial}{\partial x_i} \qquad x_i \to \hat{x}_i = x_i. \tag{6.13}$$

Operators are marked by a hat on top of the letter. One has to keep in mind the commutation rules, in particular

$$[\hat{x}_i, \hat{p}_j] = \hat{x}_i \hat{p}_j - \hat{p}_j \hat{x}_i = i\hbar\delta_{ij} \tag{6.14}$$

from which one readily deduces

$$[\hat{x}_i, \hat{H}] = \hat{x}_i \hat{H} - \hat{H}\hat{x}_i = i\frac{\hbar}{m}\hat{p}_i. \tag{6.15}$$

For the Hamiltonian of an electron in an atom perturbed by an electromagnetic field we write

$$\hat{H} = \frac{e}{2m_e c}(\hat{\mathbf{p}} \cdot \hat{\mathbf{A}} + \hat{\mathbf{A}} \cdot \hat{\mathbf{p}}) + \frac{e^2}{2m_e c^2}\hat{\mathbf{A}} \cdot \hat{\mathbf{A}}. \tag{6.16}$$

This follows from (6.11), omitting the subscript 'wave'.

6.1.3.2 *Stationary solutions of the Schrödinger equation*

The wavefunction Ψ in (6.12) depends generally on space and time,

$$\Psi = \Psi(x, t).$$

Stationary solutions correspond to fixed energy eigenvalues E. In this case, the wavefunction can be written as

$$\Psi(x, t) = \psi(x)\, e^{-iEt/\hbar}. \tag{6.17}$$

$\Psi(x, t)$ contains the time merely through the factor $e^{-iEt/\hbar}$ which cancels out on forming the probability density $|\Psi|^2$. The eigenfunction $\psi(x)$ depends only on coordinate x. The time-independent Schrödinger equation of a particle of energy E in a potential $V(x)$ follows from (6.12),

$$\frac{\hbar^2}{2m}\Delta\psi + \left[E - U(x)\right]\psi = 0. \tag{6.18}$$

The eigenfunctions ψ_n, each to an eigenvalue E_n, form a complete set so that an arbitrary wavefunction $\Psi(x, t)$ can be expanded into a sum of ψ_n:

$$\Psi(x, t) = \sum_n a_n\psi_n(x) \cdot e^{-iE_n t/\hbar} \tag{6.19}$$

with expansion coefficients a_n.

6.1.4 The dipole moment in quantum mechanics

The Hamiltonian operator $\hat{H}$ is Hermitian, which means

$$\int \psi^*(\hat{H}\phi)\,dx = \int (\hat{H}\psi)^*\phi\,dx$$

for any ψ and ϕ and guarantees real expectation values. The asterisk denotes the complex conjugate. It follows, with the help of equation (6.15), that

$$\frac{e}{m}\int \psi_j^* \hat{p}\psi_k\,dx = i\frac{e}{\hbar}(E_j - E_k)\int \psi_j^* \hat{x}\psi_k\,dx$$

so one can substitute in the left-hand integral for the momentum operator the coordinate operator (but for a constant). If we put $\hbar\omega_{jk} = E_j - E_k$ and call

$$\mu_{jk} = e\int \psi_j^*(x)\,x\,\psi_k(x)\,dV \tag{6.20}$$

the dipole moment with respect to the states described by the eigenfunctions ψ_j and ψ_k, we get

$$\frac{e}{m}\int \psi_j^* \hat{p}\,\psi_k\,dx = i\omega_{jk}\mu_{jk}. \tag{6.21}$$

6.1.5 The quantized harmonic oscillator

We review the solution of the time-independent Schrödinger equation for the undamped free linear harmonic oscillator. The oscillator obeys the equation of motion

$$\ddot{x} + \omega^2 x = 0$$

where $\omega^2 = \kappa/m$ is the square of the frequency of oscillation, m the particle mass, and $-\kappa x$ the restoring force (see section 1.3). The total energy of the system

$$E = T + V = \frac{p^2}{2m} + \frac{1}{2}m\omega^2 x^2$$

and the time-independent Schrödinger equation is, therefore, (see (6.18))

$$\frac{\hbar^2}{2m}\frac{\partial^2 \psi_n}{\partial x^2} + \left[E_n - \frac{1}{2}m\omega^2 x^2\right]\psi_n = 0. \tag{6.22}$$

This second-order differential equation looks much better in the form

$$u_n'' + (\lambda_n - y^2)u_n = 0$$

where we have introduced the function $u_n(\alpha x) = \psi_n(x)$ with

$$\alpha^2 = \frac{m\omega}{\hbar} \tag{6.23}$$

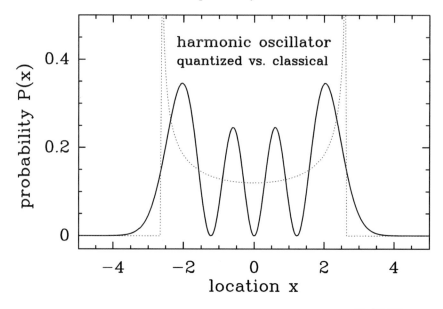

Figure 6.1. The undulating quantum mechanical probability $P(x) = |\psi(x)|^2$ for finding the particle at locus x when the oscillator is in its third energy level above ground ($n = 3$). For comparison, we show the corresponding probability of a classical oscillator (dots) of the same energy; here the particle is strictly confined to the allowed region. For high n, the dotted and full curve converge.

and put $y = \alpha x$ and $\lambda_n = 2E_n/\hbar\omega$. An ansatz for $u_n(y)$ of the form $e^{-y^2/2} H_n(y)$ yields the normalized eigenfunctions of the harmonic oscillator:

$$\psi_n(x) = N_n e^{-\alpha^2 x^2/2} H_n(\alpha x) \qquad \text{with } N_n = \sqrt{\frac{\alpha}{2^n n! \sqrt{\pi}}}. \qquad (6.24)$$

H_n are the Hermitian polynomials described in appendix A. The energy levels E_n of the harmonic oscillator in (6.22) are equidistant:

$$E_n = (n + \tfrac{1}{2})\hbar\omega \qquad n = 0, 1, 2, \ldots \qquad (6.25)$$

and the lowest level

$$E_0 = \tfrac{1}{2}\hbar\omega$$

is *above* zero. Figure 6.1 displays, as an example, the square of the eigenfunction $|\psi_n(x)|^2$ for $n = 3$ and $\alpha^2 = 1$.

With the help of (A.3) we find that the mean position and momentum of an oscillator, $\langle x \rangle$ and $\langle p \rangle$, always disappear for any energy E_n. So

$$\frac{d\langle x \rangle}{dt} = \frac{\langle p \rangle}{m}$$

is fulfilled in a trivial way but $\langle x^2 \rangle$ and $\langle p^2 \rangle$ do not vanish. For the product of the uncertainties we have

$$\Delta x \Delta p = \sqrt{\langle x^2 \rangle - \langle x \rangle^2} \sqrt{\langle p^2 \rangle - \langle p \rangle^2} = \frac{E_n}{\omega} = (n + \tfrac{1}{2})\hbar. \tag{6.26}$$

6.2 Small perturbations

6.2.1 The perturbation energy

Let us introduce, to a system with Hamiltonian $\hat{H}$ and energy eigenvalues E_n, a *small* perturbation, $\hat{H}'$, so that the Schrödinger equation (6.12) reads:

$$\left(\hat{H} + \hat{H}' - i\hbar \frac{\partial}{\partial t} \right) \psi(x, t) = 0. \tag{6.27}$$

We expand ψ after (6.19) into eigenfunctions ψ_n but now with *time-dependent* coefficients $a_n(t)$ because of the additional term $\hat{H}'$ which causes the system to change:

$$\psi(x, t) = \sum_n a_n(t) \, \psi_n(x) \, e^{-iE_n t/\hbar}. \tag{6.28}$$

Inserting this sum into (6.27) gives

$$\sum_n e^{-iE_n t/\hbar} \left(a_n \hat{H}' \psi_n - i\hbar \dot{a}_n \psi_n \right) = 0 \tag{6.29}$$

the dot over a_n means time derivative. As the eigenfunctions are orthogonal, multiplication of (6.29) with the eigenfunction ψ_f^* and integration over the space coordinate x yields

$$\sum_n e^{-iE_n t/\hbar} a_n(t) H'_{fn} - i\hbar \dot{a}_f(t) e^{-iE_f t/\hbar} = 0 \tag{6.30}$$

with the matrix coefficients

$$H'_{fn} = \int \psi_f^* \hat{H}' \psi_n \, dx. \tag{6.31}$$

For equal subindices ($f = n$), the matrix element H'_{nn} gives the perturbation energy. This means when the Hamiltonian $\hat{H}$ is replaced by $\hat{H} + \hat{H}'$, the energy eigenvalues change from E_n to $E_n + H'_{nn}$.

6.2.2 The transition probability

We consider a transition $s \rightarrow f$, from a *starting* state s to a *final* state f. Suppose that our system was at time $t = 0$ in state s,

$$\psi(x, t = 0) = \psi_s(x).$$

Then all $a_n(0)$ in (6.28) vanish, except for $a_s(0) = 1$. If the interaction operator $\hat{H}'$ is weak and the perturbation time t not too long, we can substitute in (6.30) the $a_n(t)$ by their values at $t = 0$, so

$$a_n(0) = \delta_{sn}$$

where δ_{sn} is the Kronecker symbol. If

$$\hbar\omega_{fs} = E_f - E_s$$

is the energy difference between the two states, equation (6.30) yields for the time derivative of the expansion coefficient a_f,

$$\dot{a}_f = -\frac{i}{\hbar}H'_{fn}e^{i\omega_{fs}t}. \tag{6.32}$$

After integrating from the starting value $a_f(0) = 0$ and assuming that H'_{fs} is *not* time-variable,

$$a_f(t) = \frac{i}{\hbar}H'_{fs}\frac{1 - e^{i\omega_{fs}t}}{i\omega_{fs}}. \tag{6.33}$$

The probability of finding the system at a later time t in the final state f is

$$P_{fs}(t) = |a_f(t)|^2 = \frac{2}{\hbar^2}|H'_{fs}|^2\frac{1 - \cos\omega_{fs}t}{\omega_{fs}^2}. \tag{6.34}$$

6.2.3   Transition probability for a time-variable perturbation

When the perturbation varies harmonically with time, we have to write $\hat{H}'e^{-i\omega t}$ in equation (6.29) for the perturbation operator, instead of $\hat{H}'$. The time derivative of the expansion coefficient, $\dot{a}_f$, given in (6.32), then becomes

$$\dot{a}_f = -\frac{i}{\hbar}H'_{fn}e^{i(\omega_{fs}-\omega)t}$$

and

$$P_{fs}(t) = |a_f(t)|^2 = \frac{2}{\hbar^2}|H'_{fs}|^2\frac{1 - \cos(\omega_{fs} - \omega)t}{(\omega_{fs} - \omega)^2}. \tag{6.35}$$

One just has to replace ω_{fs} by $\omega_{fs} - \omega$ and the probability $P_{fs}(t)$ now has its spike when $\omega \simeq \omega_{fs}$.

If the time t in (6.35) is large compared to the oscillation period, $t \gg \omega^{-1}$, the function

$$f(\omega) = \frac{1 - \cos(\omega_{fs} - \omega)t}{(\omega_{fs} - \omega)^2} \tag{6.36}$$

resembles the δ-function: it has values close to zero almost everywhere, except around $\omega = \omega_{fs}$ in the small interval $\Delta\omega$ from $-\pi/2t$ to $\pi/2t$. Integrated over frequency,

$$\int_{-\infty}^{\infty}\frac{1 - \cos(\omega_{fs} - \omega)t}{(\omega_{fs} - \omega)^2}\,d\omega = \pi t. \tag{6.37}$$

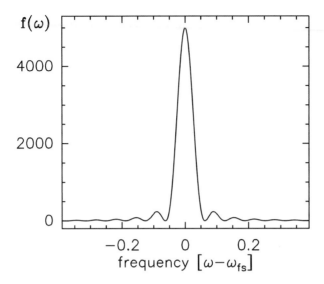

Figure 6.2. The transition probability from a starting state s to a final state f under the influence of a perturbation potential is proportional to a characteristic function given in (6.36) and shown here for $t = 100$.

The function $f(\omega)$ is plotted in figure 6.2 for $t = 100$. The probability $p_{fs}(t)$ in (6.35) is not negligible and transitions from state s to state f occur only when $|\omega - \omega_{fs}|t \leq \pi$. If one waits long enough, one has almost exactly $\omega = \omega_{fs}$.

6.3 The Einstein coefficients A and B

6.3.1 Induced and spontaneous transitions

6.3.1.1 *How A and B are defined*

We consider radiative transitions in atoms between an upper level j and a lower one i of energies E_j and E_i such that

$$E_j - E_i = h\nu.$$

Let the atoms have number densities N_j and N_i per cm^3, respectively, and be bathed in a radiation field of energy density u_ν. As atomic energy levels and lines are not infinitely sharp but have some finite width, let

$$N_i(\nu)\, d\nu = N_i \Phi(\nu)\, d\nu$$

be the number of atoms in state i that can absorb radiation in the frequency interval $[\nu, \nu + d\nu]$, where, of course,

$$\int_{\text{line}} \Phi(\nu)\, d\nu = 1.$$

Radiative transitions between level j and i may be either spontaneous or induced. Setting

$$u = \int_{\text{line}} u_\nu \Phi(\nu)\, d\nu \tag{6.38}$$

the Einstein coefficients A_{ji} and B_{ji} are defined such that per cm^3

- $N_j A_{ji} =$ rate of *spontaneous downward* transitions,
- $N_j u B_{ji} =$ rate of *induced downward* transitions and
- $N_i u B_{ij} =$ rate of *induced upward* transitions.

6.3.1.2 How A and B are related

A_{ji}, B_{ji}, B_{ij} are atomic quantities and do not depend on the radiation field. If we assume blackbody radiation ($u_\nu = 4\pi B_\nu(T)/c$) and thermodynamic equilibrium, we can derive the relations between the As and Bs. The Boltzmann formula (5.9),

$$\frac{N_i}{N_j} = \frac{g_i}{g_j} e^{-x} \qquad \text{with } x = \frac{E_j - E_i}{kT}$$

yields

$$A_{ji} + u_\nu B_{ji} = \frac{g_i}{g_j} B_{ij} u_\nu e^x.$$

If we fix ν and let T go to infinity, we are in the Rayleigh–Jeans part of the spectrum and u_ν becomes arbitrarily large so that $u_\nu B_{ji} \gg A_{ji}$. As then also $e^x \to 1$, we get the symmetry relation between the Einstein-B coefficients,

$$g_j B_{ji} = g_i B_{ij}. \tag{6.39}$$

Consequently,

$$u_\nu = \frac{A_{ji}}{B_{ji}(e^x - 1)}.$$

If we now let the temperature tend to zero, we are in the Wien part of the spectrum and

$$\frac{8\pi h \nu^3}{c^3} e^{-h\nu/kT} = \frac{A_{ji}}{B_{ji}} e^{-(E_j - E_i)/kT}.$$

As this equation holds for any small T, we obtain

$$h\nu = E_j - E_i.$$

This looks trivial but expresses the non-trivial fact that the radiation field stimulates absorption and emission at the frequency corresponding to the energy difference of the levels. It also follows that

$$A_{ji} = \frac{8\pi h \nu^3}{c^3} B_{ji} \tag{6.40}$$

which connects the A coefficients with the B coefficients.

6.3.1.3   The quantum mechanical expression for A

The quantum-mechanical correct expression for the Einstein coefficient A_{ji} for spontaneous emission from upper state j to lower state i is

$$A_{ji} = \frac{64\pi^4 \nu^3}{3hc^3} |\mu_{ji}|^2 \tag{6.41}$$

where

$$\mu_{ji} = e \int \psi_i^*(\mathbf{x}) \, \mathbf{x} \, \psi_j(\mathbf{x}) \, dV \tag{6.42}$$

is the dipole moment corresponding to the transition (see (6.20)). ψ_i and ψ_j are the eigenfunctions of the energy states. A rigorous derivation of formula (6.41) requires quantum electrodynamics; the non-relativistic limit is discussed in section 6.3.4.

6.3.1.4   A classical analogy

The essence of formula (6.41) can already be grasped using classical arguments by equating the emission rate $A_{ji}h\nu$ to the average power radiated by a harmonic oscillator (see (1.97)). If $x = x_0 e^{-i\omega t}$ is the time-variable coordinate of the electron and $p_0 = ex_0$ its dipole moment, then

$$A_{ji}h\nu = \frac{p_0^2 \omega^4}{3c^3}.$$

This yields (6.41) exactly if one puts $\mu = \frac{1}{2}p_0$ and $\omega = 2\pi\nu$. Note that according to equation (6.41), A_{ji} increases with the square of the dipole moment and with the third power of the frequency. We see from (6.41) that, cum granu salis, A_{ji} is high for optical transitions and low at radio wavelengths. To get a feeling for the numbers, we apply the formula:

(1) to an electronic transition in the hydrogen atom. Let x_0 be equal to the atomic radius in the ground state (0.5Å), the dipole moment is then $\mu = 2.4 \times 10^{-18}$ cgs $= 2.4$ Debye. At an optical frequency $\nu = 6 \times 10^{14}$ Hz, corresponding to $\lambda = 5000$ Å, the Einstein coefficient becomes $A_{ji} \sim 10^7$ s^{-1}.

(2) To the lowest rotational transition of the CO molecule. The dipole moment in this case is fairly weak, $\mu = 0.11$ Debye, and the frequency low, $\nu \simeq 1.15 \times 10^{11}$ Hz. Now $A_{ji} \sim 10^{-7}$ s^{-1}, which is 14 powers of ten smaller!

6.3.1.5   Transition probability for general forces

Equation (6.41) may also be put in the form

$$A_{ji} = \frac{8\pi^3}{3} \alpha \left(\frac{c}{\ell}\right) \left(\frac{p\ell}{h}\right)^3$$

where $\alpha = e^2/\hbar c$ is the fine structure constant, $\ell = 2x_0$ the size of the system, c/l the inverse crossing time by light, $p = h\nu/c$ the photon momentum and $(p\ell)^3$ the volume in the phase space necessary for the creation of photons by the atomic oscillator. In other words, the transition probability is the product of the coupling constant of the electromagnetic radiation, α, multiplied by the number of cells in the phase space divided by the crossing time. When the transition probability is expressed in this way, it can also be applied to other basic forces in the universe, besides electromagnetism. The coupling constant is then, of course, different.

6.3.2   Selection rules and polarization rules

Formulae (6.41) and (6.42) incorporate the selection rules because when the integral (6.42) is zero for two eigenfunctions ψ_i and ψ_j (which is usually the case), the transition is forbidden. For instance, from that part of the eigenfunctions of the hydrogen atom which is radius-independent, namely

$$P_l^m(\cos\theta)e^{im\varphi}$$

follow the selection rules

$$\Delta l = \pm 1 \qquad \Delta m = 0, \pm 1.$$

If they are not fulfilled, no emission is possible. However, a perturbation, like an electric field, can make an otherwise forbidden line to become observable because the eigenfunctions are then deformed so that the integral (6.42) no longer vanishes. One can also derive the polarization rules in the presence of a magnetic field by separating the dipole moment μ of (6.41) into its Cartesian components,

$$\mu_x = e\int\psi_i^*x\psi_j\,dV \qquad \mu_y = e\int\psi_i^*y\psi_j\,dV \qquad \mu_z = e\int\psi_i^*z\psi_j\,dV$$

and evaluating the expressions. For example, when the field is in the z-direction and $\mu_x = \mu_y = 0$ but $\mu_z \neq 0$, the radiation is linearly polarized with its electric vector parallel to the applied magnetic field.

6.3.3   Quantization of the electromagnetic field

So far we have learnt that radiative transitions occur at a rate given by the Einstein coefficients. Now we want to understand *why* they occur and how the As and Bs are computed. For this end, we have to quantize the electromagnetic field and we sketch how this can be done.

6.3.3.1   *The spectrum of standing waves in a box*

Imagine a big box of volume V and length L with an electromagnetic field inside, consisting of standing waves. The vector potential $\mathbf{A}$ is a function only of z. It oscillates in the x-direction ($A_y = A_z = 0$) and vanishes at the walls of the box, which are nodal surfaces. We expand A_x into a Fourier series:

$$A_x(z, t) = \sum_{j=1}^{\infty} q_j(t) \sin \frac{j\pi z}{L} \tag{6.43}$$

the coefficients $q_j(t)$ are time-dependent. With $\mathbf{H} = \mathrm{rot}\, \mathbf{A}$ and $\mathbf{E} = -\dot{\mathbf{A}}/c$, and the orthogonality of the sine function,

$$\int_0^{\pi} \sin ix \cdot \sin jx\, dx = \delta_{ij} \frac{\pi}{2}$$

the energy of the field in the box becomes

$$E_{\mathrm{field}} = \frac{1}{8\pi} \int (E_x^2 + H_y^2)\, dV = \frac{V}{16\pi c^2} \sum_{j=1}^{\infty} (\dot{q}_j^2 + \omega_j^2 q_j^2)$$

where the frequencies are

$$\omega_j = \frac{j\pi c}{L}.$$

If the box is very big, there is an almost continuous frequency spectrum.

6.3.3.2　Equation of motion of an elastic string

There is a strong analogy between the field and an elastic string. Suppose a string has a length L, is fixed at its ends and vibrates exactly like the vector potential A_x of (6.43). As it vibrates, it bends and becomes distorted. Hooke's law of elasticity states how strong the internal deformations are: When we pull with a force F at the end of a rectangular block of length ℓ and cross section σ, the change in length $d\ell/\ell$ is proportional to the force per unit area, $F/\sigma = Y\, d\ell/\ell$, where the proportionality factor Y is Young's elasticity module. This law implies that the potential energy stored in the deformation of the volume element goes with $(d\ell)^2$.

　　If the shape of the bent string is described by the curve $y(z)$, its local deformation $d\ell$ is proportional to the derivative of the curve, $d\ell \propto y'(z)$. So the potential energy of the string that vibrates like the magnetic field A_x is

$$U = \frac{1}{2}\kappa \int_0^L \left(\frac{\partial A_x}{\partial z}\right)^2 dz = \frac{\kappa\pi^2}{4L} \sum_{j=1}^{\infty} j^2 q_j^2$$

where κ is some elastic constant. If the string has a mass density ρ, its kinetic energy equals

$$T = \tfrac{1}{2}\rho \int_0^L \dot{A}_x^2\, dz = \frac{1}{4} L\rho \sum_{j=1}^{\infty} \dot{q}_j^2.$$

We can form the Lagrange function $L = T - U$ and find, from (6.1), the equation of a loss-free harmonic oscillator

$$\ddot{q}_j + \tilde{\omega}_j^2 q_j = 0$$

with frequency

$$\tilde{\omega}_j = \frac{j\pi}{L}\sqrt{\frac{\kappa}{\rho}}.$$

The total energy equals the sum of kinetic plus potential energy of all oscillators,

$$E_{\text{string}} = \frac{L\rho}{4}\sum_{j=1}^{\infty}(\dot{q}_j^2 + \tilde{\omega}_j^2 q_j^2).$$

6.3.3.3 *Mathematical identity between field and string*

The expressions for the energy of the string and the field, E_{string} and E_{field}, become identical when we put $c = \sqrt{\kappa/\rho}$ and

$$m \equiv L\rho = \frac{V}{4\pi c^2} \tag{6.44}$$

m may be considered to be the mass of the oscillator. The frequencies of field and string then coincide, $\tilde{\omega}_j = \omega_j$, as well as their energies. The q_js of the string are coordinates of an harmonic oscillator of frequency ω_j. Such an oscillator has, in quantum mechanics, discrete energy levels

$$E_{nj} = \hbar\omega_j(n + \tfrac{1}{2}) \qquad \text{with } n = 0, 1, 2, \ldots.$$

The quantization of the string implies that the electromagnetic field is quantized, too. Naturally, we associate the oscillator of frequency ω_j with radiation at that frequency and identify the quantum number n with the number of photons in the box at that frequency, more precisely, with the occupation number $\mathcal{N}$, which is related to the radiative energy density u_ν through formula (5.33). Because the field is, from (6.43), completely described by the coefficients $q_j(t)$, in quantum mechanics one replaces the field A_x by the amplitudes q_j; the q_j are called normal coordinates.

6.3.4 **Quantum-mechanical derivation of A and B**

6.3.4.1 *Coupling between oscillator and field*

Emission or absorption must result from the coupling between the oscillator and the electromagnetic field but how? The trick is to perceive the atom and the field as *one* system consisting of two subsystems:

- the atom described by coordinates (1) with Hamiltonian $\hat{H}_1$, eigenvalue E_1 and eigenfunction ψ_1, so that $\hat{H}_1\psi_1(1) = E_1\psi_1(1)$; and

- the electromagnetic field for which we use the subscript 2; here $\hat{H}_2\psi_2(2) = E_2\psi_2(2)$.

The total uncoupled system has the Hamiltonian $\hat{H} = \hat{H}_1 + \hat{H}_2$, the wavefunction

$$\psi(1, 2) = \psi_1(1) \cdot \psi_2(2)$$

and the energy $E_1 + E_2$,

$$\hat{H}\psi = (E_1 + E_2)\psi_1\psi_2.$$

In reality, the two subsystems couple. For the perturbation operator H' we use the Hamiltonian derived in (6.16),

$$\hat{H}' = -\frac{e}{mc}\hat{\mathbf{p}} \cdot \hat{\mathbf{A}} \tag{6.45}$$

and neglect the quadratic term $\hat{\mathbf{A}}^2$. Note that $\hat{\mathbf{p}} \cdot \hat{\mathbf{A}}$ commutes with $\hat{\mathbf{A}} \cdot \hat{\mathbf{p}}$; this follows from the definition of the momentum operator $\hat{\mathbf{p}}$ in (6.13) and the Lorentz gauge div $\mathbf{A} = 0$ (see (1.85)).

6.3.4.2 The matrix element H'_{fs}

Let us consider the process of emission or absorption of the atom as a transition of the *whole* system: atom plus field. The transition probability from the starting state s to the final state f as a function of time is given by (6.35). To compute it, we have to evaluate the matrix element H'_{fs} of (6.31) which contains the operator $\hat{H}' = -(e/mc)\hat{\mathbf{p}} \cdot \hat{\mathbf{A}}$. The integral that appears in the matrix element H'_{fs} can be separated into

$$I_1 = \frac{e}{mc} \int \psi_{1f}^* \hat{\mathbf{p}} \psi_{1s} \, dx \quad \text{and} \quad I_2 = \int \psi_{2f}^* \hat{\mathbf{A}} \psi_{2s} \, dq$$

so that $H'_{fs} = I_1 \cdot I_2$. For the first, we get from (6.21)

$$|I_1|^2 = \frac{\omega_{fs}^2}{c^2} |\boldsymbol{\mu}_{fs}|^2.$$

To evaluate the second integral I_2, for ψ_2 we insert the eigenfunctions of the electromagnetic field and thus of an harmonic oscillator. They read $\psi_{2j} = N_j H_j$, where H_j are the Hermitian polynomials of (6.24) and N_j their normalization coefficients. The field operator $\hat{\mathbf{A}}$ is replaced by the normal coordinate q. In this way, we get

$$I_2 = N_f N_s \int H_f(\alpha q) \, q \, H_s(\alpha q) e^{-\alpha^2 q^2} \, dq = \frac{N_f N_s}{\alpha^2} \int H_f(y) y H_s(y) e^{-y^2} \, dy$$

$$= \frac{N_s}{2\alpha N_f}\{\delta_{f,s+1} + 2s\delta_{f,s-1}\}$$

$$= \begin{cases} \dfrac{N_s}{2\alpha N_f} = \dfrac{\sqrt{s+1}}{\alpha\sqrt{2}} & \text{if } f = s+1 \text{ (emission)} \\[2mm] \dfrac{N_s}{2\alpha N_f} = \dfrac{\sqrt{s}}{\alpha\sqrt{2}} & \text{if } f = s-1 \text{ (absorption).} \end{cases} \tag{6.46}$$

Here we have exploited formula (A.3). In the event of atomic emission, the energy of the *atom* makes a downward jump $\hbar\omega_{fs}$. For the *field* it is the other other way round, its final state is higher than the initial state. Therefore in (6.46), which refers to the field, atomic emission corresponds to $f = s+1$ and absorption to $f = s - 1$. During the transition there is an exchange of energy between the two subsystems by the amount $\hbar\omega_{fs}$ but the overall energy stays constant. With respect to the parameter $\alpha^2 = m\omega/\hbar$ from equation (6.23), we have to insert the field mass m after (6.44), so

$$\alpha^2 = \frac{V\omega}{4\pi c^2 \hbar}. \tag{6.47}$$

This gives

$$|I_2|^2 = \frac{s+1}{2\alpha^2} = \frac{2\pi c^2 \hbar}{\omega V} \cdot (s+1)$$

and finally

$$|H'_{fs}|^2 = |I_1|^2 \cdot |I_2|^2 = \frac{\omega_{fs}^2}{c^2}|\mu_{fs}|^2 \cdot \frac{2\pi c^2 \hbar}{\omega V} \cdot (s+1).$$

6.3.4.3 *The transition probability*

The transition probability $p_{fs}(t)$ as a function of time from one discrete state to another, $s \rightarrow f$, is given in (6.35). Because of the presence of the field, the energy levels of the *total* system are not discrete, like those of the atom alone, but closely packed, like a continuum. Furthermore, we do not have *one* final state but many. If $\rho(\omega)$ is the density of these states, there are $\rho(\omega_{fs})\,d\omega$ of them in the frequency interval $d\omega$ around the energy $\hbar\omega_{fs}$. The density is equal to the number of unit cells h^3 in phase space, from (5.30),

$$\rho(\omega) = \frac{V\omega^2}{\pi^2 c^3}.$$

To find the total transition probability of our coupled system, we, therefore, have to multiply $p_{fs}(t)$ by $\rho(\omega_{fs})\,d\omega$ and integrate over frequency. The integral is according to (6.37), proportional to $\pi t\rho(\omega_{fs})$. When we make the time t large, the total transition probability *per unit time* becomes

$$P = \frac{2\pi}{3\hbar^2}|H'_{fs}|^2\rho(\omega_{fs}). \tag{6.48}$$

We introduced the correction factor $\frac{1}{3}$ to account for the fact that the orientation of the dipole with respect to the field is arbitrary. It only remains to insert the expressions for $\rho(\omega_{fs})$ and $|H'_{fs}|^2$. When we do this, the volume V of the box luckily disappears. Here are the final results for the transition probability per unit time of an atom in a radiation field:

- If exciting radiation is absent $(s = 0)$, the transition probability P equals the Einstein A coefficient of (6.41),

$$P = A_{ji} = \frac{4\omega^3}{3\hbar c^3}|\boldsymbol{\mu}_{ji}|^2.$$

It is puzzling that a seemingly unperturbed atom in an eigenstate (above the ground state) with energy E_j should emit radiation at all. One might expect it to stay there forever because the only time dependence in the eigenfunction is, from (6.17), an oscillation of the form $e^{-iE_jt/\hbar}$. However, there *are* downward transitions $(A_{ji} > 0)$, even when the atom sits in the dark. The reason is that the perturbation operator (where the field is represented by eigenfunctions of an harmonic oscillator) does not vanish when photons are absent.

- In a radiation field, the radiative density is, from (5.33),

$$u_\nu = \frac{2\hbar}{\pi c^3}\omega^3 s.$$

When there are many photons flying around, $s \simeq s + 1$ and the induced emission rate is

$$P = \frac{4\omega^3}{3\hbar c^3}|\boldsymbol{\mu}_{ji}|^2(s + 1) \simeq B_{ji}\, u_\nu.$$

The same expression holds for induced absorption (if the initial and final states of the atom have the same statistical weight). Note that the symmetry of $|H'_{fs}|^2$ with respect to s and f accounts for the symmetry of the B coefficients, $B_{sf} = B_{fs}$.

The transitions described here involve *one* field photon because they are rooted in the perturbation operator $\hat{\mathbf{p}}\cdot\hat{\mathbf{A}}$, which is linear in $\hat{\mathbf{A}}$. When we include the much weaker operator $\hat{\mathbf{A}}^2$ in the Hamiltonian of (6.16), it follows, again from the orthogonality of the Hermitian polynomials which we employed for the derivation of (6.46), that the $\hat{\mathbf{A}}^2$-term describes processes involving two photons at a time, either in emission or absorption, or one in emission and the other in absorption.

6.4   Potential wells and tunneling

6.4.1   Wavefunction of a particle in a constant potential

Consider in one dimension a free particle of mass m and energy E in a potential $V(x)$. To facilitate writing, in this section we often put

$$U(x) = \frac{2m}{\hbar^2} V(x) \qquad \varepsilon = \frac{2m}{\hbar^2} E. \qquad (6.49)$$

In this notation, the wavefunction of a stationary state satisfies the equation

$$\psi'' = \big[U(x) - \varepsilon\big]\psi. \qquad (6.50)$$

If U is constant, the general solution reads

$$\psi(x) = A_1 e^{i\alpha x} + A_2 e^{-i\alpha x} \qquad \alpha = \sqrt{\varepsilon - U} \qquad (6.51)$$

with complex A_i. There are two cases:

- $\varepsilon - U > 0$. Then $\alpha > 0$ and the wave has an oscillatory behavior. It could be represented by a combination of a sine and a cosine.
- $\varepsilon - U < 0$. Then $i\alpha$ is real and the two solutions correspond to an exponentially growing and declining function. The former is only allowed in finite intervals because the integral $\int \psi^* \psi \, dx$ must be bounded.

If the potential $U(x)$ is a step function, i.e. if it is constant over intervals $[x_i, x_{i+1}]$ but makes jumps at the connecting points x_i, with $x_1 < x_2 < x_3 < \ldots$, the wavefunction ψ and its derivative ψ' must be continuous at these points; otherwise the Schrödinger equation would not be fulfilled. The boundary conditions of continuity for ψ and ψ' at all x_i fully determine the wavefunction. To find it is straightforward but often tedious.

6.4.2   Potential walls and Fermi energy

If the particle is trapped in an infinitely deep well of length L, as depicted on the left-hand side of figure 6.3, the wavefunction must vanish at the boundaries. It cannot penetrate beyond the walls, not even a bit. The situation is reminiscent of a vibrating string fixed at its ends. If $\lambda = h/p$ is the de Broglie wavelength and p the momentum, the condition $n\lambda/2 = L$ leads to the discrete energies

$$\varepsilon_n = \frac{2m}{\hbar^2} E_n = \frac{n^2 \pi^2}{L^2} \qquad n = 1, 2, \ldots \qquad (6.52)$$

and sine-like wavefunctions

$$\psi_n(x) = A \sin\left(\frac{n\pi}{L}x\right).$$

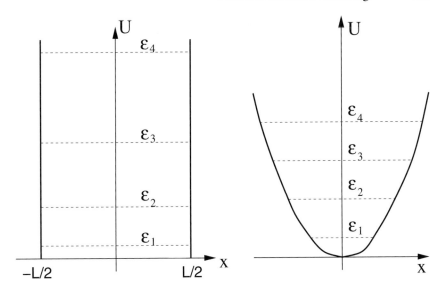

Figure 6.3. The left side shows an electron in a square potential with walls of infinite height. The stationary states correspond to energies $E_n \propto n^2$ (see (6.52)). The right side shows, for comparison, the parabolic potential and the equidistant energy levels of an harmonic oscillator.

There is a zero-point energy ($\varepsilon > 0$ for $n = 1$) because the particle is spatially localized in the interval from $-L/2$ to $L/2$, and so its momentum cannot vanish. The spacing between energy levels increases quadratically with quantum number n. This is very different from an harmonic oscillator which has a parabolic potential and equally spaced energy levels (see figure 6.3).

If a system with N electrons is in its lowest state, all levels $n \leq N/2$ are filled. The threshold or Fermi energy of the topmost filled level is

$$E_F = \frac{\hbar^2 \pi^2 N^2}{8 m_e L^2} \qquad (6.53)$$

and the velocity of those electrons is $v_F = \sqrt{2E_F/m_e}$.

If the walls of the potential well are finite and of height U and the particle energy $\varepsilon < U$, there is some chance, exponentially decreasing with distance, to find it outside the well. The energies of the eigenstates depend then in a more complicated manner on n.

In three dimensions, for a cube of length L and infinite potential barrier, the eigenfunctions are

$$\psi(\mathbf{x}) = A \sin\left(\frac{n_x \pi}{L} x\right) \sin\left(\frac{n_y \pi}{L} y\right) \sin\left(\frac{n_z \pi}{L} z\right)$$

with positive integers n_x, n_y, n_z, and the Fermi energy is

$$E_F = \frac{\hbar^2}{2m_e} \left(\frac{3\pi^2 N}{L^3} \right)^{2/3}. \tag{6.54}$$

6.4.2.1 Examples of Fermi energy and Fermi statistics

- White dwarfs are burnt-out stars with extreme densities up to 10^6 g cm^{-3}. But only the electrons are degenerate, the more massive atomic nuclei are not because their degeneracy parameter is much smaller (see (5.29)). When the Fermi energy is very high as a result of a high density $\rho = N/L^3$, the electrons are relativistic and

$$E_F = \hbar c (N/V)^{1/3}. \tag{6.55}$$

If one expresses the Fermi energy as a temperature,

$$T_F = \frac{E_F}{k}$$

one finds $T_F \sim 10^9$ K for $\rho \sim 10^6$ g cm^{-3}.
- An even more spectacular case is presented by neutron stars with densities up to 10^{15} g cm^{-3}. The neutrons do not disintegrate into protons and electrons, as one would expect, only a few do, because the energy liberated in the decay n $\to$ p+e produces only $\sim 10^6$ eV, which is miles below the Fermi threshold of (6.55) in a fictitious sea of electrons and protons at density 10^{15} g cm^{-3}.
- The conduction electrons in a metal have values T_F between 10^4 and 10^5 K, so at temperatures $T \sim 100$ K, the energy distribution of the conduction electrons given by (5.26) is highly degenerate, only few of them are thermalized. For graphite particles of interstellar dust grains, the Fermi temperature is somewhat lower, $T_F \sim 10^3$ K.
- Observing astronomers use both isotopes of liquid helium to cool their detectors. The ^{3}He atoms are made of five elementary particles (two protons, one neutron and two electrons), they have half-integer spin and are fermions. The main isotope ^{4}He, however, contains one more neutron, has zero total angular momentum and follows Bose statistics. The two isotopes show radically different behavior at low temperatures: ^{4}He becomes a superfluid below 2.2 K, whereas ^{3}He does not. Superfluidity is explicable by Bose statistics because Bose statistics allows many (even all) atoms to be at the lowest energy state.

6.4.3 Rectangular potential barriers

6.4.3.1 A single barrier

Let a particle travel from the left to the right and encounter a potential wall, as in figure 6.4. If its energy ε is greater than U, it can, of course, overcome

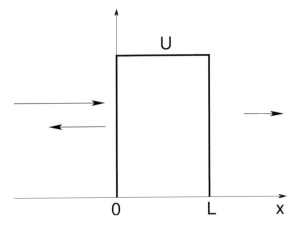

Figure 6.4. Rectangular potential barrier for a particle coming from the left. Part of the beam is reflected and part goes through the barrier.

the barrier but contrary to classical physics, part of the wave is reflected. Most relevant is the case when $\varepsilon < U$. Then the particle can tunnel through the barrier which is classically forbidden. What happens, as the particle approaches the barrier, obviously depends on time and therefore stationary solutions may not seem possible in this scenario. However, one may interpret the infinite wave $e^{\frac{i}{\hbar}(px-Et)}$ of a particle with definite energy E as a stationary particle beam, and then the time drops out.

The wavefunction is constructed piece-wise:

$$\psi(x) = \begin{cases} Ae^{i\alpha x} + Be^{-i\alpha x} & x < 0 \\ Ce^{\beta x} + De^{-\beta x} & 0 < x < L \\ Fe^{i\alpha x} + Ge^{-i\alpha x} & x > L. \end{cases}$$

The coefficient $G = 0$, as there is no wave coming from the right, and

$$\alpha = \sqrt{\varepsilon} > 0 \qquad \beta = \sqrt{U - \varepsilon} > 0.$$

Exploiting the boundary conditions and keeping in mind that $(e^x + e^{-x})^2/4 = \cosh^2 x = 1 + \sinh^2 x$, the transmission coefficient T, which is the fraction of the particle beam that penetrates the barrier, becomes

$$T = \frac{4\varepsilon(U - \varepsilon)}{4\varepsilon(U - \varepsilon) + U^2 \sinh^2(L\sqrt{U - \varepsilon})}. \tag{6.56}$$

When the barrier is high compared to the kinetic energy of the particle ($U \gg \varepsilon$) and sufficiently broad ($LU^{1/2} > 1$), the tunneling probability can be approximated by

$$T = \frac{16\varepsilon}{U} e^{-2L\sqrt{U}}. \tag{6.57}$$

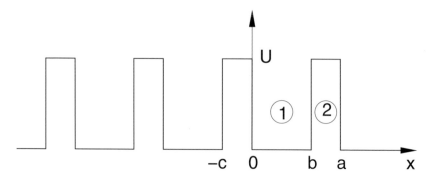

Figure 6.5. A periodic square potential of a one-dimensional lattice. The length of the period is a, the length of one barrier is $c = a - b$. In region 1, where $0 \le x \le b$, the potential is zero, in region 2, where $b \le x \le a$, it has the value U.

6.4.3.2 Periodic potential

Regularly arranged atoms in a solid (crystal) produce a periodic potential. We discuss this important case in one dimension. Let the potential be periodic with period a, as depicted in figure 6.5. The wavefunction ψ of a particle with energy $\varepsilon < U$ is then given by

$$\psi(x) = \begin{cases} Ae^{i\alpha x} + Be^{-i\alpha x} & 0 < x < b \\ Ce^{\beta x} + De^{-\beta x} & b < x < a \end{cases}$$

with

$$\alpha = \sqrt{\varepsilon} \qquad \beta = \sqrt{U - \varepsilon}.$$

Two equations for determining A–D are found from the condition that ψ and ψ' must be continuous at $x = 0$. Another two equations follow from a result detailed in section 7.2.3: In a periodic potential, any eigenfunction ψ_k, with respect to wavenumber k and energy $\varepsilon = k^2$, must have the form

$$\psi_k = u_k e^{ikx}$$

where u_k is periodic such that

$$u_k(x) = u_k(x + a).$$

Therefore, $u_k(b) = u_k(-c)$ and $u'_k(b) = u'_k(-c)$, where $u_k(x) = \psi_k(x)e^{-ikx}$. Altogether, this gives

$$A + B = C + D$$
$$i\alpha(A - B) = \beta(C - D)$$
$$e^{-ikb}\left[Ae^{i\alpha b} + Be^{-i\alpha b}\right] = e^{ikc}\left[Ce^{-\beta c} + De^{\beta c}\right]$$
$$ie^{-ikb}\left[Ae^{i\alpha b}(\alpha - k) - Be^{-i\alpha b}(\alpha + k)\right] = e^{ikc}\left[Ce^{-\beta c}(\beta - ik) - De^{\beta c}(\beta + ik)\right].$$

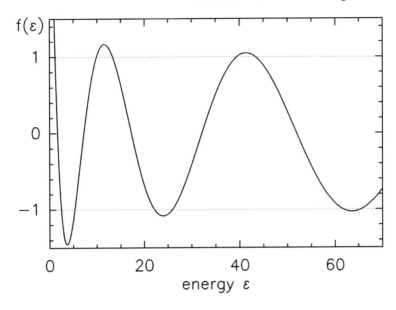

Figure 6.6. The quasi-sinusoidal variation of the right-hand side of equation (6.58) as a function of particle energy ε. It is calculated for the periodic potential of figure 6.5 with $U = 100$, $a = 2$, $c = 0.04$. Energies for which $|f(\varepsilon)| > 1$ are forbidden. This is, for instance, the case when $\varepsilon \simeq 4, 11$ or 24.

Non-trivial solutions of this linear system of equations, those for which $A = B = C = D$ does *not* equal zero, have a vanishing determinant. The condition Det $= 0$, after some algebra, leads to

$$\cos ka = \frac{\beta^2 - \alpha^2}{2\alpha\beta} \sinh \beta c \cdot \sin \alpha b + \cosh \beta c \cdot \cos \alpha b. \qquad (6.58)$$

For fixed values a, b, U of the potential in figure 6.5, the right-hand side of (6.58) is a function of particle energy ε only; we denote it by $f(\varepsilon)$. Because of the cosine term on the left-hand side of (6.58), only energies for which $f(\varepsilon) \in [-1, 1]$ are permitted and those with $|f(\varepsilon)| > 1$ are forbidden. The function $f(\varepsilon)$ is plotted in figure 6.6 for a specific set of values U, a, c.

In the potential well on the left-hand side of figure 6.3, which is not periodic and where the walls are infinite, the energy ε rises quadratically with the wavenumber, $\varepsilon = k^2$. If the length L of the well in figure 6.3 is large, there will be a continuous spectrum as depicted by the continuous line in figure 6.7. In a periodic potential, however, the function $\varepsilon(k)$ is only aproximately quadratic and there are now jumps at integral values of $|ka/\pi|$. Figure 6.7 is the basis for explaining why electrons in crystals are arranged in energy bands.

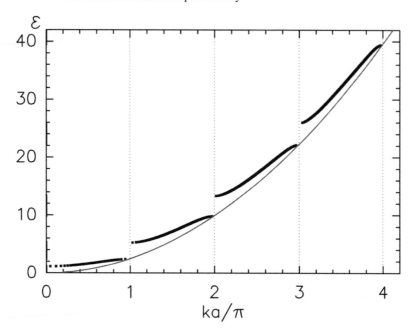

Figure 6.7. The energy spectrum $\varepsilon(k)$ of particles in a periodic potential as in figure 6.5 calculated for $U = 100$, $a = 2$ and $c = 0.04$. The function $\varepsilon(k)$ has discontinuities at integer values of ka/π. The curve without discontinuities is explained in the text. See also figure 6.6.

6.4.4   The double potential well

6.4.4.1   *Splitting of energy levels*

Next we consider an atom of mass m and energy ε in a potential $U(x)$ consisting of two symmetric adjacent wells with a barrier U_0 between them as depicted in figure 6.8. Classical motion is allowed in the intervals $[-b, -a]$ and $[a, b]$, where $U(x) \leq \varepsilon$. When the barrier U_0 is infinitely large, the particle can only be in the left- or right-hand well. The corresponding wavefunctions are denoted by ψ_- and ψ_+, respectively, and satisfy the Schrödinger equation (see (6.18)):

$$\psi_+'' + \left[\varepsilon - U(x)\right]\psi_+ = 0 \qquad \psi_-'' + \left[\varepsilon - U(x)\right]\psi_- = 0. \qquad (6.59)$$

If the barrier U_0 is finite, there is a certain chance of tunneling from one well to the other. The particle is now not localized in either of them and its wavefunction is a symmetric and anti-symmetric combination of ψ_+ and ψ_-,

$$\psi_1 = \frac{\psi_+}{\sqrt{2}} + \frac{\psi_-}{\sqrt{2}} \qquad \psi_2 = \frac{\psi_+}{\sqrt{2}} - \frac{\psi_-}{\sqrt{2}}.$$

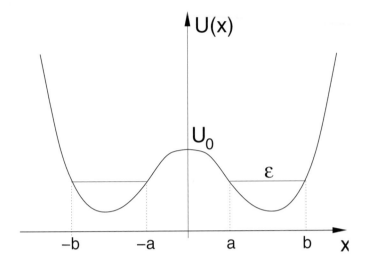

Figure 6.8. A particle in a double potential well with a barrier U_0 between the minima. Because of tunneling, an energy level ε splits into two with a separation $\Delta\varepsilon$ given by (6.61).

The probability of finding the particle either in the left or right well is unity and the integral of $|\psi_i|^2$ over the whole x-axis equals one. The wavefunctions ψ_1 and ψ_2 also obey the Schrödinger equations,

$$\psi_1'' + [\varepsilon_1 - U(x)]\psi_1 = 0 \qquad \psi_2'' + [\varepsilon_2 - U(x)]\psi_2 = 0 \qquad (6.60)$$

but with slightly different energies ε_1 and ε_2. To obtain the difference $\varepsilon_2 - \varepsilon_1$, we multiply the left equation in (6.59) by ψ_1 and the left equation in (6.60) by ψ_+, subtract the two products, and integrate from $x = 0$ to $x = \infty$. Exploiting $\psi_1(0) = \psi_+(0)$, $\psi_1'(0) = 0$ and assuming that the tunneling probability is small such that $\psi_1 \simeq \psi_+$ for $x > 0$ gives $\varepsilon_1 - \varepsilon = -2^{1/2}\psi_+'(0)\psi_+(0)$. Likewise we find $\varepsilon_2 - \varepsilon$, and finally get

$$\varepsilon_2 - \varepsilon_1 = \psi_+(0) \cdot \psi_+'(0).$$

So the energy difference is determined by the wavefunction ψ_+ in the classically forbidden region at $x = 0$. If $p = \hbar\sqrt{\varepsilon - U(x)}$ denotes the momentum and p_0 its value at $x = 0$, one finds

$$\psi_+(0) = \frac{\hbar}{p_0}\psi_+'(0) = \sqrt{\frac{v_0 m}{p_0}}\exp\left[-\frac{1}{\hbar}\int_0^a |p|\,dx\right]$$

where v_0 is the classical oscillation frequency, i.e. v_0^{-1} is the time for the particle to go from a to b and back, and, therefore,

$$\varepsilon_2 - \varepsilon_1 = \frac{4mv_0}{\hbar}\exp\left[-\frac{1}{\hbar}\int_{-a}^a |p|\,dx\right] \simeq \frac{4mv_0}{\hbar}\exp\left[-\frac{2a\sqrt{2mV_0}}{\hbar}\right]. \qquad (6.61)$$

The possibility that the particle can tunnel to the adjacent well leads to two wavefunctions, ψ_1 and ψ_2. Their space probabilities $|\psi_i|^2$ are different, although only in the forbidden region but there they couple differently to the potential $U(x)$ resulting in the splitting of the energy ε into two levels ε_1 and ε_2.

6.4.4.2   Tunneling time

To find the approximate time t_{tun} that the particle needs to tunnel from the left to the right well or back, we turn to the time-dependent wavefunction

$$\Psi(t) = \psi_1 e^{-\frac{i}{\hbar}E_1 t} + \psi_2 e^{-\frac{i}{\hbar}E_2 t}.$$

Here we have used again $E = (\hbar^2/2m)\varepsilon$ and $V = (\hbar^2/2m)U$ (see (6.49)). At time $t = 0$, the wavefunction has the value $\Psi(0) = \sqrt{2}\psi_+$, so the particle starts in the right well. It reaches the left well when $\Psi(t) = \psi_-$, which happens after a tunneling time

$$t_{tun} = \frac{\pi\hbar}{|E_1 - E_2|}. \qquad (6.62)$$

This equation expresses the uncertainty principle $\Delta E \cdot \Delta t \simeq \hbar$. A precisely defined energy means that the particle is fixed in one well and does not tunnel. A broad energy band ΔE, however, implies a high mobility, the particle is then not localized anywhere.

From (6.61) and (6.62) we see that with increasing potential barrier V_0, the energy splitting $E_2 - E_1$ becomes smaller and the tunneling time longer. Inserting the energy difference of (6.61) into (6.62), we find

$$t_{tun} \simeq v_0^{-1} \exp\left[\frac{2a\sqrt{2mV_0}}{\hbar}\right]. \qquad (6.63)$$

Chapter 7

Structure and composition of dust

We study in section 7.1 the way in which atoms are arranged in a solid and in section 7.2 what holds them together. Then we make our first serious acquaintance with observations and examine the interstellar extinction curve (section 7.3). Its interpretation requires a basic knowledge of stellar photometry and the reddening law. In view of the overall cosmic abundance of elements, the extinction curve rules out major dust constituents other than silicate and carbon grains. We, therefore, proceed in section 7.4 to discuss their atomic structure and bonding. The shape of the extinction curve can only be explained if the grains are not of uniform dimension but display a size distribution. What kind of distribution and how it can be achieved in grain–grain collisions is sketched in section 7.5. Equipped with a quantitative idea about size and chemical composition of interstellar dust, we present typical and likely grain cross sections based on a reasonable set of optical constants.

7.1 Crystal structure

7.1.1 Translational symmetry

7.1.1.1 The lattice and the base

Interstellar grains probably contain crystalline domains of sizes 10–100 Å. In a crystal, the atoms are regularly arranged in a lattice which means that the crystal can be built up by periodic repetition of identical cells. A cell is a parallelepiped defined by three vectors $\mathbf{a}$, $\mathbf{b}$ and $\mathbf{c}$ such that for any two points $\mathbf{r}$, $\mathbf{r}'$ in the crystal whose difference can be written as

$$\mathbf{r}' - \mathbf{r} = u\mathbf{a} + v\mathbf{b} + w\mathbf{c} \qquad (7.1)$$

with integral u, v, w, the environment is exactly the same. If *any* two points $\mathbf{r}$ and $\mathbf{r}'$ with an identical environment are connected through (7.1), the vectors $\mathbf{a}, \mathbf{b}, \mathbf{c}$ are said to be primitive and generate the *primitive cell*. This is the one with the smallest possible volume. There is an infinite number of ways how to construct cells or vector triplets (**a**, **b**, **c**), even the choice of the primitive cell is not unique; figure 7.3 gives a two-dimensional illustration. Among the many possibilities, one prefers those cells whose sides are small, whose angles are the least oblique and which best express the symmetry of the lattice. These are called conventional, elementary or unit cells.

For a full specification of a crystal, one needs besides the lattice points $u\mathbf{a} + v\mathbf{b} + w\mathbf{c}$, a description of the three-dimensional cell structure, i.e. of the distribution of charge and matter in the cell. The cell structure is called the base. The spatial relation between the lattice points and the base is irrelevant. So lattice points may or may not coincide with atomic centers. A primitive cell may contain many atoms and have a complicated base but it always contains only one lattice point.

7.1.1.2 Physical consequences of crystalline symmetry

Geometrically, the existence of a grid always implies spatial anisotropy because in various directions the structures are periodic with different spacings. Furthermore, the periodic arrangement of atoms are manifest by the morphological shape of the body; in non-scientific language: crystals are beautiful. The physical relevance of crystalline order reveals itself, among others, by

- the appearance of long-range, i.e. intensified, forces which lead to sharp vibrational resonances observable at infrared wavelengths;
- bonding strengths that are variable with direction: in one or two directions, a crystal cleaves well, in others it does not; and
- a generally anisotropic response to fields and forces. For example, the way crystals are polarized in an electromagnetic field or stretched under mechanical stress depends on direction.

7.1.1.3 Miller's and other indices

To find one's way around a crystal and to identify points, axes or planes, one uses the following notation:

- If the origin of the coordinate system is at a cell corner, a point $\mathbf{r} = x\mathbf{a} + y\mathbf{b} + z\mathbf{c}$ within the cell is designated by xyz with $x, y, z \leq 1$. For example, the cell center is always at $\frac{1}{2}\frac{1}{2}\frac{1}{2}$.
- Suppose a plane intercepts the axes of the crystal coordinate system at the three lattice points $u00$; $0v0$ and $00w$. One then denotes, after Miller, the plane by the triplet in *round brackets* (hkl) where h, k, l are the smallest possible integers such that

$$h : k : l = \frac{1}{u} : \frac{1}{v} : \frac{1}{w}.$$

Table 7.1. Lengths and angles in the unit cells of the seven crystal systems. See figure 7.1 for definition of sides and angles. The monoclinic, orthorhombic, tetragonal and cubic system are further split into lattice types that are either primitive or centered. For the cubic crystal system, this is shown in figure 7.2. The cells of the triclinic and rhombohedral system are always primitive, without centring. Altogether there are 14 lattice types or Bravais cells, only seven of them are primitive.

1. Triclinic:	$a \neq b \neq c$	$\alpha \neq \beta \neq \gamma$	Lowest symmetry
2. Monoclinic:	$a \neq b \neq c$	$\alpha = \gamma = 90°, \beta \neq 90°$	Not shown in figure 7.1
3. Orthorhombic:	$a \neq b \neq c$	$\alpha = \beta = \gamma = 90°$	
4. Tetragonal:	$a = b \neq c$	$\alpha = \beta = \gamma = 90°$	
5. Hexagonal:	$a = b \neq c$	$\alpha = \beta = 90°, \gamma = 120°$	
6. Rhombohedral:	$a = b = c$	$\alpha = \beta = \gamma \neq 90°$	Not shown in figure 7.1
7. Cubic:	$a = b = c$	$\alpha = \beta = \gamma = 90°$	Highest symmetry

For instance, the plane (436) runs through the points 300, 040 and 002. A plane that is parallel to **b** and **c** and cuts only the **a**-axis at a distance a from the coordinate center is identified by (100). The plane (200) is parallel to (100) but cuts the **a**-axis at $\frac{1}{2}a$.

- A direction within the lattice may be specified by an arrow that goes from the origin of the coordinate system to the point $\mathbf{r} = x\mathbf{a} + y\mathbf{b} + z\mathbf{c}$ where x, y, z are integers. There is an infinite number of such vectors pointing in the same direction. Among them, let $u\mathbf{a} + v\mathbf{b} + w\mathbf{c}$ be the one with the smallest integers u, v, w. The direction is then denoted by the triplet $[uvw]$ in *square brackets*. In the cubic system (see later), the direction $[uvw]$ is always perpendicular to the plane (hkl) with $h = u, k = v$ and $l = w$.

7.1.2 Lattice types

7.1.2.1 *Bravais cells and crystal systems*

In the general case, the lengths of the vectors **a, b, c** in the conventional cell and the angles between them are arbitrary. When lengths or angles are equal, or when angles have special values, like 90°, one obtains grids of higher symmetry. There are 14 basic kinds of translation lattices (*Bravais* lattices) which correspond to 14 elementary cells. These 14 lattice types can be grouped into seven crystal systems which are described in table 7.1 and shown in figure 7.1.

The cubic system has the highest symmetry: **a, b, c** are orthogonal and their lengths are equal, $a = b = c$. The polarizability is then isotropic, whereas for other crystal systems, it depends on direction. Figure 7.2 shows the three lattice types of the *c*ubic family: simple (sc), *b*ody centred (bcc) or *f*ace centred (fcc) of which only the sc cell is primitive. The bcc cell is two times bigger than the primitive cell and the fcc cell four times; furthermore, primitive cells do not have a cubic shape (see the two-dimensional analog in figure 7.5).

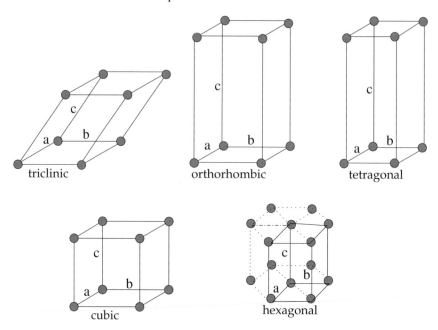

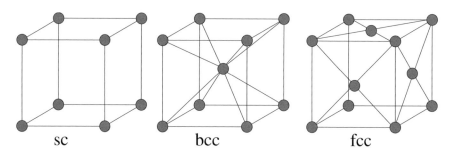

Figure 7.1. There are seven crystal systems of which five are shown here. The angle between the vector pair $(\mathbf{b}, \mathbf{c})$ is denoted α. If, symbolically, we write $\alpha = (\mathbf{b}, \mathbf{c})$, we define likewise $\beta = (\mathbf{c}, \mathbf{a})$ and $\gamma = (\mathbf{a}, \mathbf{b})$. See table 7.1.

Figure 7.2. The three lattices types of the cubic crystal system: simple cubic (sc), body-centered cubic (bcc) and face-centered cubic (fcc). To avoid confusion, for the fcc lattice the face-centring atoms are shown only on the three faces directed towards the observer.

7.1.2.2 *Microscopic and macroscopic symmetry*

Under a symmetry operation, a crystal lattice is mapped onto itself. A linear translation by one of the vectors $\mathbf{a}, \mathbf{b}, \mathbf{c}$ is one such operation but it becomes apparent only on an atomic scale because the shifts are of order 1 Å and thus much smaller than the dimensions of a real (even an interstellar) crystal. The translational symmetry is revealed from X-ray images and is described by the Bravais lattices. However, there are symmetry operations that also are evident on a macroscopic scale, such as the following ones.

- Inversion at a center. If the center is in the origin of the coordinate system, the operation may be symbolized by $\mathbf{r} = -\mathbf{r}'$.
- Rotation about an axis. One can readily work out that the only possible angles compatible with translational symmetry, besides the trivial 360°, are 60°, 90°, 120° and 180°. However, for a single molecule (not a crystal), other angles are possible, too.
- Reflection at a plane.

In each of these symmetry operations, there is a symmetry element that stays fixed in space (the inversion center, the rotation axis or the mirror plane, respectively).

Macroscopically, a crystal has the shape of a polyhedron, like a gem in an ornament or the grains in the salt shaker. When one subjects it to a symmetry operation, the normals of the faces of the polyhedron do not change their directions.

All crystals can be divided into 32 classes where each class represents a mathematical group whose elements are the symmetry operations described earlier. These 32 classes completely define the morphological (macroscopic) appearance of crystals. Macroscopic symmetry is, of course, the result of the microscopic structure, although the latter is not explicitly used in the derivation of the 32 crystal classes. Without proof, when the 14 Bravais lattices of the atomic world are combined with these symmetry elements, including two further symmetry elements (screw axis and glide plane), one arrives at 230 space groups.

7.1.2.3 Two-dimensional lattices

The classification scheme is much simpler and easier to understand in two-dimensions. The basic unit of a two-dimensional crystal is a parallelogram. Now there are only five (translational) Bravais lattices and 10 crystal classes. When the Bravais lattices are combined with the symmetry elements permitted for a plane (rotation axis, mirror plane and glide plane), one obtains 17 planar space groups. The parallelograms of Bravais lattices may be

- oblique ($a \neq b$, $\gamma \neq 90°$, figure 7.3);
- hexagonal ($a = b$, $\gamma = 120°$);
- quadratic ($a = b$, $\gamma = 90°$);
- rectangular ($a \neq b$, $\gamma = 90°$, figure 7.4); and
- centered rectangular ($a \neq b$, $\gamma = 90°$, figure 7.5).

Any two vectors $\mathbf{a}'$, $\mathbf{b}'$ with lengths $a' = b'$ and angle γ between them describe a centered cell of a rectangular lattice but the cell with vectors $\mathbf{a}$, $\mathbf{b}$

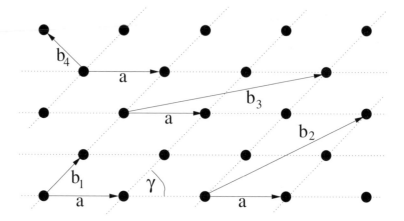

Figure 7.3. A two-dimensional lattice. There is an infinite number of ways to generate cells. Here it is done by the vector pairs $(\mathbf{a}, \mathbf{b}_i)$ with $i = 1, 2, 3, 4$. All define primitive cells, except $\mathbf{a}, \mathbf{b}_2$ which has an area twice as large.

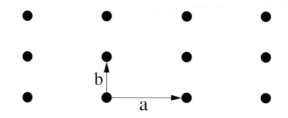

Figure 7.4. A simple rectangular lattice is generated by the primitive vectors $\mathbf{a}$ and $\mathbf{b}$.

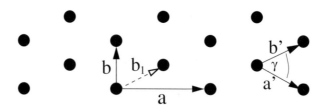

Figure 7.5. In a centered rectangular lattice, a primitive cell is generated by $\mathbf{a}, \mathbf{b}_1$ and has half the area of a conventional cell generated by $\mathbf{a}, \mathbf{b}$. Note that $\mathbf{a}', \mathbf{b}'$ with $a' = b'$ are also primitive vectors.

as in figure 7.5 seems to better convey to us the symmetry. Likewise in three dimensions, the body-centered (bcc) and face-centered (fcc) cubic type can be generated from primitive rhombohedra (where all three lattice constants are equal, $a = b = c$) but then again, one loses the advantage of a rectangular coordinate system because the angles are no longer $90°$.

7.1.3   The reciprocal lattice

Because the structure in a crystal is periodic, it is natural to expand the density $\rho(\mathbf{x})$ in a Fourier series. If $\mathbf{a}, \mathbf{b}, \mathbf{c}$ form the primitive vectors of the lattice according to (7.1), then

$$\rho(\mathbf{x}) = \sum_{\mathbf{G}} \rho_{\mathbf{G}}\, e^{i\mathbf{G}\cdot\mathbf{x}}. \tag{7.2}$$

The $\rho_{\mathbf{G}}$ are the Fourier coefficients and the sum extends over all vectors

$$\mathbf{G} = u\mathbf{a}^* + v\mathbf{b}^* + w\mathbf{c}^* \tag{7.3}$$

where u, v, w are integers and $\mathbf{a}^*$, $\mathbf{b}^*$, $\mathbf{c}^*$ are the primitive vectors of the reciprocal lattice. The first, $\mathbf{a}^*$, is defined by

$$\mathbf{a}^* = 2\pi \frac{\mathbf{b} \times \mathbf{c}}{\mathbf{a} \cdot (\mathbf{b} \times \mathbf{c})} \tag{7.4}$$

so that $\mathbf{a}^*$ is perpendicular to $\mathbf{b}$ and $\mathbf{c}$ and $\mathbf{a}^* \cdot \mathbf{a} = 2\pi$. The unit of $\mathbf{a}^*$ is one over length. There are corresponding definitions and relations for $\mathbf{b}^*$ and $\mathbf{c}^*$.

Suppose a plane wave of wavevector $\mathbf{k}$ falls on a crystalline grain. Any small subvolume in the grain, let it be at locus $\mathbf{x}$, scatters the incoming light. The amplitude of the scattered wave is proportional to the electron density $\rho(\mathbf{x})$. Two outgoing beams with the same wavenumber $\mathbf{k}'$ which are scattered by subvolumes that are a distance $\mathbf{r}$ apart have a phase difference $e^{i\Delta\mathbf{k}\cdot\mathbf{r}}$ with

$$\Delta\mathbf{k} = \mathbf{k}' - \mathbf{k}.$$

The vectors $\mathbf{k}$ and $\mathbf{k}'$ have the same length but point in different directions. To find the total scattered amplitude E_s, one must integrate over the whole grain volume V, so modulo some constant factor

$$E_s = \int_V \rho(\mathbf{x}) e^{-i\Delta\mathbf{k}\cdot\mathbf{x}}\, dV.$$

With respect to an arbitrary direction $\mathbf{k}'$, the integral usually vanishes as a result of destructive interference. It does, however, not vanish if $\Delta\mathbf{k}$ is equal to a reciprocal lattice vector $\mathbf{G}$ of (7.3), i.e.

$$\Delta\mathbf{k} = \mathbf{k}' - \mathbf{k} = \mathbf{G}. \tag{7.5}$$

Equation (7.5) is another way of formulating the Bragg law for reflection in crystals.

7.2   Binding in crystals

Solids are held together through electrostatic forces. Five basic types of bonds exist and they are discussed here. All five types are realized in cosmic dust.

7.2.1   Covalent bonding

In a covalent or homopolar bond between atoms, a pair of electrons of anti-parallel spin, one from each atom, is shared. The more the orbitals of the two valence electrons in the pair overlap, the stronger the atoms are tied together. The bonding is directional and follows the distribution of the electron density which has a significant high concentration *between* the atoms. One finds covalent binding in solids and in molecules.

The wavefunctions in covalent bonding are not extended and only nearest neighbors interact. We sketch the simplest case which is presented by the H_2^+-molecule, although it contains only one electron, and not an electron pair. A stable configuration with two protons and one electron is possible only if the protons have the right distance and the electron cloud is properly placed between them. The motion of the protons can be neglected for the moment; it may be included afterwards as a correction and results in quantized vibrational and rotational states. The Hamiltonian $\hat{H}$ of the H_2^+-molecule then represents the kinetic energy of the electron (mass m, momentum p) and the electrostatic potentials among the three particles (figure 7.3),

$$\hat{H} = \frac{p^2}{2m} + e^2 \left(\frac{1}{R} - \frac{1}{R_1} - \frac{1}{R_2} \right). \tag{7.6}$$

When one assumes the distance R between the protons to be fixed, the Hamiltonian of the electron, $\hat{H}_e(\mathbf{r})$, has the property that $\hat{H}_e(\mathbf{r}) = \hat{H}_e(-\mathbf{r})$ with $\mathbf{r} = (x, y, z)$. Any eigenfunction ψ is then even or odd if there is no degeneracy, and a linear combination of even and odd functions in the case of degeneracy. One therefore seeks ψ in the form of the *symmetric* or *anti*-symmetric functions

$$\psi_s = \psi_1 + \psi_2 \qquad \psi_a = \psi_1 - \psi_2.$$

Here ψ_1, ψ_2 are the usual hydrogenic wavefunctions with respect to proton 1 and 2 with energies $E_1 = E_2$. The electron is either at proton 1 or 2, one cannot say at which. When the distance R between the protons is very large, there is degeneracy and the system has zero binding energy ($E_b = 0$). For smaller R, the degeneracy is removed. The electron is then in a double potential well and this leads to a splitting of energy levels. The analysis is completely analogous to the discussion of the double well in section 6.4.4. There is a binding (ψ_s) and an unbinding state (ψ_a) and their separation increases as R becomes smaller. The lowering of the potential energy as a function of R is [Lan74]

$$\Delta E = 4\chi_H R e^{-R-1}$$

when R is expressed in Bohr radii, a_0, and χ_H denotes the ionization potential of hydrogen. The qualitative reason for the existence of a binding state (of lower energy) follows from the uncertainty principle: a hydrogen atom in its 1s ground state has a radius $\Delta x = a_0$ and a momentum $\Delta p = m_e c/137$, so $\Delta x \Delta p = \hbar$. The

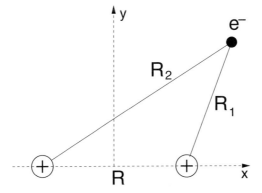

Figure 7.6. A sketch of the configuration between the two protons and one electron in the H_2^+-molecule. The protons are labeled 1 (right) and 2 (left).

binding energy of the hydrogen atom is $E_b = \chi_H = p^2/2m_e$. If the electron can spread out and be at two protons, as in the molecule H_2^+, more space is available to the electron and its momentum and the binding energy will be lower.

The electronic wavefunctions and the electron energies are schemtically drawn in figure 7.7 for intermediate R. When one includes the repulsion among the protons and considers the full Hamiltonian (7.6), the binding energy of the H_2^+ molecule has its minimum $E_b = 2.65$ eV at a distance of about two Bohr radii. This is the stable configuration of the molecule and E_b is the dissociation energy. When R is further reduced towards zero, repulsion between the protons dominates and the system is again unstable.

7.2.2 Ionic bonding

In ionic or heteropolar bonds, the adhesion is due to long-range Coulomb forces. In a crystal, the ions of opposite sign attract and those of the same sign repel each other but in the end, attraction wins. At short distances, there is an additional kind of repulsion but it acts only between adjacent atoms. It results from Pauli's exclusion principle which restricts the overlap of electron clouds.

Let us take a sodium chloride crystal (NaCl) as an example of ionic bonding. One partner, the alkali metal, is relatively easy to ionize (5.14 eV) and the other, the halogen Cl, has a high electron affinity (3.61 eV). Although the transfer of the electron from Na to Cl requires $5.14 - 3.61 = 1.53$ eV, the Na^+ and Cl^- ions can supply this deficit. They can provide even more than that in the form of potential energy by coming close together. Indeed, at the equilibrium distance of 2.8 Å, the binding energy per molecule equals 6.4 eV.

As the ions of heteropolar bonds have acquired full electron shells, like inert gases, their electron clouds are fairly spherical and the charge distribution between the ions is low. Ionic bonding is, therefore, non-directional and usually

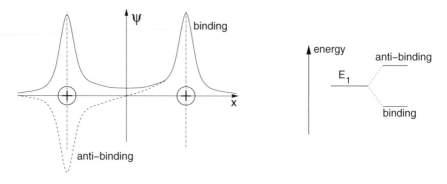

Figure 7.7. Left: A sketch of the wavefunction ψ in the H_2^+ molecule. It may be either *anti*-symmetric (ψ_a) and anti-binding, then the electron density is zero midway between the protons, or it may be *symmetric* (ψ_s) and binding, then the electron density is enhanced at $x = 0$ reducing the repulsive potential between the protons. Right: As the protons come closer, the degeneracy is removed and the atomic levels split into an unbinding and binding state.

found only in solids.

To estimate the binding energy of an ionic crystal, we write the potential φ_i at the locus of atom i resulting from all other atoms j in the form

$$\varphi_i = \sum_{j \neq i} \frac{\pm q^2}{r_{ij}} + \sum_{j \neq i} \lambda e^{-r_{ij}/\rho}. \tag{7.7}$$

In the first term representing the interaction between ions of charge q, the plus sign refers to like, the minus sign to unlike charges and r_{ij} is the distance between them. The second term heuristically describes the repulsive potential at very close distance. Because its range ρ is only about one-tenth of the equilibrium separation r_{eq} between adjacent atoms, one can replace the second sum by $n\lambda e^{-r/\rho}$, where n is the number of nearest neighbors and r the nearest-neighbor separation. The constants λ and ρ have to be determined from quantum mechanical calculations or from experiment. We put $p_{ij} = r_{ij}/r$ and introduce the *Madelung constant*

$$A = \sum_{j \neq i} \frac{(\pm)}{p_{ij}}. \tag{7.8}$$

If atom i is negative, there is a plus sign when atom j is positive and a minus sign when it is negative. For a crystal of N ion *pairs*, the total potential energy becomes

$$\Phi_{\text{tot}} = N\varphi_i = N\left(n\lambda e^{-r/\rho} - \frac{q^2}{r} A\right).$$

The equilibrium distance r_{eq} follows from $d\Phi_{tot}/dr = 0$, which yields

$$\Phi_{tot} = -\frac{Nq^2 A}{r_{eq}}\left(1 - \frac{\rho}{r_{eq}}\right).$$

The Madelung constant A is evaluated in special routines. It must obviously be positive as Φ_{tot} is negative; for NaCl, $A = 1.748$.

Pure ionic bonding is, however, an idealization; one finds it neither in molecules nor in crystals. Even when compounding alkali metals with halogenes, there is some homopolar component or overlap of orbitals. In the extreme case of CsF, homopolar binding still amounts to 5%. There are also no pure covalent bonds, even in H_2 or diamond, because of the fluctuations of the electron clouds. In the real world, there are only blends between the two bonding types.

7.2.3 Metals

Another type of strong binding exists in metals. One may imagine a metal as a lattice of positive ions bathed in a sea of free electrons; metallic binding is therefore non-directional.

7.2.3.1 Sodium

As as an example, we consider the alkali metal sodium (Na), the 11th element of the Periodic Table with an electron configuration $1s^2 \, 2s^2 \, 2p^6 \, 3s^1$. The two inner shells are complete and the 3s valence electron is responsible for binding. In sodium metal, the 3s electrons form the conduction band.

In an *isolated* sodium atom, the 3s electron is bound to the Na^+ ion by an ionization energy $\chi_{Na} = 5.14$ eV. The electron is localized near the atomic nucleus and its wavefunction ψ goes to zero at large distances from the nucleus. When, however, the 3s electron is in the potential of a lattice of Na^+-ions, as in a metal, its wavefunction ψ is very extended and subject to a different boundary condition: symmetry requires that the gradient of the wavefunction vanishes midway between atoms. The quantum mechanical calculations now yield a binding energy of the electron to the ionic lattice of 8.2 eV.

Because sodium has a density of 1.01 g cm^{-3}, the concentration of 3s electrons is 2.65×10^{22} cm^{-3} and the Fermi energy after (6.54) $E_F = 3.23$ eV. From (6.54) also follows the density of states,

$$\rho(E) = \frac{dN}{dE} = \frac{V}{2\pi^2}\left(\frac{2m_e}{\hbar^2}\right)^{3/2} E^{1/2}.$$

The *average* energy $\langle E \rangle$ of a conduction electron, is therefore,

$$\langle E \rangle = N^{-1}\int \rho(E)\,dE = \tfrac{3}{5}E_F \simeq 1.9 \text{ eV}.$$

Consequently, in a metal each sodium atom is bound by $8.2 - \chi_{Na} - 1.9 \simeq 1.16$ eV.

7.2.3.2 Conductivity and heat capacity

The free and mobile electrons of a metal are responsible for the high electric conductivity. In a constant electric field E, they are accelerated but collide after an average time τ with phonons (thermally oscillating atoms). The resulting current density J is equal to the charge density ne, where n is the number of free electrons per cm^3, multiplied by their drift velocity v. An electron experiences an acceleration eE/m_e over an average time τ, so that $v = \tau e E/m_e$. Because Ohm's law (1.112) asserts $J = \sigma E$, the conductivity σ is

$$\sigma = \frac{ne^2}{m_e}\tau.$$

At very low temperatures, collisions of the electrons with lattice imperfections are more important than collisions with phonons.

Because of free electrons, the heat capacity has a peculiar behavior at very low temperatures. Below the Debye temperature, which is typically several hundred K and thus not low, the specific heat C due to the lattice falls rapidly, like T^3 (see (8.39)). The heat capacity of the electron gas, however, declines less swiftly, it is proportional to T and, therefore, dominates at very low temperatures. Crudely speaking, when there are n free electrons per cm^3, heating of the metal from zero Kelvin to temperature T thermally excites those conduction electrons which lie T degrees below the Fermi temperature T_F, altogether a fraction T/T_F. Their total mean energy is then $U \sim nkT$ and, therefore,

$$C = \frac{\partial U}{\partial T} \sim \frac{nkT}{T_F} \propto T.$$

For an exact relation, one has to use the Fermi distribution (5.37).

7.2.3.3 Energy bands and Bloch functions

It is a general property of crystals that the electronic states are clustered in bands, with energy gaps between them. The basics of their formation was explained in section 6.4.3 on one-dimensional periodic potentials. To derive the energy states and eigenfunctions of an electron in the three-dimensional potential $U(\mathbf{x})$, we first note that as $U(\mathbf{x})$ comes from the lattice ions and is strictly periodic, it may, from (7.2), be written as

$$U(\mathbf{x}) = \sum_{\mathbf{G}} U_{\mathbf{G}}\, e^{i\mathbf{G}\cdot\mathbf{x}} \tag{7.9}$$

where the $\mathbf{G}$ are the reciprocal lattice vectors of (7.3). The general solution for the wavefunction $\psi(\mathbf{x})$ is sought in a Fourier expansion,

$$\psi(\mathbf{x}) = \sum_{\mathbf{k}} C_{\mathbf{k}}\, e^{i\mathbf{k}\cdot\mathbf{x}} \tag{7.10}$$

where the wavenumber $\mathbf{k}$ fulfils the periodic boundary conditions and is connected to the momentum $\mathbf{p}$ by

$$\mathbf{p} = \hbar \mathbf{k}.$$

The coefficients $C_{\mathbf{k}}$ are found by inserting (7.10) into the Schrödinger equation:

$$\hat{H} \psi(\mathbf{x}) = \left[\frac{\hat{\mathbf{p}}^2}{2m_e} + \hat{U}(\mathbf{x}) \right] \psi(\mathbf{x}) = E\psi(\mathbf{x}) \qquad (7.11)$$

and solving the resulting set of algebraic equations. In (7.11), the interaction among electrons is neglected. After some algebra, one is led to *Bloch's theorem*, namely that the eigenfunction $\psi_{\mathbf{k}}(\mathbf{x})$ for a given wavenumber $\mathbf{k}$ and energy $E_{\mathbf{k}}$ must have the form

$$\psi_{\mathbf{k}}(\mathbf{x}) = u_{\mathbf{k}}(\mathbf{x})\, e^{i\mathbf{k}\cdot\mathbf{x}}$$

where $u_{\mathbf{k}}(\mathbf{x})$ is a function periodic with the crystal lattice. So if $\mathbf{T}$ is a translation vector as given in (7.1), then

$$u_{\mathbf{k}}(\mathbf{x}) = u_{\mathbf{k}}(\mathbf{x} + \mathbf{T}).$$

The wavefunction $\psi_{\mathbf{k}}(\mathbf{x})$ itself does not generally display the periodicity. It is straight forward to prove Bloch's theorem in a restricted form, for a one-dimensional ring of N atoms, where after N steps, each of the size of the grid constant, one is back to the starting position.

It is instructive to consider the formation of bands from the point of view of the tightly bound inner electrons of an atom. When we imagine compounding a solid by bringing N free atoms close together, the inner electrons are afterwards still tightly bound to their atomic nuclei, only their orbits have become disturbed. Because of the disturbance, a particular energy state E_i of an isolated atom will be split into a band of N substates due to the interaction with the other $N - 1$ atoms. The shorter the internuclear distance, the stronger the overlap of the wavefunctions and the broader the band; the energetically deeper the electrons, the narrower the band.

There will be a 1s, 2s, 2p, ... band. One can determine the energy states of a tightly bound electron by assuming that its wavefunction is a linear superposition of the atomic eigenfunctions. In this way, one finds the width of the band and the average energy in a band. The latter is lower than the corresponding energy in a free atom, which implies binding of the crystal.

7.2.4   van der Waals forces and hydrogen bridges

The van der Waals interaction is much weaker than the bonding types discussed earlier and is typically only 0.1 eV per atom. It arises between neutral atoms or molecules and is due to their dipole moments. The dipoles have a potential U that rapidly falls with distance ($U \propto r^{-6}$, see section 9.1).

If the molecules do not have a permanent dipole moment, like CO_2, the momentary quantum mechanical fluctuations of the electron cloud in the neighboring atoms induces one; its strength depends on the polarizability of the molecules. This *induced* dipole moment is usually more important even for polar molecules. Its attraction holds CO_2 ice together and binds the sheets in graphite.

Hydrogen can form bridges between the strongly electron-negative atoms N, O or F; the binding energy is again of order ~ 0.1 eV per atom. For example, when a hydrogen atom is covalently bound to oxygen, it carries a positive charge because its electron has been mostly transferred to the oxygen atom. The remaining proton can thus attract another negatively charged O atom.

The double helix of the DNA is bound this way but so also is water ice. In solid H_2O, every oxygen atom is symmetrically surrounded by four others with hydrogen bonds between them. Because of these bonds, the melting and evaporation point of water is unusually high compared, for instance, to H_2S, which otherwise should be similar as sulphur stands in the same Group VI of the Periodic Table right below oxygen. This thermodynamic peculiarity of water is crucial for the existence of life.

7.3   Reddening by interstellar grains

Our knowledge of interstellar dust comes from the following measurements:

- *Interstellar extinction.* The interstellar extinction or reddening curve specifies how dust weakens the light from background stars as a function of wavelength.
- *Dust emission.* We receive this from all kinds of objects: protostars, old stars with shells, interstellar clouds and whole galaxies.
- *Infrared resonances.* These are features observed in absorption and emission which allow us to identify the chemical components of the dust material.
- *Polarization.* It refers to light from stars behind dust clouds but also to scattered radiation and dust emission.
- *Scattered starlight.* Examples are reflection nebulae, the diffuse galactic light or the zodiacal light of the solar system.

The items in italics are our five observational pillars. In this section, we are concerned with interstellar extinction and the clues it provides to dust properties, the other pillars are discussed elsewhere. We first describe the principles of stellar photometry because it is the basis for the reddening curve.

7.3.1   Stellar photometry

The interstellar extinction curve is obtained from stellar photometry. The wavelength resolution in such measurements is usually poor, typically $\lambda/\Delta\lambda \simeq 10$ but the broadness of the observational bands enhances the sensitivity. The standard photometric system is due to *H L Johnson* and *W W Morgan* and rooted

Table 7.2. The standard photometric system. The center wavelengths λ_c and the conversion factors w_λ between magnitudes and Jansky after (7.14) are averages gleaned from the literature. The last column gives the conversion factors for a blackbody of 9500 K, see text.

Band	Historical meaning	λ_c (μm)	w_λ	w_λ (bb)
U	Ultraviolet	0.365	1810	2486
B	Blue	0.44	4260	2927
V	Visual	0.55	3640	3084
R	Red	0.70	3100	2855
I	Infrared	0.90	2500	2364
J		1.25	1635	1635
H		1.65	1090	1116
K		2.2	665	715
L		3.7	277	294
M		4.8	164	184
N		10	37	46
Q		20	10	12

in optical astronomy. It was later expanded into the infrared; there the choice of wavelengths was dictated by the transmission of the atmosphere of the Earth. The observational bands are designated by letters: U, B, V, R, I, ... Table 7.2 lists their approximate center wavelengths λ_c and the conversion factors, w_λ, for translating magnitudes into Jansky and back (see (7.14)). The precise effective observation wavelength follows only after convolving the spectrum of the source with the transmission of the instrument and the atmosphere.

In photometry, especially at shorter wavelengths, it is customary to express the brightness of an object in apparent magnitudes. These are logarithmic quantities and they were appropriate units in the days when the human eye was the only detector because according to the *psycho-physical rule* of *W Weber* and *G T Fechner*: the subjective impression of the eye changes proportionally to the logarithm of the physical flux. But to the pride of many astronomers, magnitudes are still in use today. As they are laden with five millenia of history, only he who has thoroughly studied the 5000 year period can fully appreciate their scientific depth. An exhaustive explanation of magnitudes is, therefore, beyond the scope of this book.

The brightest star of all, Sirius, has a visual *apparent magnitude* $m_V = -1.58$ mag. Capella, Rigel and Vega are around zeroth and Spica (α Virginis) is only of first magnitude. Generally, a step of one up in magnitude means a factor of 2.5 down in observed flux. Whereas the apparent magnitude m_λ at wavelength

λ depends on the stellar distance D, the *absolute magnitude* M_λ, defined by

$$m_\lambda - M_\lambda = 5\log_{10}\left(\frac{D}{\text{pc}}\right) - 5 \tag{7.12}$$

does not and thus measures the intrinsic brightness of a star. At a distance of 10 pc, the apparent and absolute magnitude are, by definition, equal. When there is intervening dust, one adds, to the right-hand side of equation (7.12) a term A_λ to account for the weakening of starlight through interstellar extinction:

$$m_\lambda - M_\lambda = 5\log_{10}\left(\frac{D}{\text{pc}}\right) - 5 + A_\lambda. \tag{7.13}$$

A_λ is related to the optical depth τ_λ through

$$A_\lambda = 1.086\tau_\lambda.$$

The strange conversion factor of 1.086 between the two quantities arises from their definitions: stellar magnitudes are scaled by a factor of 2.5, the optical depth by $e \simeq 2.718$. The term $m - M$ in (7.12) is called the *distance modulus*. So ingenious astronomers have contrived a means to express length in stellar magnitudes.

To convert a flux F_λ in Jy into an apparent magnitude m_λ or vice versa, we use the relation

$$m_\lambda = 2.5\log_{10}\left(\frac{w_\lambda}{F_\lambda}\right). \tag{7.14}$$

F_λ is expressed in Jy, w_λ is given in table 7.2. The formula is simple, the difficulty lies in the calibration factors w_λ. The values in the literature scatter by about 10% and, furthermore, they do not refer to identical center wavelengths because of the use of different filters.

The magnitude difference between two bands is called the *color*. It is equivalent to the flux ratio at the corresponding wavelengths and thus determines the gradient in the spectral energy distribution. By definition, the colors of main sequence A0 stars are zero. The most famous A0V star is α Lyr (= Vega, $D = 8.1$ pc, $m_V = 0.03$ mag, $L = 54\,L_\odot$). Such stars have an effective temperature of about 9500 K and they can be approximated in the infrared by a blackbody. Indeed, if we calibrate the emission of a blackbody at 9500 K in the J band and put all colors to zero, we find the conversion factors $w_\lambda(\text{bb})$ listed in table 7.2. In the infrared, they are not very different from w_λ (adjacent column); however, the discrepancies are large at U, B and V because of the many absorption lines in the stellar spectrum that depress the emission relative to a blackbody.

7.3.2 The interstellar extinction curve

7.3.2.1 *The standard color excess E_{B-V}*

The interstellar extinction curve is obtained from photometry at various wavelengths λ on two stars of identical spectral type and luminosity class, one of which is reddened (star No. 1), whereas the other is not (star No. 2). In the case of pure extinction, without dust emission along the line of sight, one receives from the stars the flux

$$F_i(\lambda) = \frac{L(\lambda)}{4\pi D_i^2} \cdot e^{-\tau_i(\lambda)} \qquad i = 1, 2.$$

$L(\lambda)$ and $\tau_i(\lambda)$ are the spectral luminosity and optical thickness. Because the apparent photometric magnitude m is a logarithmic derivative of $F_i(\lambda)$,

$$m_i(\lambda) = \ln L(\lambda) - \tau_i(\lambda) + 2 \ln D_i + \text{ constant}$$

and because $\tau_2 = 0$, the difference in magnitude $\Delta m(\lambda)$ between the two stars becomes

$$\Delta m(\lambda) = -\tau_1(\lambda) - 2 \ln\left(\frac{D_1}{D_2}\right).$$

The difference in Δm at two wavelengths, λ and λ', gives the *color excess*

$$E(\lambda, \lambda') = \Delta m(\lambda) - \Delta m(\lambda') = -[\tau_1(\lambda) - \tau_1(\lambda')]$$

which no longer contains the distance D. At $\lambda = 0.44\ \mu\text{m}$ and $\lambda' = 0.55\ \mu\text{m}$, the center wavelengths of the B and V band, the color excess is denoted by

$$E_{B-V} = E(B, V)$$

and called the standard color excess. The band symbol is also used for the apparent magnitude, for example, $V = m_V$ or $B = m_B$. The intrinsic color is denoted $(B - V)_0$ in contrast to the observed color $B - V$, which includes the effect of the selective weakening by interstellar dust. With this notation, we can write

$$E_{B-V} = (B - V) - (B - V)_0 \tag{7.15}$$

and likewise for any other pair of wavelengths.

7.3.2.2   *How the extinction curve is defined*

Fixing the wavelength λ' in the color excess $E(\lambda, \lambda')$ at $0.55\ \mu\text{m}$ and normalizing $E(\lambda, V)$ by E_{B-V}, one arrives at the extinction curve in its traditional form,

$$\text{Ext}(\lambda) = \frac{E(\lambda, V)}{E_{B-V}} = \frac{A_\lambda - A_V}{A_B - A_V} = \frac{\tau_\lambda - \tau_V}{\tau_B - \tau_V}. \tag{7.16}$$

A_λ is the extinction in magnitudes at wavelength λ; it is the result of absorption plus scattering but observationally one cannot separate the two. The normalization of the extinction curve by E_{B-V} is important because it allows a comparison of the wavelength dependence of extinction towards stars which suffer different amounts of reddening. Obviously,

$$\text{Ext}(B) = 1 \qquad \text{Ext}(V) = 0.$$

The quantity

$$R_V = \frac{A_V}{E_{B-V}} = -\text{Ext}(\lambda = \infty) \qquad (7.17)$$

is called the *ratio of total over selective extinction* or, simply, R_V. The visual extinction in magnitudes A_V, which also appears in the distance modulus of (7.13), is the most common quantity to characterize the opaqueness of a cloud. It is not a directly observable quantity. To obtain A_V from photometry, one measures E_{B-V}, assumes an R_V value and then uses (7.17).

One can also express the reddening by dust through the mass extinction coefficient K_λ normalized at the V band. Because $K_\lambda/K_V = \tau_\lambda/\tau_V$, one gets

$$\frac{\tau_\lambda}{\tau_V} = \frac{\text{Ext}(\lambda)}{R_V} + 1 \qquad (7.18)$$

τ_λ/τ_V and $\text{Ext}(\lambda)$ are mathematically equivalent. Still other forms, containing the same information, are sometimes more suitable for displaying certain trends, for example, A_λ/A_J (the J band is at 1.25 μm) or $(A_\lambda - A_J)/(A_V - A_J)$.

7.3.2.3   Remarks on the reddening curve

The interstellar extinction curve (figure 7.8) is observationally determined from about 0.1 to 10 μm. At longer wavelengths, the optical depth is small and the extinction of background light weak; moreover, the dust begins to emit itself. The reddening curve displays the following salient features:

- In the diffuse interstellar medium, $\text{Ext}(\lambda)$ of equation (7.16) is, at optical and infrared wavelengths, fairly uniform over all directions in the sky and R_V equals 3.1; deviations occur only in the UV.
- Towards clouds, however, the shape of the curve $\text{Ext}(\lambda)$ varies from one source to another, even in the infrared. The reduction in UV extinction for large R_V suggests that the small grains have disappeared. Photometric data exist only for cloud edges, their cores are too obscured.
- The extinction curve refers to broad-band observations but nevertheless it is remarkably smooth. If one plots instead of $\text{Ext}(\lambda)$ the ratio τ_λ/τ_V then, to first order, *all* extinction curves look alike for $\lambda \geq 0.55$ μm; at shorter wavelengths, they can differ substantially. The curves are largely (but not totally) fixed over the *entire* wavelength range by the one parameter R_V of equation (7.17). The observed variations of R_V range from 2.1 to well over 5, the larger values being found towards molecular, the smaller ones toward high latitude clouds.
- The only resonance in the extinction curve is the broad bump at 4.6 μm^{-1} or 2175 Å which is always well fit by a Drude profile. The central position of the bump stays constant from one star to another to better than 1% implying that the underlying particles are in the Rayleigh limit (sizes $<$ 100 Å); however, its width varies ($\Delta\lambda^{-1} \simeq 1 \pm 0.2$ μm^{-1}). There is also scatter

in the strength of the bump. It is correlated with R_V in the sense that the
resonance is weak in clouds where R_V is large. A plausible explanation is
coagulation of dust particles in clouds as a result of which the small grains
disappear.

- From the infrared to the hump at 2175 Å, the extinction curve rises
  continuously. Over this range, stellar light is reddened because extinction
  decreases with wavelength. Beyond 2175 Å, the ratio τ_λ/τ_V first declines
  and here the light is made bluer(!). In the far UV ($\lambda^{-1} > 6\,\mu m^{-1}$), the curve
  rises again.
- If one had good albedo measurements (which one does not have), one could
  subtract scattering from extinction to obtain a pure absorption curve. This
  would provide a touchstone for any dust model.
- The two extinction curves in figure 7.8 suggest that the grains are modified
  when they change their environment. This should happen several times
  during their lifetime as they pass from dense into diffuse medium and back.
  The modifications may be due to energetic photons, condensation of gas
  atoms or collisions among grains. The latter would imply fluffy particles
  built up of smaller subunits (section 2.6).

At higher spectral resolution, one finds superposed on the extinction curve
*d*iffuse *i*nterstellar *b*ands (DIBs but they are not shown in figure 7.8). There are
altogether more than a hundred, mainly between 0.4 and 0.9 μm, of variable
strength, shape and width, the broadest being about 2 Å wide. Although
discovered 80 years ago, the identification of the DIBs is still unclear.

7.3.3 Two-color diagrams

In *t*wo-*c*olour-*d*iagrams (TCDs), one plots one color against another. TCDs are
a simple and efficient tool with which to separate distinct or identify similar
astrophysical objects. As an example of a TCD, we show in figure 7.9 the UBV
diagram of unreddened main sequence stars. All luminous stars of type B0 or
earlier cluster in a tight strip with B–V between −0.30 and −0.32 mag. The
position in the UBV diagram of blackbodies with temperatures from 3500 to
40 000 K is always well above the main sequence. So compared to a blackbody
of the same color B–V, a stellar photosphere is much weaker at U. The reason
for this behavior is that photoionization of the $n = 2$ level in hydrogen falls into
the U-band. As this is the dominant process for the photospheric opacity at this
wavelength, it greatly supresses the stellar flux. The characteristic wiggle in the
main sequence line in figure 7.9 reflects the variation of this supression (called
the Balmer decrement) with effective temperature.

Interstellar reddening with the standard extinction law ($R_V = 3.1$) shifts any
star in the UBV diagram from the main sequence in the direction parallel to the
arrow of figure 7.9. The length of the shift is proportional to the foreground A_V
and in figure 7.9, it corresponds to $A_V = 5$ mag. In the case of another extinction
law, say, with $R_V = 5$, the slope and length of the arrow would be different

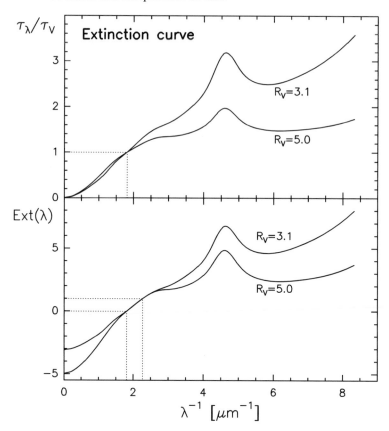

Figure 7.8. The observed interstellar extinction curve in the form τ_λ/τ_V and as Ext(λ) according to equation (7.16). The ratio of total over selective extinction, $R_V = 3.1$, refers to the diffuse medium, $R_V = 5$ to the edges of molecular clouds. Whereas τ_λ/τ_V is normalized at V, Ext(λ) is zero at V and equals one at B.

(for the same A_V). If we observe an object somewhere in the right-hand part of figure 7.9 and know, for some reason, that it is a main sequence star, we can determine the amount of extinction by moving it up, parallel to the arrow, until it reaches the main sequence. This dereddening is unique as long as the star is of type B5 or earlier so that the reddening path does not intersect the wiggle of the main sequence curve.

7.3.4 Spectral indices

Akin to a color is the spectral index, commonly defined in the infrared as

$$\alpha = \frac{d \log(\lambda F_\lambda)}{d \log \lambda}.$$

(7.19)

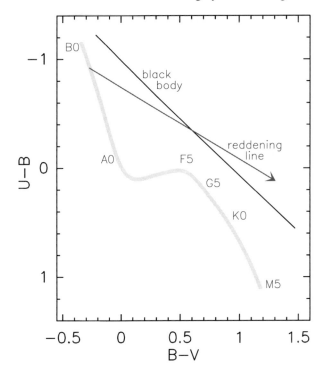

Figure 7.9. UBV diagram for unreddened main sequence stars (thick line) and for blackbodies (thin line). The arrow shows the direction and distance by which a star is displaced from the main sequence under the standard reddening law with $R_V = 3.1$ (figure 7.8) and $A_V = 5$ mag of foreground extinction. The main sequence locations of a few spectral types are indicated. The lower tip of the blackbody line refers to a temperature of 3500 K, the upper to 40 000 K. The data for the main sequence are taken from the literature, the blackbody and reddening curve have been calculated. The figure is, therefore, quite accurate and may be used to find the position of a main sequence star that suffers a certain amount of extinction A_V; see text.

Hence, a spectral index depends on wavelength. In practice, α is calculated not as a derivative but as the slope of λF_λ between two wavelengths λ_1 and λ_2:

$$\alpha = \frac{\log(\lambda_2 F_{\lambda_2}) - \log(\lambda_1 F_{\lambda_1})}{\log \lambda_2 - \log \lambda_1}. \qquad (7.20)$$

For a blackbody obscured by a foreground extinction $\tau(\lambda)$, one has, in the limit of high temperatures,

$$\alpha = -3 - \frac{\tau(\lambda_2) - \tau(\lambda_1)}{\ln(\lambda_2/\lambda_1)}. \qquad (7.21)$$

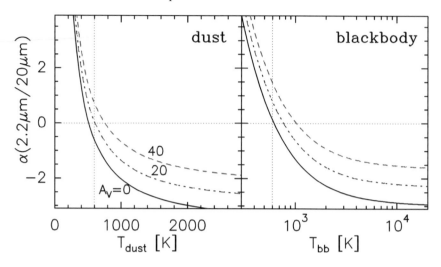

Figure 7.10. The spectral index α in the wavelength interval from 2.2 to 20 μm after (7.20) of a source at temperature T that is observed through a foreground of visual extinction A_V. In the right box, the object is a blackbody or, to first approximation, a star. In the left box it is an optically thin dust cloud with the standard mixture of silicate and carbon grains (see section 12.4). The curves show how the spectral index is modified as the foreground extinction increases from 0 to 20 mag and 40 mag. Because α is sensitive to both T and A_V, it is used in the classification of protostars. Without extinction, α of a blackbody approaches -3 at high temperatures. The weak dotted lines are drawn to alleviate comparison between the two frames. Notice the different temperature ranges right and left.

An important example is the spectral index between 2.2 and 20 μm which serves as a classifier for protostellar objects. In this case, for standrad dust (see figure 7.20 or 12.10)

$$\alpha \rightarrow -3 + A_V \cdot 0.036 \qquad \text{for } T \rightarrow \infty.$$

Of course, other wavelength intervals fulfil the same purpose; for instance, one also uses the index between 2.2 and 5 μm. Figures 7.10 and 7.11 demonstrate how the indices change as a function of source temperature.

7.3.5 The mass absorption coefficient

The reddening curve yields only a normalized extinction coefficient, for instance, K_λ / K_V. To obtain the absolute value of the cross section per gram of interstellar matter, K_λ, one has to know K_V. For this purpose, one measures towards a star the standard color excess E_{B-V} and the total hydrogen column density N_H to

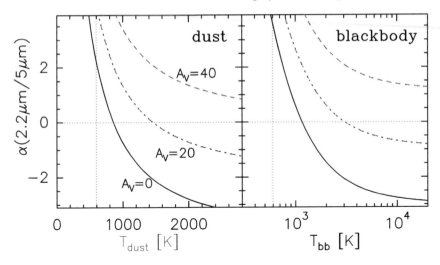

Figure 7.11. As figure 7.10 but for the spectral index α between 2.2 and 5 μm.

which both atomic and molecular hydrogen contribute:

$$N_H = N(\text{HI}) + 2N(\text{H}_2).$$

The observations are performed in the far UV from a satellite and consist of absorption measurements towards early-type stars in the Lyα line of HI and the Werner and Lyman bands of H_2. For technical reasons, one chooses stars that are only slightly reddened and located in the diffuse interstellar medium. The average value over a large sample of stars is

$$N_H \simeq 5.8 \times 10^{21} E_{B-V} \; \text{cm}^{-2} \tag{7.22}$$

or with (7.17) and $R_V = 3.1$,

$$N_H \simeq 1.9 \times 10^{21} A_V \; \text{cm}^{-2}. \tag{7.23}$$

As $A_V = 1.086 \times N_H K_V$, one finds $K_V \simeq 4.9 \times 10^{-22}$ cm^2 per H-atom and a mass extinction coefficient

$$K_V \simeq 200 \; \text{cm}^2 \; \text{per g of interstellar matter.} \tag{7.24}$$

This value is easy to remember. In combination with the interstellar reddening curve, we know K_λ in the range from about 0.1 to 5 μm. Equation (7.23) also allows us to estimate the dust-to-gas mass ratio R_d if one makes reasonable assumptions about the composition of the grains, their shape and sizes. The amount of dust relative to the gas follows then from the condition (7.22) that a hydrogen column density of 5.8×10^{21} cm^{-2} produces a standard color excess

E_{B-V} of 1 mag. Any dust model compatible with the interstellar reddening curve gives, quite independently of its particular choice,

$$R_d = \frac{M_{dust}}{M_{gas}} \simeq 0.5 \ldots 1\%. \tag{7.25}$$

The lower limit may be appropriate for the diffuse interstellar medium, the higher for dense clouds where grains are ice-coated. Because at least half a percent of interstellar matter is in the solid phase, the dust can only consist of the most abundant elements, like C, N, O, Fe, Si, or Mg.

According to equations (7.23) and (7.25), the total dust volume in a column of 1 cm^2 cross section with $A_V = 1$ mag is $V_{dust} \sim 1.2 \times 10^{-5}$ cm^{-3}, assuming a bulk density of 2.5 g cm^{-3}. Considering the uncertainties, this is in agreement with the dust-to-gas ratio derived from the Kramers–Kronig relation in (2.95) but the method employed here is probably superior.

How dusty is the interstellar medium? Very dusty! If the atmosphere of the Earth, which has a density $\rho \sim 10^{-3}$ g cm^{-3}, had the same relative dust content, the optical depth $\tau = \ell \rho K_V$ would become unity over a distance of only $\ell = 10$ cm and we could not see our feet!

7.4 Carbonaceous grains and silicate grains

7.4.1 Origin of the two major dust constituents

Interstellar grains, at least their seeds, cannot be made in the interstellar medium, they can only be modified there or destroyed. Observational evidence points towards the wind of red giants on the asymptotic branch of the Hertzsprung–Russell diagram as the place of their origin (section 9.4). At first, tiny refractory nuclei are created in the inner part of the circumstellar envelope which then grow through condensation as they traverse the cooler outer parts (section 9.5). The type of dust that is produced depends on the abundance ratio of carbon to oxygen:

- If there is less carbon than oxygen, [C]/[O] < 1, as in M-type and in OH/IR stars, no carbon is available for the solid phase because all C atoms are locked up in CO which is a diatomic molecule with an exceptionally strong bond.

  In such an environment, mainly silicates form, besides minor solid constituents like MgS, MgO and FeO. The observational characteristic of silicates is the presence of an Si–O stretching and an O–Si–O bending mode at 9.8 μm and 18 μm, respectively. Silicon has a cosmic abundance [Si]/[H] = 3×10^{-5}, it is heavily depleted in the gas phase (depletion factor $f_{depl}(Si) \sim 30$), so almost all of it resides in grains.

- If [C]/[O] > 1, as in C-type stars (these are carbon-rich objects like C, R, N stars), carbonaceous grains form.

Carbon has a cosmic abundance $[C]/[H] = 4 \times 10^{-4}$, or somewhat less, and a depletion factor $f_{\text{depl}}(C) \sim 3$. Two out of three atoms are built into solids, one is in the gas phase. There are several forms of carbonaceous grains: mainly amorphous carbon, graphite and PAHs.

Carbon and silicates are the major constituents of interstellar dust. Other types of grains, for whose existence there are either observational or theoretical indications, include metal oxides (MgO, FeO, Fe_3O_4, Al_2O_3), MgS and SiC. The last substance, silicon carbide, is only seen in C-stars through its emission signature at 11.3 μm but it is not detected in the interstellar medium.

Generally, interstellar grains are beyond the reach of spacecraft, because of their distance, so they cannot be brought to Earth and studied in the laboratory. However, as the Sun moves relative to the local interstellar cloud (it is of low density, $n_H \sim 0.3$ cm^{-3}), dust particles of this cloud sweep through the solar system. Some of them were detected by spacecraft at the distance of Jupiter and identified as belonging to the local interstellar cloud by the direction and value ($\sim$26 km s^{-1}) of their velocity vector [Grü94]. In the process of detection they were destroyed. Their masses, which could also be determined, correspond to μm-sized grains which we think are rather atypical for the interstellar medium. Smaller grains are probably prevented from penetrating deep into the solar system by radiation pressure and, if they are charged, by the interplanetary magnetic field.

In interplanetary dust particles and in primitive meteorites which have been collected on Earth, one can identify inclusions which distinguish themselves by their isotopic pattern from the solar system. Because the isotopic abundance ratios are explainable in terms of nucleosynthesis in AGB stars or supernovae, these subparticles are hypothesized to be of interstellar origin. If so, they have probably undergone considerable reprocessing in the solar nebula.

7.4.2 The bonding in carbon

Carbon, the sixth element of the periodic system, has four electrons in its second ($n = 2$) uncompleted shell: two s-electrons denoted as $2s^2$ with angular momentum quantum number $l = 0$ and two unpaired p-electrons ($2p^2$) with $l = 1$, in units of $\hbar$.

The ground state of carbon is designated $^3P_0$. It is a triplet (prefix 3) with total spin $S = 1$, total orbital angular momentum $L = 1$ (letter P) and total (including spin) angular momentum $J = 0$ (suffix 0). In the ground state, there are only two unpaired p-electrons. However, one 2s-electron may be promoted to a 2p orbital and then all four electrons in the second shell become unpaired; this is the chemically relevant case where carbon has four covalent bonds. The promotion requires some energy (4.2 eV) but it will be more than returned in the formation of a molecule.

The bonding of carbon to other atoms is of the covalent type and there are two kinds:

- In a σ-bond, the electron distribution has rotational symmetry about the internuclear axis.
- In a π-bond, the wavefunction of the two electrons has a lobe on each side of the internuclear axis.

The s and p wavefunctions can combine linearly resulting in hybrid orbitals. The combination of one s and three p electrons is denoted as an sp^3 hybrid, likewise there are sp^2 and sp hybrids. Here are examples:

- In methane, CH_4, the coupling of carbon is completely symmetric to all four H atoms and the difference between s and p valence electrons has disappeared. In the identical hybridized sp^3 orbitals, the C–H binding is through σ-bonds. They point from the C atom at the center of a tetrahedron towards the H atoms at its corners. The angle α between the internuclear axes follows from elementary geometry, one gets $\cos \alpha = -\frac{1}{3}$, so $\alpha = 109.5°$.
- The situation is similar in ethane, H_3C–CH_3, although the symmetry is no longer perfect. Each of the two C atoms has again four sp^3 hybrid orbitals connecting to the other carbon atom and to the neighboring three hydrogen atoms. Because of the σ-bond between the two Cs, the CH_3 groups can rotate relative to each other (see figure 7.12).
- An sp^2 hybrid is present in ethylene, H_2C–CH_2, which is a planar molecule, where the angle from C over C to H equals 120°. There are σ-bonds between C–H and C–C. But the two remaining 2p orbitals in the carbon atoms, besides the sp^2 hybrids, overlap to an additional π-bond, so altogether we have a C=C double bond (one σ and one π), which makes the CH_2 groups stiff with respect to rotation (see figure 7.13).
- In the linear molecule acetylene, HC–CH, the C atoms are *sp* hybridized yielding C–C and C–H σ-bonds. There are two more π-bonds, whose orbitals are perpendicular to each other and to the internuclear axis; in total, the carbon atoms are held together by a triple bond, C≡C. Acetylene is believed to be the basic molecule for the nucleation of carbonaceous grains in the wind of giant stars (see figure 7.14).
- Benzene, C_6H_6, is an especially simple and symmetric hydrocarbon (a so called PAH, see later). In this planar ring (aromatic) molecule, each carbon atom has three sp^2 hybrid orbitals connecting through σ-bonds to the adjacent two C atoms and one H atom. There is still one unhybridized 2p electron per C atom left. The carbon atoms are sufficiently close together so that neighboring 2p orbitals can pair to form an additional π-bond; the p-orbitals are perpendicular to the plane. Because of the complete symmetry of the ring, the unhybridized electrons in the 2p-orbitals are not localized but equally shared (resonant) in the ring. So, on average, two adjacent C atoms are coupled by a σ- and half a π-bond. The binding is therefore intermediate in strength between a single and a double bond. The more bonds between two C atoms there are, the stronger they are bound and the shorter their internuclear distance (see figure 7.15).

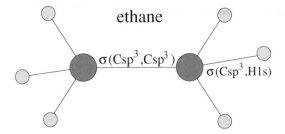

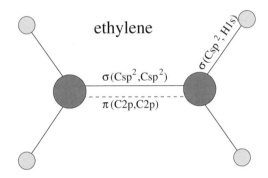

Figure 7.12. Four types of carbon binding are exemplified in figures 7.13–7.15. Here a single bond in ethane (C_2H_6); big circles, C atoms; small ones, H atoms. σ-bonds are drawn with full lines, π-bonds with broken lines. In the description of the bond, the type (σ or π) is followed by the atomic symbol and the participating orbitals, for instance, $\sigma(Csp^2, H1s)$.

Figure 7.13. Double bond in ethylene (C_2H_4).

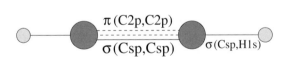

Figure 7.14. Triple bond in acetylene (C_2H_2).

7.4.3 Carbon compounds

7.4.3.1 *Diamond*

In diamond, a crystalline form of carbon, the bonding is similar to methane. Each C atom is linked tetrahedrally to its four nearest neighbors through sp^3 hybrid orbitals in σ-bonds. The distance between the atoms is 1.54 Å, the angle between the internuclear axes again 109.5°. As a C atom has only four nearest (and 12

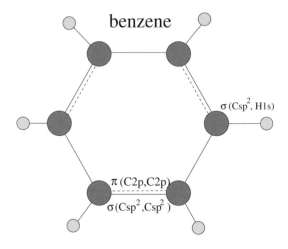

Figure 7.15. Resonance structure in benzene (C_6H_6) with three π-bonds shared among six C atoms.

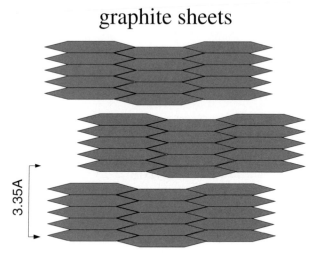

Figure 7.16. Graphite is formed by sheets of carbon atoms, each with an hexagonal honeycomb structure. The side length of the hexagons equals 1.42Å, the distance between neighboring sheets is 3.35 Å. In analogy to close packed spheres (figure 7.17), besides the sheet sequence ABABAB . . . also ABCABC . . . is possible and combinations thereof.

next nearest) neighbors, the available space is filled to only 34% *assuming* that the atoms are hard spheres. This is to be compared with a close-packed structure, either face-centered cubic or hexagonal close-packed (see figure 7.17), where in both cases each atom has 12 nearest neighbors and the volume filling factor amounts to 74%.

next higher layer B

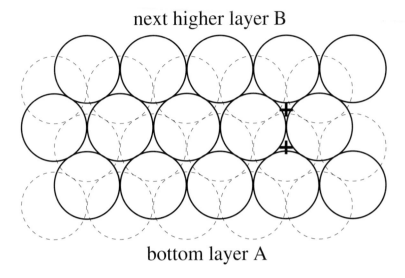

bottom layer A

Figure 7.17. A ground layer A of equal balls (dashed), each touching its six nearest neighbors, is covered by an identical but horizontally shifted layer B (full). There are two ways to put a third layer on top of B; one only needs to specify the position of one ball in the new layer, all other locations are then fixed. (a) When a ball is centered at the lower cross, directly over a sphere in layer A, one obtains by repetition the sequence ABABAB . . . . This gives an *h*exagonal *c*lose *p*acked structure (hcp). (b) When a ball is over the upper cross, one gets by repetition a sequence ABCABC . . . and a face-centered cubic lattice.

The bonding in diamond is strong (7.3 eV per atom) and due to its overall bonding isotropy, diamond is extremely hard with hardness number 10; diamond is used for cutting and drilling. The stiffness of the bonds also explains the exceptionally high thermal conductivity (several times greater than that of copper). As there are no free electrons, diamond is electrically insulating.

On Earth, diamonds are formed under high pressure and temperature at a depth of about 200 km. They are mined in Kimberlite pipes which are vents associated with volcanoes where the deep material is transported upwards so that it can be reached by man. The upward transport during the volcano eruption occurs sufficiently rapid without a phase transition. In fact, diamond is not in equilibrium at the conditions at the surface of the Earth, it is metastable in our environment but well separated by a large activation barrier from graphite, which is the energetically lower phase. So diamonds are long lasting presents that will outlive the affluent donors as well as the pretty recipients of the gems. Big raw diamonds are cleft parallel to octahedral crystal faces. In the subsequent cutting, however, into brilliants, which is the standard gem with a circular girdle and 33 facets above and 25 facets below the girdle, the angles between the facets are not related to the crystalline structure but to the refractive index n. The angles are chosen to maximize the fire (brilliance) of the stone.

Diamond grains are also found in meteorites but they are extremely small ($\sim$30 Å), so that not a negligible fraction of C atoms is at the surface. They contain many H atoms and their density ρ is considerably lower than in perfect crystals for which $\rho = 3.51$ g cm^{-3}. The origin of the meteoritic diamonds is unclear but probably interstellar (carbon stars, supernovae, interstellar shocks?).

7.4.3.2   Graphite

Carbon atoms combine also to planar structures consisting of many C_6 hexagonals. The distance between adjacent C atoms is 1.42 Å, the angle between them 120°. The bonding is similar to that in benzene as again the π-bonds are not localized in the rings and the electrons of the unhybridized 2p orbitals are free to move in the plane. This mobility endows graphite with an electric conductivity within the sheets. Graphite consists of such C_6 sheets put on top of each another. The stacking is arranged in such a way that above the center of an hexagon in the lower sheet there is always a C atom in the upper one. The distance from one sheet to the next is 3.35 Å (figure 7.16). The sheets are only weakly held together by van der Waals forces. Therefore, graphite cleaves well along these planes and owing to this property, its name is derived from the Greek word for writing: when gently pressing a pen over a piece of paper, tiny chunks peel off to form sentences of rubbish or wisdom, depending on who is pressing.

Graphite has a density of 2.23 g cm^{-3} and a very high sublimation temperature ($T_{ev} > 2000$ K at interstellar pressures). We reckon that in interstellar space about 10% of the carbon which is in solid form is graphitic.

7.4.3.3   PAHs

A *polycyclic aromatic hydrocarbon*, briefly PAH, is a planar compound made up of not too large a number of aromatic (C_6) rings (chapter 12). At the edge, hydrogen atoms are attached but only sporadically; the coverage need not be complete, as in benzene. Possibly one finds at the periphery also other species, besides hydrogen, like OH or more complex radicals. PAHs are unambiguously identified in the interstellar medium and account for a few percent by mass of the interstellar dust.

A fascinating kind of PAHs are fullerenes. The most famous, C_{60}, has the shape of a football. It is made of 12 pentagons (C_5), which produce the curvature, and 20 hexagons (C_6). The diameter of the spheres is $\sim$7.1 Å. Fullerines do not contain hydrogen. Their existence in interstellar space is hypothetical.

7.4.3.4   Amorphous carbon

Diamond and graphite are very regular (crystalline). If there are many defects and if regularity extends only over a few atoms, the material is said to be amorphous.

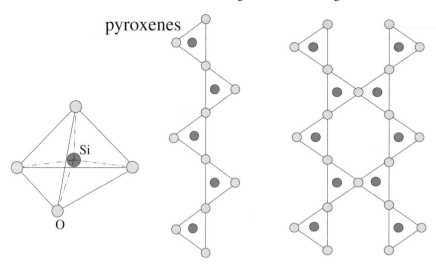

Figure 7.18. Left: The building blocks of silicates are SiO_4 tetrahedra. Middle: In pyroxenes, which have a chain structure, two adjacent tetrahedra share one oxygen atom; an example is bronzite, $(Mg,Fe)SiO_3$. The triangles represent the SiO_4 units. In interstellar grains, the chains are probably not very regular and linear, as in crystals but may more resemble worms because of widespread disorder. Right: Other chain types are realized in silicates, too; here a double chain where more than one O atom per tetrahedron is shared (amphiboles).

Most interstellar *c*arbon grains are probably *a*morphous (aC). These amorphous particles should also contain hydrogen because it is present in the environment where they are created (wind of mass loss giants); pure carbon grains could only come from the hydrogen deficient atmospheres of WC stars or R CrB stars. We may tentatively think of such *h*ydrogenated *a*morphous *c*arbon, abbreviated HAC, as a potpourri of carbon compounds with all types of the bonds discussed earlier, or as a disordered agglomerate of PAHs, some of them stacked, with many unclosed rings and an endless variety of carbon–hydrogen connections.

7.4.3.5 *The 2175 Å bump in the interstellar extinction curve*

The bump at 4.6 μm^{-1} in the interstellar extinction curve (figure 7.8) is usually attributed to small graphite-like grains (section 4.4.0.1). As we have seen in section 7.2.3, the σ- and π-electrons in such grains form the σ- and π-band, respectively. Since the π-states are weakly bound, the π-band lies energetically above the σ-band. When the electric vector of the radiation field is parallel to the plane of the carbon rings, the π-electrons can be excited into the π^*-band, and this may cause the 4.6 μm^{-1} hump. The splitting into a π- and π^*-band is analogous to the splitting into a binding and anti-binding state as illustrated in figure 7.7 for the H_2^+ molecule, only in a solid the levels are broadened because of the interaction with many lattice atoms. Note that whereas the π-band is filled, the π^*-band is empty and responsible for electric conduction. The assignment of the 4.6 μm^{-1} hump to $\pi \rightarrow \pi^*$ transitions is, however, controversial because of the overall messy character of the binding in carbon with its multitude of hard-to-identify bonds.

7.4.4   Silicates

7.4.4.1   *Silicon bonding and the origin of astronomical silicates*

Silicon stands right below carbon in Group IV of the Periodic Table. Its inner two shells are filled, in the third ($n = 3$) it has, like carbon, two s and two p electrons ($3s^2\, 3p^2$ configuration). The ground state is also $^3P_0$. There are two basic reasons why silicon is chemically different from carbon. First, it takes more energy to promote an s electron in order that all four electrons become available for covalent bonding. Second, silicon atoms are bigger, so their p orbitals cannot come together closely enough to overlap and form a π-bond. Therefore, there are no silicon ring molecules and double bonding (Si=Si) is rare (it involves d-electrons).

The dust particles formed in oxygen-rich atmospheres are predominantly silicates. Silicates consist of negatively charged SiO_4 groups in the form of tetrahedra with a side length of 2.62 Å. They are the building blocks which form an ionic grid together with positively charged Mg and Fe cations. We may idealize an SiO_4 tetrahedron as an Si^{4+} ion symmetrically surrounded by four O^{2-} ions; however, there is also substantial covalent bonding in the tetrahedron.

7.4.4.2   *Coordination number*

It is intuitively clear that the relative size of adjacent atoms in a crystal is important for the crystal structure. Let us define

$$r = \text{ratio of the radii of neighboring atoms.}$$

In silicates, the radius of the central Si^{4+} ion is 0.41 Å and much smaller than the radius of the O^{2-} ion which is 1.40 Å, therefore, $r = 0.41/1.4 \simeq 0.3$. This implies, first, that oxygen fills basically all the volume and, second, that the coordination number N of a Si^{4+} ion, which is the number of its nearest equal neighbors, is four.

In other crystals, r is different. The coordination number and the coordination polyhedron for the limiting values of r, which assumes that the ions are hard spheres touching each other, are summarized in table 7.3. For crystals with $0.23 < r \leq 0.41$, the coordination number N equals 4 and the nearest neighbors form a tetrahedron; for $0.41 < r \leq 0.73$, one has $N = 6$ and an octahedron as coordination polyhedron. The case $r = 1$ is depicted in figure 7.17.

Table 7.3. The limiting ratio of the radius of the central ion to that of its nearest neighbors determines the coordination number and the kind of polyhedron around it.

Ratio r of ionic radii	Coordination number	Polyhedron around central ion
0.15	3	flat triangle
0.23	4	tetrahedron
0.41	6	octahedron
0.73	8	cube
1	12	hexagonal or cubic close-packed

7.4.4.3 Silicate types

The (negative) $[SiO_4]$ anions and the (positive) metal cations can be regularly arranged in various ways. The basic silicate types are determined by the following criteria:

- No oxygen atom is common to two tetrahedrons. The latter are then isolated islands. The prototype is olivine $(Mg,Fe)_2SiO_4$ which belongs to the orthorhombic crystal family. By writing (Mg,Fe), one indicates that magnesium is often replaced by iron. To visualize the three-dimensional structure of this mineral, imagine that each Mg^{2+} or Fe^{2+} cation lies between six O atoms of two independent tetrahedra. These six O atoms are at corners of an octahedron around the cation. The ratio r of the ionic radius of the metal cations to that of O^{2-} is about 0.5. Because the O atoms are so big, they form approximately an hexagonal close packed structure (figure 7.17).
- One, two, three or even all four oxygen atoms are shared with neigboring tetrahedra. We mention bronzite, $(Mg,Fe)SiO_3$, also of the orthorhombic crystal family (the form without Fe is called enstatite, $MgSiO_3$). In this mineral, there are chains of SiO_4 tetrahedra in which two O atoms are common to neighboring units. The repeating unit has the formula $[SiO_3]^{2-}$.

7.4.5 A standard set of optical constants

We have learnt in this section what dust is made of. We can now proceed to present a reasonable set of optical constants $m(\lambda)$ for the two major interstellar solid substances, amorphous carbon and silicate. The wavelength ranges from 0.03 to 2000 μm, which is a sufficiently broad interval for most applications. We will call these data our *standard set* and it may be used as a benchmark for comparison between dust models. For some applications, one also needs $m(\lambda)$ for the two minor solid constituents: graphite and PAHs (section 12.4).

Optical constants vary with time; at least, in the literature they are. The question whether the choice displayed in figure 7.19 is up-to-date or out-of-date

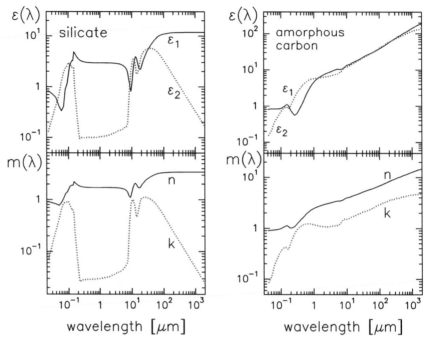

Figure 7.19. Optical constant $m = (n, k)$ and dielectric permeability $\varepsilon = (\varepsilon_1, \varepsilon_2)$ for silicate material after [Lao93] and amorphous carbon after [Zub96] (their type BE). The data fulfil the Kramers–Kronig relations of section 2.5.

degrades to secondary importance in view of our general ignorance about grain properties.

The absorption and extinction efficiencies resulting from figure 7.19 for a typical grain size are shown in figure 7.20. For amorphous carbon, the extinction is basically flat up to $\lambda \simeq 0.7$ μm and then falls steadily at longer wavelengths; in the far infrared, it declines proportionally to $\nu^{1.55}$. For silicate, the wavelength dependence is more complicated. There is also a flat part in the ultraviolet up to 0.2 μm, very low absorption in the near infrared, two resonances at 10 and 18 μm and a far IR decline proportional to ν^2. For both substances, the albedo is of order 0.5 in the UV and usually negligible beyond 3 μm.

7.5 Grain sizes and optical constants

7.5.1 The size distribution

Because the extinction curve covers a large frequency range, particles of different sizes must be invoked to explain its various sections. The infrared part requires large grains, the optical region medium-sized and the UV, the 2175 Å bump and

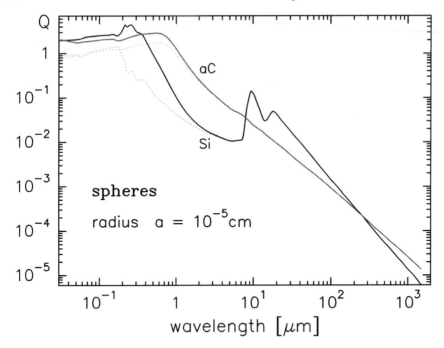

Figure 7.20. The absorption (dots) and extinction (full line) efficiency for spherical dust grains of 0.1 μm radius with the standard set of optical constants from figure 7.19. The scattering efficiency can be found as the difference between extinction and absorption. Amorphous carbon = aC, silicate = Si.

the far UV small particles. Qualitatively speaking, we expect that at wavelength λ, the radii a of the relevant grains are given by $2\pi a/\lambda \sim 1$, smaller ones are inefficient as absorbers or scatterers. To quantify this supposition, one can ask how the grains have to be distributed in size in order that their wavelength-dependent extinction follows the observed interstellar reddening curve. Assuming that they are spheres and defining the size distribution $n(a)$ such that

$$n(a)\,da = \text{number of grains per cm}^3 \text{ with radii in the interval } [a, a + da]$$

one obtains a power law:

$$n(a) \propto a^{-q} \qquad \text{with } q \simeq 3.5 \qquad (7.26)$$

which is called the MRN size distribution [Mat77]. The result is widely accepted and follows from binning the radii into intervals $[a_i, a_{i+1}]$ with $a_{i+1} > a_i$ and optimizing the number of grains n_i in each bin until agreement with the extinction curve is achieved. For the dust composition, one assumes silicate and carbon material with optical constants similar to those in figure 7.19; composite grains made of these materials are also possible.

To fix the size distribution, one must additionally specify its upper and lower limit, a_+ and a_-, roughly,

$$a_- \simeq 10\,\text{Å} \qquad \text{and} \qquad a_+ \simeq 0.3\,\mu\text{m} \tag{7.27}$$

but the outcome is not very sensitive to either boundary. Not to a_+, because extinction becomes gray for large grains, and even less to a_- because small grains are in the Rayleigh limit ($a_- \ll \lambda$) where only the total volume of the grains counts, and not their individual sizes. It is, therefore, no wonder that the smallest grains ($a < 100\,\text{Å}$) were detected through their emission and not from studies of the extinction curve.

For an MRN distribution, which has an exponent $q = 3.5$, the total mass or dust volume V is supplied by the big particles and the geometrical surface F by the small ones:

$$V \propto \int_{a_-}^{a_+} n(a)a^3\,da \propto (\sqrt{a_+} - \sqrt{a_-}) \sim \sqrt{a_+} \tag{7.28}$$

$$F \propto \int_{a_-}^{a_+} n(a)a^2\,da \propto \left(\frac{1}{\sqrt{a_+}} - \frac{1}{\sqrt{a_-}}\right) \sim \frac{1}{\sqrt{a_-}}. \tag{7.29}$$

If q were greater than four, the small grains would also contain most of the volume.

7.5.2 Collisional fragmentation

One can theoretically obtain a size distribution $n(a) \propto a^{-3.5}$ by studying grains undergoing destructive collisions [Hel70, Dor82]. Consider a *time-dependent* mass distribution $N(m, t)$ such that

$N(m, t)\,dm$ = number of particles in the interval $m \ldots m + dm$;

$P(m, t)\,dt$ = probability that particle of mass m is collisionally
destroyed during time dt;

$\xi(\mu, m)\,d\mu$ = number of fragments in the interval μ to $\mu + d\mu$ resulting
from the collision of a body of mass m.

If fragmentation is the only process that determines the mass distribution, $N(m, t)$ is governed by the formula

$$\frac{\partial N(m, t)}{\partial t} = -N(m, t)P(m, t) + \int_m^M N(\mu, t)P(\mu, t)\xi(\mu, m)\,d\mu. \tag{7.30}$$

M is the maximum mass of the particles and the total mass of the particles is conserved. To solve (7.30), one separates the variables:

$$N(m, t) = \eta(m) \cdot \zeta(t).$$

The probability $P(m, t)$ for destruction of a particle of mass m is obviously proportional to an integral over relative velocity v_{rel} of the colliding particles multiplied by the collisional cross section. If we assume, as a first approximation, that v_{rel} is constant, we may write

$$P(m, t) = Q \int_{m'}^{M} N(\mu, t) \left[\mu^{\frac{1}{3}} + m^{\frac{1}{3}} \right]^2 d\mu = \zeta(t) \cdot \gamma(m) \qquad (7.31)$$

with

$$\gamma(m) \equiv Q \int_{m'}^{M} \eta(\mu) \left[\mu^{\frac{1}{3}} + m^{\frac{1}{3}} \right]^2 d\mu.$$

Q is some proportionality factor. The lower boundary on the integral, m', is the minimum mass for destructive collisions, very small particles are not destroyed. Equation (7.30) now becomes

$$\frac{\dot{\zeta}}{\zeta^2} = -\gamma(m) + \frac{1}{\eta(m)} \int_m^M \eta(\mu) \gamma(\mu) \xi(\mu, m) d\mu. \qquad (7.32)$$

The equation is completely separated as there is a pure time dependence on the left and pure mass dependence on the right. Both sides may be set to a constant. The solution for $\zeta(t)$ is

$$\zeta(t) = \frac{\zeta_0}{1 + tC\zeta_0}$$

where ζ_0 is the value of ζ at $t = 0$ and C is a positive constant. To solve (7.31) with $\dot{\zeta}/\zeta^2 = -C$ for $\eta(m)$, we need to know the mass distribution of the debris, $\xi(\mu, m)$, after each shattering event. One expects that the number of debris, $\xi(\mu, m)$, is related to the ratio of initial mass over fragment mass, m/μ. When one puts

$$\xi(\mu, m) = \theta \frac{m^x}{\mu^{x+1}}$$

empirical evidence suggests for the exponent that $0.5 \le x \le 2$. The constant θ follows from the condition of mass conservation:

$$\theta \int_0^m \frac{m^x}{\mu^{x+1}} \mu \, d\mu = m.$$

With a power law ansatz

$$\eta(m) = m^{-y}$$

one finds with a little algebra that $y = 1.83$ over a wide range of x values ($0.5 \le x \le 3$). Because for spheres $m \propto a^3$ and $dm \propto a^2 da$, we recover, in terms of sizes, the MRN distribution $n(a) \propto a^{3y-2} \simeq a^{-3.5}$.

There are two dissatisfactory aspects to this scenario when applied to interstellar dust. First, the particles have to be born uncomfortably big, for example, in the outflow of giants [Bie80], before they can be ground down by collisions or shocks to the saturation distribution $n(a) \propto a^{-3.5}$. Second, the scenario does not include further modifications in the interstellar medium, such as accretion or coagulation, which do exist as one observes, for instance, varying values of R_V or ice mantles.

Chapter 8

Dust radiation

We treat, in this chapter, the physical background of dust emission. A single grain radiates according to Kirchhoff's law and we strictly derive this law in section 8.1 under the assumption that a grain consists of an ensemble of coupled harmonic oscillators in thermal equilibrium. In section 8.2, we show how to compute the temperature to which a grain is heated in a radiation field and add a few common examples as well as useful approximation formulae. Section 8.3 illustrates the emission of grains for a typical dust type; it also presents a first example of radiative transfer, for a dust cloud of uniform temperature. For evaluating the emission of very small grains, which are those whose temperature does not stay constant, we have to know the specific heat and the internal energy of the dust material. We, therefore, investigate, in section 8.4, the calorific properties of dust. Finally, section 8.5 presents examples of the emission of very small grains.

8.1 Kirchhoff's law

8.1.1 The emissivity of dust

A grain bathed in a radiation field acquires an equilibrium temperature T_d which is determined by the condition that it absorbs per second as much energy as it emits. According to Kirchhoff, in local thermodynamic equilibrium (LTE) the ratio of the emission coefficient ϵ_ν to the absorption coefficient K_ν^{abs} is a function of temperature and frequency only. This discovery had already been made in the 19th century; later it was found that $\epsilon_\nu / K_\nu^{abs}$ equals the Planck function.

Of course, a grain in interstellar space is not at all in an LTE environment. But whenever it is heated, for example by a photon, the excess energy is very rapidly distributed among the very many energy levels of the grain and the resulting distribution of energy states is given by the Boltzmann function and depends only on the dust temperature, T_d. Therefore, grain emission is also only a function of T_d and, as in LTE, given by

$$\epsilon_\nu = K_\nu^{abs} \cdot B_\nu(T_d). \tag{8.1}$$

The emission over all directions equals $4\pi\epsilon_\nu$. The quantity ϵ_ν is called the emissivity. It can refer to a single particle, a unit volume or a unit mass. The dimensions change accordingly. For example, the emissivity ϵ_ν per unit volume is expressed in erg s^{-1} cm^{-3} Hz^{-1} ster^{-1}.

8.1.2 Thermal emission of grains

To understand how and why an interstellar grain emits according to equation (8.1), we assume that it consists of N atoms and approximate it by a system of $f = 3N - 6$ (see section 8.4.2) weakly-coupled one-dimensional harmonic oscillators, each of mass m_i and resonant frequency ω_i. Classically, the total energy E of the system is

$$E = E_{\text{kin}} + V + H' \simeq \frac{1}{2}\sum_{i=1}^{f}\left(\frac{p_i^2}{m_i} + m_i\omega_i^2 q_i^2\right) \tag{8.2}$$

and consists of kinetic (E_{kin}), potential (V) and interaction (H') energy. The latter is small compared to E_{kin} and V but it must not be neglected altogether. Otherwise the oscillators would be decoupled and unable to exchange energy at all: starting from an arbitrary configuration, equilibrium could never be established. In (8.2), q_i and p_i denote coordinates and momenta.

In the thermal (*canonical*) description, the level population of the quantized harmonic oscillators is described by a temperature T. The ith oscillator has levels $v_i = 0, 1, 2, \ldots$ and can make transitions of the kind $v_i \rightarrow v_i - 1$. If the energy difference is lost radiatively, a photon of frequency v_i escapes. The probability $P_{i,v}$ of finding the ith oscillator in quantum state v is given by the Boltzmann equation (5.8):

$$P_{i,v} = \frac{e^{-\beta h v_i (v+\frac{1}{2})}}{Z(T)} = \left[1 - e^{-\beta h v_i}\right]e^{-v\beta h v_i}$$

with the partition function $Z(T)$ from (5.10) and $\beta = 1/kT$. The sum over all probabilities is, of course, one:

$$\sum_{v=0}^{\infty} P_{i,v} = 1.$$

To obtain $4\pi\epsilon_{\text{th}}(i, v)$, the radiation integrated over all directions from oscillator i in the transition $v \rightarrow v - 1$ at temperature T, we have to multiply $P_{i,v}$ by the Einstein coefficient $A^i_{v,v-1}$ (here the superscript i is not an exponent) and the photon energy $h v_i$,

$$4\pi\epsilon_{\text{th}}(i, v) = P_{i,v} A^i_{v,v-1} h v_i = \left[1 - e^{-\beta h v_i}\right]e^{-v\beta h v_i} A^i_{v,v-1} h v_i. \tag{8.3}$$

The Einstein coefficient $A^i_{v,v-1}$ is calculated according to the general formulae (6.41) and (6.42). For a harmonic oscillator, the eigenfunctions ψ are the

Hermitian polynomials $H_n(y)$ of (A.1). We, therefore, have to evaluate the integral

$$\int H_v(y)\, y\, H_{v-1}(y)\, e^{-y^2}\, dy$$

and its solution is given in (A.3). The Einstein coefficient is thus proportional to the quantum number v and we may put

$$A^i_{v,v-1} = v A^i_{1,0} \qquad (v \geq 1). \tag{8.4}$$

All one has to know is $A^i_{1,0}$, the coefficient for the ground transition. Note that for an ideal harmonic oscillator there is no upper limit to its maximum energy and v_i is the same for all levels v, whereas a real oscillator has a maximum energy above which the system is unbound and only its lowest levels are approximately energetically equidistant; towards the boundary their spacing decreases. To compute the total emission per solid angle of oscillator i, we sum up over all levels $v \geq 1$,

$$\epsilon_{\text{th}}(i) = \sum_{v \geq 1} \epsilon_{\text{th}}(i, v).$$

Because of the mathematical relations (A.16), we obtain

$$\epsilon_{\text{th}}(i) = h v_i A^i_{1,0} \sum_{v \geq 1} v \left[1 - e^{-\beta h v_i}\right] e^{-v \beta h v_i} = \frac{h v_i A^i_{1,0}}{e^{\beta h v_i} - 1}. \tag{8.5}$$

This is the total emission of the grain at frequency v_i. For the integrated radiation ϵ_{th} of the particle over all frequencies, one must add up all oscillators,

$$\epsilon_{\text{th}} = \sum_i \epsilon_{\text{th}}(i).$$

8.1.3 Absorption and emission in thermal equilibrium

According to the theory of line radiation, the absorption coefficient in the transition $v \to v - 1$ of the oscillators with frequency v_i can be expressed through the Einstein B coefficients and the level populations (see (13.61)),

$$K^v_{v_i} = \left[P_{i,v-1} B^i_{v-1,v} - P_{i,v} B^i_{v,v-1}\right] \frac{h v_i}{c}.$$

As the statistical weight of the levels is one, $B^i_{v,v-1} = B^i_{v-1,v}$ and a change from B to A coefficients, according to (6.40), yields

$$K^v_{v_i} = \frac{c^2}{8\pi v_i^2} A^i_{v,v-1} \left[1 - e^{-\beta h v_i}\right] \left[e^{-(v-1)\beta h v_i} - e^{-v\beta h v_i}\right].$$

Summing over all levels $v \geq 1$ gives the absorption coefficient K_{v_i} of the grain at frequency v_i,

$$K_{v_i} = \sum_v K_{v_i}^v = \frac{c^2 A_{1,0}^i}{8\pi v^2}. \tag{8.6}$$

We might drop the subscript i in equations (8.5) and (8.6), but we have to be aware that the frequency v_i refers to a certain oscillator. When we combine the two formulae, we retrieve Kirchhoff's law (8.1).

Note that the absorption coefficient $K_{v_i}^v$, which is taken with respect to a particular pair of levels $(v, v - 1)$, still contains the temperature via $\beta = 1/kT$. However, the radiation of frequency v_i interacts with all level pairs and in the *total* absorption coefficient K_{v_i} the temperature has disappeared. Also note that K_{v_i} does include stimulated emission because it is incorporated into the formula for $K_{v_i}^v$; however, the typical factor $(1 - e^{-hv_i/kT})$ has canceled out.

8.1.4   Equipartition of energy

We conclude with remarks on the distribution of energy under equilibrium conditions. Let us first consider a gas where the atoms are unbound. A gas of N atoms possesses $3N$ degrees of freedom and each degree of freedom has an average energy $\langle E \rangle = \frac{1}{2}kT$. As the atoms have only kinetic energy, altogether $E_{tot} = E_{kin} = \frac{3}{2}NkT$.

When the atoms are not free but bound to their neighbors by electric forces, as in a solid, there is, rather interestingly, the same average kinetic energy $\frac{1}{2}kT$. Furthermore, there is also potential energy V. The total energy E_{tot} is now twice as large as for a gas and evenly split between V and E_{kin}:

$$E_{tot} = E_{kin} + V = 3NkT.$$

Quite generally, thermal equilibrium implies equipartition of energy when for each particle i of the system the kinetic energy is quadratic in its momentum p_i and the potential energy is quadratic in its space coordinate q_i, and when the total energy of the rest of the system, E', depends neither on p_i nor q_i. Then the total energy E of the system, including particle i, can be written as

$$E = ap_i^2 + bq_i^2 + E'.$$

The mean kinetic energy, $\langle ap_i^2 \rangle$ of particle i is found by averaging ap_i^2 over the phase space with the Boltzmann distribution as the weight:

$$\langle ap_i^2 \rangle = \frac{\int e^{-\beta(ap_i^2 + E')} ap_i^2 \, dq_1 \ldots dp_f}{\int e^{-\beta(ap_i^2 + E')} \, dq_1 \ldots dp_f} = \frac{\int e^{-\beta ap_i^2} ap_i^2 \, dp_i}{\int e^{-\beta ap_i^2} \, dp_i} = \frac{1}{2\beta} = \frac{1}{2}kT.$$

Likewise, one finds $\langle bq_i^2 \rangle = \frac{1}{2}kT$. This derivation of equipartition presupposes, however, that energy between the members of the system can be exchanged continuously. Therefore, it holds only for *classical* systems or, equivalently, at high temperatures. When T is small and the energy levels are quantized, $\langle E \rangle \neq \frac{1}{2}kT$. Different oscillators, although in thermal equilibrium, then have different mean energies equal to $h\nu/(e^{h\nu/kT} - 1)$ after (5.11).

8.2 The temperature of big grains

8.2.1 The energy equation

Equation (8.1) permits us, in a straightforward way, to calculate the dust temperature T_d from the balance between radiative heating and radiative cooling:

$$\int K_\nu^{\text{abs}} J_\nu \, d\nu = \int K_\nu^{\text{abs}} B_\nu(T_d) \, d\nu. \tag{8.7}$$

K_ν^{abs} is the mass (or volume) absorption coefficient of dust at frequency ν, and J_ν is the average of the radiation intensity over all directions. For spherical grains, our standard case, of radius a and absorption efficiency $Q_\nu^{\text{abs}}(a)$,

$$\int Q_\nu^{\text{abs}}(a) \cdot J_\nu \, d\nu = \int Q_\nu^{\text{abs}}(a) \cdot B_\nu(T_d) \, d\nu. \tag{8.8}$$

If other forms of heating or cooling are relevant, for instance, by collisions with gas particles, they have to be added in equations (8.7) and (8.8). We neglect them for now.

It is evident from (8.8) that it is not the absolute value of Q_ν^{abs} that determines the dust temperature in a given radiation field J_ν. Two efficiencies, $Q_{1,\nu}^{\text{abs}}$ and $Q_{2,\nu}^{\text{abs}}$, that differ at all frequencies by an arbitrary, but constant factor, yield exactly the same T_d. Qualitatively speaking, a grain is hot when Q_ν^{abs} is high at the wavelengths where it absorbs and low where it emits.

8.2.2 Approximate absorption efficiency at infrared wavelengths

When one wants to determine the dust temperature from equation (8.8), one first has to compute the absorption efficiency $Q_\nu^{\text{abs}}(a)$ for all frequencies from Mie theory and then solve the integral equation for T_d. These procedures require a computer. However, there are certain approximations to Q_ν^{abs} which allow us to evaluate T_d analytically, often with only a small loss in accuracy.

For a quick and often sufficiently precise estimate, one exploits the fact that dust emits effectively only in the far infrared where the Rayleigh limit is valid. According to (3.3), the absorption efficiency Q_ν^{abs} of spheres is then proportional to the size parameter $2\pi a/\lambda$ multiplied by a function that depends only on frequency. Simplifying the latter by a power law, we get

$$Q_\nu^{\text{abs}} = \frac{8\pi a}{\lambda} \cdot \text{Im}\left\{ \frac{m^2(\lambda) - 1}{m^2(\lambda) + 2} \right\} = a Q_0 \nu^\beta. \tag{8.9}$$

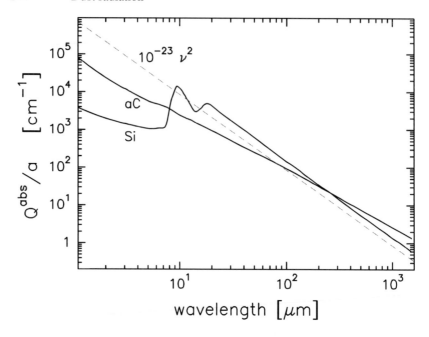

Figure 8.1. The ratio of absorption efficiency over grain radius, Q_ν^{abs}/a, for amorphous carbon (aC) and silicate (Si). The curves have been calculated with $a = 0.1\ \mu$m but are quite generally valid in the infrared. The optical constants are from figure 7.19. The broken line shows the approximation from (8.10). See also figure 7.20 which shows Q_ν^{abs} for $a = 0.1\ \mu$m.

Estimates of the exponent β range from 1 to 2. Values smaller than 1 are excluded for very long wavelengths because of the Kramers–Kronig relations (section 2.5). A blackbody would have $\beta = 0$ and $aQ_0 = 1$. When the complex optical constant $m(\lambda)$ deserves its name and is truly constant, $\beta = 1$. When the expression Im{...} in (8.9) varies inversely proportional to wavelength, $\beta = 2$. This is the value we favour.

We, therefore, propose for quick estimates at infrared wavelengths

$$Q_\nu^{abs} \simeq 10^{-23} a\nu^2 \qquad [a \text{ in cm}, \nu \text{ in Hz}]. \qquad (8.10)$$

Figure 8.1 gives a feeling for the quality of the approximation. The curves display the ratio of absorption efficiency over grain radius, Q_ν^{abs}/a, for the particular value $a = 0.1\mu$m. At wavelengths $\lambda > 10\ \mu$m, which is the relevant range of emission, they are fairly independent of radius provided the grains are not very big ($a \leq 0.3\ \mu$m). We see that the fit Q_ν^{abs}/a from (8.10) is not bad at $\lambda > 10\mu$m for both interstellar silicate and amorphous carbon dust.

When we follow the suggestion of (8.9) and use a power law for the absorption efficiency,

$$Q_\nu^{abs} \propto \nu^\beta \qquad \beta \geq 1$$

Q_ν^{abs} is, of course, overestimated at high frequencies. But this is irrelevant for the evaluation of the integral on the right-hand side of (8.8) describing emission, as long as the grain temperature stays below a few hundred K so that the Planck function at these frequencies is small. With the power law (8.9) for Q_ν^{abs}, the energy equation (8.8) becomes

$$\int_0^\infty Q_\nu^{abs} B_\nu(T)\, d\nu = a Q_0 \frac{2h}{c^2} \left(\frac{kT}{h}\right)^{4+\beta} \int_0^\infty \frac{x^{3+\beta}\, dx}{e^x - 1}.$$

The right-hand integral is evaluated in (A.17). Approximate values are:

$$\int_0^\infty \frac{x^{3+\beta}\, dx}{e^x - 1} \simeq \begin{cases} 6.494 & \text{if } \beta = 0 \\ 24.89 & \text{if } \beta = 1 \\ 122.08 & \text{if } \beta = 2. \end{cases} \qquad (8.11)$$

8.2.3 Temperature estimates

8.2.3.1 Blackbodies

Easiest of all is to find the temperature of a blackbody, a perfect absorber with $Q_\nu^{abs} = 1$ everywhere. Equation (8.8) then reduces to

$$\int J_\nu\, d\nu = \frac{\sigma}{\pi} T^4 \qquad (8.12)$$

σ being the radiation constant of (5.80). The temperature is now *independent of particle size*. In a given radiation field, a blackbody is usually colder than any other object. But this is not a law, one can construct counter examples. If a blackbody is heated by a star at distance r and bolometric luminosity L, its temperature follows from

$$\frac{L}{4\pi r^2} = 4\sigma T^4. \qquad (8.13)$$

For example, we find that a blackbody r astronomical units from the Sun heats up to

$$T = 279 \left(\frac{r}{\text{AU}}\right)^{-1/2} \text{K}.$$

For $r = 1$, this value is close to the average temperature of the surface of the Earth, although our planet does not emit like a perfect blackbody and its clouds reflect some 30% of insolation.

8.2.3.2 *Interstellar grains directly exposed to stars*

For real grains, we may use, for the absorption efficiency Q_ν^{abs}, the approximation (8.10) together with the integral (8.11) for $\beta = 2$. The energy equation (8.8) then transforms into

$$\int_0^\infty Q_\nu^{\text{abs}} J_\nu \, d\nu \simeq 1.47 \times 10^{-6} a T_{\text{d}}^6 \qquad \text{(cgs units).} \qquad (8.14)$$

In cgs units, a is expressed in cm, ν in Hz, T_{d} in K, and J_ν in erg cm^{-2} s^{-1} Hz^{-1} ster^{-1}. The equation is applicable as long as the grain diameter stays below ~ 1 μm, which is usually true in the interstellar medium.

According to (8.14), the heating rate of a grain and, in view of equilibrium, also its total emission rate are proportional to T_{d}^6.

Consequently, a moderate temperature difference implies a vast change in the energy budget. For example, to warm a particle, or a whole dust cloud, from 20 to 30 K requires a tenfold increase in flux.

To determine T_{d} in (8.14), it remains to evaluate the integral on the left. There is a further simplification if the radiation field is hard. For an early-type star or even the Sun, one may put, at all wavelengths *relevant for absorption*,

$$Q_\nu^{\text{abs}} \simeq 1$$

so that

$$\int_0^\infty J_\nu \, d\nu \simeq 1.47 \times 10^{-6} a T_{\text{d}}^6 \qquad \text{(cgs units).} \qquad (8.15)$$

If, additionally, J_ν is the diluted blackbody radiation of temperature T_*, the integral over J_ν is directly proportional to T_*^4. Consider a grain at distance r from a hot star of luminosity L, effective temperature T_* and radius R_*. From the position of the grain, the star subtends a solid angle $\pi R_*^2/r^2$ and the intensity towards it is $B_\nu(T_*)$. The average intensity over all directions is, therefore,

$$J_\nu = \frac{B_\nu(T_*) \pi R_*^2}{4\pi r^2}.$$

When we insert this expression of J_ν into (8.15) and use $L = 4\pi\sigma R_*^2 T_*^4$ after (5.82), the stellar radius disappears and we find (a, r in cm, L in erg s^{-1})

$$\frac{L}{16\pi^2 r^2} \simeq 1.47 \times 10^{-6} a T_{\text{d}}^6 \qquad \text{(cgs units).} \qquad (8.16)$$

So the grain temperature falls off with distance from the star like

$$T_{\text{d}} = 4.0 \left(\frac{L}{a}\right)^{\frac{1}{6}} r^{-\frac{1}{3}} \qquad \text{(cgs units).} \qquad (8.17)$$

If there is extinction along the line of sight, one must replace the luminosity L by a frequency average over $L_\nu e^{-\tau_\nu}$. In view of the usual uncertainties regarding radiation field, grain properties and geometry of the configuration, estimates of this kind are often no less trustworthy than sophisticated computations.

8.2.4 Relation between grain size and grain temperature

In the estimate (8.17), the dust temperature changes with particle radius a like

$$T_{\mathrm{d}} \propto a^{-\frac{1}{6}}.$$

However, this relation is valid only for grains of intermediate size. If the particles are very small so that the Rayleigh limit is applicable to *both emission and absorption*, i.e. to either side of (8.8), the radius a in the absorption efficiency $Q_\nu^{\mathrm{abs}}(a)$ cancels out (see (3.3)) and all tiny grains of the same chemical composition have the same temperature. If the grains are big, much bigger than the particles found in interstellar space, $Q_\nu^{\mathrm{abs}}(a)$ in (8.8) becomes independent of size and the grain temperature levels off to some limit, T_{lim}. This value is similar but not identical to that of a blackbody, T_{bb}, because Q_ν^{abs} stays under the integrals in (8.8) and is not constant.

As an example, we compute grain temperatures near a luminous star and in the interstellar radiation field (ISRF). By the latter we mean the environment in the solar neighborhood but outside clouds and not close to any star, rather in the space between the stars. The radiation field expected there averaged over a solid angle is depicted in figure 8.2. Its main sources are stars of spectral type A and F, giants and interstellar dust. The ISRF should be uniform in the galactic disk on a scale of 1 kpc. One can discern, in figure 8.2, several components. The two major ones are of comparable strength in terms of $\nu J_\nu^{\mathrm{ISRF}}$, one is from starlight peaking at ~ 1 μm, the other from interstellar dust with a maximum at ~ 200 μm. Integrated over frequency and solid angle,

$$4\pi \int J_\nu^{\mathrm{ISRF}} \, d\nu \sim 0.04 \text{ erg s}^{-1} \text{ cm}^{-2}.$$

Figure 8.3 demonstrates the size dependence of the dust temperature for these two environments. Generally speaking, small grains are warmer than big ones. The figure contains none of our previous approximations, the absorption efficiencies are calculated from Mie theory with optical constants from figure 7.19. The temperature increases from tiny to huge particles by a factor of about three. For the size range supposed to prevail in the interstellar medium ($a \leq 0.3$ μm), the maximum temperature ratio is smaller, of order 1.5.

We will see in section 8.6 that really tiny grains may not attain an equilibrium temperature at all, instead their internal energy fluctuates in response to the quantum character of the absorbed photons. Then a plot of the kind of figure 8.3 becomes meaningless.

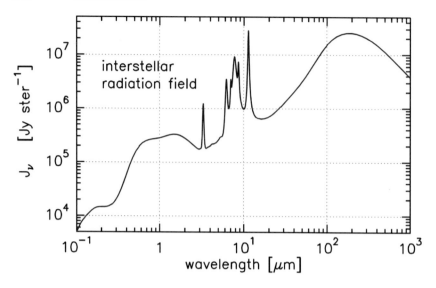

Figure 8.2. The approximate mean intensity J_ν^{ISRF} of the interstellar radiation field (ISRF) in the Milky Way at the locus of the Sun. The curve is based on data from [Per87] but includes the emission of PAHs (spikes). Interstellar dust exposed to a radiation field J_ν^{ISRF} emits at wavelengths $> 3\ \mu$m a flux proportional to J_ν^{ISRF}. The underlying dust model is described in section 12.2.

8.2.5 Temperature of dust grains near a star

The variation in dust temperatures in a reflection nebula, which is typically excited by a B1V star, is depicted in figure 8.4. The grains have radii of 0.01 and 0.1 μm and consist either of silicate or amorphous carbon. In the logarithmic plot, the curves are very smooth, almost linear in the case of carbon, and there is a wiggle around $r = 10^{15}$ cm for silicates. Near the star, silicates are warmer, further away colder than carbon grains. This behavior can be explained with the help of figure 8.1: at short distances, grains are very hot and emit also below 10 μm. At these wavelengths, the emissivity of silicate particles is small, so they have to be hotter than the carbon grains in order to get rid of their energy.

- Writing the dependence of the temperature on distance as $T_d \propto r^{-\alpha}$, the slope α is not far from 1/3. Carbon has a very constant α of 0.375.
- Small grains are always warmer than big ones. An increase in grain radius from 0.01 to 0.1 μm lowers the temperature by $\sim 30\%$.
- When we check the accuracy of the approximation formula (8.17) with the help of figure 8.4, it turns out to be satisfactory in both variables, grain radius a and distance r.

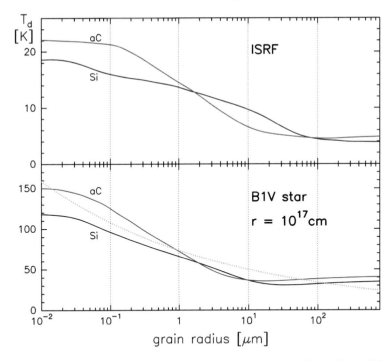

Figure 8.3. The temperature of spheres of amorphous carbon (aC) or silicate (Si) as a function of grain radius: Top, in the interstellar radiation field with mean intensity J_ν^{ISRF} after figure 8.2. Bottom, at a distance of 10^{17} cm from a B1V star with $L = 10^4 \, L_\odot$ and $T_* = 2 \times 10^4$; here the integrated mean intensity of the radiation field is about 7000 times stronger. The dotted line represents T_d from the approximate formula (8.17). In both boxes, the optical constants are from figure 7.19. For comparison, a blackbody would acquire a temperature $T_{bb} = 34.2$ K near the B1 star and $T_{bb} = 3.8$ K in the ISRF.

8.2.6 Dust temperatures from observations

Whereas gas temperatures can often be obtained rather accurately from properly chosen line ratios, the experimental determination of dust temperatures is always dubious. After the basic equation of radiative transport (see section 13.1), the intensity I_ν towards a uniform dust layer of optical thickness τ_ν and temperature T_d is

$$I_\nu(\tau) = B_\nu(T_d) \cdot [1 - e^{-\tau_\nu}] \tag{8.18}$$

so that an observer receives from a solid angle Ω the flux

$$F_\nu = B_\nu(T_d) \cdot [1 - e^{-\tau_\nu}] \cdot \Omega \tag{8.19}$$

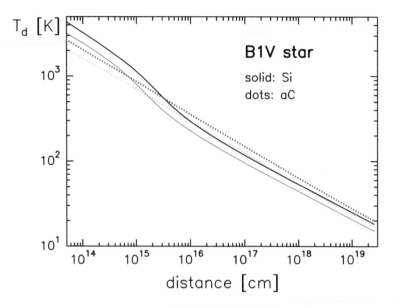

Figure 8.4. Temperature of amorphous carbon (aC, dotted curve) and silicate (Si, full curve) grains as a function of distance from a B1 main sequence star with $10^4 \, L_\odot$ and effective temperature $T_* = 2 \times 10^4$ K. Particle radii are 0.01 μm for the upper full and dotted curve and 0.1 μm for the lower ones. Optical constants are from figure 7.19. The optical depth towards the star is zero, i.e. there is no absorption between the star and the grain.

ν denotes the frequency. The flux ratio at two frequencies, marked by the subscripts 1 and 2, becomes

$$\frac{F_1}{F_2} = \frac{B_1(T_d)\,[1 - e^{-\tau_1}]\,\Omega_1}{B_2(T_d)\,[1 - e^{-\tau_2}]\,\Omega_2}. \tag{8.20}$$

It is customary to extract from (8.20) the dust temperature under the following premises:

- The emission is optically thin ($\tau \ll 1$), which is usually true in the far infrared.
- The observational frequencies, ν_1 and ν_2, do not lie both in the Rayleigh–Jeans limit of the Planck function. If they did, T_d would cancel out because $B_\nu(T_d) \propto T_d$.
- The observations at the two frequencies refer to the same astronomical object, for instance, a certain region in a galactic cloud. This apparently trivial condition is sometimes hard to fulfil when the source is extended relative to the telescope beam (see section 13.1.4) and the observations have to be carried out with different spatial resolution.

- The dust temperature in the source is uniform.
- The ratio of the dust absorption coefficients, K_1/K_2, over the particular frequency interval from v_1 to v_2 is known or, equivalently the exponent β in (8.9); K_v itself is not needed.

If all requirements are fulfilled, (8.20) simplifies to

$$\frac{F_1}{F_2} = \frac{K_1 \cdot B_1(T_d)}{K_2 \cdot B_2(T_d)} \tag{8.21}$$

and yields T_d in a straightforward manner. Should the assumed value for K_1/K_2 be wrong, one gets a purely formal color temperature, without any physical correspondence. We recall here that the surface temperature of a star can also be obtained through photometry from an equation like (8.21) but with $K_1/K_2 = 1$ because a star is, to first order, a blackbody.

In view of the numerous restrictions and because the exponent β in (8.9) is debatable, we should always be sceptical about absolute values of T_d. However, we may trust results of the kind

$$T_d \text{ (source A)} > T_d \text{ (source B)}$$

and their implications for the energy budget because they depend only weakly on the ratio K_1/K_2.

The flux ratio also determines the spectral index α of the energy distribution; it is defined by

$$\frac{F_1}{F_2} = \left(\frac{v_1}{v_2}\right)^{\alpha}.$$

The exponent β in (8.9) and the slope α are, *at very long wavelengths*, related through

$$\alpha = \beta + 2$$

independent of temperature. In a double-logarithmic plot of F_v versus v and when λ is large, all curves are parallel for any T. The number 2 in this equation comes from the Rayleigh–Jeans part of the Planck function. A blackbody ($\beta = 0$), like a planet, has a spectral index $\alpha = 2$.

8.3 The emission spectrum of big grains

8.3.1 Constant temperature and low optical depth

As a first illustration of the spectral energy distribution emitted by dust, figure 8.5 displays, for silicate grains of 600 Å radius, the product $Q_v^{abs} B_v(T)$. This quantity is proportional to the emissivity ϵ_v of (8.1). The blackbody intensity ($Q_v^{abs} = 1$) is shown for comparison.

The temperatures chosen in figure 8.5 represent three astronomical environments:

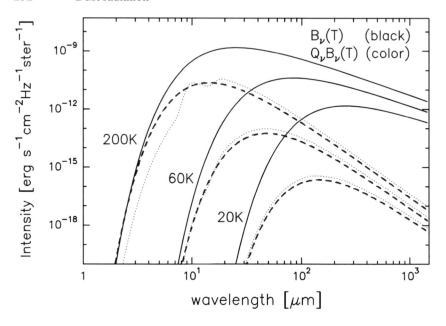

Figure 8.5. Emission of a blackbody (full curve) and of a silicate grain of 600 Å $= 0.6\,\mu$m radius at different temperatures. The efficiency Q_ν^{abs} of the silicate grain is calculated from approximation (8.10) (broken curve) and from Mie theory with optical constants after figure 7.19 (dotted curve). The Planck function $B_\nu(T_d)$ as well as $Q_\nu^{abs} B_\nu(T_d)$ have the units of an intensity.

- 20 K for the bulk of the dust in the Milky Way,
- 60 K for warm dust in star-forming regions and
- 200 K for hot dust close to stars.

Several points deserve a comment:

- A dust particle emits at all wavelengths less than a blackbody, especially in the far infrared.
- A seemingly unspectacular temperature change by a factor of three, say from 20 to 60 K, can boost emission by orders of magnitude.
- Although the dust optical constants of figure 7.19 have some structure, the dust emission is very smooth. For instance, only in the hot dust do we see the 10 μm resonance.
- Unless the dust is hot ($\gtrsim$100 K), the power law approximation for the absorption efficiency Q_ν^{abs} from (8.10) gives acceptable results.
- The spectrum of a dust grain can be crudely characterized by the wavelength of maximum emission. This is similar to Wien's displacement law for a blackbody (see discussion after (5.77)).

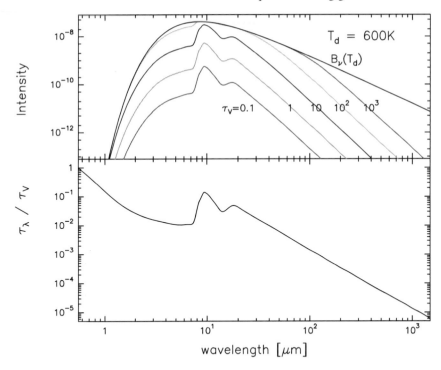

Figure 8.6. Bottom, the normalized optical depth of silicate grains with 600 Å radius. The wavelength scale starts at 0.55 μm where $\tau_\lambda/\tau_V = 1$; Top, The intensity (in units erg s^{-1} cm^{-2} Hz^{-1} ster^{-1}) towards a cloud of temperature $T = 600$ K filled with such grains. The visual optical thickness τ_V varies between 0.1 and 10^3. The Planck function $B_\nu(T_d)$ corresponds to $\tau_V = \infty$.

8.3.2 Constant temperature and arbitrary optical depth

If the dust in a cloud is at constant temperature and the emission optically thin, the curves $Q_\nu^{abs}B_\nu(T_d)$ in figure 8.5 are proportional to the observed intensity I_ν as given by equation (8.18). To obtain the *absolute* value of I_ν, one needs the optical depth τ_ν. For small τ_ν, the spectral distribution of the intensity is

$$I_\nu = \tau_\nu B_\nu(T).$$

In figure 8.6, the intensity has been calculated as a function of wavelength and arbitrary optical depth (from $\tau_V = 0.1$ to ∞) for a cloud consisting of identical silicate spheres of 600 Å radius at a temperature of 600 K. The variation of the optical depth with wavelength is depicted in the lower part of the figure; the curve there is normalized at 0.55 μm.

The line with $\tau_V = 0.1$ approximately reflects the optically thin case and may be compared with the plots in figure 8.5 but the temperature is now higher. As

τ_V increases, I_ν asymptotically approaches the Planck function. For instance, the cloud becomes optically thick at 25 μm and emits like a blackbody for $\lambda \leq 25\,\mu$m when $\tau_V > 100$. Any information about the emitters is then lost. This is exemplified by the disappearance of the 10 μm feature. At long wavelengths, the intensity curves for all τ run parallel. They are equidistant and have the same spectral slope $I_\nu \propto \nu^\alpha$ with $\alpha \simeq 4$.

8.4 Calorific properties of solids

To investigate the calorific properties of a a dust particle, we treat the conglomeration of atoms in the grain as an ensemble of harmonic oscillators, determine their eigenfrequencies and density of states and evaluate the energy of the grain under the assumption of thermal equilibrium.

8.4.1 Normal coordinates

In a simple-minded model, the atoms in a grain are replaced by mass points connected through springs. To find the eigenfrequencies of the oscillators, one uses normal coordinates. If there are N mass points in the grain, we describe their positions by *one* vector in Cartesian coordinates,

$$\mathbf{x} = (x_1, \ldots, x_n) \qquad n = 3N.$$

At the equilibrium position, designated $\mathbf{x}_0$, all forces exactly balance, so the derivatives of the potential $V(\mathbf{x})$ vanish,

$$\left(\frac{\partial V}{\partial x_i} \right)_{\mathbf{x}_0} = 0.$$

We introduce coordinates relative to equilibrium,

$$\eta_i = x_i - x_{i0}$$

and put $V(\mathbf{x}_0) = 0$, which is always possible. If the oscillations are small, one can approximate the potential by a Taylor expansion. Because V and its first derivatives are zero, the first non-vanishing term is

$$V(\mathbf{x}) = \frac{1}{2} \sum_{ij} \left(\frac{\partial^2 V}{\partial x_i \partial x_j} \right)_0 \eta_i \eta_j = \frac{1}{2} \sum_{ij} V_{ij} \eta_i \eta_j. \qquad (8.22)$$

As the kinetic energy T is quadratic in $\dot{\eta}_i$,

$$T = \tfrac{1}{2} \sum_i m_i \dot{\eta}_i^2 \qquad (8.23)$$

one derives from the Lagrange function $L = T - V$ of (6.1) the equations of motion

$$m_i \ddot{\eta}_i + \sum_j V_{ji} \eta_j = 0 \qquad i = 1, \ldots, n. \tag{8.24}$$

Trying, as usual, a solution of the kind

$$\eta_i = a_i e^{-i\omega t}$$

yields a set of n linear equations for the amplitudes a_i,

$$\sum_j (V_{ji} - \delta_{ji} m_i \omega^2) a_i = 0.$$

For non-trivial values of a_i, the determinant must vanish,

$$|V_{ji} - \delta_{ji} m_i \omega^2| = 0. \tag{8.25}$$

This is an algebraic equation of nth order in ω^2. It has n solutions and one thus finds n frequencies. In fact, six solutions will be zero if there are more than two (nonlinear) atoms but the remaining frequencies refer only to internal atomic oscillations. The motion in any coordinate x_i will be a superposition of n harmonic oscillations of different frequencies ω_i and amplitudes a_i.

It is a standard procedure to obtain from the relative coordinates, η_i, through a linear transformation new, so called *normal coordinates* y_i, with the property that each normal coordinate corresponds to a harmonic motion of just one frequency, and that the kinetic and potential energy are quadratic in y_i and $\dot{y}_i$, respectively,

$$T = \tfrac{1}{2} \sum_i \dot{y}_i^2 \qquad V = \tfrac{1}{2} \sum_i \omega_i^2 y_i^2.$$

So a grain of N atoms may be substituted by $f = 3N - 6$ independent oscillators, each with its personal frequency.

8.4.1.1 *Oscillators in a linear chain*

Instructive and well-known examples are provided by linear chains which present the simplest configuration in which atoms can be arranged. Let all atoms be of the same mass m, with a constant distance d between them when they are at rest, and connected through springs of force constant κ. Let us consider a longitudinal wave. The force on any mass point depends then only on the distance to its nearest neighbors on the left and right. It is straightforward to derive the dispersion relation

$$\omega^2 = \frac{2\kappa}{m} \big[1 - \cos(kd)\big] \tag{8.26}$$

where $k = 2\pi/\lambda$ is the wavenumber and λ the wavelength. When there are two kinds of atoms in the chain of mass m and $M > m$, in alternating sequence, the spectrum splits into what is called an acoustical and an optical branch. For each wavenumber k, there are now two eigenfrequencies. In the acoustical mode, adjacent atoms swing in phase, in the optical mode where frequencies are higher, oppositely and out of phase. The dispersion relation reads

$$\omega_{\pm}^2 = \frac{\kappa}{mM} \left[M + m \pm \sqrt{M^2 + m^2 + 2mM \cos(2kd)} \right] \tag{8.27}$$

where the minus sign refers to the acoustic and the plus sign to the optical branch.

8.4.2 Internal energy of a grain

A point-like unbound atom has three degrees of freedom corresponding to the three independent spatial coordinates along which it can move. A solid body consisting of N atoms has $3N$ degrees of freedom, although there are now forces between the atoms. When $N \geq 3$, the number of vibrational modes in a grain is

$$f = 3N - 6$$

because one has to subtract $3 + 3$ to account for the translatory and rotational motion of the grain as a whole. But in most applications, we can put $f \simeq 3N$.

An atom in a solid can be considered as a three-dimensional harmonic oscillator. If the body has N atoms, there are $f = 3N - 6$ one-dimensional oscillators. We bin oscillators of identical frequencies. Let there be s different frequencies altogether. If n_i oscillators have the frequency ν_i, then

$$\sum_{i=1}^{s} n_i = f$$

and the possible energy levels are

$$E_{i\nu} = h\nu_i (\nu + \tfrac{1}{2}) \qquad \nu = 0, 1, 2, \ldots \qquad i = 1, \ldots, s.$$

At temperature T, the mean energy of the oscillators of frequency ν_i is, from (5.11),

$$\langle E_i \rangle = \frac{1}{2} h\nu_i + \frac{h\nu_i}{e^{h\nu_i/kT} - 1}. \tag{8.28}$$

The zero-point energy $\frac{1}{2}h\nu_i$ is not necessarily small compared to the second term in (8.28) but drops out when calculating the specific heat and is, therefore, disregarded. The total energy content U of the solid is obtained by summing over all oscillators:

$$U(T) = \sum_{i=1}^{s} \frac{h\nu_i}{e^{h\nu_i/kT} - 1} n_i. \tag{8.29}$$

When N is large, the eigenfrequencies ν_i are closely packed and one may replace the sum by an integral:

$$U(T) = \int_0^{\nu_D} \frac{h\nu}{e^{h\nu/kT} - 1} \rho(\nu) \, d\nu \tag{8.30}$$

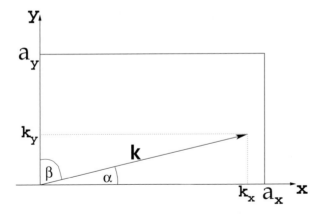

Figure 8.7. Modes have nodes at the walls of the crystal. Here a two-dimensional sketch for a rectangle with sides of length a_x and a_y. The wave propagates in the direction of the wavevector **k**. The conditions for a standing wave are $l\lambda/2 = a_x \cos\alpha$ and $m\lambda/2 = a_y \cos\beta$ for integer l, m.

where $\rho(\nu)$ denotes the number density of oscillator frequencies and ν_D the upper integration limit. One has to be on the alert when applying (8.30) to very small grains, like PAHs (section 12.2), where N is only of order 100.

8.4.3 Standing waves in a crystal

The density of states, $\rho(\nu)$, is found by determining first the low-frequency part (*sound waves*). This is easy and similar to finding modes in acoustics. Then $\rho(\nu)$ is extrapolated to high frequencies (*optical waves*) assuming the same functional dependence of $\rho(\nu)$. The extrapolation is convenient but far from accurate.

Consider a rectangular crystal of N atoms with sides a_x, a_y, a_z along the coordinate axes and one corner at the origin $\mathbf{r} = (x, y, z) = 0$ (see figure 8.7). A plane wave in the crystal of wavelength λ has the form

$$e^{i(\mathbf{k}\cdot\mathbf{r}-\omega t)} = \Psi(\mathbf{r})e^{-i\omega t}$$

$$\mathbf{k} = (k_x, k_y, k_z) = k(\cos\alpha, \cos\beta, \cos\gamma) \qquad k = |\mathbf{k}| = \frac{2\pi}{\lambda}.$$

At the walls, the wave must have nodes because the atoms cannot move; for instance, $\Psi(0, y, z) = \Psi(a_x, y, z) = 0$. This implies

$$\pm 1 = e^{-ik_x a_x} = \cos k_x a_x - i \sin k_x a_x$$

which leads to

$$l\frac{\lambda}{2} = a_x \cos\alpha \qquad m\frac{\lambda}{2} = a_y \cos\beta \qquad n\frac{\lambda}{2} = a_z \cos\gamma \qquad l, m, n = 0, 1, 2, \ldots.$$

Each triple (l, m, n) of positive integers specifies *one* mode and can be represented by a point in a Cartesian coordinate system. The direction into which the wave travels is given by the angles α, β, γ. Because

$$\cos^2 \alpha + \cos^2 \beta + \cos^2 \gamma = 1$$

it follows that

$$\left(\frac{l}{2a_x/\lambda}\right)^2 + \left(\frac{m}{2a_y/\lambda}\right)^2 + \left(\frac{n}{2a_z/\lambda}\right)^2 = 1.$$

Thus all points (l, m, n) corresponding to wavelengths greater than or equal to λ lie inside an ellipsoid of volume $V = 32\pi a_x a_y a_z/3\lambda^3$. Their total number is approximately

$$Z \approx \frac{4\pi}{3 \cdot 8} \frac{8 a_x a_y a_z}{\lambda^3} = \frac{4\pi a_x a_y a_z}{3c^3} v^3.$$

We divided by 8 because l, m, n are positive numbers and fall into the first octant; $c = \lambda v$ denotes the sound velocity.

8.4.4   The density of vibrational modes in a crystal

8.4.4.1   *Longitudinal and transverse waves*

A disturbance can travel in a crystal in two ways:

- as a longitudinal (compressional) wave where the atoms move in the direction of wave propagation or opposite to it; and
- as a transverse wave in which the atomic motion is perpendicular to wave propagation.

Because the transverse wave has two kinds of polarization corresponding to motions with perpendicular velocity vectors, there are altogether three types of waves and we have to make separate counts for all three of them. If c_l and c_t are the sound velocities for the *l*ongitudinal and for both *t*ransverse waves, and if we define the average $\bar{c}$ by

$$\frac{1}{\bar{c}^3} = \frac{1}{c_l^3} + \frac{2}{c_t^3}$$

we find for the frequency density in a crystal

$$\rho(v) = \frac{dZ}{dv} = \frac{4\pi a_x a_y a_z}{\bar{c}^3} v^2.$$

The number of eigenfrequencies in the interval v to $v + dv$ is then

$$dZ = \frac{4\pi a_x a_y a_z}{\bar{c}^3} v^2 \, dv. \tag{8.31}$$

8.4.4.2   The Debye temperature

The cutoff ν_D that appears in (8.30) for the internal energy of a crystal is the highest possible frequency. There is a limit because Z obviously cannot exceed $3N$, the number of all modes. Therefore the maximum of Z is $Z_{max} = 3N$. If

$$n = \frac{N}{a_x a_y a_z}$$

is the density of atoms, we get

$$\nu_D = \bar{c} \cdot \sqrt[3]{\frac{9n}{4\pi}}. \tag{8.32}$$

This is the Debye frequency. The Debye temperature θ is related to ν_D through

$$\theta = \frac{h\nu_D}{k}. \tag{8.33}$$

The frequency density may now be written as

$$\rho(\nu) = \frac{9N}{\nu_D^3}\nu^2. \tag{8.34}$$

Actually, the crystal does not have to be rectangular but may have any shape. What the true frequency density might look like compared to the Debye approximation is qualitatively illustrated in figure 8.8.

8.4.5   Specific heat

The specific heat at constant volume of a system of f one-dimensional oscillators follows from (8.29),

$$C_v = \left(\frac{\partial U}{\partial T}\right)_V = k \sum_{i=1}^{f} \frac{x_i^2 e^{x_i}}{[e^{x_i} - 1]^2} n_i \qquad x_i = \frac{h\nu_i}{kT}. \tag{8.35}$$

It specifies how much energy is needed to raise the temperature of a body by 1 K. For a continuous distribution of modes, we insert $\rho(\nu)$ from (8.34) into (8.30) to find the internal energy and specific heat,

$$U(T) = \frac{9N}{\nu_D^3} \int_0^{\nu_D} \frac{h\nu^3}{e^{h\nu/kT} - 1} \, d\nu \tag{8.36}$$

$$C_v(T) = \left(\frac{\partial U}{\partial T}\right)_V = 9kN \left(\frac{T}{\theta}\right)^3 \int_0^{\theta/T} \frac{x^4 e^x}{(e^x - 1)^2} \, dx. \tag{8.37}$$

There are two important limits:

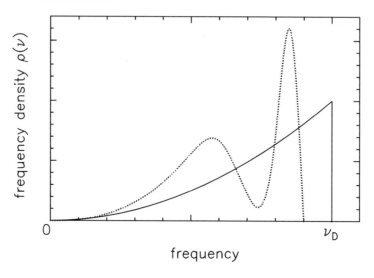

Figure 8.8. The frequency density $\rho(\nu)$ of crystal vibrations. A true distribution might look like the dotted curve, which is to be compared with the Debye approximation of (8.34) (full curve). The latter is correct only for sound waves with wavelengths much larger than the grid size of the crystal ($\sim$1 Å). It yields at low temperatures the famous T^3 dependence of C_V. There is a cutoff at ν_D because the number of degrees of freedom of all atoms is finite.

- At low temperatures, the upper bound θ/T approaches infinity and the total energy of the grain and its specific heat are (see (A.17) for the mathematics):

$$U = \frac{3\pi^4}{5} NkT \left(\frac{T}{\theta}\right)^3 \tag{8.38}$$

$$C_V = \frac{12\pi^4}{5} Nk \left(\frac{T}{\theta}\right)^3 \tag{8.39}$$

  The characteristic feature in C_V is the proportionality to T^3.
- At high temperatures, the formulae approach the classical situation where a three-dimensional oscillator has the mean energy $\langle E \rangle = 3kT$, so

$$U = 3NkT \tag{8.40}$$

and the specific heat is constant and given by the rule of *Dulong–Petit*,

$$C_V = 3Nk. \tag{8.41}$$

Any improvement over the Debye theory has to take the force field into account to which the atoms of the body are subjected.

8.4.6 Two-dimensional lattices

When the analysis of section 8.4.4 is repeated for a two-dimensional crystal of N atoms, one finds a density of modes

$$\rho(v) = \frac{6N}{v_D^2} v \tag{8.42}$$

which increases only linearly with frequency, not as v^2. Therefore now $C_V \propto T^2$ and

$$U(T) = \frac{6N}{v_D^2} \int_0^{v_D} \frac{hv^2}{e^{hv/kT} - 1} \, dv$$

$$C_V = 6kN \frac{1}{y^2} \int_0^y \frac{x^3 e^x}{(e^x - 1)^2} \, dx \qquad y = \frac{\theta}{T}. \tag{8.43}$$

The two-dimensional case is important for PAHs because they have a planar structure. It also applies to graphite which consists of PAH sheets; however, only at temperatures above 20 K. Although the coupling between the sheets is relatively weak, when the temperature goes to zero, C_V eventually approaches the T^3 dependence.

For graphite sheets, one can improve the model of the specific heat by using two Debye temperatures:

$$C_V(T) = kN \left[f(\theta_z) + 2f(\theta_{xy}) \right] \tag{8.44}$$

$$f(y) = \frac{2}{y^2} \int_0^y \frac{x^3 e^x}{(e^x - 1)^2} \, dx.$$

With $\theta_z = 950$ K referring to out-of-plane bending and $\theta_{xy} = 2500$ K to in-plane stretching vibrations [Kru53], there is good agreement with experimental data [Cha85].

Very low vibrational frequencies cannot be excited in small grains. One can estimate the minimum frequency v_{min} from the condition that in the interval $[0, v_{min}]$ the PAH has just one frequency,

$$1 = \int_0^{v_{min}} \rho(v) \, dv.$$

With $\rho(v)$ from (8.42), this gives

$$v_{min} = \frac{k\theta}{h\sqrt{3N}}. \tag{8.45}$$

The bigger the PAH, the smaller v_{min} becomes. Collisional excitation of the modes by gas atoms is possible provided $kT_{gas} \geq hv_{min}$. Under normal interstellar conditions, only the lowest levels are populated.

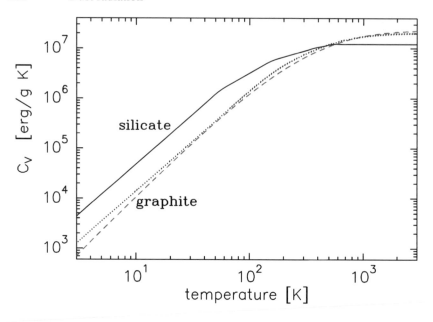

Figure 8.9. The specific heat per gram of dust for silicate (full curve, similar to [Guh89]), graphite (broken curve, after [Cha85]) and PAHs without hydrogen atoms (dotted curve, after [Kru53] using (8.44)).

Figure 8.9 summarizes the specific heats $C_V(T)$ which we use in the calculations of the emission by small grains. At low temperatures, graphite and silicate have a T^3 dependence, for PAHs C_V changes like T^2. As the grains get warmer, the curves flatten towards the Dulong–Petit rule (8.41), $C_V = 3Nk$. Graphite has 5.0×10^{22} and silicate about 2.6×10^{22} atoms per gram.

8.5 Temperature fluctuations of very small grains

When a dust particle is very small, its temperature will fluctuate. This happens because whenever an energetic photon is absorbed, the grain temperature jumps up by some not negligible amount and subsequently declines as a result of cooling. We will speak of very small grains (vsg for short) when we have in mind particles whose temperature is time variable because they are tiny. To compute their emsission, we need their optical and thermal properties. The optical behavior depends in a sophisticated way on the two dielectric functions $\varepsilon_1(\omega)$ and $\varepsilon_2(\omega)$ and on the particle shape. The thermal behavior is determined more simply from the specific heat.

8.5.1 The probability density $P(T)$

Consider a large ensemble of identical grains in some interstellar environment. Let us arbitrarily pick out one of them and denote by $P(T)\,dT$ the chance that its temperature lies in the interval from $T \ldots T + dT$ and call $P(T)$ the probability density. It is, of course, normalized:

$$\int_0^\infty P(T)\,dT = 1. \tag{8.46}$$

In normal interstellar grains of average size, the temperature oscillates only a little around an equilibrium value $T_{\rm eq}$ and in the limit of large grains, $P(T)$ approaches the δ-function $\delta(T_{\rm eq})$, where $T_{\rm eq}$ follows from the steady-state balance between emission and absorption from (8.8),

$$\int Q_\nu^{\rm abs} J_\nu \, d\nu = \int Q_\nu^{\rm abs} B_\nu(T_{\rm eq}) \, d\nu.$$

Even for a very small particle we will assume that its radiation, at any time, obeys Kirchhoff's law (8.1), so in the case of a sphere of radius a, we can express its average monochromatic emission per solid angle by

$$\epsilon_\nu = \pi a^2 Q_\nu^{\rm abs} \int B_\nu(T) P(T) \, dT. \tag{8.47}$$

Although the emission of such a single grain is not constant over time, the whole ensemble radiates at any frequency at a steady rate. We are faced with the problem of finding $P(T)$ and we describe below its solution (an elaborate treatment can be found in [Guh89] and [Dra01]).

8.5.2 The transition matrix

When a grain absorbs or emits a photon, its internal energy $U(T)$, which is a function of temperature only, changes. We bin $U(T)$ into N states U_j of width ΔU_j with $j = 1, 2, \ldots, N$. Each state U_j corresponds to a temperature T_j or frequency $\nu_j = U_j/h$ with corresponding spreads ΔT_j and $\Delta \nu_j$. The probability P_j of finding an arbitrary grain in a large ensemble of $\mathcal{N}$ particles in state U_j is equal to the number of all grains in level j divided by $\mathcal{N}$.

An absorption or emission process implies a transition in U from an *initial* state i to a *final* state f. They occur at a rate $\mathcal{N} P_i A_{fi}$, where the matrix element A_{fi} denotes the transition probability that a single grain changes from state i to f. In equilibrium, for each level j the number of populating and depopulating events, $N_{\rm pop}$ and $N_{\rm depop}$, must be equal:

$$N_{\rm pop} = N_{\rm depop}.$$

Therefore, for each level j,

$$N_{\text{pop}}/\mathcal{N} = \sum_{k>j} P_k A_{jk} + \sum_{k<j} P_k A_{jk} = \sum_{k\neq j} P_k A_{jk}$$

$$N_{\text{depop}}/\mathcal{N} = \underbrace{P_j \sum_{k<j} A_{kj}}_{\text{cooling}} + \underbrace{P_j \sum_{k>j} A_{kj}}_{\text{heating}} = P_j \sum_{k\neq j} A_{kj}.$$

In both formulae, the first sum after the first equal sign refers to processes that cool and the second to those which heat the grain. With the purely mathematical definition

$$A_{jj} = -\sum_{k\neq j} A_{kj}$$

we may write the condition $N_{\text{pop}} = N_{\text{depop}}$ for all j as

$$\sum_k A_{jk} P_k = 0. \tag{8.48}$$

Only $N-1$ of these N equations are linearly independent. To find the probability density $P(T)$ required in equation (8.47), one may first put $P_1 = 1$, solve (8.48) for $P_2, \ldots, P_N$ and then rescale the P_j by the obvious condition that all probabilities must add up to one (see (8.46)),

$$\sum_i P_j = 1. \tag{8.49}$$

A matrix element A_{kj} referring to dust heating ($j < k$) is equal to the number of photons of frequency $\nu_k - \nu_j$ which a grain absorbs per Hz and second multiplied by the width of the final bin $\Delta\nu_k$,

$$A_{kj} = \frac{4\pi C_\nu^{\text{abs}} J_\nu}{h\nu} \Delta\nu_k \qquad \nu = \nu_k - \nu_j \qquad j < k. \tag{8.50}$$

J_ν stands for the mean intensity of the radiation field and C_ν^{abs} is the absorption cross section of the grain. As it should be, the number of transitions $P_j A_{kj}$ from $j \to k$ is thus proportional to the width of the initial (via P_j) and final energy bins. Likewise we have, for dust cooling from state j to a lower one k,

$$A_{kj} = \frac{4\pi C_\nu^{\text{abs}} B_\nu(T_j)}{h\nu} \Delta\nu_k \qquad \nu = \nu_j - \nu_k \qquad k < j. \tag{8.51}$$

Above the main diagonal stand the cooling elements, below those for heating. The energy balance requires for the cooling and heating rate of each level j,

$$\sum_{k<j} A_{kj} v_{kj} = \underbrace{\int_0^\infty 4\pi C_\nu^{abs} B_\nu(T_j) \, d\nu}_{\text{cooling}}$$

$$\sum_{k>j} A_{kj} v_{kj} = \underbrace{\int_0^\infty 4\pi C_\nu^{abs} J_\nu \, d\nu}_{\text{heating}}$$

with $v_{kj} = |v_j - v_k|$. As cooling proceeds via infrared photons which have low energy, their emission changes the grain temperature very little. This suggests that in cooling from state j one needs to consider only the transitions to the levels immediately below. In fact, in practical applications it suffices to ignore cooling transitions with $j \to k < j - 1$. One can, therefore, put all matrix elements A_{fi} above the main diagonal to zero, except $A_{j-1,j}$. But the latter, in order to fulfil the energy equation, have to be written as

$$A_{j-1,j} = \int_0^\infty 4\pi C_\nu^{abs} B_\nu(T_j) \, d\nu \cdot \left[h(v_j - v_{j-1}) \right]^{-1}. \tag{8.52}$$

The total matrix A_{fi} has thus acquired a new form where, above the main diagonal, only the elements A_{fi} with $f = i - 1$ are non-zero. One now immediately obtains from (8.48) the computationally rapid recursion formula (but see the simple trick described in [Guh89] to safeguard against numerical rounding errors).

$$P_{j+1} = -\frac{1}{A_{j,j+1}} \sum_{k \le j} A_{kj} P_k \qquad j = 1, \ldots, N-1. \tag{8.53}$$

We mention that *heating* may not be reduced to transitions $j \to j + 1$. This would ignore the big energy jumps of the grain after UV photon absorption which are important for the probability function $P(T)$. Although heating elements of the form

$$A_{j+1,j} = \int_0^\infty 4\pi C_\nu^{abs} J_\nu \, d\nu \cdot \left[h(v_{j+1} - v_j) \right]^{-1}$$

do not violate energy conservation, they would result in an unrealistically small spread around the equilibrium temperature T_{eq} of (8.8).

8.5.3 Practical considerations

Calculating the emission of a grain with temperature fluctuations is not straightforward. Therefore, it is good to know when such calculations are necessary. Generally speaking, they should be carried out whenever the time interval between the absorption of photons with an energy comparable to the heat capacity of the grain is larger than the cooling time. Unfortunately, this is not very practical advice. However, we will present examples that give some feeling for when temperature fluctuations are important and how they modify the spectrum. Qualitatively, there are three requirements for temperature fluctuations to occur:

- The radiation field is hard to ensure the presence of energetic photons. We are purposely vague about what is meant by *hard* but one may think of the UV range.
- The grains are tiny so that their heat capacity is small.
- The radiation field is weak to make the intervals between capture of energetic photons long. This point is counterintuitive but is exemplified in figure 8.13.

Evaluating the probability density $P(T)$ when temperature fluctuations are small is not only unnecessary but can also incur numerical difficulties. Whenever the function $P(T)$ becomes very sharp, it suffices to calculate the radiation of the particle from (8.1) with a constant temperature from (8.8).

To minimize the computational effort, especially in problems of radiative transfer, one has to avoid a large transition matrix A_{fi} and properly select the energy bins. As $P(T)$ varies strongly with hardness and strength of the radiation field as well as with particle size, a good grid is sometimes not easy to find. The boundaries of the grid, U_{min} and U_{max}, are hard to determine beforehand and intuitive values like $U_{min} = 0$ and $U_{max} = h\nu_{max}$ are often not adequate. An iterative procedure is, therefore, recommended: starting with a first choice that crudely brackets the maximum of $P(T)$, one can find a better adapted grid by using as new boundaries those values of T where $P(T)$ has dropped by some large factor (for instance, 10^{12}) from its maximum.

An indicator for the computational accuracy, but not a precise one, is the time-averaged ratio of emitted over absorbed energy which must, of course, equal one. Generally, a grid of 100 energy bins is sufficient provided they are properly selected. The simplest grid has a constant mesh size, either in temperature or energy. Because $U \propto T^{\alpha}$ with $0 \leq \alpha \leq 3$, a grid of constant ΔT seems to be better suited than one with constant ΔU, as the former has at low temperatures a finer spacing in energy to handle the infrared photons.

8.5.4   The stochastic time evolution of grain temperature

The probability density $P(T)$ describes the steady-state temperature distribution of a large ensemble of grains. The temperature of an individual particle, however, is time variable. Radiative cooling is not balanced at every instant by heating and so the internal energy U changes for a spherical grain of radius a according to the first-order differential equation:

$$\frac{dU}{dt} = 4\pi a^2 \left\{ \int Q_\nu^{abs} J_\nu(t) \, d\nu - \int Q_\nu^{abs} B_\nu(T(t)) \, d\nu \right\}.$$

The right-hand side describes the difference between the power absorbed from the radiation field J_ν and the cooling rate. Replacing U by the specific heat

$C(T)$ (see (8.35)), we may write

$$\frac{dT}{dt} = \frac{4\pi \, \pi a^2}{C(T)} \left\{ \int Q_\nu^{abs} J_\nu(t) \, dv - \int Q_\nu^{abs} B_\nu(T(t)) \, dv \right\}. \tag{8.54}$$

Given a starting value T_0 at $t = 0$, this ordinary differential equation yields the time evolution of the temperature, $T(t)$. One must, however, take into account that the electromagnetic field is quantized, although its time-averaged flux is constant. To make equation (8.54) easier, we assume that the cooling rate, i.e. the flux of emitted low-energy infrared photons, is continuous. Most of the grain heating, however, occurs sporadically through absorption of individual energetic photons (>1 eV) and leads to temperature jerks.

To solve (8.54) numerically from $t = 0$ until $t = \tau$, we proceed as follows. The total period τ is divided into small time steps Δt and the total frequency range into intervals $\Delta v_j = v_{j+1/2} - v_{j-1/2}$. The number of photons with frequency Δv_j that are absorbed within each time step equals

$$N_j = \Delta t \cdot 4\pi \, \pi a^2 \frac{J(v_j) Q^{abs}(v_j)}{h v_j} \Delta v_j.$$

For short time steps Δt, the N_j are much smaller than one and represent also the probability for photon capture. Let U_i and T_i be the energy and temperature of the grain at time t_i, at the end of the ith time step. At the beginning of the new time step $i + 1$, the particle energy is raised by an amount ΔU_i that accounts for the stochastically absorbed photons during the new interval Δt; the temperature is increased accordingly by ΔT_i. To incorporate the quantum character of the photons, ΔU_i is written as

$$\Delta U_i = \sum_j a_j h v_j$$

where the a_j are either one or zero. They are computed from N_j with a random number generator. The chance that $a_j = 1$ is N_j, the probability that $a_j = 0$ equals $1 - N_j$. Then we solve for the $(i + 1)$th time step lasting from t_i until $t_{i+1} = t_i + \Delta t$ the differential equation

$$\frac{dT}{dt} = -\frac{4\pi^2 a^2}{C(T)} \int Q_\nu^{abs} B_\nu(T(t)) \, dv$$

with the boundary condition $T(t_i) = T_i + \Delta T_i$; this gives T_{i+1}. For each time step, one thus gets a temperature.

When the evolution of the temperature, $T(t)$, has been laboriously evaluated from (8.54), the probability density $P(T)$ follows from counting how many times during the period τ the temperature attains values in the interval $T \ldots T + dT$. These numbers are then normalized according to (8.46). The emissivity ϵ_ν can be computed as an average over time τ,

$$\epsilon_\nu = \frac{\pi a^2}{\tau} \int_0^\tau Q_\nu^{abs} B_\nu(T(t)) \, dt.$$

8.6   The emission spectrum of very small grains

We illustrate the stochastic temperature fluctuations and their effect on the spectrum for grains in two environments:

- Near a B1V star. It is luminous ($L = 10^4 \, L_\odot$), has a hot atmosphere ($T_{\text{eff}} = 2 \times 10^4$ K) and guarantees a copious amount of energetic quanta. Its monochromatic flux, L_ν, is proportional to the Planck function,

$$L_\nu = \frac{\pi L}{\sigma T_{\text{eff}}^4} B_\nu(T_{\text{eff}}),$$

and satisfies the condition

$$\int L_\nu \, d\nu = L.$$

- In the interstellar radiation field (ISRF) with a spectral shape as depicted in figure 8.2.

In figures 8.10–8.13, the emission coefficient ϵ_ν refers to one grain and is plotted in the unit erg s^{-1} Hz^{-1} ster^{-1}. All examples deal with silicate grains because of their 10 μm resonance which occasionally shows up. The true emission (from (8.47)) is displayed as a solid line. For comparison, the emission under the false supposition of constant temperature from (8.1) is shown dotted.

The probability density $P(T)$ in figures 8.10–8.13 is also computed in two ways, either from the stochastic time evolution of the temperature $T(t)$ described in section 8.5.4 or, more simply, from the formalism developed in section 8.5.2. In the first case, we depict $P(T)$ by dots, otherwise by a solid line.

8.6.1   Small and moderate fluctuations

The top panel of figure 8.10 displays the temperature variation $T(t)$ of a single silicate grain of 40 Å radius at a distance $r = 10^{17}$ cm from the star. Altogether we followed $T(t)$ over a time $\tau = 2 \times 10^4$ s partitioned into 10^5 steps of fixed length $\Delta t = 0.1$ s. We show an arbitrary section of 400 s. The temperature excursions are small and amount to some 10% around a mean value.

The probability density $P(T)$ in the lower left panel of figure 8.10 gives the chance of finding the grain within a temperature interval of 1 K width centered on T. The full and dotted curves for $P(T)$ (see previous explanation) practically coincide; the area under the curves equals one. In the emission spectrum of the lower right panel the 10 μm feature is indicated. In this example, it does not matter whether ones takes the temperature variations into account or assumes a (constant) equilibrium temperature T_{eq} after (8.8). Here $T_{\text{eq}} = 116.9$ K, which is close to but not identical to the value $T_{\text{max}} = 115.1$ K, where $P(T)$ attains its maximum.

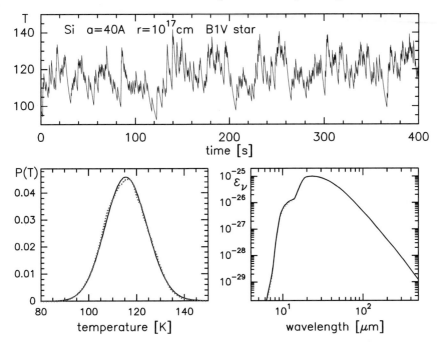

Figure 8.10. Small stochastic temperature excursions: top, time evolution of the temperature of a silicate grain of 40 Å radius at a distance of 10^{17} cm from a B1V star; lower left, probability distribution $P(T)$ of the temperature shown on a linear scale (the dotted line is computed from section 8.5.4, the solid from section 8.5.2); lower right, emissivity ϵ_ν of the grain in erg s^{-1} Hz^{-1} ster^{-1}. The dots (hardly discernible) plot the emission assuming a constant temperature; the full curve includes temperature fluctuations, however, its effect is negligible here. See text.

In figure 8.11, the same grain is placed at a distance ten times greater from the star ($r = 10^{18}$ cm), so the rate of impinging photons is reduced by a factor 100 and the particle is, on average, colder. As one can estimate from the figure, it is hit by an energetic photon only once every 100 s and the temperature excursions are no longer small. There are now discernible diffferences in the probability $P(T)$ depending on the way in which it is computed (see previous explanation). Because the total time interval over which we integrated, although large, was finite, temperatures far from the mean never occurred. One would have to wait very long to see the grain, say, at 170 K. Therefore, the dotted line displays a scatter, although close to the full one, and is determined well only around the maximum of $P(T)$ and does not extend to probabilities below $\sim 10^{-5}$. In the bottom frame on the right, when the emission is evaluated under the false supposition of temperature equilibrium (dotted), the spectrum is a good approximation only at far infrared wavelengths. In the mid infrared, the errors

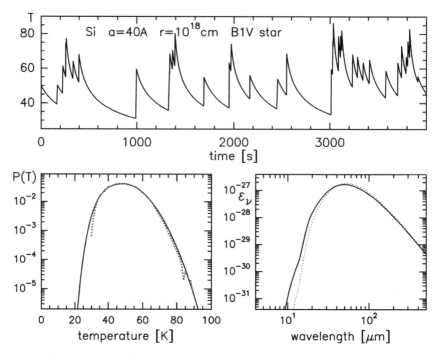

Figure 8.11. Moderate fluctuations. As in figure 8.10 but the same grain is ten times further away from the star. The temperature fluctuations have increased. The ordinate for $P(T)$ is now logarithmic.

are large (two powers of ten at 10 μm) because the grain is occasionally at temperatures far above the average.

8.6.2 Strong fluctuations

In figure 8.12, the grain is again at a distance $r = 10^{17}$ cm from the star but has a radius of only 10 Å. Compared to the two preceding examples where $a = 40$ Å, the heat capacity of the particle is now $4^3 = 64$ times lower and photon absorption thus induces a much larger relative change in energy. The absorption cross section is also 4^3 times smaller than before (the Rayleigh limit) and photon capture is, accordingly, less frequent.

When looking at the temperature evolution $T(t)$, which gives a better feeling for the scatter than the probability function $P(T)$, one hesitates to assign an average temperature at all, although mathematically this can be done. There are now two disparate regimes: most of the time the grain is cold and cooling is slow but occasionally the grain is excited to a high temperature from which it rapidly cools. The probability density $P(T)$ has turned asymmetric; the maximum is at $T_{\text{max}} = 51.9$ K and far from the equilibrium temperature $T_{\text{eq}} = 116.2$ K after

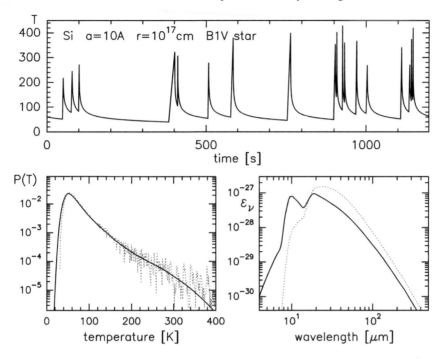

Figure 8.12. Strong fluctuations. As in figure 8.10 but the grain is now small ($a = 10$ Å).

(8.8). The dotted line, representing $P(T)$ as determined from $T(t)$, is jerky and inaccurate above 120 K because of the finite time over which we calculated the evolution; a longer integration time would smooth it. To evaluate the emission without taking into account the hot excursions no longer makes sense, even in the far infrared. The dotted line in the bottom right-hand box bears no resemblance to the real emission (full curve), although the frequency integrated emission is, in both cases, the same and equal to the absorbed flux. Note the strong 10 μm silicate feature.

In figure 8.13, the grain has a radius of 40 Å and is heated by the weak light of the ISRF. Absorption here is a rare event, even with regards to low-energy photons and occurs approximately every few hours. Note the much larger time scale in this figure. The probability density $P(T)$ tapers off gradually from its maximum at $T_{max} = 8.3$ K to ~55 K; beyond 55 K a sharp drop sets in. The dotted line shows the probability $P(T)$ as calculated from the time evolution $T(t)$.

In these examples, we can follow in detail individual absorption events and the subsequent cooling. The cooling rate varies after (8.14) with the sixth power of the temperature. The energy reservoir $U(T)$ of a grain is proportional to T^4 when it is cold, and above the Debye temperature $U \propto T$ (see (8.38), (8.40) and

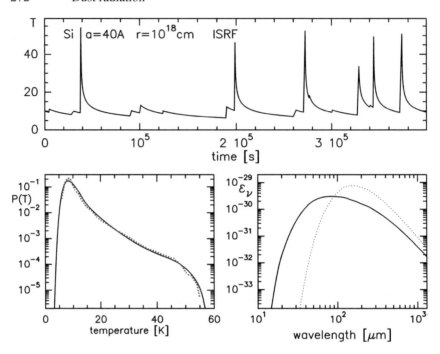

Figure 8.13. Strong fluctuations. A silicate grain of 40 Å radius in the ISRF.

also figure 8.9). The cooling time is roughly

$$\tau_{\text{cool}} \propto \frac{U(T)}{T^6}.$$

and, therefore, falls rapidly as T increases. In the spikes of the present examples, τ_{cool} is of order 10 s.

8.6.3 Temperature fluctuations and flux ratios

Very small grains display in their emission another peculiar feature. It concerns their color temperature or, equivalently, the flux ratio at two wavelengths, λ_1 and λ_2. In a reflection nebula, a grain of normal size (>100 Å) becomes colder when the distance to the exciting star is increased, obviously because it receives less energy. For a very small grain, the situation is more tricky. Of course, it also receives less photons farther from the star but its color temperature at shorter wavelengths is only determined by the hot phases corresponding to the spikes in figure 8.12. The interesting point is: no matter whether these spikes are rare or common, as long as they do not overlap, the flux *ratio* is constant; only the emitted *power* in the wavelength band from λ_1 to λ_2 diminishes with distance. When multi-photon events occur, i.e. when an energetic photon is absorbed before

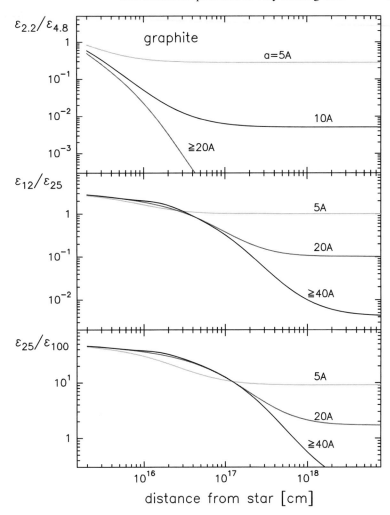

Figure 8.14. Flux ratios for graphite grains of various sizes as a function of distance to a B1V star. The wavelengths in micrometres are indicated as subscripts in the emissivity. Small particles undergo strong temperature fluctuations which lead to flux ratios that are almost independent of the distance from the star.

the grain has had time to cool from a preceding capture, such as that illustrated in figure 8.10, the ratio of the emission coefficients, $\epsilon(\lambda_1)/\epsilon(\lambda_2)$, is again distance-dependent.

Figure 8.14 shows the flux ratio for several wavelengths as a function of distance from the star. The grains are made of graphite and have radii between 5 and 40 Å. For all grain sizes, the near infrared colors, like K–M corresponding to $\epsilon_{2.2\ \mu m}/\epsilon_{4.8\ \mu m}$, first fall as one recedes from the star. But already at distances smaller than the typical dimension of a reflection nebula, they level off. Near IR color temperatures do not change across a reflection nebula if the grain radius is $\sim$10 Å or smaller. For mid and far IR colors, the critical particle size is pushed up a bit and the flux ratios are constant only at larger distances. They continue to stay so far away from the star where the stellar UV radiation field resembles that of the diffuse interstellar medium.

Chapter 9

Dust and its environment

9.1 Grain surfaces

9.1.1 Gas accretion on grains

In the interstellar medium, gas atoms and molecules continually collide with the dust particles. They may either rebounce from the grain surface or stick. In equilibrium, the number of accreted atoms per unit time is equal to the number of atoms leaving the grain. Gas accretion has great astronomical consequences as the following examples demonstrate.

- The transfer of energy from gas to dust and back which may strongly influence the temperature of the components.
- A change in the optical properties of the grains and thus their emission characteristics.
- The depletion of molecular species in the gas phase as they freeze out on the grains. Such species may not then be observable, whereas otherwise they would be strong emitters and important coolants.
- The formation of new molecules. Accretion of H atoms is the only way in which molecular hydrogen can be made at all in relevant quantities.

Consider a gas of temperature T_{gas} containing a certain species (subscript i) with particle mass m_i, number density density n_i and mean velocity (see (5.16))

$$v_i = \sqrt{\frac{8kT_{gas}}{\pi m_i}}. \qquad (9.1)$$

A grain of geometrical cross section πa^2, which can be considered to be at rest, accretes the gas species i at a rate

$$R_{acc} = n_i \, \pi a^2 \, v_i \, \eta_i \qquad (9.2)$$

where η_i denotes the sticking probability. When the grains have a size distribution $n(a) \propto a^{-3.5}$ with lower and upper limit a_- and a_+, their total cross section F for gas capture per cm^3 is

$$F = \frac{3 n_H m_H R_d}{4 \rho_{gr} \sqrt{a_- a_+}}.$$

Here R_d is the dust-to-gas mass ratio, $n_H = n(HI) + n(H_2)$ the total number density of hydrogen, either in atomic or molecular form and ρ_{gr} the density of the grain material, and it was assumed that $a_+ \gg a_-$. The lifetime τ_{acc} of a gas atom before it is swallowed by dust may be defined by $\tau_{acc} R_{acc} = 1$ and thus becomes

$$\tau_{acc} = \frac{1}{F v_i \eta_i} = \frac{4 \rho_{gr}}{3 m_H R_d \eta_i} \frac{\sqrt{a_- a_+}}{n_H v_i}. \tag{9.3}$$

Inserting reasonable numbers ($R_d = 0.007$, $\rho_{gr} = 2.5$ g cm^{-3}, $a_- = 100$ Å, $a_+ = 3000$ Å, $v_i = 0.3$ km s^{-1}), we find that an atom that sticks on collision ($\eta_i = 1$) freezes out after a time

$$\tau_{acc} \sim \frac{2 \times 10^9}{n_H} \text{ yr}. \tag{9.4}$$

τ_{acc} is shorter than the lifetime of a molecular cloud ($\tau_{cloud} \sim 10^7 \ldots 10^8$ yr), because such clouds always have densities above 10^2 cm^{-3}. If the gas density is high ($n_H > 10^4$ cm^{-3}), as in clumps, depletion and mantle formation proceed more quickly ($\tau_{acc} \leq 10^5$ yr) than the dynamical processes which are characterized by the free-fall time scale ($t_{ff} \sim 2 \times 10^7 / \sqrt{n_H}$ yr).

The growth rate da/dt of the grain radius should be independent of a, as the impinging atoms do not see the grain curvature. Therefore, the matter will mostly accrete on the small dust particles as they have the larger total surface area. Consequently, large grains hardly grow any bigger through accretion, although the mean grain size, which is the average over the total distribution, increases.

9.1.2 Physical adsorption and chemisorption

9.1.2.1 The van der Waals potential

To describe the collision between an atom on the grain surface and an approaching gas atom, we first derive their interaction potential. Both atoms are electrically neutral but have dipole moments: let $\mathbf{p}_1$ refer to the surface and $\mathbf{p}_2$ to the gas atom.

A point charge q in an external electrostatic potential ϕ, which is given by some fixed far away charges, has at position $\mathbf{x}_0$ the electrostatic potential energy $U = q\phi(\mathbf{x}_0)$. If instead of the point charge q there is, localized around $\mathbf{x}_0$, a charge distribution $\rho(\mathbf{x})$, the energy becomes

$$U = \int \rho(\mathbf{x}) \, \phi(\mathbf{x}) \, dV.$$

Expanding $\phi(\mathbf{x})$ to first order around $\mathbf{x}_0$,

$$\phi(\mathbf{x}) = \phi(\mathbf{x}_0) + \mathbf{x} \cdot \nabla\phi(\mathbf{x}_0) = \phi(\mathbf{x}_0) - \mathbf{x} \cdot \mathbf{E}(\mathbf{x}_0)$$

and substituting the expansion into the volume integral for U, we obtain

$$U = q\phi(\mathbf{x}_0) - \mathbf{p} \cdot \mathbf{E}(\mathbf{x}_0)$$

where we have used the general definition (1.2) of a dipole $\mathbf{p}$. In our case of colliding atoms, the effective charge q vanishes. Suppose the field $\mathbf{E}(\mathbf{x})$ is due to the dipole $\mathbf{p}_1$ on the grain surface; $\mathbf{E}(\mathbf{x})$ is then described by (3.10). Denoting by $\mathbf{r}$ the vector from $\mathbf{p}_1$ to $\mathbf{p}_2$, we get

$$U(r) = \frac{\mathbf{p}_1 \cdot \mathbf{p}_2}{r^3} - \frac{3(\mathbf{p}_1 \cdot \mathbf{r})(\mathbf{p}_2 \cdot \mathbf{r})}{r^5}. \tag{9.5}$$

Let us further assume that the dipole $\mathbf{p}_2$ of the gas atom is induced by $\mathbf{p}_1$ of the surface atom. The strength of the former depends then on its polarizability α (see (3.10) and (1.8)):

$$p_2 = \frac{2\alpha p_1}{r^3}$$

and the dipole moments will be parallel $(\mathbf{p}_1 \cdot \mathbf{p}_2 = p_1 p_2)$ so that

$$U(r) = -\frac{4\alpha p_1^2}{r^6}. \tag{9.6}$$

9.1.2.2   *The full potential including repulsion*

The absolute value of the van der Waals potential increases rapidly at short distances, much quicker than for monopoles. But when the two dipoles are very close, a repulsive interaction sets in because the electrons tend to overlap and *Pauli*'s exclusion principle forbids them to occupy the same quantum state. This potential bears a more empirical character and changes with distance even more abruptly, like $1/r^{12}$. The net result of the combination between repulsion and attraction is

$$U(r) = 4D\left[\left(\frac{\sigma}{r}\right)^{12} - \left(\frac{\sigma}{r}\right)^6\right] \tag{9.7}$$

with $D\sigma^6 = \alpha p_1^2$. The parameter D defines the strength of the potential and σ its range. $U(r)$ becomes infinite for $r \to 0$, changes sign at $s = \sigma$, reaches its minimum value $-D$ at $r = \sqrt[6]{2}\sigma$ and remains negative as it approaches zero for $r \to \infty$ (figure 9.1).

In a collision of an atom with the grain surface, one has to include the contribution from all force centers, i.e. atoms on the grain surface. On a regular surface, the potential attains minima at various locations privileged by symmetry. For a simple or body centered cubic lattice, these minima are at the mid-points between four neighboring surface atoms (see figure 7.2).

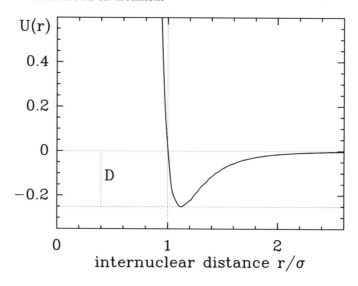

Figure 9.1. The potential $U(R)$ between dipoles after (9.7); here with $D = 0.25$ and $\sigma = 0.5$. The curve is relevant for physical adsorption. The minimum is at $r = \sqrt[6]{2}\sigma$ and $U(r) = -D$.

9.1.2.3 Physical and chemical binding

Physical adsorption is rather weak; it needs typically only 0.1 eV to remove the atom or molecule from the grain surface, corresponding to 1000 K if the binding energy E_b is expressed as a temperature $T = E_b/k$. The exact value depends on the composition and structure of the surface and on the type of gas species.

A much tighter coupling of a gas atom colliding with the grain surface is possible through chemical binding or chemisorption. It involves a profound change in the electron structure of the binding partners. The interaction is therefore much stronger (from 0.5 to 5 eV) and the range of the chemical potential shorter than in the case of physical adsorption.

As a precondition for chemisorption, the surface must contain chemically active sites and the gas species should not consist of saturated molecules, like H_2 or H_2O but atoms, particularly hydrogen. In principle, radicals like OH could also chemically bind to the grain surface but prior to the reaction an activation barrier must be surmounted. This is not possible on a cold grain so that the process is probably unimportant there.

Chemically active sites will have disappeared once the grain surface is covered by physical adsorption by one monolayer of, for example, H_2 or H_2O. Therefore, the mean binding energy of atoms and molecules in an ice mantle several Å or more thick is less than 0.5 eV. Quite generally, we expect grains in cold and dense clouds to be covered by a sheet of physically adsorbed ice. The sheet may be envisaged to be made either of pure water ice or to be a composite of different ices, such as H_2O, CO_2, NH_3 and others, mixed in a certain proportion.

9.1.3 The sticking probability

We estimate the probability η that an impinging gas atom stays on the grain surface; obviously, $0 \leq \eta \leq 1$. For the gas atom to stick, it must transfer to the dust particle in the collision an energy ΔE_s greater than its kinetic energy E_k far from the surface; otherwise it will rebound. One can approximately evaluate the process by studying the interaction of the gas atom (subscript g) with just *one* surface atom (subscript s). We imagine the surface atom to be attached to a spring of force constant κ so that it oscillates at frequency

$$\omega_s = \sqrt{\kappa/m_s}.$$

For the interactive potential $U(r)$ between the two atoms we use (9.7), where $r = x_g - x_s > 0$ is their distance and D the binding energy. Such a simple mechanical system is governed by the equations of motion

$$m_s \ddot{x}_s = -\kappa x_s - F \tag{9.8}$$

$$m_g \ddot{x}_g = F \tag{9.9}$$

if the surface atom of mass m_s has its equilibrium position at $x = 0$. The force associated with the potential, $F(r) = -U'(r)$, is attractive and negative when r is large and repulsive and positive when r is small. The gas atom passes in the collision first through the attractive potential where its kinetic energy increases, then encounters the repulsive part, which is almost like a wall (see figure 9.1), and rebounces at the distance r_{min} where $U(r_{min}) = E_k$.

One finds the energy ΔE_s which is transferred to the surface atom during the impact by numerically integrating the equations of motion. An analytical solution under simplifying assumptions is illustrative. Let us neglect the attraction in the potential and approximate its repulsive part by a parabola [Wat75],

$$U(r) = \begin{cases} -D + \dfrac{D + E_k}{4b^2}(r - 2b)^2 & 0 \leq r \leq 2b \\ -D & r > 2b. \end{cases}$$

The repulsive potential U is often also written as $U(r) = -D + (E_k + D)e^{-r/b}$; the constant b is then called the slope parameter and is of order 0.4 Å. In the *quadratic* equation, the force F with which the gas atom is rejected becomes

$$F(r) = \begin{cases} F_0 - \dfrac{F_0}{2b}r & 0 \leq r \leq 2b \\ 0 & r > 2b \end{cases} \qquad \text{with } F_0 = \dfrac{D + E_k}{b}.$$

To fix the initial conditions, let the gas atom at time $t = 0$ pass the point $r = 2b$ with a velocity $\dot{r} = -\sqrt{2(D + E_k)/m_g}$. Its equation of motion, $F = m_g \ddot{r}$, then has, for times $t > 0$, the solution

$$r = 2b \cdot (1 - \sin \omega_0 t)$$

where the frequency

$$\omega_0 = \sqrt{F_0/2bm_{\mathrm{g}}} \tag{9.10}$$

is of order $10^{14} \ \mathrm{s}^{-1}$. Its inverse defines the collision time,

$$\omega_0^{-1} = \tau_{\mathrm{coll}}.$$

The infall of the gas atom is reversed at the turning point $r = 0$ at time $t = \pi/2\omega_0$ where also $\dot{r} = 0$. Altogether the force varies with time like

$$F(t) = \begin{cases} F_0 \sin \omega_0 t & 0 < t < \dfrac{\pi}{\omega_0} = t_1 \\ 0 & \text{else .} \end{cases} \tag{9.11}$$

Now consider the collision from the point of view of the surface atom which is subjected to a force $f(t)$ of equal strength but opposite sign. During the impact, it undergoes a forced oscillation described by

$$m_{\mathrm{s}} \ddot{x}_{\mathrm{s}} + \kappa x_{\mathrm{s}} = f(t). \tag{9.12}$$

Putting

$$\xi = \dot{x}_{\mathrm{s}} + i\omega_{\mathrm{s}} x_{\mathrm{s}}$$

where $\omega_{\mathrm{s}} = \sqrt{\kappa/m_{\mathrm{s}}}$ is the vibration frequency of the surface atom, one can replace the second-order differential equation (9.12) by one of first order,

$$\dot{\xi} - i\omega_{\mathrm{s}} \xi = \frac{1}{m_{\mathrm{s}}} f(t)$$

which has the solution

$$\xi(t) = e^{i\omega_{\mathrm{s}} t} \left\{ \int_0^t \frac{1}{m_{\mathrm{s}}} f(t) e^{-i\omega_{\mathrm{s}} t} \, dt + \xi_0 \right\}.$$

In our case, $\xi_0 = \xi(t = 0) = 0$. Because the energy of the surface atom is

$$E_{\mathrm{s}} = \tfrac{1}{2} m_{\mathrm{s}} \{\dot{x}_{\mathrm{s}}^2 + \omega_{\mathrm{s}}^2 x_{\mathrm{s}}^2\} = \tfrac{1}{2} m_{\mathrm{s}} |\xi|^2$$

the energy which it receives in the collision becomes

$$\Delta E_{\mathrm{s}} = \frac{1}{2m_{\mathrm{s}}} \left| \int_0^{t_1} f(t) e^{-i\omega_{\mathrm{s}} t} \, dt \right|^2.$$

With t_1 and $f(t) = -F(t)$ from (9.11), the integrations can be evaluated analytically because for $\omega_0 \neq \omega_{\mathrm{s}}$,

$$\int e^{-i\omega_{\mathrm{s}} t} \sin \omega_0 t \, dt = \frac{e^{-i\omega_{\mathrm{s}} t}}{\omega_{\mathrm{s}}^2 - \omega_0^2} (\omega_0 \cos \omega_0 t + i\omega_{\mathrm{s}} \sin \omega_0 t).$$

This can be verified immediately by taking the derivative. For the limiting cases, we obtain

$$\Delta E_s = \frac{m_g}{m_s}(D + E_k) \begin{cases} 4 & \omega_0 \gg \omega_s \\ \pi^2/4 & \omega_0 = \omega_s \\ \dfrac{2\omega_0^4}{(\omega^2 - \omega_0^2)^2} & \omega_0 \ll \omega_s. \end{cases} \qquad (9.13)$$

In the evaluation of the integral for $\omega_0 \ll \omega_s$, it was assumed that the term $\cos(\pi \omega_s/\omega_0)$, which then appears, vanishes on average.

- When $\omega_0 \gg \omega_s$, the collision time τ_{coll} is very short and $e^{-i\omega_s t}$ is practically constant.
- When $\omega_0 \ll \omega_s$, the collision time is long compared to the oscillation period ω_s^{-1} of the crystal. Other atoms will then also absorb energy and the assumption of just two colliding bodies is inadequate. Nevertheless, one can get a feeling from (9.13) for whether the energy loss of the gas atom, ΔE_s, is bigger than its initial kinetic energy E_k, implying sticking, or not.
- Two numerical examples for the case $\omega_s = \omega_0$ are presented in figure 9.2.

As under most interstellar conditions $E_k \ll D$, we learn from formula (9.13) that the sticking efficiency is largely determined by the binding energy D for physical adsorption and by the mass ratio of gas-to-surface atom; this ratio is generally smaller than one, $m_g/m_s \ll 1$.

This simple theory can be refined in various ways: one can take into account the reaction of the crystal; use a more suitable interaction potential because a gas atom generally hits a spot somewhere between several surface atoms; consider oblique incidence and the possibility that the gas atom repeatedly recoils from the grain surface. Elaborate estimates suggest that under most astronomically important circumstances sticking of the atoms is likely. To use a sticking coefficient between 0.1 and 1 presents an educated guess.

9.1.4 Thermal hopping, evaporation and reactions with activation barrier

The more or less evenly spaced atoms on the surface of a grain act as force centers and the surface potential has a semi-regular hilly structure. A physically adsorbed gas atom or molecule finds itself in a potential minimum. It cannot travel freely along the surface because to move, it has to overcome the surrounding potential barrier of height U_0. The value of U_0 is a few times smaller than the binding energy E_b of the gas atom, typically $U_0 \sim 0.3E_b$. The atom can jump over the barrier classically by means of thermal excitation; alternatively, it may quantum mechanically tunnel through the barrier (see [Tie87] for details).

- The adsorbed atom vibrates and thereby swaps its kinetic and potential energy with a characteristic frequency ν_0. For two states separated by an energy difference U_0, the Boltzmann distribution gives the population ratio

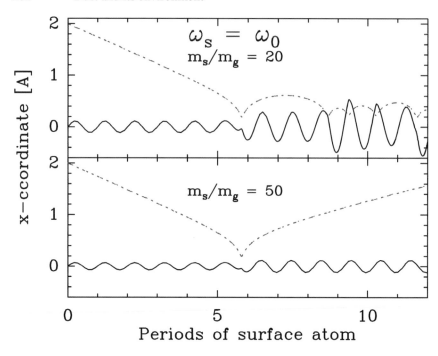

Figure 9.2. Two examples of the time evolution of a one-dimensional collision along the x-axis between a gas atom (subscript g) and a surface atom (subscript s). The forces between the two particles follow from the U12–6 potential of equation (9.7). The vibrational frequency of the surface atom, ω_s, is 1.58×10^{14} s^{-1} and equal to the collisional frequency ω_0. The inverse of the latter can be considered as the effective collision time, $\tau_{coll} = \omega_0^{-1}$. The collisional frequency is defined by $\omega_0 = \sqrt{2(D + E_k)/\sigma^2 m_g}$, slightly different from (9.10) which contains the range parameter b, whereas here we use the range parameter σ after (9.7). Full curves denote the surface atom, enlarged 20 times; the dash-dots, the gas atom. It is evident from the figure that when the gas atom loses energy in the collision, the elongation of the surface atom increases. Bottom: Rebounce with $m_g = m_H$, $m_s/m_g = 50$ and $\Delta E_s/E_k = 1.2$. Top: Sticking with $m_s/m_g = 20$ and $\Delta E_s/E_k = 3.0$. Both numbers for $\Delta E_s/E_k$ agree qualitatively with the approximate formula (9.13) according to which the gas atom rebounds for $\Delta E_s/E_k < 1$, or otherwise sticks. Further parameters are: range of the repulsive potential $\sigma = 0.2$ Å, binding energy $D = 0.05$ eV, temperature of grain and gas: $T_s = 20$ K, $T_g = 50$ K. Curves are obtained by integrating the two second-order differential equations (9.8) and (9.9) by a Runge–Kutta method. The time coordinate on the abscissa is in units of the vibration period of the surface atom.

of upper over lower level or, equivalently, the time ratio of how long the system resides, on average, in these two states. For the lower level, the residence time is also the time t_{hop}, which is needed for the atom to overcome the energy barrier and move to a neighboring site. If the mean residence time in the upper level is v_0^{-1}, the time scale for thermal hopping is

$$t_{\text{hop}} \sim v_0^{-1} e^{U_0/kT} \tag{9.14}$$

where T is the grain temperature. For v_0, one may use the frequency of lattice vibration of the grain, typically a few times 10^{13} s^{-1}. A more accurate expression in the case of a symmetric harmonic potential is

$$v_0 = \sqrt{2n_s E_b/\pi^2 m}$$

where n_s denotes the surface density of sites, m and E_b are the mass and binding energy of the adsorbed atom. If $U_0 < kT$, thermal hopping is very quick and $t_{\text{hop}} \sim v_0^{-1}$, otherwise t_{hop} increases exponentially.

When an adsorbed atom moves on the surface of a grain by thermal hopping, it has no preferential direction when jumping from one site to the next and its path is a random walk (see figures 9.4 and 9.5). If the atom takes N steps of constant length a equal to the distance between two surface atoms, the root mean square deviation from the starting position is $\sigma = N^{1/2} a$. Therefore, the adsorbed atom needs a time $N^2 t_{\text{hop}}$ to move a distance Na from the starting point.

- Replacing in (9.14) U_0 by the binding energy E_b, one gets the evaporation time scale

$$t_{\text{evap}} \sim v_0^{-1} e^{E_b/kT}. \tag{9.15}$$

- One can argue, as before, that a reaction in which a molecule on the grain surface is chemically transformed but which requires an activation energy E_a, occurs on average after a time

$$t_{\text{chem}} \sim v_0^{-1} e^{E_a/kT}. \tag{9.16}$$

9.1.5 Tunneling between surface sites

Because of its exponential dependence on grain temperature in (9.14), thermal hopping does not work when the grain is cold and kT falls below U_0. However, an atom has the possibility to quantum mechanically tunnel through the potential barrier as discussed in section 6.4. When the potential barrier U_0 is infinite, the levels are degenerate (the U here is denoted V in section 6.4). For finite U_0, there is energy splitting with an amount $E_2 - E_1$ that increases as U_0 is lowered (see (6.61)). According to (6.63), the tunneling time

$$t_{\text{tun}} \sim v_0^{-1} \exp\left(\frac{2a}{\hbar}\sqrt{2mU_0}\right)$$

is short for small U_0. The gas atom on the grain surface then has a high mobility and is not fixed to any particular site. The time t_{tun} is very sensitive to all numbers in the exponent, i.e. to the lattice spacing $2a$, the mass of the gas atom m and the barrier height U_0. Obviously, tunneling can only be important for the lightest atoms, in practice for hydrogen or deuterium. The mobility of other species is restricted to thermal hopping.

Let us evaluate the separation of the energy levels, $E_2 - E_1$, and the tunneling time t_{tun} from (6.61) and (6.63) for a hydrogen atom ($m = 1.67 \times 10^{-24}$ g) physically adsorbed on an ice layer with a spacing $2a = 2$ Å between force centers and vibrating at a frequency $\nu_0 = 3 \times 10^{12}$ s^{-1}. Putting the potential barrier U_0 to 0.3 times the binding energy E_b, i.e. $U_0/k \simeq 100$ K, we obtain $t_{tun} \simeq 10^{-11}$ s and $E_2 - E_1 \simeq 10$ K. Because of the exponential terms, the estimates are coarse.

The case of a one-dimensional double potential well can be generalized to a perfectly regularly structured grain surface, where the potential is periodic in two dimensions [Hol70]. In the earlier expression for t_{tun}, the energy difference between the two split states $|E_1 - E_2|$ is replaced by the width ΔE of the lowest energy band. It consists of as many sublevels as there are interacting wells. The full analysis gives a very similar residence time for an atom in a particular well,

$$t_{tun} \simeq \frac{4\hbar}{\Delta E}. \tag{9.17}$$

When it is applied to a hydrogen atom on ice, the calculations yield $\Delta E/k \simeq 30$ K and thus $t_{tun} \simeq 10^{-12}$ s, an order of magnitude shorter than the crude estimate given here, but this discrepancy does not worry us considering the uncertainties and simplifications.

9.1.6 Scanning time

For an *ideal* surface, tunneling to a site n grid spacings away is not a random walk, but the well is reached after a time $t_n \simeq n \cdot t_{tun}$. A real grain has defects on its surface, i.e. sites surrounded by a high potential which the atom cannot penetrate and where it is deflected in its motion, thus leading to a random walk. The defect sites invariably arise because the grain is bombarded by soft X-rays and cosmic rays.

Suppose the grid constant of the atomic lattice is of unit length. Let ℓ be the mean free path before the atom is scattered. The time to reach a site n spacings away increases from $n \cdot t_{tun}$ without scattering to $(n^2/\ell)t_{tun}$ when defects are taken into account because one needs $(n/\ell)^2$ steps of length ℓ. If the surface contains N sites altogether, its linear dimension is of order $N^{1/2}$ and the mobile particle covers in a random walk such a distance in a time

$$\tau_{scan} \simeq \frac{N}{\ell} t_{tun}. \tag{9.18}$$

Provided that $\ell \ll n$, one calls τ_{scan} the scanning time. Inserting numbers for a grain of 1000 Å radius, we find $t_{scan} \lesssim 10^{-6}/\ell$ s, so scanning is quick.

9.2   Grain charge

A grain in interstellar space is not likely to be electrically neutral. Mechanisms are at work that tend to alter its charge, notably

* the impact of an electron
* the impact of a positive ion and
* the ejection of an electron by a UV photon.

In equilibrium, the processes that make the grain positive and negative exactly balance.

9.2.1   Charge equilibrium in the absence of a UV radiation field

First we neglect the radiation field. Consider a spherical grain of radius a and charge Z. Its cross section for capturing electrons of mass m_e and velocity v is

$$\sigma_e(v) = \pi a^2 \left(1 + \frac{2Ze^2}{a m_e v^2} \right).$$

The term in the brackets determines the change over the pure geometrical cross section $\sigma_{geo} = \pi a^2$. The bracket has a value greater than one when the dust is positively charged $(Z > 0)$ and less or equal unity otherwise. The enhancement in the cross section follows immediately from the two equations describing conservation of energy and angular momentum for an electron with grazing impact,

$$\frac{1}{2} m_e v^2 = \frac{1}{2} m_e V^2 - \frac{Ze^2}{a} \tag{9.19}$$

$$a_{eff} v = a V. \tag{9.20}$$

V is the actual impact velocity and $\sigma_e = \pi a_{eff}^2$ the effective cross section. The number of electrons striking and then staying on the grain:

$$Y_e n_e \langle v \sigma_e(v) \rangle = Y_e n_e \pi a^2 4\pi \left(\frac{m_e}{2\pi kT} \right)^{3/2} \int_{v_0}^{\infty} v^3 \left(1 + \frac{2Ze^2}{a m_e v^2} \right) e^{-m_e v^2/2kT} \, dv \tag{9.21}$$

n_e is their density, Y_e their sticking probability and the bracket $\langle \ldots \rangle$ denotes an average over the Maxwellian velocity distribution (5.13) at the temperature T of the plasma. In the case of negative grain charge $(Z < 0)$, some electrons are repelled and only those whose velocity exceeds a critical value v_0 reach the surface at all. This critical value is determined by

$$\frac{m_e v_0^2}{2} = \frac{Ze^2}{a}.$$

Otherwise, if $Z \geq 0$, we must put $v_0 = 0$. There are corresponding formulae for ion capture where the quantities m, σ, n, Y bear the subscript i. For a plasma with equal density and sticking probability for electrons and singly charged ions, $n_e = n_i$ and $Y_e = Y_i$. In equilibrium, we can write for the brackets that appear in (9.21),

$$\langle v\sigma(v)\rangle_e = \langle v\sigma(v)\rangle_i. \tag{9.22}$$

On the left, the average refers to a Maxwell distribution for electrons and on the right for ions. The impact rate of electrons on a grain of charge Z is $n_e \langle v \rangle \sigma_{eff}$, where

$$\sigma_{eff} = \sigma_{geo} \begin{cases} (1 + Ze^2/akT) & \text{if } Z > 0 \\ e^{Ze^2/akT} & \text{if } Z < 0 \end{cases} \tag{9.23}$$

is the effective cross section and $\langle v \rangle = (8kT/\pi m_e)^{1/2}$ the mean electron velocity. There is a corresponding equation for ions.

Without photoemission, the grain must be negatively charged ($Z < 0$) because the electrons are so much faster than ions. When we evaluate the integrals $\langle v\sigma(v)\rangle$ using the relations (A.26), (A.28) for the ions and (A.30) and (A.32) for the electrons, we get

$$\exp\left(\frac{Ze^2}{akT}\right) = \left(\frac{m_e}{m_i}\right)^{\frac{1}{2}}\left[1 - \frac{Ze^2}{akT}\right].$$

Solving this equation for $x = Ze^2/akT$ and assuming that the ions are protons, one finds

$$Z \simeq -2.5\frac{akT}{e^2}. \tag{9.24}$$

Interestingly, the degree of ionization and the density of the plasma do not appear in this formula. Heavier ions make the grain more negative but only a little. We see in figure 9.3 that for a fixed temperature, large grains bear a greater charge than small ones but the potential

$$U = \frac{Ze}{a}$$

does not change. As one would expect, Z adjusts itself in such a way that, by order of magnitude, the mean kinetic energy of a gas atom, kT, equals the work Ze^2/a necessary to liberate one unit charge. We notice that in a hot plasma, the charge can become very large.

9.2.2 The photoelectric effect

9.2.2.1 The charge balance

In the presence of a hard radiation field of mean intensity J_ν, one has to include a term in the charge equilibrium equation (9.22) that accounts for the fact that UV

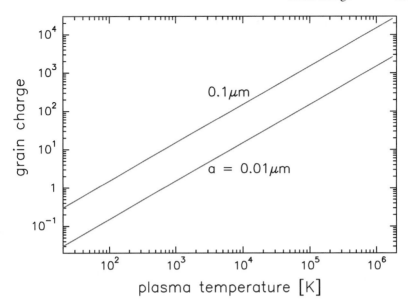

Figure 9.3. The equilibrium charge of a grain (in units of the charge of an electron) in a hydrogen plasma of temperature T for two particle radii. Photoelectric processes are absent.

photons can chip off electrons from the bulk material of the grain. This certainly happens near a star of early spectral type but also in diffuse clouds permeated by the average interstellar radiation field. If photoemission is strong, the grain will be positively charged; one can then neglect the impact of positive ions. Putting the sticking probability Y_e of electrons equal to one, we find for the charge balance with respect to impinging electrons and absorbed photons that

$$n_e \langle v \rangle \left(1 + \frac{Ze^2}{akT}\right) = 4\pi \int_{\nu_t}^{\infty} \frac{J_\nu Q_\nu^{abs}}{h\nu} y_\nu \, d\nu. \qquad (9.25)$$

We divided under the integral by $h\nu$ because we wanted the *number* of absorbed photons. One can immediately solve (9.25) for the charge Z, provided one knows the parameters ν_t and y_ν. The first, ν_t, represents a threshold frequency for photon absorption leading to electron emission; the energy $h\nu_t$ is of order 10 eV and includes the work to liberate an electron from the solid ($\sim$4 eV) and to overcome the potential U of the positively charged grain. The second parameter, y_ν, is the yield for photoemission.

9.2.2.2   The photon yield

The yield y_ν for photoemission may be estimated from the following physical picture based on classical electrodynamics [Pep70]. The energy absorbed by a subvolume dV of a grain is given by (2.52). When one applies the Gauss theorem (A.13), one gets

$$dW_a = -\operatorname{div} \mathbf{S}\, dV$$

where $\mathbf{S}$ is the Poynting vector. Let the subvolume be at a depth x below the surface. The likelihood P that the absorption leads to emission of a photoelectron is assumed to have the form

$$P = C \exp(-x/l_e).$$

The factor C incorporates the following two probabilities: for excitation of an electron to a 'free' state and, when such an electron has reached the surface, for penetrating to the outside and not being reflected. The exponential term $\exp(-x/l_e)$ gives the probability that the electron reaches the surface at all and is not de-excited in any of the scattering processes on the way. The deeper the subvolume dV below the grain surface, the higher the chance for de-excitation. The mean free path of electrons in the bulk material, l_e, is of order 30 Å, possibly shorter for metals and longer in dielectrics. Because of the factor e^{-x/l_e}, it is evident that small grains (radii $a \sim 50$ Å) are much more efficient in photoemission than big ones ($a \sim 1000$ Å). From Mie theory, one can compute the internal field of the grain (the relevant formula is (2.45)) and thus the Poynting vector.

The yield y_ν for electron emission induced by photons of frequency ν is now defined via the equation

$$y_\nu \int_V \operatorname{div} \mathbf{S}\, dV = C \int_V e^{-x/l_e} \operatorname{div} \mathbf{S}\, dV \qquad (9.26)$$

where the integrals extend over the whole grain volume V. The material constants are uncertain but various evidence points towards $y_\nu \sim 0.1$ [Wat72].

9.2.2.3   The photoelectric effect and gas heating

Photoemission can also be important for heating the interstellar gas. As the mean kinetic energy of a photoejected electron, E_ν, exceeds the average thermal energy $\frac{3}{2}kT$ of a gas particle, the excess energy, after subtraction of the electrostatic grain potential U, is collisionally imparted to the gas. In the end, the electron is thermalized and its own average energy will then also be $\frac{3}{2}kT$. The heating rate due to one dust grain is therefore

$$H = 4\pi\, \pi a^2 \int_{\nu_t}^{\infty} \frac{J_\nu Q_\nu^{\text{abs}}}{h\nu} y_\nu \big[E_\nu - U - \tfrac{3}{2}kT\big]\, d\nu. \qquad (9.27)$$

The photoeffect has to be invoked to explain the fairly high temperatures of 50 ... 100 K observed in HI regions; there it is the dominant heating mechanism. Of course, the temperature does not follow from the rate H in (9.27) alone but only from the balance with the cooling processes.

9.3 Grain motion

9.3.1 Random walk

Suppose one makes, in three-dimensional space, starting from position $\mathbf{R}_0 = \mathbf{0}$, a sequence of steps defined by the vectors $\mathbf{L}_i$. After N steps, one has arrived at the position $\mathbf{R}_N = \mathbf{R}_{N-1} + \mathbf{L}_N$. If the vectors $\mathbf{L}_i$ are of constant length L but in an arbitrary direction, the mean square of the distance, $\langle R_N^2 \rangle$, grows like

$$\langle R_N^2 \rangle = NL^2. \tag{9.28}$$

It is straightforward to prove (9.28) by induction because the average of $\mathbf{R}_N \cdot \mathbf{L}_{N-1}$ obviously vanishes. Figure 9.4 shows a two-dimensional random walk and figure 9.5 the verification of formula (9.28) in a numerical experiment.

9.3.2 The drag on a grain subjected to a constant outer force

Let us consider the one-dimensional motion of a heavy test particle (grain) through a fluid or gas in more detail. Suppose one applies a constant outer force F on the test particle of mass M. It then experiences a drag from the fluid or gas molecules that is proportional to its velocity $\dot{x} = V$ and the equation of motion of the test particle reads:

$$F = M\ddot{x} + \mu\dot{x}. \tag{9.29}$$

On the right stands an accelerational term $M\ddot{x}$ and a dissipational term. The coefficient μ in the latter depends on the properties of both the fluid and the test particle. In a steady state, $\ddot{x} = 0$ and $F = \mu V$.

We determine the force F in (9.29) needed to move a spherical grain of geometrical cross section πa^2 at a constant velocity V through a gas. The gas has a number density N at temperature T and its atoms have a mass $m \ll M$. Let the grain advance in the positive x-direction. The velocity distribution $N(v_x)$ of the gas atoms along this direction *with respect to the grain* is no longer given by equation (5.14) but has an offset in the exponent,

$$N(v_x)\,dv_x = N\left(\frac{m}{2\pi kT}\right)^{1/2} e^{-m(v_x+V)^2/2kT}\,dv_x.$$

The y- and z-axes are, of course, unaffected and purely Maxwellian. The maximum of $N(v_x)$ is now at $v_x = -V$, whereas in the rest frame of the gas it is at zero velocity. The number of atoms in the velocity range $[v_x, v_x + dv_x]$ that hit the grain head-on ($v_x < 0$) equals $-\pi a^2 v_x N(v_x)\,dv_x$ and each atom imparts in an

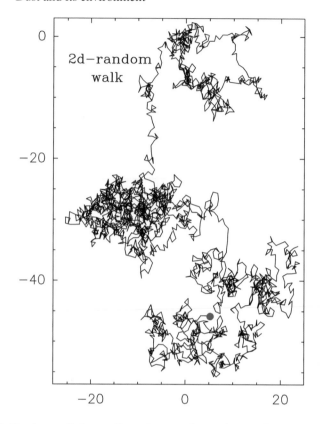

Figure 9.4. Random walk in two dimensions starting at the coordinate center $(0, 0)$ and ending after 2700 steps of equal length $L = 1$ at the big dot near $(5, -46)$.

inelastic collision, the momentum mv_x. To obtain the total momentum transfer, one has to integrate over all velocities $v_x < 0$, which leads to the expression

$$\pi a^2 Nm \left(\frac{m}{2\pi kT}\right)^{1/2} \int_{-\infty}^{0} v_x^2 e^{-m(v_x+V)^2/2kT} \, dv_x. \tag{9.30}$$

- Suppose the grain moves much slower than sound. Substituting in (9.30) $w = v_x + V$ and taking the momentum difference between front and back, one gets the retarding force on the grain. The term that does not cancel out under this operation is

$$F = \pi a^2 4Nm V \left(\frac{m}{2\pi kT}\right)^{1/2} \int_{0}^{\infty} w e^{-mw^2/2kT} \, dw$$

and therefore,

$$F = \pi a^2 Nm V \left(\frac{8kT}{m\pi}\right)^{1/2}. \tag{9.31}$$

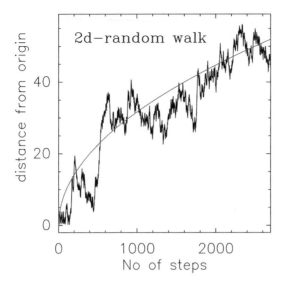

Figure 9.5. According to (9.28), in a random walk one travels after N steps a mean distance $\sqrt{N}L$ (smooth line). The actual distance in the numerical experiment of figure 9.4 is the jittery line.

The drag is proportional to the geometrical cross section of the grain, πa^2, and the drift velocity V. With the mean gas velocity $\langle v \rangle = \sqrt{8kT/m\pi}$, the friction coefficient becomes

$$\mu = \pi a^2 Nm\langle v \rangle.$$

- When the grain moves highly supersonically, there is only a force acting on the front side and the momentum transfer is simply

$$F = \pi a^2 NmV^2. \tag{9.32}$$

F is now proportional to the square of the velocity. In elastic collisions, the forces would be larger by about a factor of two.

An astronomically important case of an outer force acting on dust particles is provided by radiation pressure (see formulae (2.7) and (2.8) but there the letter F means flux). In the case of direct illumination by a star, we may approximately set the cross section for radiation pressure, C^{rp}, equal to the geometrical cross section $\sigma_{\mathrm{geo}} = \pi a^2$, corresponding to an efficiency $Q^{\mathrm{rp}} = 1$. If L_* is the stellar luminosity and r the distance of the grain to the star, the drift velocity, for subsonic motion, is then

$$V = \frac{L_*}{4\pi cr^2 Nm} \left(\frac{m\pi}{8kT}\right)^{1/2}.$$

N, m and T refer to the gas. Evaluating this formula for typical numbers of L_*, N and T, one finds that in stellar environments supersonic drift speeds are easily achieved. However, in many configurations grains are not directly exposed to starlight but irradiated by infrared photons to which the starlight has been converted through foreground matter. In such circumstances, C^{rp} is much smaller than the geometrical cross section and V accordingly smaller, too.

Around main sequence stars, the gas density is always low and the gas drag unimportant. A grain is attracted towards the star of mass M_* by gravitation and repelled by radiation pressure. From the balance between the two,

$$\frac{GM_*M}{r^2} = \frac{L_*\sigma_{\mathrm{geo}}}{4\pi c r^2} \qquad (9.33)$$

one obtains a critical grain radius independent of the distance:

$$a_{\mathrm{cr}} = \frac{3L_*}{4\pi c \, G\rho_{\mathrm{gr}} M_*} \qquad (9.34)$$

or

$$\frac{a_{\mathrm{cr}}}{\mu\mathrm{m}} \simeq 0.24 \left(\frac{L_*}{L_\odot}\right)\left(\frac{M_*}{M_\odot}\right)^{-1}. \qquad (9.35)$$

Grains smaller than a_{cr} are expelled. For instance, a 10 $M_\odot$ star has $a_{\mathrm{cr}} \simeq 2.4$ mm, so all particles with sizes of interstellar grains (~ 0.1 μm) are blown away by radiation pressure. For low mass stars $L_* \propto M_*^{3.5}$ and the critical radius falls with $M_*^{2.5}$. The removal is always rapid as one can show by integrating the outward acceleration $\dot{v} = 3L_*/16\pi c\rho_{\mathrm{gr}} a r^2$ (see (9.33)).

9.3.3 Brownian motion of a grain

Under equipartition between dust particles and gas, the grains perform a Brownian motion. It has a translatory and a rotational part but for the moment we neglect rotation. It is a fundamental result of thermodynamics that in equilibrium the mean kinetic energy of a gas atom is equal to the mean kinetic energy E_{kin} of a grain. If M and V denote the mass and velocity of the dust particle,

$$\tfrac{3}{2}kT_{\mathrm{gas}} = E_{\mathrm{kin}} = \tfrac{1}{2}MV^2.$$

The Brownian velocity V follows from

$$V = \sqrt{\frac{3kT_{\mathrm{gas}}}{M}}. \qquad (9.36)$$

Even a small grain of $a = 100$ Å with bulk density $\rho_{\mathrm{gr}} = 2.5$ g cm^{-3} has in a 20 K gas only $V = 28$ cm s^{-1}: it is practically at rest. The kinetic energy of the grain, E_{kin}, is, on average, equally distributed among the three degrees of freedom, so

$$MV_x^2 = MV_y^2 = MV_z^2.$$

The one-dimensional equation (9.29) is also correct with respect to the Brownian motion of interstellar grains when the retardation is not due to an external force but to internal friction. The latter always involves dissipation of energy. In a quasi-stationary state at constant velocity V, the associated heat loss FV must be compensated by a permanent acceleration $\Delta p^2/2M$, where Δp^2 is the average change in the square of the momentum, p^2, per unit time due to collisions with gas atoms. Thus we require $FV = \Delta p^2/2M$ and because $MV^2 = kT$ in equilibrium at temperature T, we get $\Delta p^2 = 2\mu kT$. In three dimensions, when all directions are equal,

$$\Delta \mathbf{p}^2 = 6\mu kT. \tag{9.37}$$

To find the mean distance that a grain travels when there is no external force, we multiply equation (9.29) by x,

$$Fx = Mx\ddot{x} + \mu x\dot{x} = \frac{M}{2}\frac{d}{dt}\frac{dx^2}{dt} - M\dot{x}^2 + \frac{\mu}{2}\frac{dx^2}{dt}.$$

On taking averages, we note, first, that $\langle Fx \rangle = 0$ because x and F are unrelated and the force F due to collisions with gas atoms is completely stochastic in its direction. Second, the rate of change of the mean square distance, $d\langle x^2 \rangle/dt$, must be constant in time because the particle has no memory about its past. On averaging, this equation shortens, therefore, to $\frac{1}{2}\mu\, d\langle x^2 \rangle/dt = M\langle \dot{x}^2 \rangle = kT$. Therefore, in three dimensions, the square of the distance r grows linearly with time,

$$r^2 \equiv \langle x^2 + y^2 + z^2 \rangle = \frac{6kT}{\mu}t. \tag{9.38}$$

9.3.4 The disorder time

Suppose a grain has, at a certain instant, the momentum p_0. After a disorder or damping time t_{dis}, it will have completely lost its memory about p_0. One impinging gas atom of mass $m \ll M$ and velocity v changes the momentum of the grain statistically by mv, and Z atoms by $mv\sqrt{Z}$, as in a random walk. So in equipartition, when

$$MV^2 = mv^2$$

the momentum of the grain is profoundly altered after M/m collisions, i.e. when the mass of the colliding atoms equals the mass of the grain. Therefore, in a gas of number density N, the disorder or damping time is

$$t_{\mathrm{dis}} = \frac{M}{Nmv\pi a^2}. \tag{9.39}$$

Because $F = \mu V = M\,dV/dt \simeq MV/t_{\mathrm{dis}}$, the friction coefficient can be expressed through the damping time t_{dis},

$$\mu = \frac{M}{t_{\mathrm{dis}}}. \tag{9.40}$$

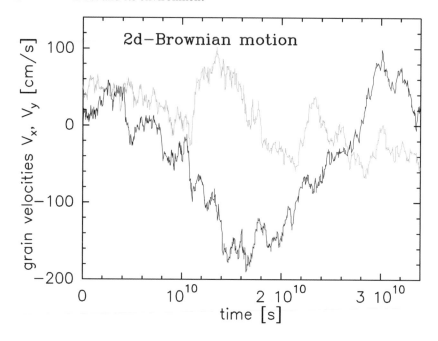

Figure 9.6. Numerical experiment of a two-dimensional Brownian motion: gas parameters, hydrogen atoms at $T = 100$ K with number density 10^4 cm^{-3}; grain parameters, radius 100 Å, mass $M = 1.05 \times 10^{-17}$ g. The figure shows the x- and y-component of the velocity; initially $\frac{1}{2}MV_x^2 = kT$ and $V_y = 0$. Because of the large time interval displayed in the plot, the actual jitter is graphically not fully resolved.

Figure 9.6 displays a numerical experiment of a two-dimensional Brownian motion assuming inelastic collisions. The parameters are such that a gas atom impacts the grain about once every 1000 s. To follow the stochastic evolution, we bin the velocity distribution of the gas atoms $N(v_x)$ of (5.14) into 100 velocity intervals, and likewise for $N(v_y)$. Choosing a time step of 10^3 s, the chance that during one time step the grain is hit by an atom within a certain velocity interval, either from the front or back, is small. The actual occurrence of such an event is prompted by a random number generator. The disorder time according to (9.39) equals 1.5×10^9 s. It is roughly the interval after which there is a change in the direction of the velocity vector of the grain by 90° or in its kinetic energy by more than 50%.

The computations pertaining to figure 9.6 also yield, of course, the time evolution of the mean square deviation $r^2 = \langle x^2 + y^2 \rangle$. Therefore one can check whether $r^2(t)$ fulfils the theoretical prediction of equation (9.38) with the dissipation constant μ from (9.40). One finds good agreement between numerical experiment and (9.38) only after integrating over many disorder times t_{dis} because one disorder time corresponds to one step in the random walk of figure 9.4.

9.3.5 Laminar and turbulent friction

In this section, we consider gases of sufficiently high density such that the mean free path ℓ of the atoms after (15.90) is smaller than the dimension of the body moving through the gas. So the discussion does not apply to grains in the interstellar medium but is relevant to dust particles in accretion disks (section 15.4) or planetary atmospheres.

- A laminar flow of velocity v around a sphere of radius a exerts, according to Stokes law, a force

$$F = 6\pi \eta a v \tag{9.41}$$

where $\eta = \rho v$, and ρ and v are the density and kinematic viscosity coefficient of the gas (see the empirical relation (15.74) and equation (15.89)). For the coefficient μ in (9.29) we find, by comparison with (9.41), that

$$\mu = 6\pi \eta a.$$

The retarding force F in (9.41) is linearly proportional to the velocity v but also to the radius a. This somewhat strange result can be understood from (15.74), if one writes for a sphere:

$$F = v\rho \frac{\partial v}{\partial a} 2\pi a^2$$

and replaces the derivative $\partial v/\partial a$ by v/a. To obtain the coefficient 6π rigorously is tiresome. A sphere rotating at velocity v at the circumference is retarded by a torque $\tau = Fa$ with F also approximately given by (9.41).
- When the flow is turbulent, the decelerating force becomes

$$F = c_W \pi a^2 \rho v^2$$

as in (9.32). F is now proportional to the cross section πa^2, the gas density $\rho = Nm$ and the velocity squared. The coefficient c_W depends on the Reynolds number (15.93) and is of order one.

But even in a turbulent flow there is a laminar surface layer of thickness D, in which the gas velocity increases from zero, on the surface of the body, to v at the boundary of the layer. For a rectangle of area $A = bl$ and with normal perpendicular to the velocity $\mathbf{v}$, the retarding force is

$$F = \eta A \frac{v}{D}. \tag{9.42}$$

The work required to renew the kinetic energy of the surface layer continually is

$$Fv = \eta bl \frac{v^2}{D} = \frac{1}{2}v^2 bv D\rho.$$

It is assumed that the side l of the rectangle runs parallel to the flow and b perpendicular to it. When one takes into account that the velocity in the surface layer increases linearly from zero to v, one finds by integration for the thickness of this so called *Prandtl* layer

$$D = \sqrt{\frac{6\eta l}{v\rho}}.$$

Inserting D into (9.42) gives a force that is in between the purely laminar and turbulent cases.

9.3.6   A falling rain drop

As an example, we study a falling rain drop. This has the advantage over an interstellar grain that its properties are known and that it is shaped like a perfect sphere, almost. Not to be a sphere, i.e. to have (for the same mass) a surface larger than necessary, would be energetically quite disadvantageous. To create an area dA of surface takes the work

$$dW = \zeta dA \tag{9.43}$$

where $\zeta \simeq 75$ erg cm^{-2} is the surface tension of water. It is slightly temperature dependent; a more precise formula is

$$\zeta(T) = 116.82 - 0.151T \quad \text{erg cm}^{-2}. \tag{9.44}$$

One roughly estimates that to lift a molecule from the interior of the drop, where it is surrounded by other water molecules on all sides, to the surface requires a few hundredths eV. By equating the $p\,dV$ work (here $dV = $ infinitesimal volume) to the work required to create new surface, it also follows that the pressure p inside a drop of radius a is given by

$$p = \frac{2\zeta}{a}. \tag{9.45}$$

The pressure can become high, it is more than ten times the atmospheric pressure at sea level for a rain drop 0.1 μm in radius. Equation (9.45) is needed when evaluating the evaporation of small grains.

The rain drop is pulled by gravity with the force $F = 4\pi a^3 g\rho_w/3$ (gravitational acceleration $g = 981$ cm s^{-2}, water density $\rho_w = 1$ g cm^{-3}). With the appropriate numbers for air (kinematic viscosity $v \simeq 0.15$ cm^2 s^{-1}, density $\rho \simeq 1.2 \times 10^{-3}$ g cm^{-3}, $\eta = \rho v \simeq 1.8 \times 10^{-4}$ g cm^{-1} s^{-1}), the turbulent and laminar deceleration become equal, $6\pi \eta a V = \pi a^2 \rho V^2$, when the drop has a radius $a \sim 0.1$ mm and sinks at a velocity $V \sim 1$ m s^{-1}. The Reynolds number Re $= aV/v$ of (15.93) is then slightly below 10. For small Reynolds numbers, the terminal velocity of the falling drop changes like $V \propto a^2$, for large ones, $V \propto a^{1/2}$.

9.3.7 The Poynting–Robertson effect

Consider a grain of mass m, radius a and geometrical cross section $\sigma_{geo} = \pi a^2$ circling a star at distance r with frequency ω and velocity $v = \omega r$. The angular momentum of the grain is

$$\ell = mr^2\omega = mrv.$$

Let M_* and L_* be the mass and luminosity of the star. Because there is mostly optical radiation, the absorption coefficient of the grain is $C^{abs} \simeq \sigma_{geo}$ and the particle absorbs per unit time the energy

$$\Delta E = \frac{L_*\sigma_{geo}}{4\pi r^2}.$$

In thermal balance, the same amount is re-emitted. Seen from a non-rotating rest frame, the stellar photons that are absorbed travel in a radial direction and carry no angular momentum, whereas the emitted photons do because they partake in the circular motion of the grain around the star. If we associate with the absorbed energy ΔE a mass $m_{phot} = \Delta E/c^2$, the angular momentum of the grain decreases per unit time through emission by

$$\frac{d\ell}{dt} = -m_{phot}rv = -\frac{\Delta E}{c^2}rv = -\frac{L_*\sigma_{geo}}{4\pi c^2 r^2}\frac{\ell}{m}.$$

Because for a circular orbit

$$\frac{v^2}{r} = \frac{GM_*}{r^2}$$

we get $v = \sqrt{GM_*/r}$ and

$$\frac{d\ell}{dt} = \frac{\ell}{2r}\frac{dr}{dt}.$$

Due to the loss of angular momentum, the distance of the grain to the star shrinks according to

$$\frac{dr}{dt} = -\frac{L_*}{2\pi r}\cdot\frac{\sigma_{geo}}{mc^2}. \tag{9.46}$$

When we integrate the equation $dt = -(2\pi mc^2/L_*\sigma_{geo})r\,dr$ from some initial radius r to the stellar radius $R_* \ll r$, we find the time τ_{PR} that it takes a grain to fall into the star,

$$\tau_{PR} = \frac{m\pi c^2}{L_*\sigma_{geo}}r^2 \tag{9.47}$$

or, in more practical units, assuming a density of $\rho_{gr} = 2.5$ g cm^{-3} for the grain material,

$$\frac{\tau_{PR}}{yr} = 1700\left(\frac{a}{\mu m}\right)\left(\frac{L_*}{L_\odot}\right)^{-1}\left(\frac{r}{AU}\right)^2. \tag{9.48}$$

The Poynting–Robertson effect is an efficient way to remove dust in the solar system. For instance, during the lifetime of the Sun ($\sim 5 \times 10^9$ yr) only bodies with a diameter greater than 6 m (!) can have survived within the orbit of the Earth. Particles existing today smaller than 6 m must have been replenished from comets or asteroids.

The Poynting–Robertson effect may also be described by an observer in a frame corotating with the grain. In such a reference frame, the stellar photons approach the grain not exactly along the radius vector from the star but hit it slightly head-on in view of the aberration of light. This phenomenon arises because one has to add the velocity of the photon and the grain according to the rules of special relativity. The photons thus decrease the angular momentum of the grain and force it to spiral into the star. The Poynting–Robertson effect also works when the photons are not absorbed but isotropically scattered.

9.4   Grain destruction

9.4.1   Mass balance in the Milky Way

Grains do not live forever but have a finite lifetime. As part of the dynamic, continually changing interstellar medium, they are born, modified and destroyed in processes such as

Birth:	mainly in old (evolved) stars with extended shells but possibly also in novae and supernovae (SN).
Accretion:	in dark clouds, gas atoms condense onto grain surfaces to form mantles.
Coagulation:	in dark clouds, grains collide, stick and form larger particles.
Destruction:	partial or complete, mainly in star formation, hot gas or shocks.

In a steady state, dust formation is balanced by dust destruction. The rates at which this happens are quite uncertain. Therefore the numbers given here are not precise at all but they do show the order of magnitude.

As gas and dust are intimately linked, we begin with a few remarks about the mass balance of *all* interstellar matter in the Milky Way; four important numbers are listed in table 9.1. The star formation rate implies that within one billion years, a short period compared to the Hubble time, all interstellar matter, including dust, is processed in stars. The continuous mass loss of gas in star formation must be offset by an equal mass input. It comes mainly from planetary nebulae which supply about 80% of the gas, red giants ($\sim$20%) and supernovae ($\sim$10%?).

Let us denote the input rates of *dust* due to planetary nebulae, red giants and supernovae by I_{PN}, I_{RG} and I_{SN}, respectively, and express them in solar masses per year. Most grains probably form in the wind of red giants (like M stars, OH/IR and carbon stars) as high temperature condensates. The total gas input in the Milky Way from red giants is about 1 $M_\odot$ yr^{-1} and about 1% of this is expected to be in the form of dust. So the input rate of dust from red giants is

$$I_{RG} \sim 0.01 \ M_\odot \ \text{yr}^{-1}.$$

Table 9.1. Global parameters of the Milky Way relevant to the mass balance of dust.

Total gas mass	$M_{gas} \sim 5 \times 10^9$ $M_\odot$
Total dust mass	$M_{dust} \sim 3 \times 10^7$ $M_\odot$
Star formation rate	$\tau_{SFR} \sim 5$ $M_\odot$ yr^{-1}
Supernova rate	$\tau_{SN} \sim 0.03$ yr^{-1}

Planetary nebulae, although they inject more gas into the interstellar medium, are less effective than giants in forming grains so that probably $I_{PN} < I_{RG}$. To estimate the contribution of supernovae to the dust balance, suppose that 3 $M_\odot$ of heavy elements are ejected per SN explosion. Although this is all potential dust material, a considerable fraction is locked up in CO, H_2O or otherwise and is unable to form grains. If one assumes that each explosion creates 0.3 $M_\odot$ of dust and if there are three supernova events per century in the Milky Way (see table 9.1), one gets a dust input rate from supernovae comparable to that of red giants of

$$I_{SN} \sim 0.01 \ M_\odot \ yr^{-1}.$$

For each input rate, one can define a formation time scale t by

$$t = \frac{M_{dust}}{I}.$$

9.4.2 Destruction processes

Grains are destroyed in various processes (recommended reading [Sea87] and [McK89]) such as the following ones.

- *Evaporation.* The solid is heated up to the condensation temperature. This happens near luminous stars, to a certain extent in HII regions (section 14.4), also in the diffuse medium (section 14.3) when the grains are extremely small or have volatile mantles. But by far the greatest sink is star formation and the destruction rate t_{SF} gives the maximum lifetime of a grain,

$$t_{SF} = \frac{M_{gas}}{\tau_{SFR}} \sim 10^9 \ yr. \tag{9.49}$$

  If grains should live a billion years, they are cycled at least 10 times between cloud and intercloud medium. The loss rate of dust associated with astration equals $I_{SF} = M_{dust}/t_{SF} \sim 0.03 \ M_\odot \ yr^{-1}$.
- *Sputtering.* Atoms are ejected from a grain by colliding gas particles, either neutrals or ions. The threshhold impact energy for liberating an atom from the solid phase is a few times larger than the binding energy of an atom (~ 5 eV). But even for impact energies above the threshhold, atoms are chipped off only with an efficiency η smaller than one, possibly $\eta \sim 0.1$.

- *Grain–grain collisions* after acceleration in a magnetic field.
- *Shattering* which is the destruction of a grain into smaller units, usually in grain–grain collisions.
- *Photodesorption* which is the ejection of grain atoms by photons.

9.4.2.1　Destruction in shocks

Besides star formation, grains are mostly destroyed in shocks associated with supernova remnants (SNR). The observational evidence comes from clouds of high velocity ($v > 100$ km s^{-1}) where more than half of Si and Fe is in the gas phase, whereas in normal clouds only a fraction of order 1% of these atoms is in the gas phase. High-velocity clouds are interpreted to be fragments of an expanding and shocked supernova shell. The mechanism of grain destruction depends on the type of shock:

- In a fast shock, cooling is slow. The shock is adiabatic or non-radiative and has a moderate density jump of four. Sputtering is thermal because the velocities of the gas atoms are thermal. A shock with $v \sim 300$ km s^{-1} corresponds to a temperature $T \sim 3 \times 10^6$ K. If one determines the rate of impinging protons using (9.1) and adopts a gas density $n = 1$ cm^{-3}, one finds that at least small grains are likely to be eroded.
- In a low-velocity shock, the cooling time is smaller than the expansion time of the supernova remnant. The shock is radiative and the density jump is much greater than four. A charged grain of mass m will gyrate in the magnetic field **B**. As $B_{\parallel}$, the component of **B** parallel to the shock front, is compressed, the grain velocity v_{gr} increases because its magnetic moment $\mu = m v_{\text{gr}}^2 / 2B_{\parallel}$ is conserved and the grain is accelerated (betatron acceleration). The collisional velocities between the grain and the gas atoms are now non-thermal and non-thermal sputtering dominates. The final velocity v_{gr} is limited by drag forces from the gas. The deceleration is proportional to the inverse of the grain radius, so big grains become faster than small ones and are more easily destroyed; small grains may survive. Grains may also collide with one another. A velocity of 100 km s^{-1} corresponds to $\sim$1000 eV kinetic energy per atom, whereas the binding energy is only $5 \ldots 10$ eV for refractory material; grains will be vaporized if the efficiency for knocking off atoms is greater than 1%.

To estimate the importance of shock destruction, we turn to the theory of supernova blasts. During the first (Sedov) stage, the remnant of mass M_{SNR} and velocity v is adiabatic and its kinetic energy $E = \frac{1}{2} M_{\text{SNR}} v^2$ conserved; at later times, the momentum $v M_{\text{SNR}}$ is constant. According to [McK89], the shock becomes radiative when $v \leq 200 (n^2 / E_{51})^{1/14}$ km s^{-1}; E_{51} is the energy expressed in units of 10^{51} erg. In a gas of density $n = 1$ cm^{-3} and with the standard explosion energy $E_{51} = 1$, one finds for the mass of the supernova shell $M_{\text{SNR}} \sim 10^3$ M$_\odot$ at the onset of the radiative phase. A supernova rate

$\tau_{SN} = 0.03$ yr^{-1} then implies that all interstellar material of the Milky Way is processed in a time

$$t_{SNR} = \frac{M_{gas}}{M_{SNR}\tau_{SN}} \sim 2 \times 10^8 \text{ yr.}$$

As the high-velocity clouds indicate an efficiency for dust destruction of 50%, the associated timescale ($\sim 4 \times 10^8$ yr) is substantially shorter than the joint dust injection time scale from supernovae and red giants ($\sim 1.5 \times 10^9$ yr). In a steady state, this requires an additional dust source for which accretion in the interstellar medium, especially in dark clouds, is usually invoked.

9.4.2.2 *Sputtering in the coronal gas*

Sputtering of refractory grains also occurs in the hot coronal gas ($T = 10^6$ K) where all atoms are very energetic (~ 100 eV). Assuming there a gas density $n = 10^{-3}$ cm^{-3}, grains of 1000 Å size are eroded in $\sim 10^9$ yr. For the total dust in the Milky Way, destruction in the coronal gas is, however, negligible because it contains only $\sim 0.1\%$ of all interstellar matter.

9.4.2.3 *Sputtering of grain mantles*

Similarly simple computations show that volatile grain mantle material which has low binding energies per atom (~ 0.1 eV) is effectively sputtered in the diffuse interstellar medium ($T \sim 10^3 \ldots 10^4$ K, $n > 10$ cm^{-3}). Low-velocity shocks are also efficacious. Therefore the mantle material with a low binding energy is confined to dark clouds, in agreement with observations.

9.5 Grain formation

We derive approximate equations that describe the nucleation of monomers into large clusters. The examples refer to water because its properties are experimentally well established but the physics apply, with some modifications, also to interstellar grains. There, however, the material constants are poorly known.

9.5.1 Evaporation temperature of dust

When two phases, like liquid and gas, are in equilibrium at temperature T, the vapor pressure P changes with T according to the *Clausius–Clapeyron* equation,

$$\frac{dP}{dT} = \frac{Q}{(V_g - V_f)T} \tag{9.50}$$

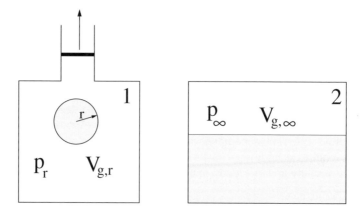

Figure 9.7. A reversible cycle employing two vessels to find the vapor pressure of a drop of radius r. The shaded area marks the fluid; see text.

where V_g and V_f are the volume per mol in the gas and fluid phase, respectively, and Q is the heat necessary to evaporate 1 mol. We remember that 1 mol consists of

$$L = 6.02 \times 10^{23}$$

molecules. L is Loschmidt's number, and the gas constant R, the Boltzmann constant k and L are related through

$$k = \frac{R}{L}. \tag{9.51}$$

The Clausius–Clapeyron equation is, with obvious modifications, also valid with respect to sublimation, which is the phase transition between the solid and gas phases, and thus applies to the evaporation of interstellar grains and their mantles. The fundamental formula (9.50) is derived in a thought experiment from a Carnot cycle:

- First, one totally vaporizes at constant temperature T one mol of a liquid in a vessel closed by an adjustable weight (of the kind as depicted on the left-hand of figure 9.7); this requires heat Q. During the phase transition from fluid to gas, the substance expands from V_f to V_g and the vapor pressure P does the work $P (V_g - V_f)$ by lifting a piston.
- Then the vapor is condensed in another heat bath at slightly lower temperature $T - dT$ and pressure $P - dP$. In this way, one returns the latent heat Q' and the work $(P - dP)(V_g - V_f)$. Finally, one brings the system back to its initial state (P, T).

As the internal energy of the system is the same at the beginning and the end of the cycle, the total effective work done is, in view of the first law of thermodynamics (see (5.49) with $dU = 0$),

$$dP \cdot (V_g - V_f) = Q - Q'.$$

The efficiency of converting heat Q into work is

$$\eta = \frac{dP \cdot (V_g - V_f)}{Q} = \frac{dT}{T} \tag{9.52}$$

and, according to the second law of thermodynamics, no heat engine can do better. Because $V_f \ll V_g$, we can approximate (9.50) by

$$\frac{d \ln P}{dT} = \frac{Q}{RT^2} \tag{9.53}$$

which has the solution

$$P = P_0 \, e^{-T_0/T} \qquad T_0 \equiv Q/R. \tag{9.54}$$

The affinity to the Boltzmann distribution is not only formal but physical because atoms in the gas are in an energetically higher state than atoms in a solid. Equation (9.54) is very general and does not say anything about the details of a specific phase transition. Therefore, P_0 and T_0 have to be determined experimentally. Laboratory data for water ice from $0\,°C$ down to $-98\,°C$ are displayed in figure 9.8. They can be neatly fit over the whole range by equation (9.54) with properly chosen constants.

One gram of H_2O, in whatever form, has $N = 3.22 \times 10^{22}$ molecules. If it is ice, it takes 79.4 cal to liquify it at $0\,°C$, then 100 cal to heat it up the water to $100\,°C$ and another 539.1 cal to vaporize the water. In cooling the vapor by $\Delta T = 100$ K from 100 to $0\,°C$, assuming six degrees of freedom per molecule, one gains $3Nk\Delta T = 1.37 \times 10^9$ erg. Because

$$1 \text{ cal} = 4.184 \times 10^7 \text{ erg}$$

the total sublimation energy per H_2O molecule at $0\,°C$ is, therefore, $kT = 8.64 \times 10^{-13}$ erg corresponding to a temperature of 6260 K. This agrees nicely with the fit parameter $T_0 = 6170$ K in figure 9.8.

The evaporation temperature T_{evap} in interstellar space of a species X frozen out on grain mantles follows by setting its partial pressure in the gas phase, $n_X kT$, equal to the evaporation pressure,

$$P_0 e^{-T_0/T_{evap}} = n_X k T_{evap}. \tag{9.55}$$

Because of the exponential dependence of the vapor pressure on temperature, the result is rather insensitive to the gas density. At hydrogen densities of order $\sim 10^6$ cm^{-3}, water ice evaporates at $T \sim 120$ K as one can read from figure 9.8 assuming an H_2O abundance of 10^{-7}. Ammonia and methane go into the gas phase already at about 90 and 70 K, respectively. A rise in density from 10^6 to 10^{12} cm^{-3} increases the evaporation pressure for H_2O, NH_3 and CH_4 by ~ 30 K. Typical evaporation temperatures of other astronomically important species are: hydrogen (3 K), CO, CO_2 (20 K), silicates (1800 K) and graphites (2500 K).

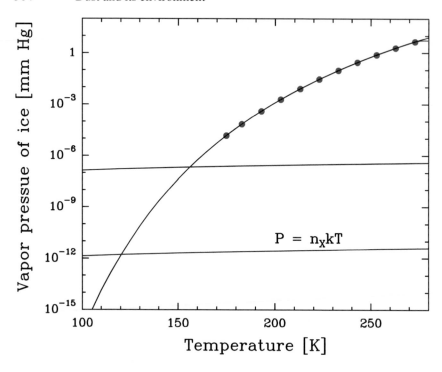

Figure 9.8. Vapor pressure of ice. Experimental points (dots) from [Wea77], the full curve is a fit according to (9.54) with $T_0 = 6170$ K and $P_0 = 4.0 \times 10^{13}$ dyn cm^{-2} or 3.0×10^{10} mm of Hg column. The fit is extrapolated to lower temperatures. The lower horizontal line shows the partial gas pressure, $P = n_X kT$ in mm Hg column for a number density $n_X = 0.1$ cm^{-3}. Such an H_2O pressure prevails in interstellar space when the total gas density $n_H = 10^6$ cm^{-3} and the water abundance (in the gas phase) is 10^{-7}. The upper horizontal line is for $n_X = 10^5$ cm^{-3}.

9.5.2 Vapor pressure of small grains

Condensation and evaporation of small interstellar grains has the peculiarity that the transition surface between the solid and the gas is not flat but curved. As molecules leave the grain, the surface area shrinks by an amount dA thereby creating the energy $dW = \zeta dA$ according to (9.43). Evaporation is, therefore, easier in grains of small radius r and their vapor pressure p_r is higher than that over a flat surface, p_∞. This has fundamental consequences.

To compute p_r, one performs an isothermal reversible cycle of four steps. Here, we write down the work W_i that is being done in each step; the formulae for W_i are elementary.

(1) In vessel 1 is a small liquid sphere of radius r at temperature T in equilibrium with its vapor of pressure p_r (figure 9.7). The volumes per mol of gas and

fluid are $V_{g,r}$ and V_f, respectively. We evaporate the droplet but in such a way that while evaporating we inject with a syringe into the sphere against the internal pressure $p = 2\varsigma/r$ of (9.45) and the outer pressure p_r the same amount of liquid as is being lost to the gas, so all the time $r =$ constant. When 1 mol has become gaseous,

$$W_1 = p_r V_{g,r} - \left(p_r + \frac{2\varsigma}{r}\right) V_f.$$

(2) Isothermal expansion from pressure p_r to p_∞ of the 1 mol of gas that has been vaporized in step (1),

$$W_2 = RT \ln(p_r/p_\infty).$$

(3) The 1 mol of vapor is pressed into vessel 2 where the liquid has a plane surface and is under its vapor pressure p_∞,

$$W_3 = -RT.$$

(4) To complete the cycle, we take from vessel 2 with a syringe 1 mol of liquid,

$$W_4 = V_f p_\infty.$$

As the cycle was isothermal ($dT = 0$), the sum $\sum W_i$ must be zero. Let v_0 denote the volume of one molecule in the liquid phase. Because $RT = pV$ and because the pressure in the drop, $2\varsigma/r$, is much bigger than $p_r - p_\infty$, we obtain

$$\ln \frac{p_r}{p_\infty} = \frac{2\varsigma v_0}{kTr}. \qquad (9.56)$$

This tells us that the vapor pressure $p_r = p_\infty \cdot e^{2\varsigma v_0/kTr}$ equals p_∞ for big spheres but increases exponentially when the radius becomes very small. Condensation of the first seed grains is thus possible only when the partial pressure p exceeds p_∞ or when the saturation parameter s (capital S for the entropy) defined by

$$s = \frac{p}{p_\infty} \qquad (9.57)$$

is greater than one. The vapor is then said to be supersaturated. At pressure p_r given by (9.56), a sphere of diameter $2r$ is in equilibrium with the vapor, a bigger drop will grow indefinitely, a smaller one will evaporate, so r is the critical cluster radius.

9.5.3 Critical saturation

To estimate the critical saturation parameter $s_{cr} > 1$ at which clusters are created, we calculate the work W needed to form in a reversible cycle a drop of radius

r. The drop consists of n atoms and is formed in a vessel with vapor at pressure p and temperature T. The work in each of the four steps is denoted by W_i, so $W = \sum W_i$.

(1) Remove n gas atoms from the vessel,

$$W_1 = -nkT.$$

(2) Isothermally expand these n atoms from p to p_∞,

$$W_2 = -nkT \ln(p/p_\infty).$$

(3) Press the n atoms into another vessel, also at temperature T but containing a liquid of flat surface under vapor pressure p_∞,

$$W_3 = nkT.$$

(4) Form there from the liquid a drop of radius r (and bring it back to the first vessel),

$$W_4 = 4\pi \zeta r^2.$$

For the sum we get

$$W = -nkT \ln(p/p_\infty) + 4\pi \zeta r^2 = \tfrac{1}{3}\zeta \cdot 4\pi r^2. \tag{9.58}$$

The term $-nkT \ln(p/p_\infty)$ is the potential energy of the drop. It is proportional to the number of molecules n because the molecules possess short-range forces and, therefore, only connect to their nearest neighbors. For long-range forces, the term would be proportional to n^2. The other term $4\pi \zeta r^2$, which goes as $n^{2/3}$, introduces a correction because atoms on the surface are attracted only from one side.

As the energy of the system stays constant over the cycle, one has, according to the first law of thermodynamics, to subtract from the system the heat $Q = W$. Therefore, during formation of the droplet the entropy of the system falls by (see (5.51))

$$S = Q/T = 4\pi \zeta r^2/3T.$$

S decreases because a liquid represents a state of higher order than a gas. Although the entropy of a macroscopic system can only increase, on a microscopic level all processes are reversible and S fluctuates. Formation of a seed grain comes about by such an entropy fluctuation. As the entropy is after (5.43) equal to Boltzmann constant k multiplied by the logarithm of the number Ω of states, $S = k \ln \Omega$, the probability $P = 1/\Omega$ for the formation of a seed is given by $e^{-S/k}$. When we insert the saturation parameter $s = p/p_\infty$ of (9.57), simple algebra yields

$$P = \frac{1}{\Omega} = e^{-S/k} = \exp\left(-\frac{4\pi \zeta r^2}{3kT}\right) = \exp\left(-\frac{16\pi \zeta^3 v_0^2}{3k^3 T^3 \ln^2 s}\right). \tag{9.59}$$

For water ($\zeta = 75$ erg cm^{-2}, $v_0 = 3.0 \times 10^{-23}$ cm^3) at temperature $T = 275$ K, the exponent $S/k \sim 116/\ln^2 s$. For carbon compounds ($\zeta \sim 1000$ erg cm^{-2}, $v_0 = 9 \times 10^{-24}$ cm^3) at $T = 1000$ K, $S/k \sim 500/\ln^2 s$.

The probability P is most sensitive to ζ/T and, if ζ/T is fixed, to s, changing from practically impossible to highly likely in a narrow interval Δs. It is exactly this property which allows us to estimate the critical value s_{cr} as follows.

If N, v and σ denote the number density, mean velocity and collisional cross section of the vapor atoms, there are, per second, $N^2 v\sigma$ atomic collisions each leading with a probability P to the formation of a seed. So $J = N^2 v\sigma P$ is the rate at which seeds form and the condition

$$N^2 v\sigma P \sim 1$$

yields s_{cr}. The outcome of the simple calculation depends entirely on the exponent in (9.59) which has to be of order one; the factor $N^2 v\sigma$ has very little influence. One finds typical values of the critical saturation parameter for water at room temperature around 10 which implies in view of (9.56) a critical seed radius r_{cr} of a few ångströms.

9.5.4 Equations for time-dependent homogeneous nucleation

Let us study the creation of seed grains in a kinetic picture. If the gas has only one kind of atoms or molecules of mass m, one speaks of homogeneous nucleation, in contrast to heterogeneous nucleation when different molecular species or ions are present. A small cluster or droplet of n molecules is an n-mer with concentration c_n. The number density of the gas molecules (they are monomers) is, therefore, denoted c_1. A cluster has a radius r_n and a surface area $A_n = 4\pi r_n^2$. When v_0 is the volume of one molecule, n and r_n are related through

$$n v_0 = \frac{4\pi}{3} r_n^3.$$

Consider a gas of constant temperature T and pressure $p = c_1 kT$ with saturation parameter $s = p/p_\infty > 1$ after (9.57). If atoms impinge on a drop at a rate $\beta_n A_n$ and evaporate from it at a rate $\alpha_n A_n$, the concentration c_n changes with time like

$$\dot{c}_n = -c_n\big[\beta_n A_n + \alpha_n A_n\big] + c_{n-1}\beta_{n-1}A_{n-1} + c_{n+1}\alpha_{n+1}A_{n+1}. \tag{9.60}$$

This is a very general equation, all the physics is contained in the coefficients for evaporation and accretion, α_n and β_n. If we assume that atoms are added to the grain at a rate $\pi r_n^2 \langle v\rangle c_1$ and leave at a rate $\pi r_n^2 \langle v\rangle p_n/kT$, where $\langle v\rangle$ is the mean velocity of (9.1) and p_n/kT from (9.56), then

$$\alpha_n = \frac{p_n}{\sqrt{2\pi mkT}} \qquad \beta = \frac{p}{\sqrt{2\pi mkT}}. \tag{9.61}$$

In this case, β is constant for all n and the sticking coefficient is one; p_n is the vapor pressure of an n-mer according to (9.56). One defines a particle flux or current from n-mers to $(n+1)$-mers by

$$J_n = c_n \beta A_n - c_{n+1} A_{n+1} \alpha_{n+1}. \tag{9.62}$$

We can now transform equation (9.60) into the deceptively simple looking form

$$\dot{c}_n = J_{n-1} - J_n.$$

Equation (9.60) disregards collisions between clusters, only molecules impinge on a drop but that is a valid assumption as one can easily verify.

9.5.5 Equilibrium distribution and steady-state nucleation

When the system is in phase equilibrium, detailed balance holds. We will flag equilibrium concentrations by the superscript 0. In equilibrium, $J_n = 0$ and

$$c_n^0 \beta A_n = c_{n+1}^0 \alpha_{n+1} A_{n+1}.$$

The last relation also enables us to write, for non-equilibrium conditions,

$$J_n = -c_n^0 \beta A_n \left[\frac{c_{n+1}}{c_{n+1}^0} - \frac{c_n}{c_n^0} \right]. \tag{9.63}$$

The equilibrium concentrations c_n^0 are given by the Boltzmann formula

$$c_n^0 = c_1^0 e^{-\Delta G_n / kT} \tag{9.64}$$

where ΔG_n is the work to be expended for creating an n-mer out of monomers (see (9.58)),

$$\Delta G_n = 4\pi \zeta r_n^2 - nkT \ln(p_n/p_\infty). \tag{9.65}$$

In view of the discussion in section 5.3.5 on the equilibrium conditions of the state functions, ΔG_n is, for a system at constant temperature and pressure, the difference in free enthalpy. The function ΔG_n is depicted in figure 9.9; it has its maximum at n_* where $\partial \Delta G_n / \partial n = 0$. All values at n_* will, in the following, be marked by an asterisk. Clusters of size smaller than n_* are inside a potential well and fight an up-hill battle when growing. They have to overcome the barrier of height ΔG_*. When they have climbed the barrier, their size is n_* and for $n > n_*$, they will grow unrestrictedly. So n_* is the critical size. Figure 9.10 plots the dependence of n_* on the saturation parameter s.

In a steady state (subscript s), nothing changes with time and $\partial/\partial t = 0$. The conditions for steady state are less stringent than for equilibrium because J_n need not vanish, it only has to be a positive constant,

$$J_n = J^s = \text{constant} > 0.$$

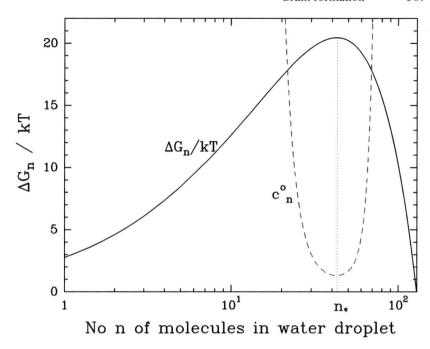

No n of molecules in water droplet

Figure 9.9. The full curve shows $\Delta G_n/kT$ where $\Delta G_n = 4\pi \zeta r_n^2 - nkT \ln p/p_\infty$ is the change in free enthalpy when n water molecules condense into a drop of radius r_n (see (9.65)). The curve reaches its maximum at n_*. The broken curve depicts the equilibrium concentration c_n^0 of n-mers on an arbitrary linear scale; the minimum is at $n = n_*$. In these plots, $T = 300$ K, $s = p/p_\infty = 4$ and the surface tension $\zeta = 75$ erg cm^{-2}.

Under this condition, we get, from (9.63),

$$\frac{J^s}{c_n^0 \beta A_n} = \frac{c_n^s}{c_n^0} - \frac{c_{n+1}^s}{c_{n+1}^0} \simeq -\frac{\partial}{\partial n}\left(\frac{c_n}{c_n^0}\right). \tag{9.66}$$

To realize a steady state in a thought experiment, one has to invoke a Maxwellian demon who removes large clusters with $n \geq L > n_*$ by gasifying them. The demon supplies the boundary condition for the cluster distribution,

$$c_n^s = 0 \qquad \text{for } n \geq L.$$

One does not have to be very particular about L, it just has to be somewhat greater than n_*. As a second boundary condition we impose

$$c_1^s/c_1^0 = 1$$

which means that only a very small fraction of the total mass is in clusters.

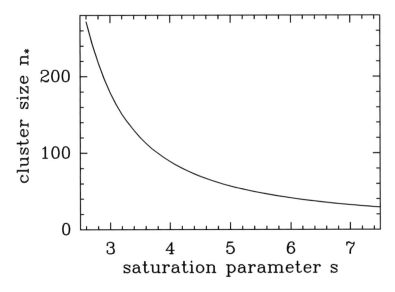

Figure 9.10. The dependence of the critical cluster size n_*, defined in figure 9.9, on the saturation parameter $s = p/p_\infty$ for water vapor at 273 K. For clusters of size n_*, the difference in the Gibbs potential between gas and droplet, ΔG_n, has its maximum and $\partial \Delta G_n / \partial n = 0$.

Consequently, by performing a sum over (9.66),

$$J^s \sum_{n=1}^{L-1} \frac{1}{c_n^0 \beta A_n} = \frac{c_1^s}{c_1^0} - \frac{c_L^s}{c_L^0} = 1 \tag{9.67}$$

we can directly compute the steady-state nucleation rate J^s. The latter may also be appproximated analytically from the relations:

$$1 \simeq \frac{J^s}{\beta} \int_1^L \frac{dn}{A_n c_n^0} \simeq \frac{J^s}{\beta c_1^0} \int_1^L \frac{e^{\Delta G_n / kT}}{A_n} \, dn \simeq \frac{J^s}{\beta A_* c_1^0} \int_1^L e^{\Delta G_n / kT} \, dn$$

which follow from (9.67). Because ΔG_n has a fairly sharp peak at n_*, the surface area $A_n = 4\pi r_n^2$ can be taken out from under the integral and replaced by A_*. A Taylor expansion of ΔG_n around the maximum n_*,

$$\Delta G_n \simeq \Delta G_* + \frac{1}{2} \frac{\partial^2 \Delta G_n}{\partial n^2} (n - n_*)^2$$

with

$$\Delta G_* = \frac{4\pi \zeta}{3} r_*^2 = \frac{16\pi \zeta^3}{3} \left(\frac{v_0}{kT \ln s} \right)^2 \tag{9.68}$$

and

$$\frac{\partial^2 \Delta G_n}{\partial n^2} = -\frac{\zeta v_0^2}{2\pi r_*^4}$$

yields, upon replacing the integration limits $(1, L)$ by $(-\infty, \infty)$ to obtain the standard integral (A.25), the steady-state nucleation rate J^s. The number of clusters that grow per unit time and volume from size n to $n+1$ is then independent of n and one readily finds

$$J^s = v_0 \left(\frac{2\zeta}{\pi m}\right)^{1/2} (c_1^0)^2 \exp\left[-\frac{4\mu^3}{27 \ln^2 s}\right]. \tag{9.69}$$

The exponent in square brackets equals $-\Delta G_* / kT$ and μ is defined by

$$\mu = \frac{4\pi \zeta a_0^2}{kT} \tag{9.70}$$

where a_0 denotes the radius of a monomer, so $v_0 = 4\pi a_0^3/3$. The expression in (9.69) is quite a satisfactory approximation to the sum in (9.67). The magnitude of J^s is, of course, entirely determined by the exponent. Figure 9.11 illustrates the immense change in J^s, over many powers of ten, with the saturation parameter s. Figure 9.12 shows the deviation of the equilibrium from the steady-state concentrations, the latter being calculated from

$$c_n^s = c_n^0 J^s \sum_{i=n}^{L-1} \frac{1}{c_i^0 \beta A_i}.$$

We learn from figure 9.12 that for small n-mers, equilibrium and steady-state concentrations are the same. Around the critical cluster size n_*, the ratio c_n^s/c_n^0 declines with an approximately constant slope $b = \partial(c_n^s/c_n^0)/\partial n$ evaluated at n_* over a region of width $\Delta n = -1/b$. For large n, the ratio c_n^s/c_n^0 tends to zero.

9.5.6 Solutions to time-dependent homogeneous nucleation

9.5.6.1 *The evolution towards the steady state*

The time-dependent system of equations for nucleation presented in (9.60) can easily be solved when the coefficients for evaporation and accretion, α_n and β_n, are as simple as in (9.61). Let us write the time derivative of the concentration of n-mers as

$$\dot{c}_n = \frac{c_n - c_n^{\text{old}}}{\tau}$$

where c_n^{old} is the old value of the concentration a small time step τ ago; we want to find the new one, c_n. Equation (9.60) can immediately be brought into the form

$$A_n c_{n-1} + B_n c_n + C_n c_{n+1} + D_n = 0 \tag{9.71}$$

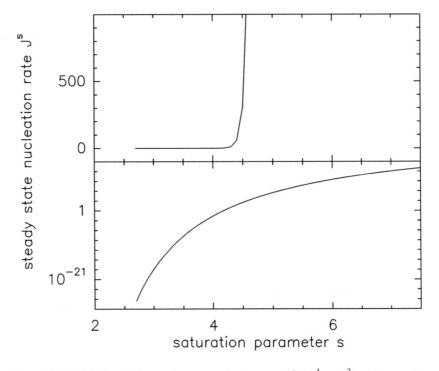

Figure 9.11. Variation of the steady-state nucleation rate J^s (s^{-1} cm^{-3}) with saturation parameter $s = p/p_\infty$ for water vapor at $T = 273$ K; surface tension $\zeta = 75$ erg cm^{-2}. J^s is computed from (9.69) and displayed on a linear (top) and logarithmic scale (bottom). The pressure p_∞ is derived from (9.54) and equals 1.92×10^{17} dyn cm^{-2}. Below $s \simeq 5$, nucleation is negligible, above catastrophic and $s \simeq 5$ represents the critical value of the saturation parameter.

where A_n, B_n, C_n, D_n are known coefficients (so for the moment, A_n is not the surface area of an n-mer). Putting

$$c_n = \gamma_n c_{n-1} + \delta_n \tag{9.72}$$

and inserting it into (9.71) yields

$$\gamma_n = -\frac{A_n}{B_n + \gamma_{n+1} C_n} \qquad \delta_n = -\frac{\delta_{n+1} C_n + D_n}{B_n + \gamma_{n+1} C_n}.$$

We determine the concentrations c_n in the size range

$$g \leq n \leq L$$

with $1 \ll g \ll n_*$ (as far as possible) and $L > n_*$. The boundary conditions are suggested by figure 9.12: At the upper end, we take $c_L = 0$ because $c_L/c_L^0 = 0$.

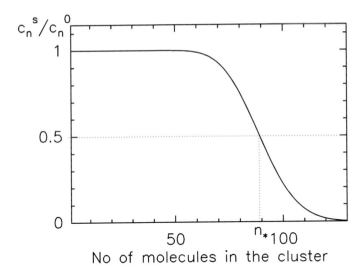

Figure 9.12. The ratio of the cluster abundance in a steady state to the cluster abundance in equilibrium as a function of cluster size n; here for water vapor at $T = 273$ K and saturation parameter $s = p/p_\infty = 4$. At $n = n_*$, the ratio is about one half.

Therefore, $\gamma_L = \delta_L = 0$, and this allows to compute $\gamma_n = \delta_n = 0$ for $n = L - 1, \ldots, g + 1$. At the lower end, the abundance of clusters is very close to equilibrium and we put $c_g = c_g^0$. By choosing g considerably greater than one, we avoid using the enthalpy ΔG_n of (9.65) for the smallest clusters where it cannot be correct. We then find c_n for $n = g + 1, \ldots, L - 1$ from (9.72).

Figure 9.13 presents a numerical experiment for water vapor at constant temperature $T = 263$ K and saturation parameter $s = 4.9$. These two values imply a monomer concentration $c_1^0 = 3.5 \times 10^{17}$ cm^{-3} and a critical size $n_* = 71$. At time $t = 0$, we start with a configuration where only very small clusters exist and put $c_n = 0$ for $n > g$. We see that the fluxes J_n, as defined in (9.63), converge towards their steady state values J^s given in (9.69). The steady state is reached after a relaxation time $\tau_{\rm rel}$ that can be shown by an analysis of equation (9.66) to be of order

$$\tau_{\rm rel} \sim \frac{r_*^2}{4\beta\zeta v_0^2}. \tag{9.73}$$

In the particular example of figure 9.13, $\tau_{\rm rel} \sim 2 \times 10^{-7}$ s and $J_n^s \simeq 1$ s^{-1} cm^{-3}. The asymptotic values of the concentration ratios c_n/c_n^0 are qualitatively the same as in the steady-state distribution of figure 9.12 (although the latter is computed for somewhat different values of T and s).

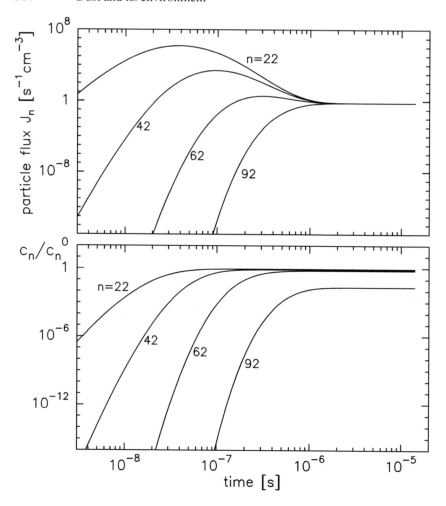

Figure 9.13. Water vapor at 263 K supersaturated by a factor $s = 4.9$. Under these conditions, the critical cluster size $n_* = 71$ and the concentration of the monomers $c_1 = 3.5 \times 10^{17}$ cm^{-3}. The smallest clusters for which we compute the time evolution consist of $g = 12$ molecules, the largest of $L = 120$. At time $t = 0$, the smallest grains are in equilibrium, $c_n/c_n^0 = 1$ for $n \leq g = 12$, and bigger clusters are absent ($c_n = 0$ for $n > g$). The bottom plot shows how the concentrations of various cluster sizes, normalized to the equilibrium value, evolve in time. The top panel depicts the corresponding particle fluxes J_n of (9.62).

9.5.6.2   *Time-dependent nucleation using steady-state fluxes*

As the relaxation time τ_{rel} in (9.73) is short compared to the time it takes to grow big stable drops, the steady-state flux J^s represents a good approximation to all fluxes J_n after the time τ_{rel} when transient effects have died out. The J_n may then be replaced by J^s and this greatly simplifies the further analysis in which we study nucleation in a cooling gas. Such a situation prevails in the outward flowing wind of mass loss giants where most interstellar grains are formed.

When the partial gas pressure of the condensible component, $P = c_1 kT$, reaches the vapor pressure P_∞ over a flat surface, the saturation parameter s equals one. At this instant, the gas temperature is denoted by T_e and the time t is set to zero. As the gas cools further, s increases. At first, the steady-state nucleation rate J^s is negligible because all clusters are below the critical size n_* (see figure 9.9) and the time it takes to form one critical cluster, roughly given by the inverse of J^s, is unrealistically long. Only when the saturation parameter approaches the critical value s_{cr} at time t_{cr} do clusters become bigger than the critical size n_* and catastrophic nucleation sets in.

Let $N(t)$ be the number of monomers in clusters that were formed at time t_{cr} and and let $R(t)$ be their radius, so $4\pi R^3/3 = Nv_0$. For $t \geq t_{cr}$, these clusters grow at a rate

$$\dot{R} = \pi a_0^2 \langle v \rangle c_1(t) \tag{9.74}$$

where a_0 and $\langle v \rangle = \sqrt{8kT/\pi m}$ are the radius and mean velocity of a monomer. The rate at which monomers are depleted through steady-state nucleation is, therefore,

$$\dot{c}_1(t) = -J^s(t) \cdot N(t) \tag{9.75}$$

which leads to a monomer depletion since time zero of

$$c_1(t) - c_1(0) = -\int_0^t J^s(t')N(t')\,dt'. \tag{9.76}$$

If the cooling is due to an adiabatic expansion, $K = T/c_1^{\kappa-1}$ is a constant, where $\kappa = C_p/C_v$ is the ratio of specific heats. One then has to add to the density decrease $\dot{c}_1(t)$ of (9.75) the term $\gamma \dot{T} T^{\gamma-1}/K^\gamma$ where $\gamma = 1/(\kappa - 1)$. The time dependence of the saturation parameter s follows via P_∞ as given by (9.54),

$$s(t) = \frac{T(t)}{T_e} e^{T_0/T(t)-T_0/T_e}.$$

To illustrate the formation of grains, we integrate equations (9.74) and (9.75) with Runge–Kutta asssuming an adiabatic expansion with a cooling rate

$$T = T_e - wt$$

with $T_e = 292$ K and $w = 100$ K s^{-1}. Results are plotted in figure 9.14. Until the moment of crtitical supersaturation, grains do not form and the decrease in the gas density and temperature and the rise of the saturation parameter are due only to the adiabatic expansion at the prescribed rate. Then at $t_{cr} = 0.247$ s, nucleation rises in a spike as rapidly as shown in figure 9.11, depriving the gas of its atoms

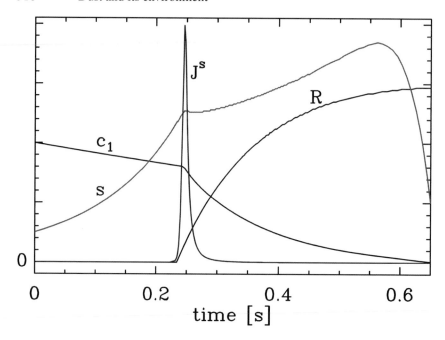

Figure 9.14. Nucleation in an adiabatically cooling water vapor. At $t = 0$, the gas is at $T_e = 292$ K and just saturated ($s(0) = 1$) with a monomer concentration $c_1(0) = 6.6 \times 10^{17}$ cm^{-3}. The gas expands adiabatically cooling by 100 K every second. We plot as a function of time the saturation parameter s, the number density of the gas molecules c_1, the steady-state nucleation rate J^s and the radius R of the largest drops. The ordinate is for all four variables linear with the zero point indicated. The maximum values are 7.3 for s, 10^3 s^{-1} cm^{-3} for J^s and 280 μm for R. At the maximum of the steady-state flux J^s, the saturation parameter s is near its critical value of 5.0 and the temperature has fallen to 268 K. At $t = t_{cr} = 0.247$ s, $\mu = 9.7$ and $\Lambda \sim 6 \times 10^6$.

at an accelerated pace. From that moment on, clusters grow beyond n_* according to (9.74) whence they are stable. The radius at which their growth levels off (at $t = 0.65$ s) represents the typical final grain size (one can also determine a size distribution). The saturation continues to increase after t_{cr} because of cooling but eventually falls for lack of gas atoms.

9.5.7 Similarity relations

The number of variables that enters the nucleation calculations is quite large ($v_0, m, c_1, \zeta, T, \langle v \rangle, s, p_0, T_0$) but one can drastically reduce the parameter space to only two variables: μ from (9.70) and Λ defined below [Yam77]. Therefore, the models presented here for the condensation of water near room temperature can be carried over to grain formation in late-type stars ($T \sim 1000$ K, $c_1 \sim$

10^9 cm^{-3}, $\zeta \sim 10^3$ dyn cm^{-2}) as long as μ and Λ are comparable.

As the gas temperature changes only a little during the brief period of catastrophic nucleation when $T \sim T_{cr}$, it may be approximated by a linear function in time,

$$T(t) = T_{cr} - at.$$

When one inserts this $T(t)$ into the evaporation pressure $p_\infty = p_0 e^{T_0/T}$ of (9.54), one finds for the time dependence of the saturation parameter

$$s(t) = \frac{c_1(t)}{c_1(0)} e^{t/\tau_{sat}}$$

where

$$\frac{1}{\tau_{sat}} \simeq \frac{T_0}{T_{cr}} \left| \frac{d \ln T}{dt} \right|.$$

The cooling rate $d \ln T/dt$ is evaluated at T_{cr}. By defining

$$y = c_1(t)/c_1(0)$$
$$x = t/\tau_{sat}$$
$$\Lambda = \tau_{sat}\, c_1(0)\, \pi a_0^2 \langle v \rangle$$
$$\rho = \frac{3R}{a_0 \Lambda}$$

the nucleation formulae (9.74) and (9.76) can be transformed into the new equations:

$$\frac{d\rho}{dx} = y \tag{9.77}$$

$$1 - y = \frac{\Lambda^4}{81} \left(\frac{\mu}{\pi}\right)^{1/2} \int_0^x y^2 \rho(x') \exp\left\{-\frac{4\mu^3}{27(x' + \ln y)^2}\right\} dx'. \tag{9.78}$$

They contain only the two independent parameters μ and Λ. The first, μ, is fixed by the ratio of surface tension over temperature and thus incorporates grain properties. The second, Λ, is determined by the conditions in the environment, such as gas density c_1 and cooling rate $\dot{T}$ but it also reflects the physics of the dust via the constant T_0 in the expression for the evaporation temperature of the grain material (see (9.54)). The final grain size is largely influenced by Λ (see definition of ρ).

9.5.7.1 Cautioning remarks

The quantitative results of homogeneous nucleation theory outlined here are fairly speculative because of the intrinsically exponential behavior of the process of grain formation and a number of dubious assumptions, oversimplifications and unsolved problems, such as (recommended reading [Fed66, Abr74]) the following ones.

- The smallest clusters are not spheres.
- The latent heat of the condensing molecules should not be neglected.
- The surface tension of the smallest clusters is unknown, the concept being even ill defined. Replacing ζ by $\zeta_r = \zeta/[1 + \delta/r]$, where δ is a curvature correction of atomic length ($\sim 1\text{Å}$) and r the grain radius, does not solve the problem.
- The thermodynamics of small clusters, in particular, the proper expression for the free enthalpy, is still controversial. A drop also has macroscopic motion but it is not clear how its translation and rotation replaces the internal degrees of freedom. Because of this uncertainty, immense fudge factors (up to 10^{17} for the concentration of the critical clusters) are cited in the literature. We mention that because of the macroscopic motion, which was neglected in our derivation of (9.57), the correct formula for the vapor pressure of a drop is not (9.57) but $\ln(p_r/p_\infty) = 2\zeta v_0/kTr - 4/n$, where n is the number of molecules in the cluster [Kuh52].

Besides such intrinsic uncertainties in the theory of homegenous nucleation, astronomical real life is even more complicated. First, nucleation in the wind of giants is not homogeneous but proceeds in a vast network of chemical reactions involving many different species for which we do not know the rate coefficients. Second, stellar photons are present in the wind that create new reaction channels and the radiation field is not in equilibrium with the gas. So when it comes to the question of how much one learns by applying homogeneous nucleation theory to grain formation in the outflow of red giants, an honest answer is sobering.

Another reason for this dissatisfactory state of affairs is the limited possibility to compare observations with theory. Observations consist of infrared spectra that do not carry detailed information about the dust. All they tell us is that silicates do form in M-type stars, as can be inferred from the presence of the 10 μm band, and where we expect carbon dust, it is there and not graphitic because of the absence of the 2200 Å feature. However, there is no information in the spectra about grain size or the dependence of growth on the physical parameters of the medium.

Chapter 10

Polarization

Electromagnetic radiation can become polarized by interstellar dust in three ways:

- Through *scattering*, which we have already discussed and illustrated. The achieved degree of polarization can be very high.
- Through *extinction*. When starlight passes through a dust cloud, the weakening of the flux depends on the orientation of the electric vector. Many reddened stars appear to be polarized, up to 10% or so, although the emission from the stellar photosphere is certainly not.
- Through *emission*. The far infrared radiation of some clouds shows a small degree of polarization.

To explain the latter two phenomena, one has to evoke the presence of non-spherical grains, for example elongated particles shaped like cigars. They must also be aligned, otherwise the net polarization would be zero.

10.1   Efficiency of infinite cylinders

As an approximation to elongated particles, we can treat dust grains as infinite circular cylinders. Although interstellar grains are certainly neither exactly cylindric nor infinitely long (at best they are elongated), such an idealization provides quantitative estimates for the real particles. The problem of computing the electromagnetic field in the interaction of light with an infinite cylinder for any ratio of cylinder radius over wavelength was first solved by Rayleigh [Ray18] for normal incidence and by Wait [Wai55] for oblique incidence. It is also possible to handle coated or multi-layered cylinders. We skip the mathematical derivation of the cross sections, the numerical code is still fairly simple.

10.1.1   Normal incidence and picket fence alignment

Consider a very long circular cylinder of length L much longer than its radius, a, or the wavelength λ. We call the cylinder infinite when on halving its length

L, the cross section C is also halved; C/L is then constant. When light impinges under normal incidence, the cylinder axis is perpendicular to the direction of wave propagation. In this case, there are two extinction cross sections depending on the polarization of the incident light:

- $C_\parallel^{\text{ext}}$: the cylinder axis is *parallel* to the electric and perpendicular to the magnetic vector of the incident electromagnetic field.
- $C_\perp^{\text{ext}}$: the cylinder axis is *perpendicular* to the electric and parallel to the magnetic vector of the incident electromagnetic field.

They are not equal and, therefore, unpolarized light becomes slightly linearly polarized after interaction with the cylinder. What counts is the difference

$$\Delta C^{\text{ext}} = C_\parallel^{\text{ext}} - C_\perp^{\text{ext}}.$$

When there are many such cylinders with their axes all parallel, one speaks of *picket fence alignment*. The efficiency Q is defined as the cross section over projected area,

$$Q = \frac{C}{2aL}. \tag{10.1}$$

In this section, we mostly talk about extinction efficiencies and sometimes omit the superscript 'ext'. In figure 10.1 we present $Q_\parallel^{\text{ext}}$ and $Q_\perp^{\text{ext}}$ for infinite cylinders with circular cross section. We adopt $m = 1.7 + i0.03$, which is about the optical constant of silicate material between 0.3 and 2 μm, i.e. over the whole visible region and part of the near IR. Carbon, the other major chemical component of interstellar dust, is often disregarded in the context of polarization because, without any magnetic inclusions, it is hard to align. However, we will also consider in this chapter examples of carbon particles. By comparing the top of figure 10.1 with figure 4.2, we see that the extinction efficiency of a cylinder mimics, in every aspect that of a sphere, except that we now have two Qs differing by an amount $\Delta Q^{\text{ext}} = Q_\parallel^{\text{ext}} - Q_\perp^{\text{ext}}$. Note that ΔQ^{ext} may be negative.

The efficiencies $Q_\parallel^{\text{ext}}$ and $Q_\perp^{\text{ext}}$ reach their first maximum at size parameter $x \simeq 2.5$, whereas the difference ΔQ^{ext} has a flat peak near $x \simeq 1$. In the bottom frame of figure 10.1, the display of ΔQ^{ext} is rearranged. It refers to the same optical constant $m = 1.7 + i0.03$ but we now keep the wavelength λ fixed and vary the radius. We choose $\lambda = 0.55$ μm, which is the center of the visual band V, because there the observed interstellar polarization typically attains its maximum. The figure informs us that in polarizing visible light, silicate grains of cylindrical shape are most efficient when their radii are about 0.1 μm and that much thinner or thicker cylinders are ineffective. Because for a given dielectric permeability, ΔQ^{ext} is only a function of the size parameter $x = 2\pi a/\lambda$, we further learn that there is a direct proportionality between the cylinder radius and the wavelength λ_{max} of maximum polarization.

To derive from figure 10.1, which shows the efficiencies $Q_\parallel$ and $Q_\perp$, the volume coefficients $K_\parallel$ and $K_\perp$, taken with respect to 1 cm^3 of dust material (see

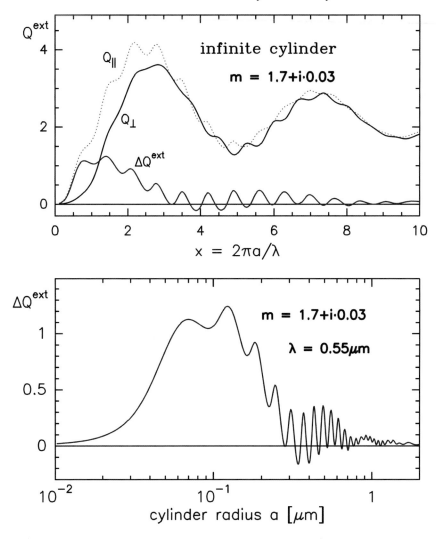

Figure 10.1. The top frame shows the extinction efficiencies $Q_\parallel$ and $Q_\perp$ and their difference, $\Delta Q^{\text{ext}} = Q_\parallel - Q_\perp$, as a function of the size parameter $x = 2\pi a/\lambda$ for infinite cylinders of radius a and optical constant $m \simeq 1.7 + i0.03$. The bottom frame shows the dependence of ΔQ^{ext} on radius a for a wavelength $\lambda = 0.55 \ \mu$m and the same m. The data for ΔQ^{ext} are exactly the same as in the top box, only rearranged.

section 2.1), one has to multiply Q by the geometrical cross section $2aL$ under vertical incidence and divide by the cylinder volume $\pi a^2 L$. This gives $K \propto a^{-1}$. Therefore, if at some wavelength ΔQ^{ext} is equal for thin and thick rods, the thin ones are more efficient polarizers (with respect to the same total dust mass).

10.1.2   Oblique incidence

If the cylinder is oriented arbitrarily, its axis can form any angle θ with the wavevector $\mathbf{k}$. Normal incidence corresponds to $\theta = 90°$, whereas at $\theta = 0$, $\mathbf{k}$ is parallel to the cylinder axis. It is also, in the case of oblique incidence, customary to call the quantity $Q = C/2aL$ the efficiency, although this is not in accord with the definition for Q in (2.9), where C is divided by the *projected* geometrical cross section $2aL \sin \theta$. Anyway, with $Q = C/2aL$,

$$Q(\theta) \to 0 \qquad \text{for } \theta \to 0.$$

Furthermore, when the cylinder has a very large radius, the efficiency reflects the geometrical contraction, so

$$Q(\theta) \propto \sin \theta \qquad \text{for } a \gg \lambda.$$

Formulae to compute Q for normal incidence are found, for instance, in [Boh83], for oblique incidence in [Lin66].

Figure 10.2 shows the efficiences $Q_{\parallel}$ and $Q_{\perp}$ as a function of orientation angle θ for size parameters $x = 2\pi a/\lambda$ in the range where polarization has its maximum. Around normal incidence down to orientation angles of 40°, the efficiencies $Q_{\parallel}$ and $Q_{\perp}$ as well as their difference ΔQ are rather constant. When the cylinder is tilted further, $Q_{\parallel}$ and $Q_{\perp}$ converge and for extremely oblique incidence they tend to zero. There is a sharp resonance for the curve with the size parameter $x = 0.6$ that peaks at an orientation angle of 0.55°. The sinusoidal dependence of Q on θ for large size parameters is already indicated in the curve with $x = 2$.

10.1.3   Rotating cylinders

Grains in interstellar space are knocked about by gas atoms and rotate. Therefore we also have to evaluate the cross section of a spinning cylinder. The incidence is generally oblique but we have just learnt how to handle this. During rotation the incident angle θ varies, so one has to form an average over a rotation cycle.

Suppose the cylinder circles in the (x, y)-plane of a Cartesian coordinate system, presented by the page of this book, and its momentary location is characterized by the angle ϕ (see figure 10.3). Let the cylinder be illuminated by an unpolarized flux whose wavevector $\mathbf{k}$ lies in the (y, z)-plane and makes an angle $90 - \Psi$ with the z-axis. At $\Psi = 90°$, light falls directly from above on the page and then, for reasons of symmetry, there will be, on average, no polarization. At an angle $\Psi = 0°$, the wavevector lies in the (x, y)-plane; then cross sections are different for electric vectors along the x- and z-axes.

While the cylinder spins, the angle θ between $\mathbf{k}$, the direction of wave propagation, and the cylinder axis changes. We can figure out in a quiet moment that θ is given by

$$\cos \theta = \cos \Psi \cos \phi. \qquad (10.2)$$

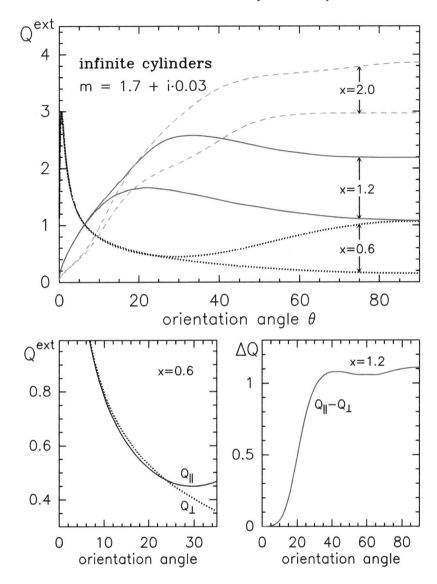

Figure 10.2. The top frame shows the extinction efficiencies for polarized light of obliquely illuminated infinite cylinders. An orientation angle $\theta = 90°$ corresponds to normal incidence. There are three pairs of curves for three size parameters; the upper curve gives $Q_\parallel$ when the electric vector **E**, the wavevector and the cylinder axis are in one plane; the lower shows $Q_\perp$ when **E** is perpendicular to the plane defined by the wavevector and the cylinder axis. The bottom left panel is a magnification for $x = 0.6$. Where $Q_\parallel$ and $Q_\perp$ (dots) cross over, the polarization reverses. The bottom right panel gives the relevant quantity for polarization, $Q_\parallel - Q_\perp$, for a size parameter $x = 1.2$.

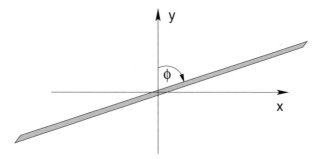

Figure 10.3. An infinite cylinder rotating in the (x, y)-plane. The direction of the incident light, given by the wavevector **k**, makes an angle $90 - \Psi$ with the z-axis which is perpendicular to the (x, y)-plane.

Now, as before, let $Q_\|(\theta)$ be the efficiency of the cylinder for oblique incidence under the angle θ, i.e. when the electric vector, the cylinder axis and **k** lie all in one plane. Let Q_x be the efficiency when the electric field swings in the plane containing **k** and the x-axis; equivalent definitions hold for $Q_\perp(\theta)$ and for Q_y. Then

$$Q_y = Q_\| \cos^2 \varphi + Q_\perp \sin^2 \varphi$$
$$Q_x = Q_\| \sin^2 \varphi + Q_\perp \cos^2 \varphi$$

so that

$$Q_y + Q_x = Q_\|(\theta) + Q_\perp(\theta)$$
$$Q_y - Q_x = [Q_\|(\theta) - Q_\perp(\theta)] \cos 2\varphi.$$

The angle φ is just the projection of ϕ onto a plane perpendicular to **k**,

$$\tan \varphi = \frac{\tan \phi}{\sin \Psi}. \tag{10.3}$$

Denoting by ΔQ the *time average* of $Q_y - Q_x$, we get

$$\Delta Q = \langle Q_y - Q_x \rangle = \frac{2}{\pi} \int_0^{\pi/2} \{ Q_\|(\theta) - Q_\perp(\theta) \} \cos 2\varphi \, d\phi.$$

ΔQ is proportional to the degree of linear polarization that the rotating cylinder produces. In the integral, θ and φ are both functions of ϕ as specified in (10.2) and (10.3).

An example is shown in figure 10.4. This again applies to silicate at optical wavelengths. We see that the *picket fence*, which refers to rigid cylinders, agrees qualitatively with the more sophisticated models that include rotation. Naturally, ΔQ^{ext} is greatest for the *picket fence*, by some 30% compared to rotation with $\Psi = 0$ and by more when Ψ is larger. Polarization will become weak when the rotation axis begins to point towards the observer.

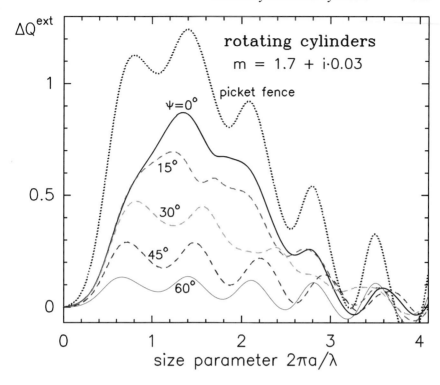

Figure 10.4. The difference in the extinction efficiency, $\Delta Q^{\text{ext}} = \langle Q^{\text{ext}}_{\parallel} - Q^{\text{ext}}_{\perp} \rangle$, of a rotating infinite cylinder with refractive index $m = 1.7 + i0.03$ as a function of size parameter for various inclinations of the rotation axis. The curve labeled *picket fence* can also be identified in the top frame of figure 10.1; here the cylinders do not rotate but are fixed in space and their axes are parallel to the electric vector and perpendicular to the direction of wave propagation given by the wavenumber **k**. When the cylinder is spinning, Ψ denotes the angle between **k** and the plane of rotation.

10.1.4 Absorption efficiency as a function of wavelength

When cylinders have a diameter $2a \simeq 0.1\ \mu$m, which is a typical size of interstellar grains, their efficiencies $Q_{\parallel}$ and $Q_{\perp}$ and the efficiency of a sphere of the same radius a are, at optical wavelengths, quite similar. Within a factor of two,

$$Q_{\parallel} \simeq Q_{\perp} \simeq Q \text{ (sphere)}.$$

In the mid and far infrared at wavelengths $\lambda \gg a$, the situation is radically different. This is demonstrated in figure 10.5, where we have calculated the Qs over a broad wavelength region for cylinders as well as spheres. The particles consist of our standard dust materials, amorphous carbon and silicate, with frequency-dependent dielectric permeabilities from figure 7.19. Whereas for both

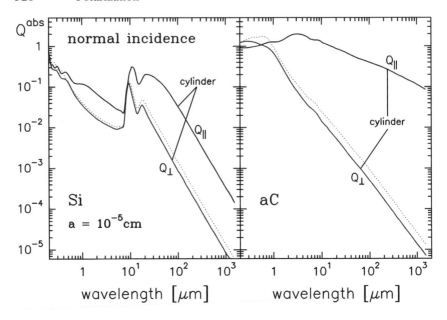

Figure 10.5. The absorption efficiencies, $Q_\|$ and $Q_\perp$, of infinite cylinders (full lines) under normal incidence of light. The cylinder radius is $a = 10^{-5}$ cm. The left panel is for silicates, the right, amorphous carbon. The efficiency of a sphere (dotted) of the same radius is shown for comparison.

substances, $Q_\perp$ and Q(sphere) are comparable at all λ, the efficiency $Q_\|$, where the electric vector of the light is along the cylinder axis, is in the far infrared much larger than Q(sphere). This has consequences for the temperature of dust, its emission and the degree to which the emission is polarized; extinction is usually irrelevant at these wavelengths.

When the cylinders in a cloud are randomly oriented or randomly rotating in space, one has to take an average over the solid angle and over the polarization direction. The mean efficiency is given by

$$\langle Q_\nu \rangle = \tfrac{1}{2} \int_0^{\pi/2} \left[Q_{\nu\|}(\theta) + Q_{\nu\perp}(\theta) \right] \sin \theta \, d\theta \tag{10.4}$$

where θ equals 90° when the light falls at a right angle on the cylinder. $\langle Q_\nu^{\text{abs}} \rangle$ is plotted in figure 10.6. It must be plugged into equation (8.8) when one wants to determine the temperature of the cylinders in a radiation field J_ν. In figure 10.7, we have computed, from equation (8.1), the emission ϵ_ν of cylinders near a B1V star and compare it with that of spheres. There are tremendous differences in the dust temperature, a factor of three (!) for amorphous carbon and a factor of 1.5 for silicate. Despite the fact that cylinders are much colder than spheres, they are much stronger emitters in the far infrared, especially if they consist

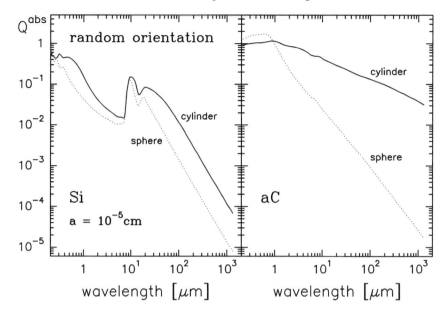

Figure 10.6. The average $\langle Q_\nu \rangle$ of (10.4) for randomly oriented rotating cylinders and for spheres (dots); otherwise as figure 10.5.

of amorphous carbon which has a large absorption efficiency because of its metallicity. Infinite cylinders are, of course, an extreme example but particles with realistic elongations also show considerable enhancement in their far infrared emission (one may consult figure 3.3).

10.2 Linear polarization through extinction

10.2.1 Effective optical depth and degree of polarization $p(\lambda)$

In a cloud with anisotropic grains that are aligned to some degree, the extinction of a linearly polarized wave depends on the orientation of the electric field vector. Let τ_{max} be the optical depth in the direction of maximum attenuation, perpendicular to it, the optical depth has its minimum, τ_{min}. When the unpolarized light from a background star traverses the cloud, the initial stellar intensity I_* is weakened to the observed intensity

$$I_{obs} = \tfrac{1}{2} I_* (e^{-\tau_{max}} + e^{-\tau_{min}}) = I_* e^{-\tau_{eff}}$$

where τ_{eff} defines the effective optical thickness,

$$\tau_{eff} = -\ln\left[\tfrac{1}{2}(e^{-\tau_{max}} + e^{-\tau_{min}})\right] > 0. \tag{10.5}$$

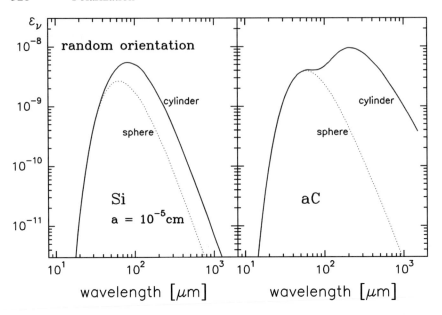

Figure 10.7. Emission in erg s^{-1} ster^{-1} Hz^{-1} of 1 cm^3 of amorphous carbon (right) and silicate dust (left). The grains are at a distance of 10^{18} cm from a B1V star ($L = 10^4$ L$_\odot$, $T_{eff} = 20\,000$ K) and are either spheres (dots) or randomly rotating cylinders (full curve) but always of radius $a = 10^{-5}$ cm. The total volume of the cylinders equals also 1 cm^3, so they are not infinite, just very long. The dust temperatures are: $T(\text{aC,sphere}) = 56.3$ K, $T(\text{aC,cyl}) = 19.3$ K, $T(\text{Si,sphere}) = 44.4$ K, $T(\text{Si,cyl}) = 30.9$ K. See also previous figure.

In the case of weak extinction ($\tau_{max} \ll 1$),

$$\tau_{eff} = \tfrac{1}{2}(\tau_{max} + \tau_{min}).$$

The (positive) degree of polarization is (see (2.71) and (2.77))

$$p = \frac{e^{-\tau_{min}} - e^{-\tau_{max}}}{e^{-\tau_{min}} + e^{-\tau_{max}}}. \tag{10.6}$$

The difference in optical depth, $\tau_{max} - \tau_{min}$, is usually small (although τ_{max} and τ_{min} themselves need not be so) and the polarization may then be approximated by

$$p = \tfrac{1}{2}(\tau_{max} - \tau_{min}) \qquad \text{for } \tau_{max} - \tau_{min} \to 0. \tag{10.7}$$

However, when $\tau_{max} - \tau_{min}$ is large, the polarization goes to unity,

$$p = 1 \qquad \text{for } \tau_{max} - \tau_{min} \to \infty.$$

The more foreground extinction there is, the higher the degree of polarization. The observed percentage of polarization rarely exceeds 10%. The upper limit is set by the difficulty of observing faint, strongly reddened stars.

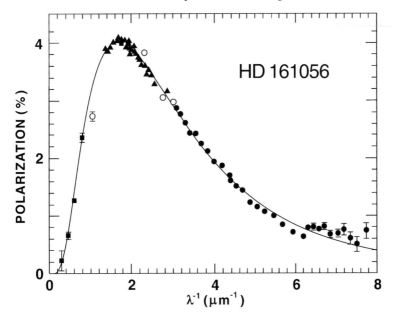

Figure 10.8. Linear polarization of the star HD 161056 [Som94]. Dots represent measurements, the full curve is the best fit Serkowski curve. Reproduced from [Som94] with permission of the American Astronomical Society.

10.2.2 The Serkowski curve

A central question is how the polarization, $p(\lambda)$, varies with wavelength as stellar light is extinguished by aligned grains in foreground clouds. Figure 10.8 presents, as an example, multi-wavelength observations of the star HD 161056 in the spectral interval from 0.13 to 3.6 μm. The data are nicely fitted by the expression

$$\frac{p(\lambda)}{p_{max}} = \exp\left[-k \ln^2 \left(\frac{\lambda}{\lambda_{max}}\right)\right]. \qquad (10.8)$$

This functional form of $p(\lambda)$ is called the Serkowski curve after one of the pioneers in the field. As the logarithm in (10.8) is squared, it does not matter whether we write λ/λ_{max} or λ_{max}/λ in the exponent. The Serkowski curve approximates polarization measurements of most stars quite well. It has three parameters: λ_{max}, p_{max} and k. The meaning of λ_{max} is obvious—it is the wavelength where the degree of polarization reaches its maximum p_{max}. It is also easy to see that the parameter k determines the width of the curve. Equation (10.8) is just a mathematical construct to fit empirical data, nothing more and certainly not a law. Nature also takes the liberty of behaving otherwise.

In section 10.1, we calculated, from Mie theory, the cross sections of infinite cylinders. We now apply these results in an attempt to give a theoretical explanation to the Serkowski curve. For this purpose, we go back to figure 10.4 which displays the difference $\Delta Q^{\text{ext}} = Q_\parallel - Q_\perp$ in the extinction efficiency of rotating cylinders as a function of size parameter $x = 2\pi a/\lambda$, and also for various inclination angles Ψ of the rotation axis. We infer from equation (10.7) that $\Delta Q^{\text{ext}}(\lambda)$ is generally proportional to the polarization $p(\lambda)$ and thus has the same wavelength dependence. The refractive index of the cylinder material in figure 10.4 is $m = 1.7 + i0.03$, representative of silicate at optical and near IR wavelengths.

When we use exactly the data of figure 10.4 but rearrange them by fixing the cylinder radius to some typical size, say $a = 0.1$ μm, so that only λ appears in the abscissa, the plots already look quite similar to a Serkowski curve. Figure 10.9 shows two alignment scenarios. On the left, the cylinders form a picket fence, on the right, they rotate with the rotation axis inclined by an angle $\Psi = 45°$. In either configuration, the Serkowski curve gives a remarkably good fit. This quick result is gratifying and guides us in understanding why interstellar polarization has the functional form of equation (10.8). It also testifies that such artificial geometrical bodies, like infinite cylinders, are, in their polarizing capabilities over a wide wavelength interval, from a tenth to a few micrometres, akin to real interstellar grains.

Of course, the agreement in figure 10.9 between model and Serkowski curve is not perfect, locally it is far from it. There are broad and deep wiggles. Their maxima are spaced by $\Delta x \simeq 0.7$ but as the difference in the extinction efficiency, ΔQ^{ext}, depends only on $x = 2\pi a/\lambda$, we know that a spread in cylinder radii by only 50% will effectively smear out the wiggles. Therefore, for a *size distribution* of cylinders, $p(\lambda)$ will be smooth and close to the Serkowski curve everywhere. Indeed, to achieve *good* agreement it takes only a few cylinder radii with proper size steps. For modeling polarization, the size distribution is not nearly as important as it is for modeling the extinction curve.

We also conclude from figure 10.9 that rotation, as opposed to the picket fence, does not change the shape of the wavelength dependence of polarization greatly. Rotation mainly reduces the height of the curve (degree of polarization), whereas the parameters λ_{max} and k are not much affected.

When one observes the wavelength dependence of polarization towards a large and diverse sample of stars, and then fits the measurements for each star by a Serkowski curve, one gets a continuous range of values for the three parameters $(p_{\text{max}}, k, \lambda_{\text{max}})$. Naturally, we expect strong fluctuations in p_{max}. Their origin is trivial as this quantity must depend on the degree of grain alignment and is at small optical depth proportional to the number of particles, N_{g}, in the line of sight.

The variations in the other two parameters, k and λ_{max}, are more subtle. We do not expect them to be related to N_{g} and the influence of alignment is uncertain and probably slight. If interstellar grains were the same everywhere, which is not a bad assumption for many purposes, we would be at a loss to explain the observed variations in k and λ_{max}. As variations are common, we conclude that

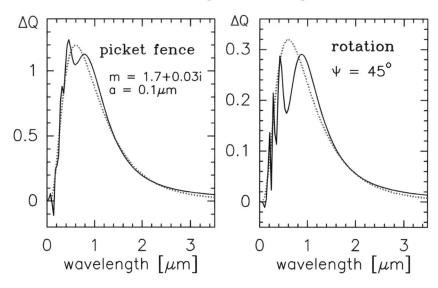

Figure 10.9. The wavelength dependence of linear polarization produced by aligned infinite cylinders with optical constant $m = 1.7 + i0.03$ (full curve). Also shown are Serkowski curves after (10.8) (dotted lines) with parameters $\lambda_{max} = 0.6\ \mu$m and $k = 1.2$; they fit more or less. In both frames, the cylinder radius $a = 0.1\ \mu$m. The left panel shows rigid cylinders under normal incidence, the right, rotating cylinders with rotation axis inclined 45° to the line of sight. The full curve is a rearrangement of the curve in figure 10.4 for $\Psi = 45°$ and is obtained by putting $a = 0.1\ \mu$m.

intrinsic differences exist among the polarizing grains in various environments.

For field stars, λ_{max} has a large scatter from approximately 0.35 to 0.9 μm but values cluster at 0.55 μm. We will argue in the discussion of figure 10.10 that, for cylinders, the maximum wavelength λ_{max} is proportional to grain size. A dependence of k on cylinder radius a is not evident. Nevertheless, there exists an interesting statistical correlation between k and λ_{max} [Whi92],

$$k \simeq 1.66\lambda_{max}.$$

A good explanation for this formula is still missing.

10.2.3   Polarization $p(\lambda)$ of infinite cylinders

Figures 10.10 and 10.11 display how, for infinite cylinders, the difference in the extinction efficiency, $\Delta Q^{ext} = Q_{\parallel} - Q_{\perp}$, which is proportional to the polarization $p(\lambda)$, changes with wavelength. The plots refer to picket fence alignment. If we allow for rotation and an inclined rotation axes, nothing essentially changes. The computations are performed for two materials, silicate and amorphous carbon,

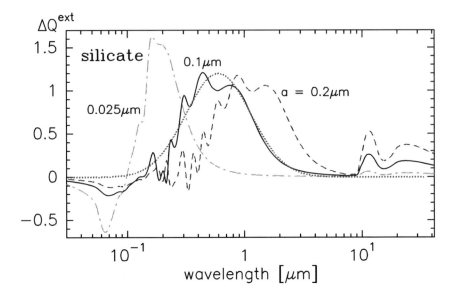

Figure 10.10. $\Delta Q^{\text{ext}}(\lambda)$ for infinite silicate cylinders in picket fence alignment. Computations were done for three cylinder radii, a, as indicated. The optical constant $m(\lambda)$ is from figure 7.19. The dotted line represents a least–square fit by a Serkowski curve with $\lambda_{\text{max}} = 0.6$ μm and $k = 1.2$.

and cover a wide wavelength range. The cylinder radii change from 0.025 to 0.2 μm, which should comprise the astrophysically relevant sizes.

10.2.3.1 Silicate cylinders

Both silicate and amorphous carbon grains have their peculiarities. Let us first look at the polarization curve of silicate cylinders with radius $a = 0.1$ μm. This is the full curve in figure 10.10. It has three bumps (at about 0.6, 12 and 23 μm). The one around 0.6 μm is very similar to the bump in the left frame of figure 10.9, only there the optical constant $m(\lambda)$ was fixed and the wavelength display linear, whereas in figure 10.10, $m(\lambda)$ is variable and the wavelength scale logarithmic. In least–square fits to $p(\lambda)$ by a Serkowski curve, the parameters k and λ_{max} are, in both cases, almost identical (dots in figures 10.9 and 10.10).

When one smooths the oscillations in the curves of figure 10.10 and determines the location of the first peak (the one at $\lambda < 2$ μm), one finds that the relation between λ_{max} and the cylinder radius a is roughly linear. Consequently, if the polarizing dust particles in interstellar space are silicate cylinders, they must have diameters of $\sim$0.2 μm, or a bit less, to yield the Serkowski curve with the standard $\lambda_{\text{max}} = 0.55$ μm.

Besides the bump at 0.6 μm, there appear in figure 10.10 two other peaks

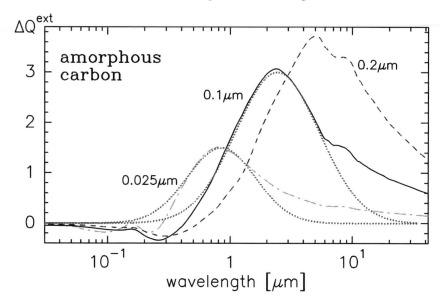

Figure 10.11. As figure 10.10, but now for cylinders of amorphous carbon (aC). For cylinder radius $a = 0.025$ μm, the optical constant $m(\lambda) = 1.85 + 1.23i$ is fixed and has the value of aC at 0.55 μm; for the other cylinder radii, $m(\lambda)$ is from figure 7.19 and variable. Dots show Serkowski curves. For cylinder radius $a = 0.025$ μm, the fit parameters are $\lambda_{max} = 0.81$ μm, $k = 1.2$; for $a = 0.1$ μm, they are $\lambda_{max} = 2.5$ μm, $k = 0.8$.

in the mid infrared which are *not predicted* by the Serkowski curve. Their origin is of an entirely different nature and due to resonances in the bulk material (see figure 7.19). If $m(\lambda)$ were constant, these bumps would vanish. The first, sharper peak is associated with the 9.7 μm silicate absorption feature but its position is substantially offset by about 2 μm to the red. The location of the polarization bump is independent of the cylinder radius; however, its strength increases with a.

10.2.3.2 Carbon cylinders

Carbon grains are depicted in figure 10.11. The bumps in the mid infrared have disappeared, nevertheless the polarization at these long wavelengths is several times stronger than for silicates. The curve $p(\lambda)$ now has only one peak that occurs, for the same cylinder diameter, at much larger λ_{max} compared to silicates. Extremely thin carbon rods ($a \leq 200$ Å) are required to produce a hump at 0.55 μm, where most polarization curves peak.

But even when carbon rods of a certain radius produce the right wavelength of maximum polarization, λ_{max}, the overall shape of $p(\lambda)$ does not agree with observations. For example, in the plot of $p(\lambda)$ for $a = 0.025$ μm, we can fit only the peak of the Serkowski curve. Likewise grains with $a = 0.1$ μm, which does not seem to be an unreasonable size, can be excluded not only because λ_{max} is far off in the near infrared but also because $p(\lambda)$ overshoots on the red side and is much too broad ($k = 0.8$). On these grounds, amorphous carbon is not likely to be the prime polarizing agent.

10.2.3.3   The ratio $C_\parallel / C_\perp$ at optical and infrared wavelengths

In the optical region, polarization is well described by the Serkowski curve which can be quantitatively reproduced by applying Mie theory to infinite cylinders. Maximum polarization occurs when the cylinder radius is comparable to the wavelength. The cylinder cross section $C_\parallel$ is about 25% greater than $C_\perp$ (see figure 10.1). So the ratio $C_\parallel / C_\perp$ is not far from unity and the cross sections are not very sensitive to the direction of the electric vector. In the infrared region, however, where the particles are small compared to wavelength, the ratio $C_\parallel / C_\perp$ can be large. For instance, silicate cylinders at 10 μm (optical constant $m = 1.36 + i0.95$) have $C_\parallel / C_\perp = 2.61$. At $\lambda = 1.3$ mm, where $m = 3.41 + i0.034$, this value goes up to 39.9 and for amorphous carbon cylinders the ratio approaches 10^4 (see figure 10.5). These numbers are independent of cylinder radius as long as $a \ll \lambda$.

10.2.4   Polarization $p(\lambda)$ of ellipsoids in the Rayleigh limit

Although infinite cylinders can reproduce the basic features of optical polarization, it is unsatisfactory that their shape is so artificial. We can improve the infinite cylinder model by treating ellipsoids as a more realistic alternative. They allow an investigation of a much greater variety of shapes. Cigars, a subclass with major axes $a > b = c$, include cylinders in the limit of infinite axial ratio a/c. In the *infrared*, the computation of cross sections is easy because the interstellar particles are then small compared to wavelength and we may apply the electrostatic approximation of section 3.3.

It is also straightforward to compute coated ellipsoids, although we will not do so. They further widen the parameter space: one can then vary the axial ratio of the ellipsoids, their mantle thickness and the optical constants of core and mantle. Coated ellipsoids will thus be superior in fitting observations. However, a fit to observations becomes astronomically relevant not by its mere agreement with the data but by its ability to predict and to constrain our conception about dust. One must, therefore, critically ask whether a refinement of the dust model in the direction of complicating shape and composition deepens or muddles our understanding of dust.

Figure 10.12 depicts how the optical depth $\tau(\lambda)$ and the polarization $p(\lambda)$ produced by small silicate cigars change with wavelength. The particles are spinning with their rotation axis perpendicular to the line of sight ($\Psi = 0$ in

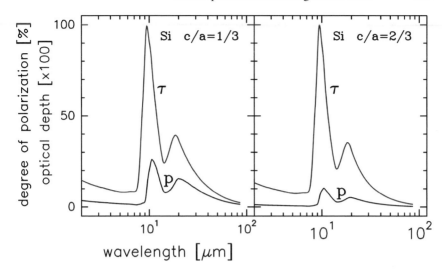

Figure 10.12. Optical thickness τ and degree of polarization p in the infrared as a function of wavelength for small rotating silicate cigars; no other grains are asumed to be present. The rotation axis is perpendicular to the line of sight. The dielectric permeability $\varepsilon_{Si}(\lambda)$ is taken from figure 7.19. The left box shows moderate elongation with axial ratio $a/c = 3$, the right box, weak elongation, $a/c = 1.5$. The optical depth is one at 9.5 μm. The curve $p(\lambda)$ gives the degree of polarization one would observe in a column with $\tau_{9.5\ \mu m} = 1$ under perfect spinning alignment.

the nomenclature of section 10.1). $\tau(\lambda)$ is the effective optical depth after (10.5) and $p(\lambda)$ is defined in (10.6). The size of the particles is irrelevant as long as they are in the Rayleigh limit. In this particular example, τ is one at 9.5 μm (curve maximum). At low optical depth, which is approximately true outside the peak at 9.5 μm, $\tau(\lambda)$ is proportional to the time-averaged cross section $C = \frac{1}{2}(C_\perp + C_\parallel)$, and $p(\lambda)$ to the difference $\Delta C = C_\perp - C_\parallel$ (see formulae (3.49) and (3.50)). The results for silicate pancakes are practically identical. When the spheroids are made of amorphous carbon (not shown), the curves are less interesting and display a monotonic decline with wavelength.

A different presentation of the polarization produced by perfectly aligned spinning spheroids is given in figure 10.13. It allows us to read off $p(\lambda)$ if the optical depth $\tau(\lambda)$ at wavelength λ equals 1 or 3. The polarization $p(\lambda)$ for other optical depths may be estimated from interpolation remembering that $p(\lambda)$ goes to zero for $\tau(\lambda) \rightarrow 0$ and approaches one for $\tau(\lambda) \gg 1$.

From figure 10.13 we conclude that if interstellar dust is made entirely of weakly elongated particles ($a/c = 2/3$, about the shape of an egg), a cloud with an optical depth of three at 20 μm, corresponding to $\tau_V \sim 100$ mag, would produce a polarization of 50% under perfect grain alignment. The optical

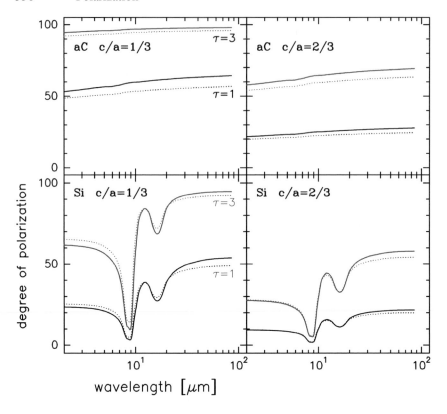

Figure 10.13. The percentage of polarization due to perfectly aligned spinning spheroids of silicate and amorphous carbon. Note that in contrast to figure 10.12, here the degree of polarization $p(\lambda)$ is plotted for fixed effective optical depth $\tau(\lambda)$, so $p(\lambda)$ always refers either to $\tau(\lambda) = 1$ (lower pair of curves in each box) or $\tau(\lambda) = 3$ (upper pair of curves). Full curves are for cigars, dotted ones for pancakes. Optical constants are from figure 7.19.

polarization $p(V)$ is then close to one but could not be observed because of the large extinction. Although alignment in real clouds will not be perfect, one nevertheless expects a considerable degree of infrared polarization in those cases where there is a suitable infrared background source.

From figures 10.12 and 10.13 and similar calculations for small amorphous carbon spheroids (not shown) we learn that polarization through extinction and extinction itself are qualitatively similar in their wavelength dependence. However, there exist significant offsets in the resonance peaks, as can be checked by scrutinizing figure 10.12, if necessary with the help of a ruler. Furthermore, one finds that a small or moderate axial ratio ($a/c \leq 3$) has little affect on extinction (see figure 3.3) but has a great influence on polarization.

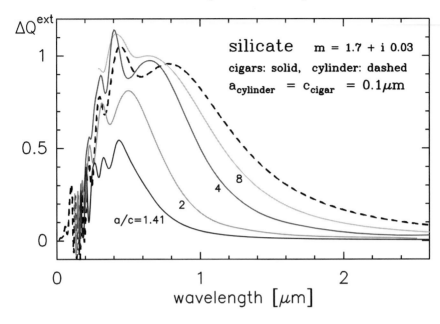

Figure 10.14. $\Delta Q^{\text{ext}}(\lambda)$ for prolate spheroids (full curves) and a cylinder (dashed curve) under normal incidence in picket fence alignment. The numerical code was kindly provided by N V Voshchinnikov [Vos93]. The cigars have principal axes $a > b = c$ with uniform $c = 0.1$ μm but various axial ratios a/c ranging from 1.41 to 8. The optical constant is fixed and appropriate for silicate in the visual band.

10.2.5 Polarization $p(\lambda)$ of spheroids at optical wavelengths

Unfortunately, in the visual band, interstellar grains of spheroidal shape are not small relative to wavelength and computing their cross sections entails some numerical effort [Asa75]. We restrict the discussion to cigars, which have principal axes $a > b = c$, and demonstrate how their axial ratio influences the polarization. Results for silicates under normal incidence are displayed in figures 10.14 and 10.15. When all axes are equal, the body is a sphere and there is no polarization. If we keep the small axes $b = c$ fixed and increase a, the cigars begin to resemble cylinders. To be specific, we assume constant values $c = 0.1$ μm and $\lambda = 0.55\mu$m (visual band) and vary the axial ratio $a/c \geq 1$. The results depend, of course, only on the size parameters a/λ, and c/λ, and on the refractive index m.

For an axial ratio of $a/c = 8$, the polarization, expressed through the difference in extinction efficiency ΔQ, is similar to that of an infinite cylinder, as it should be. When a/c decreases, the maxima of the curves shift to shorter wavelengths and the degree of polarization (ΔQ) falls; the curves also become narrower. By and large, the response in amorphous carbon particles to a change

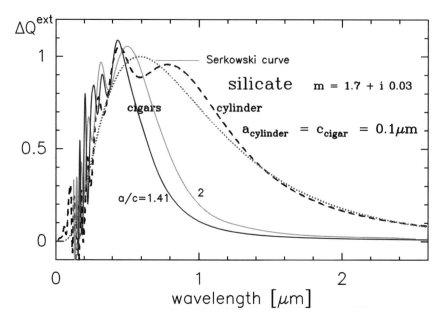

Figure 10.15. This is a derivate of figure 10.14. A Serkowski curve ($k = 1.15$, $\lambda_{max} = 0.6\,\mu\text{m}$) is added to study the shape of the polarization curves better. The ΔQs for cigars with axial ratio 4 and 8 have been omitted, those for $a/c = 1.41$ and 2 are enhanced so that the maxima of all curves are about equal.

in the ratio a/c is similar. Naturally, spheroids of variable axial ratio show a greater diversity in their polarizing properties than infinite cylinders but it also becomes harder to infer from the shape of the Serkowski curve size and chemical composition of the grains, assuming that they are spheroids. For instance, the width of the curve $p(\lambda)$ is influenced by the optical constant *and* the particle shape, and it is hard to tell one from the other.

   If we want to know how much polarization we get from a unit volume of spheroidal dust particles, we must convert ΔQ into a mass coefficient, ΔK. To determine ΔK, we multiply ΔQ by the geometrical cross section πac and divide by the volume of one grain, $4\pi ac^2/3$. So ΔK is proportional to $\Delta Q/c$. In order that the *cylinders* also refer to a unit volume, we have multiplied ΔQ_{cyl} in figure 10.14 by $8/3\pi$. This enables direct comparison of all polarizing efficiencies.

10.2.6   Polarization and reddening

It is instructive to combine photometric with polarimetric observations. The former yield the effective optical depth τ_{eff} of (10.5), the latter the degree of polarization p. Both are produced by the extinction of dust particles and both depend on wavelength.

The ratio of p_{max} over standard color excess E_{B-V} as defined in (7.15) combines photometry and polarimetry, or τ_{eff} and p, and gives the polarization per unit reddening. It is a measure of the *polarizing efficiency* at visual wavelengths along the line of sight. There is an important empirical upper limit,

$$\frac{p_{max}}{E_{B-V}} \lesssim 9\%. \tag{10.9}$$

Because the highest polarization is generally reached in the visual,

$$p_{max} \simeq p(V)$$

and the visual extinction in magnitudes, A_V, is related to E_{B-V}, from (7.17), through

$$A_V = R_V E_{B-V} = 1.086\tau_V$$

where τ_V is the corresponding optical depth, we can express the condition (10.9) also as $p(V)/\tau_V \leq 0.031$, which gives, using equation (10.7),

$$\beta \equiv \frac{\tau_{max}}{\tau_{min}} - 1 \lesssim 0.062. \tag{10.10}$$

This inequality implies that the optical depth at visual wavelengths in two perpendicular directions never differs in interstellar space by more than $\sim$6%. However, in those cases, where $p_{max}/E_{B-V} \simeq 9\%$ is observed, the quantity β must attain this maximum value of 6%.

We can determine a theoretical upper limit to β by assuming that *all* grains are cylinders and optimally aligned. Without pretending to be precise, we can read from figures 10.1 and 10.4 that the maximum value of $\beta = \Delta Q/Q$ is about 25%. This number is only four times greater than what actually is observed under favorable conditions and thus surprisingly low when one considers that some of the real grains do not polarize at all and those that do are not infinite cylinders but only weakly anisotropic; furthermore, the orientation is never perfect and possibly changing along the line of sight.

10.3   Polarized emission

Figures 10.10 to 10.12 present the theoretical wavelength dependence of polarization due to extinction by silicate cylinders and ellipsoids. These plots include the Serkowski curve but also cover the mid infrared region. When extended further into the far infrared, no detectable effect is expected because the extinction optical depth becomes too small.

10.3.1   The wavelength dependence of polarized emission for cylinders

However, dust not only absorbs but also radiates and *polarized emission* of aligned grains *can* be observed in the far infrared and millimeter spectral region. Towards a uniform cloud of optical depth τ_ν, temperature T_d and intensity

$$I_\nu = B_\nu(T_d) \cdot [1 - e^{-\tau_\nu}]$$

the degree of polarized emission, $p(\lambda)$, becomes

$$p = \frac{e^{-\tau_{\min}} - e^{-\tau_{\max}}}{2 - (e^{-\tau_{\min}} + e^{-\tau_{\min}})}. \tag{10.11}$$

We abide by the nomenclature of section 10.2 where $\tau_{\max}$ and $\tau_{\min}$ denote the optical depth at the position angle of maximum and minimum attenuation. In most cases and certainly at submillimeter wavelengths, the optical depth is small. The (positive) degree of polarization can then be written as

$$p \simeq \frac{\tau_{\max} - \tau_{\min}}{\tau_{\max} + \tau_{\min}}. \tag{10.12}$$

p depends on wavelength but not on τ. Even for arbitrarily small τ, the polarization is finite. As the optical depth increases, the polarization can only decrease. This is quite contrary to polarization by extinction which grows with τ and vanishes for $\tau \to 0$.

We start the discussion of polarized dust emission according to equation (10.12) with cylinders. Their cross sections for light when the electric vector lies along or perpendicular to the cylinder axis, $C_\parallel$ and $C_\perp$, can differ by orders of magnitude, as is evident from figure 10.5. Even for a dielectric, the ratio $C_\parallel/C_\perp$ may be above 10. We, therefore, expect a very high degree of polarization in the case of complete alignment (see figure 10.16). Rotation of the cylinders roughly halves the value of $C_\parallel/C_\perp$ with respect to picket fence alignment but $C_\parallel/C_\perp$ is still large, so $p(\lambda)$ does not change much. It also turns out that when the rotation vector is inclined towards the observer, the degree of polarization drops less swiftly than in the case of extinction (see figure 10.4).

10.3.2   Infrared emission of spheroids

At low frequencies, one can also easily compute spheroids in the small particle limit as outlined in section 3.3. Let the grains have major axes $a \geq b \geq c$ with corresponding cross sections C_a, C_b, C_c. If they rotate about their axis of greatest moment of inertia and are fully aligned, the emission in a direction perpendicular to their rotation axes is polarized to a degree:

$$p = \frac{\tau_{\max} - \tau_{\min}}{\tau_{\max} + \tau_{\min}} = \begin{cases} \dfrac{C_a - C_c}{C_a + 3C_c} & \text{cigars } (C_a > C_b = C_c) \\[3mm] \dfrac{C_a - C_c}{C_a + C_c} & \text{pancakes } (C_a = C_b > C_c). \end{cases} \tag{10.13}$$

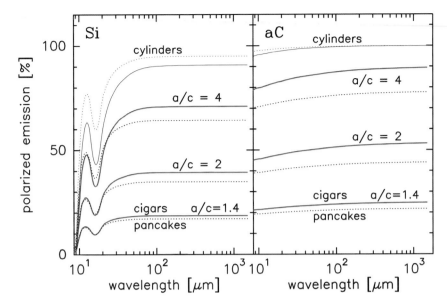

Figure 10.16. The percentage at which thermal emission by dust is polarized in the far infrared. The particles consist of silicate (Si, left box) or amorphous carbon (aC, right box) with optical constants from figure 7.19. They are either infinite cylinders (the two top curves in each frame), cigars (all other full curves) or pancakes (all other dotted curves) of varying axial ratios a/c. There is perfect spinning alignment for cylinders, cigars and pancakes and the rotation axis is perpendicular to the line of sight. Only the dotted top curves in each frame show cylinders arranged in a picket fence.

Some results are shown in figure 10.16 and we point out that

- even spheroids of small elongation ($a/c = 1.40$) produce very substantial polarization ($\sim 20\%$);
- the polarization curves are flat at long wavelengths.

10.3.3 Polarized emission versus polarized extinction

Sometimes towards a dust cloud one detects polarized submillimeter radiation and simultaneously polarized light at much shorter wavelengths from embedded or background stars. It is easy to see that the polarization vectors associated with these two processes, emission and extinction, are perpendicular to each other.

A polarization vector is defined by its length, which is equal to the degree of polarization, and by the direction where the electric field vector has its maximum; the associated angles have values between $0°$ and $180°$.

If p_{sub} is the submillimeter polarization due to emission and p_V the visual polarization due to extinction, the two are given by (10.7) and (10.11) and are related through

$$\frac{p_V}{p_{sub}} = \tau_V \frac{\beta_{sub}}{\beta_V}$$

where at both wavelengths we use the quantity $\beta = \tau_{max}/\tau_{min} - 1$ defined in equation (10.10), which specifies the maximum change in the optical depth with direction. As p_V, p_{sub} and τ_V can be determined observationally, one obtains β_{sub}/β_V. Because the influence of grain elongation on β is markedly different at visual and far infrared wavelengths, this ratio contains, in principle, information on the particle shape.

10.4 Circular polarization

In some media, the propagation of a linearly polarized wave is sensitive to the orientation of polarization because the building blocks of matter are anisotropic as, for example, in all crystals of the non-cubic lattice or substances consisting of long and aligned fibrous molecules. The optical constant, $m = n + ik$, depends then on orientation, too. This produces a number of interesting phenomena which can all be understood if the molecules are treated as harmonic oscillators. The particular structure of the molecules in the medium has the effect that the motions of the charges, which determine m or ε after (1.77), depend on the direction of the electric field vector.

- When the real part of the refractive index, n, is anisotropic, one speaks of *birefringence*. Calcite, $CaCO_3$, is the best known example. It splits an incident unpolarized beam into two polarized waves with different refraction angles (this is actually a special kind of birefringence known as *anomalous refraction*).
- When the imaginary part of the refractive index, k, differs in two directions, the phenomenon bears the name *dichroism*. The gem tourmaline is renowned for this property.

We will be dealing with birefringence only. Imagine a slab filled with a medium that has unequal indices $n_\parallel$ and $n_\perp$ in two perpendicular directions and thus unequal phase velocities. Let us shine linearly polarized light through the slab. The light is oriented in such a way that it can be decomposed into two equally strong waves, $E_\parallel$ and $E_\perp$, linearly polarized along the principal directions. At the front of the slab, $E_\parallel$ and $E_\perp$ are in phase, inside they will fall out of phase. When the phase difference amounts to $\pi/2$, the initially linearly polarized wave presented by the *sum* $E_\parallel + E_\perp$ has been converted into a circularly polarized wave; a slab with the right thickness to do just that is called a λ-quarter plate. Quite generally, when light that is linearly polarized (but not along a principal direction) enters a birefringent medium, it leaves with some degree of circular polarization.

There is another related phenomenon, *optical activity*, which we mention for completeness. Here the phase velocity of the medium is different for *circularly* polarized waves with opposite senses of rotation. A linearly polarized wave entering an optically active medium exits it still linearly polarized but with its orientation shifted by some angle. The electric vector is rotated to the right when the left-circular wave is faster and to the left otherwise.

10.4.1 The phase shift induced by grains

When a dust cloud is traversed by a plane wave, the grains not only reduce the intensity of radiation but may also change the phase of the wave. To understand this, we exploit the complex scattering function $S(\theta, \phi)$ discussed in section 2.2. Its value in the forward direction, $S(0)$, can *formally* be interpreted as the optical constant of the interstellar medium, $\overline{m} = \overline{n} + i\overline{k}$. The imaginary part, $\text{Im}\{S(0)\}$, determines, according to (2.22), the index of refraction, $\overline{n} - 1$, which is responsible for the phase velocity.

A wave of wavenumber $k = 2\pi/\lambda$ that passes through a plate of unit thickness and refractive index $\overline{m} = \overline{n} + i\overline{k}$ is phase-shifted with respect to a wave traversing a vacuum by

$$\Delta\phi = k \cdot (\overline{n} - 1).$$

Let us define the volume coefficient for phase lag, K^{pha}, by putting $K^{\text{pha}} = \Delta\phi$. If there are N particles per cm^3, each with a cross section for phase lag C^{pha}, so that $K^{\text{pha}} = NC^{\text{pha}}$, we get, from (2.22),

$$C^{\text{pha}} = \frac{\lambda^2}{2\pi} \text{Im}\{S(0)\}. \tag{10.14}$$

When the wave has crossed a distance l, the components $\mathbf{E}_\parallel$ and $\mathbf{E}_\perp$ of the electric vector $\mathbf{E}$ are out of phase by an angle

$$\Delta\phi = l(K_\parallel^{\text{pha}} - K_\perp^{\text{pha}}) = lN\Delta C^{\text{pha}}.$$

The optical thickness over this distance l equals NlC^{ext}. Therefore, an optical depth τ produces a phase lag

$$\Delta\phi = \tau \frac{\Delta C^{\text{pha}}}{C^{\text{ext}}}.$$

Near the wavelength λ_{max} of maximum linear polarization in the Serkowski curve, the ratio $\Delta Q^{\text{pha}}/Q^{\text{ext}} = \Delta C^{\text{pha}}/C^{\text{ext}}$ (see (10.1)) is for silicate cylinders, very roughly, one tenth. Consequently, the phase lag in a cloud of visual optical depth τ due to perfectly aligned cylinders is of order $\Delta\phi \approx \tau/10$. If there are also non-polarizing particles in the line of sight, $\Delta Q^{\text{pha}}/Q^{\text{ext}}$ and $\Delta\phi$ will be smaller.

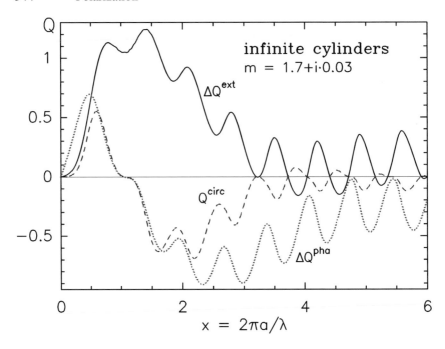

Figure 10.17. Linear and circular polarization and phase lag of infinite cylinders in picket fence alignment as a function of size parameter $x = 2\pi a/\lambda$. The optical constant m is fixed. The curve ΔQ^{ext} (full curve) is identical to the one in figure 10.1.

10.4.2   The wavelength dependence of circular polarization

Interstellar circular polarization proceeds in two stages:

- One needs linearly polarized light to start with, the more the better. This is provided by a cloud with grains aligned in one direction. The percentage of linear polarization is usually small.
- One needs another cloud where the particles are aligned in some other direction to produce the phase shift.

Therefore circular polarization is a second-order effect and weak. Its wavelength dependence, $p_{\text{circ}}(\lambda)$, is given by the product of two first-order effects,

$$p_{\text{circ}}(\lambda) \propto \left[Q_{\|}^{\text{ext}}(\lambda) - Q_{\perp}^{\text{ext}}(\lambda) \right] \cdot \left[Q_{\|}^{\text{pha}}(\lambda) - Q_{\perp}^{\text{pha}}(\lambda) \right]. \tag{10.15}$$

The first factor stands for the linear polarization, the second for the phase lag. As usual, the Qs are the efficiencies. Circular polarization contains also information how the alignment changes along the line of sight but it is buried and to find it one has to dig deep.

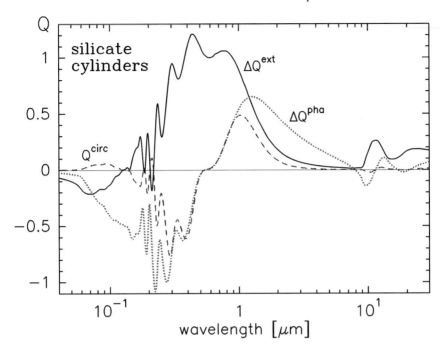

Figure 10.18. Linear and circular polarization and phase lag of infinite silicate cylinders with 0.1 μm radius in picket fence alignment. Optical constants from figure 7.19.

The indices $\bar{n}$ and $\bar{k}$ of the interstellar medium as given in (2.22) and (2.23) as well as the differences $\bar{n}_\parallel - \bar{n}_\perp$ and $\bar{k}_\parallel - \bar{k}_\perp$ are connected through the Kramers–Kronig relations. Because of the proportionalities

$$(\bar{n}_\parallel - \bar{n}_\perp) \propto (C_\parallel^{\text{pha}} - C_\perp^{\text{pha}}) \qquad (\bar{k}_\parallel - \bar{k}_\perp) \propto (C_\parallel^{\text{ext}} - C_\perp^{\text{ext}})$$

there is, in principle, nothing new in circular polarization or in its dependence on wavelength, $p_{\text{circ}}(\lambda)$. It suffices to know the linear polarization, which is given via the index $\bar{k}$. When one computes the index $\bar{n}$ with the help of the Kramers–Kronig relation, one arrives at circular polarization. This whole section might, therefore, seem superfluous. However, such a point of view is too academic because, in reality, we do not know the linear polarization at *all wavelengths*.

Figures 10.17–10.19 illustrate the behavior of circular polarization for infinite cylinders in picket fence alignment. The figures display the following quantities, either as a function of size parameter or of wavelength:

$\Delta Q^{\text{ext}} = Q_\parallel^{\text{ext}} - Q_\perp^{\text{ext}}$, the difference in the extinction efficiency which is proportional to linear polarization;

$\Delta Q^{\text{pha}} = Q_\parallel^{\text{pha}} - Q_\perp^{\text{pha}}$, the difference in the phase lag efficiency; and

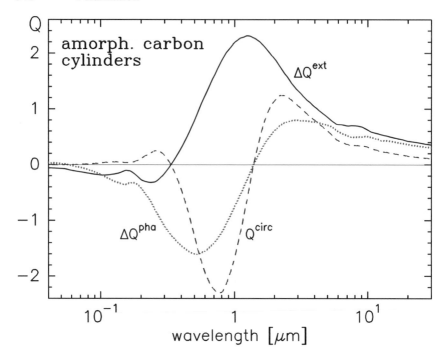

Figure 10.19. Linear and circular polarization and phase lag of infinite carbon cylinders with 0.05 μm radius in picket fence alignment. Optical constants from figure 7.19.

$Q^{\text{circ}} = \Delta Q^{\text{ext}} \Delta Q^{\text{pha}}$, the product describing after (10.15) circular polarization.

We summarize the noteworthy points about the figures:

- The efficiency for circular polarization, $Q^{\text{circ}}(\lambda)$, can have several local extrema.
- Where $Q^{\text{circ}}(\lambda)$ reverses sign, the sense of rotation changes and either ΔQ^{ext} or ΔQ^{pha} vanish. This occurs, for example, in silicate cylinders of 0.1 μm radius at 0.55 μm wavelength and in carbon cylinders of 0.05 μm radius at 1.4 μm. In either case, although the optical constants are completely different, the sign reversal of p_{circ} is close to but not exactly at the maximum of linear polarization.
- The curves for silicate are rather jittery. However, in a size distribution the spikes would disappear.
- The 10 μm silicate resonance is very weakly discernible in circular polarization, accompanied by a change of sign.

Chapter 11

Grain alignment

To produce polarization by extinction, interstellar grains must be elongated or flattened *and* aligned, at least, some of them. As they are not sitting still but receive stochastically angular momentum in collisions with gas atoms, they spin and it is their rotation axes that have to be aligned somehow, not the grains themselves. Take, for example, a long cylinder spinning about an axis perpendicular to its symmetry axis. If the rotation vector points towards the observer, there will obviously be no preferential direction and the time-averaged polarization vanishes. However, if the rotation axis is at a right angle to the line of sight, one gets maximum polarization.

The question of how alignment works has not been settled satisfactorily. One usually assumes the *Davis–Greenstein* mechanism which involves cumbersome details and is rather speculative. Although the impact of alignment studies on the rest of astronomy is humble, it is a delightful topic because of its physical richness. We summarize in the first two sections the tools necessary for understanding the mechanical and magnetic processes. Afterwards we describe them emphasizing the ideas behind them and not the final numbers which are rather vague.

11.1   Grain rotation

11.1.1   Euler's equations for a rotating body

Rotating bodies seem to misbehave in the sense that their motion evades intuition: we push a spinning top one way and it goes the other. However, a child can ride a bicycle with ease, so somehow we can cope with rotation, at least *in praxi*.

A freely rotating body, like an interstellar grain, rotates about its center of mass. The moment of inertia, I, depends on the axis about which it rotates. One can always find a rectangular coordinate system, fixed to the body, in which I is greatest about the z-axis and smallest about the x-axis. The moments along these principal axes are labeled I_x, I_y, I_z. If we plot, from the coordinate center, the moment of inertia in any possible direction as an arrow whose length is given by the value of I, the tips of the arrows form the inertia ellipsoid. This is a very regular figure, no matter how irregular the body. For example, the inertia ellipsoid of a cube or a regular tetrahedron must, for reasons of symmetry, be a sphere.

When we resolve the angular velocity with respect to the instantaneous values along the principal axes,

$$\boldsymbol{\omega} = (\omega_x, \omega_y, \omega_z).$$

the total angular momentum **L**, its square and the kinetic energy of the body are:

$$\mathbf{L} = (I_x\omega_x, I_y\omega_y, I_z\omega_z)$$
$$L^2 = I_x^2\omega_x^2 + I_y^2\omega_y^2 + I_z^2\omega_z^2 \tag{11.1}$$
$$E_{\text{rot}} = \tfrac{1}{2}(I_x\omega_x^2 + I_y\omega_y^2 + I_z\omega_z^2). \tag{11.2}$$

The equation of motion of a rotating body describes how its angular momentum **L** changes when a torque $\boldsymbol{\tau}$ is applied:

$$\boldsymbol{\tau} = \frac{d\mathbf{L}}{dt}. \tag{11.3}$$

It corresponds to the equation for translatory motion, $\mathbf{F} = \dot{\mathbf{p}}$, if we replace $\boldsymbol{\tau}$ by the force **F** and **L** by the momentum **p**. Likewise, the time variation in kinetic energy E of a freely spinning top is given by

$$\frac{dE}{dt} = \boldsymbol{\tau} \cdot \boldsymbol{\omega}. \tag{11.4}$$

This equation is analogous to $\dot{E} = \mathbf{F} \cdot \mathbf{v}$ for translational motion, where **v** is the velocity. The product $\boldsymbol{\tau} \cdot \boldsymbol{\omega}$ also gives the power needed to sustain rotation at constant angular velocity $\boldsymbol{\omega}$ against a torque $\boldsymbol{\tau}$ due to friction.

Without a torque, the angular momentum **L** stays constant in space and the tip of the vector **L** marks there a certain fixed point, let us call it Q. With respect to a coordinate system connected to the rotating body, Q is not constant but moving and the vector **L** as seen from the rotating body varies like

$$\left(\frac{d\mathbf{L}}{dt}\right)_{\text{body}} = \mathbf{L} \times \boldsymbol{\omega}.$$

If there is a torque, **L** changes in space according to (11.3). As viewed from the rotating body, **L**, therefore, changes like

$$\left(\frac{d\mathbf{L}}{dt}\right)_{\text{body}} = \mathbf{L} \times \boldsymbol{\omega} + \boldsymbol{\tau} \tag{11.5}$$

or in components

$$\tau_x = I_x \dot{\omega}_x - \omega_y \omega_z (I_y - I_z)$$
$$\tau_y = I_y \dot{\omega}_y - \omega_z \omega_x (I_z - I_x) \qquad (11.6)$$
$$\tau_z = I_z \dot{\omega}_z - \omega_x \omega_y (I_x - I_y).$$

Formulae (11.5) or (11.6) are known as *Euler's equations.*

11.1.2 Symmetric tops

The kinematics of free rotation, which means without a torque, are *not* trivial, quite unlike the translation analog, where $\mathbf{F} = 0$ implies boring uniform linear motion. For our purposes, it suffices to consider symmetric tops, i.e. bodies where the moments of inertia about two principal axes are equal, say $I_x = I_y$. The body is then symmetric about the z-axis, which is called the *figure axis.*

The solution of (11.6) is easy for $\tau = 0$ and $I_x = I_y$. Obviously, $\dot{\omega}_z$ is then zero. Taking the time derivative of the first equation of (11.6) to eliminate $\dot{\omega}_y$ gives $\ddot{\omega}_x \propto \omega_x$ and likewise for the y-component $\ddot{\omega}_y \propto \omega_y$, altogether

$$\omega_x = A \sin \Omega t \qquad \omega_y = A \cos \Omega t \qquad (11.7)$$

where A is a constant and

$$\Omega = \omega_z \frac{I_x - I_z}{I_x}. \qquad (11.8)$$

During the rotation, $\boldsymbol{\omega}$ and the figure axis describe circular cones around $\mathbf{L}$. The length of the angular velocity vector $\boldsymbol{\omega}$ and its component ω_z are constant but $\boldsymbol{\omega}$ changes its direction in space as well as in the grain. It precesses around the figure axis with angular velocity Ω. This implies time-variable stresses and, consequently, dissipation.

In the general torque-free case, where all momenta I_x, I_y, I_z are unequal, equation (11.3) can still be solved with pencil and paper but the rotation axis and the principal axes perform complicated wobbling motions around $\mathbf{L}$.

- Formula (11.8) applies to the wandering of the pole of the Earth. For our planet, $2\pi/\omega_z = 1$ day and $I_z/(I_z - I_x) = 305$. The moment of inertia I_z around the polar axis is greater than I_x around an axis that lies in the equatorial plane because the Earth has a bulge at the equator. From (11.8) one finds $\Omega \simeq 1$ yr, in qualitative agreement with the *Chandler period.*
- The axis of the Earth is inclined against the normal vector of the ecliptic by $\psi = 23.5°$. Its rotation is not really free because the Sun and the moon exert a torque τ on the equatorial bulge of the Earth which causes the polar axis to slowly precess. To estimate the effect, we use equation (11.3) as we want to derive the change of the rotation axis with respect to the stars. The precession velocity ω_{pr} follows immediately from (11.3), $\omega_{pr} = \tau/L$, where

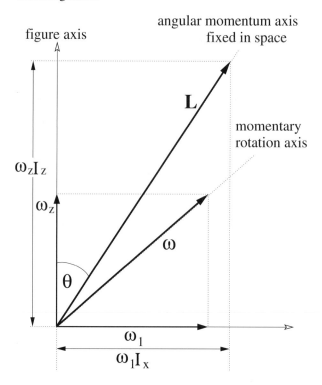

Figure 11.1. Rotation of a symmetric top with figure axis z and moments of inertia $I_x = I_y$. In the plane defined by the figure axis and the momentary rotation axis, the angular velocity vector ω can be decomposed into ω_z, which is constant, and into a perpendicular component ω_1 with moment of inertia $I_1 = I_x$. The angular velocity ω, the figure axis and the angular momentum vector **L** all lie in one plane; the latter two form an angle θ.

$L = I_z \omega_z$ is the angular momentum of the Earth. The torque is a tidal effect inversely proportional to the third power of the distance r. One easily finds

$$\omega_{\mathrm{pr}} = \frac{\tau}{L} = 2 \sin 2\psi \, \frac{G}{\omega_z} \frac{I_z - I_x}{I_z} \left[\left(\frac{M}{r^3} \right)_{\mathbb{C}} + \left(\frac{M}{r^3} \right)_{\odot} \right]$$

where G is the gravitational constant and M/r the ratio of mass over distance for the Sun or moon. When one corrects for the fact that the torque oscillates as the Sun and the moon apparently orbit the Earth, the factor $\sin 2\psi$ changes into $1.5 \cos \psi$. Inserting numbers yields for the period T of the *Platonic year*, $T = 2\pi/\omega_{\mathrm{pr}} \sim 26\,000$ yr; about two-thirds of the effect are due to the moon, the rest to the Sun.

Nowadays the polar axis is directed towards the Northern Star and gives guidance to nocturnal wanderers. In future millenia, the polar star will no longer point north and one will have to be more careful when walking away far from home at night.

11.1.3   Atomic magnet in a magnetic field

An atomic magnet in a field **B** is another example for a rotation which is not force-free. If one applies the classical formula (1.13) (in which the vector **J** is the current density) to an electron in circular motion of angular momentum **L**, one gets

$$\mathbf{m} = g\,\frac{e}{2m_e c}\,\mathbf{L} \tag{11.9}$$

with the Landé factor $g = 1$. This result is also true in quantum mechanics with respect to the orbital momentum **L** of the electron. However, in the case of pure spin momentum (**S**), $g = 2$. So the magnetic moment m associated with the spin of an electron ($S = \frac{1}{2}\hbar$) equals Bohr's magneton:

$$\mu_B = \frac{e\hbar}{2m_e c} = 9.27 \times 10^{-21} \text{ [cgs]}. \tag{11.10}$$

Quantum electrodynamics gives an upward correction; to first order in the fine structure constant α, it amounts to $0.116\% \simeq \alpha/2\pi$. When **S** and **L** couple to form the total angular momentum $\mathbf{J} = \mathbf{S} + \mathbf{L}$, the latter is usually not parallel to the net magnetic moment **m** because of the different Landé factors for **S** and **L**. We will mainly be concerned with pure spin angular momentum where $\mathbf{J} = \mathbf{S}$.

The field **B** exerts on the dipole **m** a torque

$$\boldsymbol{\tau} = \mathbf{m} \times \mathbf{B}. \tag{11.11}$$

Because $\boldsymbol{\tau}$ stands perpendicular to the angular momentum **J**, the torque changes only its direction, not its length J. Therefore **J** precesses around **B** with angular frequency

$$\omega_{\mathrm{pr}} = \frac{mB}{J}. \tag{11.12}$$

11.1.4   Rotational Brownian motion

The results of section 9.3 on translatory Brownian motion may, in a straightforward way, be carried over to rotation. When there is equipartition between dust particles and gas, the grains perform translatory and rotational Brownian motion. The average kinetic energy of a gas atom is then equal to the average rotational energy E_{rot} of a grain:

$$\tfrac{3}{2}kT_{\mathrm{gas}} = E_{\mathrm{rot}} = \tfrac{1}{2}I\omega^2$$

Table 11.1. Moment of inertia for spheres, cylinders and disks of mass M.

Sphere of radius r	$I_{\text{sphere}} = \frac{2}{5}Mr^2$
Thin cylinder of length l with rotation axis through the mass center and perpendicular to the symmetry axis	$I_{\text{cyl}\perp} = \frac{1}{12}Ml^2$
Cylinder of radius r rotating about its symmetry axis	$I_{\text{cyl}\parallel} = \frac{1}{2}Mr^2$
Disk of radius r with rotation axis through the mass center and perpendicular to the disk	$I_{\text{disk}\perp} = \frac{1}{2}Mr^2$
Flat disk with rotation axis through the mass center and parallel to the disk	$I_{\text{disk}\parallel} = \frac{1}{4}Mr^2$

and E_{rot} is, on average, equally distributed over the three degrees of rotational freedom:

$$\omega_x^2 I_x = \omega_y^2 I_y = \omega_z^2 I_z. \tag{11.13}$$

The angular momentum $\mathbf{L}$ tends to point along the axis of greatest moment of inertia. Indeed, we see from the condition of equipartition (11.13) that $I_x < I_y < I_z$ implies $L_x < L_y < L_z$. For example, in the case of a thin rod ($I_x = I_y \ll I_z$), almost all angular momentum will reside in rotation perpendicular to the symmetry axis z.

The rotational velocity of a dust particle in Brownian motion is

$$\omega_{\text{Brown}} = \sqrt{\frac{3kT_{\text{gas}}}{I}}. \tag{11.14}$$

Moments of inertia are given in table 11.1 for a few relevant configurations. Note that $I_{\text{disk}\parallel} = \frac{1}{2}I_{\text{disk}\perp}$, and that the ratio $I_{\text{cyl}\perp}/I_{\text{cyl}\parallel} = l^2/6r^2$ can take up any large value. If we adopt grains of moderate elongation, volume 10^{-15} cm^{-3} and bulk density 2.5 g cm^{-3}, we get, for their Brownian motion, from (11.14) in a gas of 20 K

$$\omega_{\text{Brown}} \sim 3 \times 10^5 \ \text{s}^{-1}.$$

For real interstellar grains, this may be a gross underestimate if they rotate suprathermally (see later). The disorder time t_{dis} of equation (9.39) stays practically the same with respect to angular momentum. The gas atoms that collide with the spinning grain retard its rotation by exerting a torque τ. In a likewise manner, one defines a friction coefficient ζ through

$$\tau = \zeta\omega$$

where τ is the torque needed to sustain a constant angular velocity ω and

$$\zeta = I t_{\text{dis}}^{-1}.$$

Again, in a stationary state, the heat loss rate $\omega\tau$ (see (11.4)) must be compensated by a spinup $\Delta L^2/2I$, where ΔL^2 is the average change in the square of the angular momentum per unit time due to collisions:

$$\omega\tau = \frac{I\omega^2}{t_{\text{dis}}} = \frac{\Delta L^2}{2I}. \tag{11.15}$$

In thermal equilibrium at temperature T, we get $I\omega^2 = 3kT$ and $\Delta L^2 = 6\zeta kT$.

We mention another randomizing process, besides grain collision, that is always at work: infrared radiation. Each photon carries off a quantum $\hbar$ in an arbitrary direction. After N emission processes, the angular momentum of the grain has changed in a random walk by an amount $\sqrt{N}\hbar$. If the grain emits per second N_{IR} infrared photons, the associated disorder time is

$$t_{\text{dis}} = \frac{8\pi kT\rho_{\text{gr}}a^5}{5\hbar^2 N_{\text{IR}}}.$$

The expression follows from $\sqrt{N_{\text{IR}}t_{\text{dis}}}\hbar = I\omega$ with $I\omega^2 = 3kT$. This process may dominate over collisional randomization for small grains. Usually the dust is heated by the UV field. In thermal balance, the number of UV photons absorbed per second, N_{UV}, is typically 50 times smaller than N_{IR}. If one can estimate the mean intensity of the radiation field, for example, from the dust temperature, N_{UV} follows from

$$N_{\text{UV}} = 4\pi\,\pi a^2 \int_{\text{UV}} \frac{J_\nu Q_\nu}{h\nu}\,d\nu.$$

Inside a molecular cloud, where only a few energetic photons fly around, the discrepancy between N_{UV} and N_{IR} is less severe.

11.1.5 Suprathermal rotation

An interstellar grain is probably not in energy equilibrium with the gas, as assumed earlier, and its rotational energy may be orders of magnitude greater than $\frac{3}{2}kT_{\text{gas}}$. It then rotates much faster than estimated in (11.14) from Brownian motion [Pur79].

To see why, imagine S-shaped grains of mass M_{gr} (figure 11.2) that are cooler than the gas,

$$T_{\text{dust}} < T_{\text{gas}}$$

and have a uniform sticking coefficient η over their surface. A fraction η of the gas atoms, where $0 < \eta < 1$, stays after the collision for a while on the grain before it leaves with a lower temperature (momentum) than it had before the impact. Because of the particular form of the grain, gas atoms approaching from the right and hitting the top of the S may collide twice and their chance for an inelastic

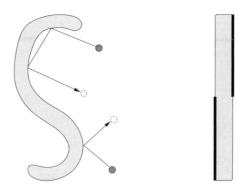

Figure 11.2. A grain in the form of the letter S, colder than the gas and with a constant sticking probability η, is accelerated clockwise (left). The same holds for a less fancy-shaped particle if η is not uniform but displays in its distribution rotational symmetry (right). The fat stretches on the bar indicate places of enhanced sticking probability.

collision is higher. As a result, they impart, on average, less momentum to the grain than those striking the bottom half of the S. Therefore, the S will rotate clockwise. Should the grain be hotter than the gas, it will rotate anti-clockwise and, without a temperature gradient, the mechanism does not work at all.

The process leads to a spinup. The dust particle rotates faster and faster and a limit is set only by friction when the lateral velocity of points on the grain surface becomes comparable to the velocity of gas atoms. Then each atom transfers approximately the angular momentum $m_a v_a a$ to a grain of radius a. If we use the formula (11.15) for the frictional loss and put $\Delta L^2 \simeq 2L\Delta L \sim 2I\omega \cdot n v_a \pi a^2 \cdot m_a v_a a$, we find indeed $\omega a \sim v_a$. Consequently, the grain's rotational energy E_{rot} is not $\frac{3}{2}kT_{gas}$ but

$$E_{rot} \simeq \frac{3}{2}kT_{gas}\frac{M_{gr}}{m_a}$$

i.e. $\simeq 10^9$ times greater. The particle is said to rotate *suprathermally*.

Suprathermal spinup does not really require the dust particles to be S-shaped. A sticking probability η that varies the way as shown on the right-hand of figure 11.2 will also do. Because of the immense factor 10^9, minor inevitable irregularities, like unevenly distributed cavities over the surface or small variations in η, are sufficient to boost E_{rot} by powers of ten.

There are two other possibilities for spinup that do not need a temperature difference between dust and gas.

- One is the formation of molecular hydrogen. If there are certain favorable (*active*) sites on the grain surface, where an H atom gets trapped until another comes by so that they can leave together as H_2, the ejection of the

H$_2$ molecule imparts angular momentum to the grain. The sites will not be exactly balanced over the surface and they endow the grain with a net rotation.

- Another mechanism is the ejection of electrons by UV photons. The kinetic energy of the ejected electron is typically ~ 1 eV and thus less than the energy $h\nu$ of the UV photon but the momentum of the electron is a hundred times greater than $h\nu/c$. Here again one gets suprathermal rotation if the yield for this process changes over the surface.

Even when the accommodation coefficient is perfectly uniform, random excitation, some clockwise and some anticlockwise, speeds up the rotation in the same way as a randomly kicked ball moves with each kick, on average, farther away from its starting position.

11.2   Magnetic dissipation

11.2.1   Diamagnetism

When a diamagnetic particle moves in a static inhomogeneous magnetic field in the direction of growing field strength, it experiences a repulsive force. According to the classical explanation, the magnetic flux through the electronic orbits of the atoms in the particle increases during the movement and thus induces an electric field. The field leads to an additional electronic current and to a magnetic moment which points, from the *Lenz law*, in the opposite direction to the outer magnetic field. The explanation applies to all atoms and, therefore, all matter is, in some way, diamagnetic; however, other effects may override it. The magnetic permeability μ, defined by the relation $\mathbf{B} = \mu\mathbf{H}$ of (1.16), is for a diamagnetic particle a little smaller than one. We will not come across diamagnetism any more.

11.2.2   Paramagnetism

Paramagnetism is relevant astrophysically and shows up when atoms have a *permanent* intrinsic magnetic moment as a consequence of unfilled inner electron shells. The electrons of an unfilled outer shell usually pair in binding and give no effect. A paramagnetic substance is pushed in the direction of increasing field strength, i.e. in the opposite way to a diamagnet and its μ is a little bit greater than one. The atoms with the strongest paramagnetism are the lanthanides.

The paramagnetism of interstellar dust is probably due to iron. This element has a high cosmic abundance and must be mostly solid because of its heavy depletion in the gas phase. It can easily be built into silicates, for example olivine, $Mg_{2-x}Fe_xSiO_4$, with x not far from one (see section 7.4.4). The amount of silicate grains is presumably limited by the cosmic abundance of Mg, which is some 20% higher than that of iron. If all iron in the interstellar medium is in

Table 11.2. Parameters of iron pertaining to the magnetic properties of interstellar grains, from [Kit96]. For the value of N_{pm} it is assumed that all iron is condensed into silicate particles.

Abundance:		
cosmic abundance		$[Fe]/[H] = 3.2 \times 10^{-5}$
depletion in gas phase		$>90\%$
number density in		$N_{pm} \simeq 6.2 \times 10^{21}$ cm^{-3}
paramagnetic silicates		
Bulk iron:		
crystal structure		cubic
γ-form	$1183K \le T \le 1674K$	face-centered (fcc)
α-form	$T \le 1183K$	body-centered (bcc)
Curie temperature		$T_c = 1043$ K
density		$\rho = 7.87$ g cm^{-3}
number density		$N_{bulk} = 8.50 \times 10^{22}$ cm^{-3}
lattice constant (bcc)		2.87 Å
nearest neighbor distance		2.48 Å
saturation magnetization		$M_{sat} = 1750$ G
effective magnetic moment		$m \simeq 2.22 \, \mu_B$ per atom
static susceptibility		$\chi_0 \simeq 12$
Electron configuration:		
Fe atoms	ground state $^5D_4$	$n = 4$ shell: 2s electrons
		$n = 3$ shell: 2s, 6p and 6d
Fe^{2+} ions	ground state $^5D_4$	eff. mag. moment: $m \simeq 5.4 \, \mu_B$
Fe^{3+} ions	ground state $^6S_{5/2}$	eff. mag. moment: $m \simeq 5.9 \, \mu_B$

silicate particles, Fe atoms constitute some 10% of all atoms in such grains and their number density there should then be $N_{pm} \simeq 6.2 \times 10^{21}$ cm^{-3}.

The electron configuration of iron, together with some of its other parameters, is summarized in table 11.2. The two inner shells of iron are complete. The third ($n = 3$, M-shell) could accommodate $2n^2 = 18$ electrons but contains only 14. There are two more electrons in the 4s subshell, totalling 26. The magnetic moments of the electrons in the third shell do not balance resulting in a net moment and thus paramagnetism. The ground state of atomic iron, as of Fe^{2+}, is $^5D_4$ (D stands for orbital angular momentum $\mathbf{L} = 2\hbar$, the multiplicity is 5 and the total angular momentum $\mathbf{J} = 4\hbar$). The ion Fe^{3+} has a ground term $^6S_{5/2}$ ($\mathbf{L} = 0$, multiplicity 6, $\mathbf{J} = \frac{5}{2}\hbar$).

An atom of magnetic dipole moment $\mathbf{m}$, oriented under an angle θ to a field $\mathbf{B}$, has a potential energy (see (1.133))

$$U = -\mathbf{m} \cdot \mathbf{B} = -mB \cos\theta. \qquad (11.16)$$

Suppose a grain has N such atoms per unit volume. Its magnetization $\mathbf{M}$ in the direction of **B** is then

$$M = NmL(x) \qquad \text{with } x = \frac{mB}{kT}. \tag{11.17}$$

The Langevin function $L(x)$ is defined in equation (1.137); for small arguments, $L(x) \simeq x/3$. Equation (11.17) corresponds to (1.136) and its derivation was given in section 1.6. In weak fields, as in interstellar space, $mB \ll kT$ and the susceptibility $\chi = M/B$ depends on temperature according to *Curie's law*,

$$\chi = \frac{M}{B} = \frac{C}{T} \qquad \text{with } C = \frac{Nm^2}{3k}. \tag{11.18}$$

So far we have neglected the quantization of space. When taken into account, only those directions for which the projection of the angular momentum along the direction of **B** differs by multiples of $\hbar$ are allowed. The magnetic susceptibility is then slightly modified. We mention two cases for a system with spin angular momentum S only. It has $2S + 1$ equally spaced energy levels.

- If $S = \frac{1}{2}$ and the field is arbitrary, the magnetic moment equals twice Bohr's magneton, $m = 2\mu_B$, and

$$M = Nm \tanh x \qquad \text{with } x = \frac{mB}{kT}. \tag{11.19}$$

- If the field is weak and S arbitrary, the constant C in (11.18) becomes

$$C = \frac{NS(S + 1)m^2}{3k}. \tag{11.20}$$

There are some minor changes if there is also orbital momentum L.

11.2.3 Ferromagnetism

Ferromagnetism appears in certain metals, like Fe, Co and Ni but also in alloys of these elements and some of their oxides. It only shows up below the Curie temperature T_C. Iron, the relevant element for us, has a Curie temperature $T_C = 1043$ K, well above any common dust temperature. Crystalline iron possesses a cubic grid. Below 1183 K, the grid is body-centered (α-form), above face-centered (γ-form) but this form is, of course, never ferromagnetic.

The magnetization M in a ferromagnet is very strong even when the applied magnetic field H_a is weak or absent. This is the result of a collective, non-magnetic effect of the electrons. A single iron atom is not ferromagnetic, only a cluster of atoms can show this property. Whereas in paramagnetism only the applied field acts on the electrons and tries to align them, in ferromagnetism there is an additional interaction among the electrons themselves *formally* described by an *exchange field* H_e. This is a fictitious field insofar as it does not appear in Maxwell's equations. It exerts on the electrons an apparent force which makes their spins parallel. Ferromagnetism can only be understood quantum mechanically. At its root is *Pauli*'s exclusion principle. One gets an idea of how the exchange field comes about from Heisenberg's theory for just two electrons: The relevant quantity is the exchange integral which occurs because the electrons are indistinguishable and which determines whether the spins are parallel or antiparallel.

The conception of an exchange field H_e is useful. Let us assume that it is proportional to the magnetization,

$$H_e = \lambda M.$$

The exchange field is much stronger than the applied field,

$$H_e \gg H_a$$

so the total effective field is

$$H_a + \lambda M \simeq \lambda M.$$

Any attempt to explain the huge magnetization M by the interaction of the magnetic dipoles associated with the electron spins fails by orders of magnitude. Indeed, at the Curie point ($\sim$1000 K), where ferromagnetism begins to fail, the kinetic energy of the atoms must be comparable to the potential energy of a Bohr magneton in the exchange field,

$$\mu_B H_e \simeq k T_c.$$

Therefore we estimate that H_e to be of order 10^7 G. However, the dipole field of a Bohr magneton at the distance $r \approx 2.5$ Å between neighboring iron atoms is $2\mu_B/r^3 \approx 10^3$ G, or ten thousand times weaker.

11.2.4 The magnetization of iron above and below the Curie point

Above the Curie point, iron is paramagnetic and the dependence of the magnetization M on the total field, $H_e + H_a$, follows Curie's law,

$$M = \frac{C}{T}(H_a + \lambda M).$$

As the magnetic susceptibility χ combines the applied field H_a with the magnetization,

$$M = \chi H_a$$

it follows that

$$\chi = \frac{C}{T - T_c} \qquad \text{with } T_c = \lambda C. \tag{11.21}$$

Here C is the same as in (11.20). We now have a relation between the Curie temperature T_c and the magnification factor λ. When the temperature approaches T_c from above, the susceptibility rises steeply. Equation (11.21), known as the *Weiss law*, is only approximate.

To evaluate the magnetization *below the Curie point*, we use (11.19), which expresses M as a function of mB/kT but replace B by $H_a + \lambda M$. We then have an equation for M. Solving it, one gets values of M which are different from zero even without an applied field ($H_a = 0$). This phenomenon is known as spontaneous magnetization. With $H_a = 0$, the spontaneous magnetization M as a function of T has its maximum at zero temperature given by the saturation value

$$M_{\text{sat}} = Nm$$

when all moments are perfectly aligned. With rising temperature, M falls, first gradually, then rapidly and goes to zero at $T = T_c$. For bulk iron, the effective magnetic moment per atom is $m \simeq 2.22 \ \mu_B$, the number density $N = 8.50 \times 10^{22} \ \text{cm}^{-3}$, so the saturation magnetization

$$M_{\text{sat}}(\text{Fe}) \simeq 1750 \ \text{G}. \tag{11.22}$$

This scenario of spatially constant magnetization applies to single domains and thus only to very small particles. In a mono-crystal of iron of only $0.1 \ \mu\text{m}$ size, the magnetization below the Curie point is not uniform over the crystal but divided into small domains. Within each domain, there is almost perfect alignment and the magnetization is close to its saturation value. But in an adjacent domain the magnetization has a different orientation so that the effective M over the whole crystal may be much lower than M_{sat}, although the susceptibility χ is still large. The crystal is partitioned into domains because its *total* energy, which includes the magnetic field outside the crystal, is then lowest; if the crystal were one big single domain, the total energy would be higher. In a ferromagnet consisting of several domains, χ has a complicated behavior depending on the applied field H_a and also on the *magnetic history*.

Akin to ferromagnetism is *ferrimagnetism*, except that the saturation magnetization is much lower. To explain it, one proposes that neighboring electron spins are not parallel, as in a ferromagnet but anti-parallel without fully compensating each other. Some iron oxides that might be found in interstellar grains, like magnetite (Fe_3O_4) or maghemite (γ–F_2O_3), are ferrimagnetic.

11.2.5 Paramagnetic dissipation: spin–spin and spin–lattice relaxation

After these preliminary remarks about magnetism, we discuss and quantify processes of magnetic damping. The one most frequently evoked for the alignment of interstellar grains is paramagnetic relaxation [Cas38, Dav51, Jon67, Spi79].

There are two thermodynamic systems in a grain:

(1) The spin system. The ensemble of magnetic moments arises from the electrons of the paramagnetic atoms. The magnetic moments may be due to electron spin as well as orbital angular momentum. In a constant magnetic field of strength H, there is some net orientation of the dipoles. Equilibrium is characterized by a Boltzmann distribution e^{-U/kT_S} at temperature T_S, where after (11.16) $U = -mH \cos\theta$ is the potential energy of a dipole of moment m aligned with the field at an angle θ.

(2) The lattice system. It consists of the vibrating atoms or ions in the grid of the solid body. It is at temperature T_L, has much more energy than the spin system, serves as a heat reservoir and is always relaxed.

In equilibrium, the temperatures of the two systems are equal, $T_S = T_L$. When the field is suddenly increased from H to $H' > H$, while keeping its direction, the spin system gets disturbed. Associated with the two systems are two damping constants:

- The spin system comes to equilibrium very quickly at a new temperature T_S' after the spin–spin (ss) relaxation time $\tau_{ss} \sim 10^{-10}$ s. This process is adiabatic; there is no heat exchange with the lattice, the level populations of the dipoles stay the same so that $mH/kT_S = mH'/kT_S'$.
- Equilibrium with the lattice system is established much more slowly after the spin–lattice (sl) relaxation time $\tau_{sl} \sim 10^{-6}$ s. The coupling between spin and lattice is mediated via phonons; they correspond to changes in the quantum numbers of the lattice vibrations.

The above numbers for τ_{ss} and τ_{sl} are very crude but certainly $\tau_{ss} \ll \tau_{sl}$. While τ_{sl} varies inversely with temperature, τ_{ss} is more or less independent of it. For $T_S \neq T_L$, heat flows from the lattice to the spin system at a rate

$$-\alpha(T_S - T_L) \qquad (\alpha > 0).$$

The heat flow is proportional to the temperature difference $dT = T_S - T_L$. It stops when the two temperatures are equal. If the spin system was initially hotter, it will lose energy. Once equilibrium has been established ($T_S = T_L$), the dipoles are better aligned in the new stronger field H' and their potential energy U is, therefore, lower than before the disturbance. The excess energy has been transferred to the lattice.

11.2.6 The magnetic susceptibility for spin–lattice relaxation

We next calculate the susceptibility for spin–lattice relaxation using the thermodynamic relations for magnetic materials of section 5.3.8. Suppose a bias field H_0 has produced a constant magnetization M_0 and superimposed onto H_0 there is a small periodic perturbation so that the total field and the total magnetization are:

$$H(t) = H_0 + he^{-i\omega t} \qquad M(t) = M_0 + h\chi e^{-i\omega t}. \qquad (11.23)$$

The perturbations $dH = he^{-i\omega t}$ and $dM = h\chi e^{-i\omega t}$ imply a heat input δQ to the spin system given by (5.66). When the time derivative of the heat input,

$$\frac{d\delta Q}{dt} = -i\omega C_H \left(\frac{\partial T}{\partial M}\right)_H dM - i\omega C_M \left(\frac{\partial T}{\partial H}\right)_M dH$$

is equated to the heat flow $-\alpha\, dT$, where

$$dT = T_S - T_L = \left(\frac{\partial T}{\partial H}\right)_M dH + \left(\frac{\partial T}{\partial M}\right)_H dM$$

one obtains the complex susceptibility

$$\chi(\omega) = \frac{dM}{dH} = \frac{-\alpha + i\omega C_M}{\alpha - i\omega C_H}\left(\frac{\partial T}{\partial H}\right)_M \cdot \left(\frac{\partial M}{\partial T}\right)_H .$$

C_H and C_M are the specific heat at constant field and constant magnetization, χ_{ad} and χ_T denote the adiabatic and isothermal value of the susceptibility. The product of the two brackets in the expression for $\chi(\omega)$ yields $-\chi_T$ according to (5.68) and (5.69). Therefore the susceptibility for spin–lattice relaxation is:

$$\chi_1(\omega) = \chi_{ad} + \frac{\chi_T - \chi_{ad}}{1 + \omega^2\tau_{sl}^2} \qquad \chi_2(\omega) = (\chi_T - \chi_{ad})\frac{\omega\tau_{sl}}{1 + \omega^2\tau_{sl}^2}. \qquad (11.24)$$

The imaginary part χ_2 stands for the dissipational loss. It is positive, as it should be, because

$$\chi_T = \chi(0) > \chi_{ad} = \chi(\infty).$$

At low frequencies, χ_2 is proportional to ω. A look at the permeability of (1.146) confirms that we are dealing with some kind of Debye polarization. After a sudden disturbance, the return of the spin system to equilibrium with the lattice proceeds exponentially on a characteristic time τ_{sl} which is related to the constant α and the specific heat C_H through

$$\tau_{sl} = \frac{C_H}{\alpha}.$$

The power W which has to be expended for the magnetization of a body in a variable field of the kind given by (11.23) follows from equation (5.63),

$$W = \tfrac{1}{2}\omega V \chi_2 h^2.$$

Here V is the grain volume. The formula is correct whether a bias field H_0 is present or not. It also agrees with the energy dissipation rate (1.55) which we derived before, because in the case of a small susceptibility, χ approaches the polarizability α.

However, the very process of dissipation by spin–lattice relaxation does not work for purely paramagnetic interstellar grains. The reason being that the bias field H_0 is absent. The spinning particles only feel the time-variable part, $dH = he^{-i\omega t}$, where h is now the interstellar magnetic field. As the grains rotate, the magnetic dipoles have to adjust their precession to a field that changes its direction but not its strength. So their potential energy U in the field and the level populations remain the same. Consequently, the spin temperature T_S does not oscillate, $dT = 0$, and there is no interaction with the lattice.

Nevertheless, our troubles are not in vain. In section 11.3 we will revive spin–lattice relaxation by proposing an internal bias field in the grain material due to ferromagnetic inclusions.

11.2.7 The magnetic susceptibility in spin–spin relaxation

The rotation of *paramagnetic* dust particles may be damped via spin–spin relaxation. The dissipation comes about because the alignment of the dipoles always lags behind the new field direction (see figure 11.5). The underlying microscopic processes are very complex and their treatment is beyond the scope of this text. We merely mention that there is:

- a magnetic dipole–dipole interaction (the paramagnetic atoms are the source of a magnetic field in which other magnetic dipoles precess);
- Stark splitting of the magnetic sublevels (arising from the electric field of the crystal);
- a quantum-mechanical exchange interaction (which tends to make spins parallel as in ferromagnetism); and
- a hyperfine splitting interaction (accounting for the influence of the magnetic moments of the nuclei of the paramagnetic atoms).

We wish to estimate the values of $\chi(\omega)$ for magnetic grain material at the frequencies at which interstellar grains may rotate, i.e. in the range from $\omega = 10^5$ s^{-1} for Brownian motion up to 10^9 s^{-1} in the case of suprathermal spinup. It is, of course, also of interest to know $\chi(\omega)$ at *all* frequencies and with respect to *all* dissipation processes so that one can check whether dust absorption or emission due to para- or ferro-magnetism can ever become important (see second term in (3.5)).

The magnetic susceptibility $\chi(\omega)$ for spin–spin relaxation must have the functional dependence of the Debye permeability after equation (1.147) because the magnetization of the grain is due to orientational alignment of the dipoles in a time-variable field, as discussed in section 1.6. The imaginary part $\chi_2(\omega)$, which is responsible for dissipation, has, therefore, the form

$$\chi_2(\omega) = \chi_0 \frac{\omega\tau}{1 + \omega^2\tau^2}. \tag{11.25}$$

To extract numbers, one must know χ_0 and τ. The former is inversely proportional to temperature and follows from Curie's law (11.18):

$$\chi_0 = \frac{N_{pm}m^2}{3kT}. \tag{11.26}$$

When the volume density of parmagnetic atoms, N_{pm}, and the effective magnetic dipole moment of an iron atom are taken from table 11.2, the static susceptibility becomes $\chi_0 = 0.045/T$. Because spin–spin relaxation is so fast, χ_0 is equal to χ_{ad} in equation (11.24) (see also (5.69)).

The electron spins precess around the applied field at frequency ω_{pr}. To get a handle on the spin–spin relaxation time τ, one sets it equal to ω_{pr}^{-1} because this is the characteristic time for a change in the magnetization perpendicular to the applied field. Note that the electrons do not precess in the weak interstellar magnetic field but in the much stronger local field B^{loc} created by all paramagnetic atoms (see section 3.4). This field is, of course, violently fluctuating because the spin directions are tossed about in collisions with phonons. In contrast to the perfect cubic lattice discussed in section 3.4, the paramagnetic atoms are dispersed irregularly over the grain. Their *average* contributions to B^{loc} cancel but the root mean square deviation does not vanish. As fields from dipoles fall off quickly and proportionally to r^{-3}, only the nearest matter. To first order, we assume the field to come from a dipole at a distance $r = N_{pm}^{-1/3}$, which is the average separation between Fe atoms (see table 11.2). When its moment is directed towards the precessing electron, it produces a field $B^{loc} \simeq 2mN_{pm} \simeq 700$ G.

As the ground state of Fe^{3+} is classified as $^6S_{5/2}$, we assume that the outermost five electrons in the 3d shell have only spin momentum and put $g = 2$ in (11.9) and (11.12) so that

$$\tau \simeq \omega_{pr}^{-1} \simeq \frac{\hbar}{2\mu_B B^{loc}} \simeq 8.4 \times 10^{-11} \text{ s.}$$

It is remarkable that in the product $\chi_0\tau$ the density of paramagnetic atoms in the grain, N_{pm}, cancels out and, therefore, does not appear in the expression for $\chi_2(\omega)$. One thus finds $\chi_0\tau = 0.49\hbar/kT$ and

$$\chi_2(\omega) = \chi_0 \frac{\omega\tau}{1 + \omega^2\tau^2} \simeq 4 \times 10^{-12}\frac{\omega}{T}. \tag{11.27}$$

The approximation on the right holds at low (rotational) frequencies. The full expression should be used for magnetic emission at wavelengths $\lambda = 1$ cm or shorter.

Interestingly, neither does the Bohr magneton show up in the product $\chi_0\tau$. So the mechanism is independent of the strength of the magnetic moments and, therefore, includes, besides electronic paramagnetism which we have had in mind so far, nuclear paramagnetism. In particular, hydrogen nuclei, which are building blocks of the otherwise non-magnetic H_2O molecules, have unsaturated nuclear moments. They are a grain constituent when the grain has an ice mantle.

Another form of the susceptibility for spin–spin relaxation uses the fact that the frequency distribution function defined by

$$f(\omega) = \frac{\chi_2(\omega)}{\omega}$$

is mathematically fully determined by all its moments:

$$\langle \omega^n \rangle = \int f(\omega)\omega^n d\omega \cdot \left[\int f(\omega)d\omega \right]^{-1}.$$

For plausible reasons, $f(\omega)$ is related through Fourier transformations to the relaxation function $\varphi(t)$, which describes how the magnetic moment M decays when a constant magnetic field H is suddenly switched off. We encountered an example of $\varphi(t)$ in equation (1.143) of section 1.6, where the decay was exponential. When, in line with experiments, one approximates the relaxation function by a Gaussian, $\varphi(t) = \chi_0 e^{-\frac{1}{2}\langle \omega^2 \rangle t^2}$, and keeps only the second moment $\langle \omega^2 \rangle$, one obtains

$$\chi_2(\omega) = \chi_0 \omega \tau \left(\frac{\pi}{2} \right)^{\frac{1}{2}} e^{-\omega^2 \tau^2 / 2} \tag{11.28}$$

with $\tau = \langle \omega^2 \rangle^{-\frac{1}{2}}$. But for a factor of order unity, $\chi_2(\omega)$ resembles Debye relaxation at frequencies $\omega \ll \tau^{-1}$.

11.3   Magnetic alignment

The polarization surveys of the Milky Way, which comprise several thousand stars both in the southern and northern hemisphere, revealed that the polarization vector from interstellar extinction runs predominantly parallel to the galactic plane. This is exemplified, for example, in the figures of [Mat70] for stars with a large distance modulus. One should also note in these figures, however, the large local excursions. As the magnetic field, derived from the polarization of radio synchrotron emission, has the same direction (projected on the sky) as the electric field vector of polarized starlight, it seemed natural to propose that magnetic fields are somehow responsible for the grain alignment.

When one compares the magnetic field lines with the optical polarization vectors in other spiral galaxies, the coincidence is less impressive. This is demonstrated in figures 11.3 and 11.4 for M 51, a galaxy which has beautiful spiral arms and is viewed face-on. The magnetic field follows the arms nicely but the optical polarization vectors, especially in the lower right of the figure, match the magnetic field vectors only poorly. The reason that their orientation is not as uniform as in the Milky Way is probably due to the fact that one also sees scattered light in external galaxies. The polarization resulting from scattering by dust is usually strong, especially when it occurs at right angles, and is completely unrelated to magnetic fields. In extinction measurements of the Milky Way, however, scattered light is absent and cannot contaminate the polarization.

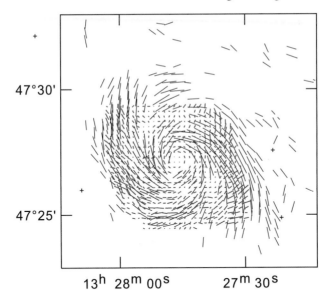

Figure 11.3. Radio and optical polarization in M51. The magnetic *B*-vectors at 2.8 cm [Nei90] are overlaid on the electric polarizarion vectors at the center wavelengths of 0.85 μm (adapted from [Sca87]). The radio polarization is fairly high ($\sim$20%) and the corresponding vectors are shown by the large dashes; they follow the spiral arm pattern. The electric polarization is small ($\sim$1%) and represented by short (unfortunately, also black) dashes.

11.3.1   A rotating dipole in a magnetic field

A particle of volume V acquires, from (1.18), in the constant interstellar magnetic field **B** a magnetization **M** and a dipole moment

$$\mathbf{m} = V\mathbf{M}.$$

The field exerts, on the dipole, a torque (see (11.11))

$$\boldsymbol{\tau} = \mathbf{m} \times \mathbf{B}$$

directed perpendicularly to **m** and **B**. If the grain spins, its magnetization **M** changes. For simplicity, we consider, as in figure 11.5, a cylinder with its angular velocity and momentum vectors, $\boldsymbol{\omega}$ and **L**, pointing the same way but perpendicular to the cylinder axis. The torque tries to line up the dipole with the field. To determine $\boldsymbol{\tau}$, we need the magnetization **M** and, because $\mathbf{M} = \chi\mathbf{B}$, the susceptibility χ taken at the rotation frequency ω, which is assumed to be small.

If χ were real, the magnetization would adjust instantaneously to the direction of **B** and the torque would vanish because of the term $\mathbf{B} \times \mathbf{B} = 0$ (see (11.11)). But χ is complex and the magnetization **M** stays a small angle χ_2/χ_1

Figure 11.4. An optical picture of the spiral galaxy M51 from the Palomar Sky Survey shown in the same orientation and on approximately the same scale. The companion on top is NGC 5195. Use of this image is courtesy of the Palomar Observatory and the Digitized Sky Survey created by the Space Telescope Science Institute, operated by Aura Inc. for NASA, and is reproduced here with permission from AURA/ST Scl.

behind the field **B** (see (1.12)). Therefore, as viewed by an observer in the rotating grain, the magnetization responds as if there were two magnetic field components: a major one, $\mathbf{B}_1 \simeq \mathbf{B}$, producing a magnetization $\mathbf{M}_1 = \chi_1 \mathbf{B}$ and a small one, $\mathbf{B}_2$, perpendicular to the first and to $\boldsymbol{\omega}$ and of length $B_2 = (\chi_2/\chi_1)B \sin\theta$. This produces the magnetization

$$\mathbf{M}_2 = \frac{\chi_2}{\omega}\boldsymbol{\omega} \times B$$

which does not disappear in the vector product with **B**. This leads to the following equations of motion for the rotating magnetic dipole:

$$\frac{d\mathbf{L}}{dt} = \boldsymbol{\tau} = V\mathbf{M} \times \mathbf{B} = V\frac{\chi_2}{\omega}(\boldsymbol{\omega} \times \mathbf{B}) \times \mathbf{B} \qquad (11.29)$$

$$\frac{dE_{\text{rot}}}{dt} = \boldsymbol{\tau} \cdot \omega = -V\frac{\chi_2}{\omega}(\boldsymbol{\omega} \times \mathbf{B})^2. \qquad (11.30)$$

The rotational energy decreases, it is wasted into heat. Because of the double vector product in (11.29), the torque $\boldsymbol{\tau}$ and the change in angular momentum $d\mathbf{L}$ are perpendicular to **B** and lie in the same plane as **B** and $\boldsymbol{\omega}$ (see figure 11.5).

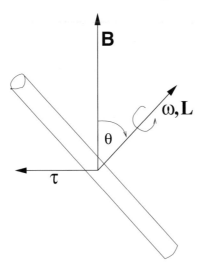

Figure 11.5. A paramagnetic cylinder rotating in a magnetic field **B** acquires a dipole moment that is not fully aligned with **B**. This leads to a torque τ on the grain. All vectors τ, ω, **B**, **L** are in the same plane, ω and **L** lie parallel to each other and perpendicular to the axis of the spinning cylinder.

As a result, the cylinder lines up with its long axis perpendicular to the field **B**. When the alignment is complete ($\omega \parallel$ **B**), there will be no more losses. This is the *Davis–Greenstein* mechanism, the standard scenario for magnetic alignment.

11.3.2 Timescales for alignment and disorder

We estimate a relaxation time t_{rel} for magnetic alignment from the condition $L/t_{\mathrm{rel}} = dL/dt$ from (11.29):

$$t_{\mathrm{rel}} = \frac{I\omega}{V\chi_2 B^2}. \tag{11.31}$$

A typical number is 10^6 yr. The inverse t_{rel}^{-1} is, by order of magnitude, the rate at which the angle θ between the field and the angular momentum vector changes. We see that magnetic alignment is favoured by a low ratio of moment of inertia over volume, I/V, and a strong magnetic field. But there are limits to these numbers. A minimal value of I/V is given by a sphere and, from diverse independent evidence, B does not exceed several times 10^{-6} G in low density clouds ($n_{\mathrm{H}} < 10^3$ cm^{-3}). Note that a higher rotation speed, although it implies enhanced dissipation, does not decrease the relaxation time t_{rel} because $\chi_2 \propto \omega$ at low frequencies.

When we compare t_{rel} with the time t_{dis} after which disorder is established through collisions with gas atoms and assume Brownian motion so that t_{dis} is given by (9.39), we find

$$\frac{t_{rel}}{t_{dis}} = \frac{3}{10}\left(\frac{8km_a}{\pi}\right)^{\frac{1}{2}}\frac{T_{gas}^{\frac{1}{2}}an}{B^2}\frac{\omega}{\chi_2} \simeq 7.3 \times 10^{-21}\frac{T_{gas}^{\frac{1}{2}}an}{B^2}\frac{\omega}{\chi_2}. \qquad (11.32)$$

m_a and n are the mass and density of the gas atoms. For alignment to be effective, t_{rel}/t_{dis} must be smaller than one. The rotation speed ω does not appear in this formula. Pure paramagnetic spin–spin relaxation with χ_2 from (11.27) yields

$$\frac{t_{rel}}{t_{dis}} \simeq 2 \times 10^{-9}an\frac{T_{dust}T_{gas}^{\frac{1}{2}}}{B^2}.$$

Because the Serkowski curve mandates a typical grain size a of order 10^{-5} cm, one can readily check that in the environment where we actually observe optical polarization, t_{rel}/t_{dis} would be substantially above unity. This implies that the grains change their angular momentum chaotically *before* paramagnetic relaxation has had time to do its job and they would not be aligned. Therefore one has to search for a more effective magnetic alignment mechanisms. Suprathermal rotation, supra–paramagnetism and ferromagnetic inclusions can bring the ratio of relaxation time over collisional disorder time, t_{rel}/t_{dis}, down and thus point a way out of the dilemma [Jon67].

11.3.2.1   *Magnetic alignment in suprathermal rotation*

Should the grains be rotating suprathermally with angular velocity $\omega \gg 10^5$ s^{-1}, as discussed in section 11.1, they have a very high kinetic energy and the time to create disorder, t_{dis}, will be much increased and no longer be given by (9.39). As the relaxation time t_{rel} of (11.31) is not altered by the spinup as long as $\chi_2 \propto \omega$, effective grain alignment according to the Davis–Greenstein mechanism with damping from paramagnetic spin–spin relaxation seems possible. Of course, the angular momentum of the fast rotating particle must stay stable in space over a time greater than t_{rel}, otherwise orientation is poor. Because inhomogeneities on the grain surface drive the suprathermal spinup, disorientation will set in when the surface properties change. This may occur, for example, when new layers are deposited or active sites are destroyed or created. Time scales for these processes are debatable.

11.3.3   Super-paramagnetism

Interstellar dust may contain not only isolated iron atoms, which would only produce paramagnetism of possibly insufficient strength but also tiny pure iron subparticles in which the collective effect of the electron spins dominate. The subparticles are possibly synthesized in supernova explosions and somehow built into interstellar grains. Even if they contain only a few tens of iron atoms, they already behave like ferromagnetic single domains, i.e. they show spontaneous magnetization. A grain with such ferromagnetic inclusions may then display super-paramagnetism. The phenomenon is halfway between ferro- and para-magnetism in the sense that the clusters are ferromagnetic single domains but their magnetization changes spontaneously by thermal perturbations so that the grain, as a whole, behaves like a paramagnet.

Suppose the grain has N_{cl} iron clusters, each consisting of n atoms of magnetic moment m_a. The magnetic moment of each cluster is at its saturation value $m_{cl} = nm_a$ but the whole particle is paramagnetic because its magnetization follows equation (11.17) with the Langevin function $L(x)$. However, in the expression for the static susceptibility (11.26) the moment m_a of a single Fe atom has to be replaced by the cluster moment m_{cl} and the volume density of magnetic atoms N_{pm} by the number of clusters N_{cl} in the grain. As obviously $N_{pm} = N_{cl}n$, the static susceptibility of the grain as a whole becomes

$$\chi_0 = n \left(\frac{N_{pm}m_a^2}{3kT} \right).$$

Therefore χ_0 increases by a factor n over the static susceptibility with isolated iron atoms. When a cluster changes its magnetization, it must overcome a potential barrier U_{bar}, because not all directions of magnetization are equal. There is an easy one and a hard one (for iron, these are the cube edges [100] and the body diagonals [111], respectively) separated by U_{bar}. The more atoms there are in the cluster, the higher the barrier, $U_{bar}(= nk\theta)$ becomes. It may be overcome by thermal excitation and the rate at which this happens is proportional to $e^{-U_{bar}/kT}$. Therefore the relaxation time has the form

$$\tau \simeq \gamma e^{U_{bar}/kT}$$

with $\theta = 0.11$ K and $\gamma = 10^{-9}$ s determined experimentally. It is very sensitive to the number n of iron atoms per cluster. As n increases, $\chi_2(\omega)$ after (11.25) can become very large for small frequencies.

11.3.4 Ferromagnetic relaxation

In a more general way, not restricted to paramagnetic relaxation, one assumes that the magnetization M is forced by a periodic outer field H to oscillate harmonically,

$$\ddot{M} + \gamma \dot{M} + \omega_0^2 M = \omega_0^2 \chi_0 H. \tag{11.33}$$

In a static field, $M = \chi_0 H$. Equation (11.33) implies with $\tau \equiv \gamma/\omega_0^2$ (see (1.77))

$$\chi_2(\omega) = \chi_0 \frac{\omega\tau}{\left[1 - (\omega/\omega_0)^2\right]^2 + \omega^2\tau^2}. \tag{11.34}$$

For small frequencies ($\omega \ll \omega_0$), the expression again approaches Debye relaxation (11.25). If τ is small, corresponding to weak damping, a pronounced resonance appears at ω_0. Equation (11.34) says nothing about the dissipation mechanism. The difficulty is to associate χ_0, τ and ω_0 with physical processes and to determine their values from a combination of experimental data and theory. If the system is critically damped, $\gamma = 2\omega_0$ after (1.63). Then there are only two parameters, χ_0 and τ, and

$$\chi_2(\omega) = \chi_0 \frac{\omega\tau}{[1 + (\omega\tau/2)^2]^2}. \tag{11.35}$$

For a grain with ferromagnetic inclusions, two scenarios exist:

- If the iron clusters are big enough to be divided into several domains, the changing magnetic field causes the transition layer between adjacent domains (*Bloch* walls) to rearrange and the wandering of the walls dissipates energy. The moving walls can phenomenologically be described by a harmonic oscillator, as in (11.33). With the static susceptibility $\chi_0 = 12$ from table 11.2 and an estimate for τ of order 10^{-8} s from the laboratory, the low-frequency limit of $\chi_2(\omega)/\omega$ becomes $\sim 10^{-7}$ which is four orders of magnitude greater than for spin–spin relaxation. The effect on the damping of the grain as a whole depends, of course, on the volume fraction f_{Fe} of the iron inclusions but even $f_{\text{Fe}} = 0.1\%$ would increase the susceptibility tenfold.

- If the iron clusters have sizes below ~ 200 Å, they are stable as ferromagnetic single domains. One can squeeze out some semi-theoretical number for χ_2 supposing, as we did for paramagnets, that dissipation comes from precession of the magnetic moments in the local field. The latter is equal to $4\pi M_{\text{sat}}/3$ (see the electric analog in (3.54)). Putting the saturation magnetization $M_{\text{sat}} = 1750\,\text{G}$ and $\chi_0 = 12$ (table 11.2), the precession frequency $\omega_{\text{pr}} \simeq \tau^{-1} \simeq (4\pi M_{\text{sat}}/3)(e/m_e c) \simeq 1.3 \times 10^{11}\,\text{s}^{-1}$. As now $\chi_2(\omega)/\omega \sim 10^{-10}$, particles with single-domain iron inclusions seem not to be terribly effective in damping.

However, the presence of single-domain ferromagnetic inclusions has the interesting consequence that it provides a bias field H_0 for the adjacent paramagnetic material [Dul78]. This enables paramagnetic dissipation by spin–lattice relaxation to operate in the otherwise paramagnetic grain material. This mechanism, which we had so far discarded, is much more efficacious in enhancing the ratio $\chi_2(\omega)/\omega$ than spin–spin relaxation, especially at low frequencies as evidenced by figure 11.6. The field produced by the inclusions must, of course, be as strong as the fluctuating field from the isolated iron atoms in the paramagnetic grain material, for which we estimated above 700 G, in order for the resulting field to be considered at a constant bias. A spherical single domain of radius a has, at a distance r from its center, a field (see table 11.2)

$$\frac{4\pi a^3}{3r^3} N_{\text{bulk}}\, 2.22\mu_B \simeq 7000 \left(\frac{a}{r}\right)^3.$$

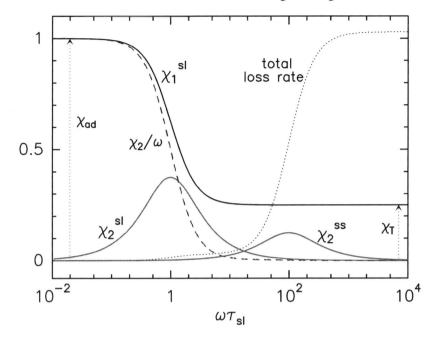

Figure 11.6. A sketch of the magnetic susceptibility, $\chi = \chi_1 + i\chi_2$, as a function of frequency in paramagnetic dissipation for spin–spin (ss) and for spin–lattice (sl) relaxation. For the relaxation times it is assumed that $\tau_{sl} = 100\tau_{sl}$; we further adopt $\chi_{ad} = \chi_2^{ss}(0) = 0.25\chi_T$. The dots show the total dissipational loss rate $W \propto \chi_2\omega$ when *both* processes are activated (here $\chi_2 = \chi_2^{sl} + \chi_2^{ss}$); the broken line depicts for this case χ_2/ω which determines the alignment efficiency after (11.31). The ordinate is arbitrary.

Therefore, each ferromagnetic inclusion provides a bias field for spin–lattice relaxation in the paramagnetic material over a volume which is about ten times bigger than its own.

11.3.5 Alignment of angular momentum with the axis of greatest inertia

There is likely to be damping within the particle independent of the outer magnetic field. The rotational energy E_{rot} of the grain then decreases and is turned into heat, while the angular momentum vector **L** stays constant in the absence of an external torque. Assuming, for example, $I_x < I_y < I_z$, it follows readily from (11.1) and (11.2) that the angular velocity ω_z grows at the expense of ω_x and ω_y. Consequently, the axis of maximum inertia (the z-axis) tends to become parallel to **L**. When such grains are then magnetically aligned, their orientation will be almost perfect and they will, therefore, be most effective polarizers.

With the help of (11.7) and using the angle θ between the figure axis and **L** (see figure 11.1), the rotational energy E_{rot} of (11.2) can be written as

$$2E_{rot} = \frac{L^2}{I_x} + L^2 A \cos^2 \theta \qquad A = \frac{I_x - I_z}{I_x I_z}.$$

If one knows the dissipation rate $W = dE_{rot}/dt$, one can work out the time $t_\parallel$ which it takes until the z-axis and the angular momentum $\mathbf{L}$ are parallel:

$$t_\parallel \sim \frac{t_\parallel}{\theta} \sim \frac{dt}{d\theta} = W^{-1} \frac{dE_{rot}}{d\theta} = -W^{-1} L^2 A \sin\theta \cos\theta \sim -W^{-1} L^2 A. \quad (11.36)$$

For the alignment to work, it is necessary that $t_\parallel$ be shorter than the disorder time t_{dis} (see (9.39) for its value for Brownian motion).

11.3.6 Mechanical and magnetic damping

We discuss two further dissipation mechanisms: one mechanical, the other gyromagnetic.

11.3.6.1 Mechanical damping via variable stresses

Interstellar dust particles are not perfectly rigid bodies and the centrifugal forces in a spinning grain change the distances among the atoms. When rotation is not about a principal axis, the rotation vector $\boldsymbol{\omega}$ moves with respect to the grain coordinates and so the stresses arising from the centrifugal forces are time variable with frequency of order ω (see (11.8)).

A perfect and cold crystal is very elastic with hardly any loss but a real grain has imperfections and mechanical deformations inevitably drain some of the rotational energy into heat. The previously introduced time $t_\parallel$ depends on the stiffness of matter, the mechanical damping constant of the grain material and, crucially, on the rotational velocity. It turns out that for suprathermally spinning inelastic grains, $t_\parallel$ is much shorter than t_{dis}, so they will always rotate about their axis of greatest moment of inertia. Interstellar grains in Brownian motion, however, will probably not show such alignment.

11.3.6.2 Magnetic damping via the Barnett effect

More efficient is a gyromagnetic mechanism which is based on the coupling between the spin system and the *macroscopic* rotation of the body. To see how it works, we first recall the famous experiments by Einstein and de Haas and its counterpart by Barnett.

- When an unmagnetized iron cylinder with inertia I is suddenly magnetized by an outer field, the atomic magnetic moments and their associated angular momenta line up. Suppose that in the process of magnetization, N electrons flip their spin direction from up to down. As the *total* angular momentum of the cylinder is zero before and after the magnetization, the cylinder acquires an angular momentum $L = N\hbar$ and rotates afterwards at a speed $\omega = N\hbar/I$ (Einstein–de Haas effect).

- When an unmagnetized cylinder is rotated at angular velocity $\boldsymbol{\omega}$, it feels a magnetic field $\mathbf{B}$, which is parallel to $\boldsymbol{\omega}$ and has the absolute value

$$B = \omega \frac{2m_e c}{e}. \qquad (11.37)$$

The field $\mathbf{B}$ then produces, in the cylinder, a magnetization $\mathbf{M} = \chi \mathbf{B}$ (Barnett effect).

To understand the linear relation (11.37) between ω and B, we mention, first, that in a rotating frame K' the acceleration $\ddot{\mathbf{x}}'$ of a point mass is different from the acceleration $\ddot{\mathbf{x}}$ as seen in an inertial frame K:

$$\ddot{\mathbf{x}}' = \ddot{\mathbf{x}} + 2\boldsymbol{\omega} \times \dot{\mathbf{x}}' + \boldsymbol{\omega} \times (\boldsymbol{\omega} \times \mathbf{x}). \qquad (11.38)$$

The prime refers to the uniformly rotating system in which there additionally appear a Coriolis and a centrifugal term (second and third on the right). When ω is small, the centrifugal force may be neglected and one may put $\mathbf{v} = \dot{\mathbf{x}} \simeq \dot{\mathbf{x}}'$. Therefore, when we take an atom and apply a magnetic field and furthermore slowly rotate the atom, each of its electrons is subjected to the additional force:

$$2m_e \mathbf{v} \times \boldsymbol{\omega} + (e/c)\mathbf{v} \times \mathbf{B}.$$

The first term is due to rotation and the other represents the Lorentz force of the magnetic field. The sum of the two is zero when $\boldsymbol{\omega} + (e/2m_e c)\mathbf{B} = 0$. Rotating the atom is thus equivalent to exposing it to a magnetic field of strength $B = (2m_e c/e)\omega$; this is Larmor's theorem.

When we apply this result to an interstellar grain of volume V and compute the dissipation rate $W = \chi_2 \omega B^2$ after (11.30) with χ_2 from (11.27) and B from (11.37), we find that the alignment time $t_{\parallel}$ in (11.36) is inversely proportional to ω^2 and much shorter than the disorder time t_{dis}.

11.4 Non-magnetic alignment

We discuss two other ways in which grains may be aligned in the interstellar medium. They are interesting in themselves but except for special configurations, not likely to be responsible for polarization.

11.4.1 Gas streaming

When gas streams with a velocity $\mathbf{v}$ past a dust particle, for example, in a stellar wind, each atom of mass m endows the grain upon collision with a small angular momentum

$$d\mathbf{L}_{gas} = \mathbf{r} \times m\mathbf{v}. \qquad (11.39)$$

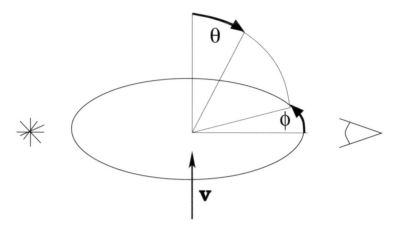

Figure 11.7. An observer (right) views a star (left) through a cloud of cylindrical grains. Perpendicular to the line of sight, gas streams at supersonic velocity **v** relative to the grains. Directions are given in the spherical polar coordinates (θ, ϕ). Cylindrical grains will rotate in planes ϕ = constant and this results in a net orientation of their rotation axes and in polarization of the starlight [Gol52].

If the grain is a rod of length l, we have, approximately,

$$dL_{\text{gas}} = \tfrac{1}{2}lmv.$$

The situation is depicted in figure 11.8. Let us choose spherical polar coordinates (θ, ϕ) so that the direction of streaming is given by $\theta = 0$ and the direction towards the observer by $\theta = \pi/2$ and $\phi = 0$ (figure 11.7). Because of the vector product, $d\mathbf{L}_{\text{gas}}$ lies in the plane $\theta = \pi/2$ which is perpendicular to **v**. If the grain is hit by N atoms, all changes in angular momentum, $d\mathbf{L}_{\text{gas}}$, lie in this plane but are randomly distributed there. So the final angular momentum $\mathbf{L}_{\text{gas}}$ is also in the plane $\theta = \pi/2$ and its length is the result of a random walk:

$$L_{\text{gas}} = \sqrt{N} \cdot dL_{\text{gas}}.$$

When the number N of collisions is large, L_{gas} will be limited by friction with the gas.

If the grain is very elongated, the vector **r** in figure 11.8 falls almost along the x-axis. Therefore the angular momentum is perpendicular to the x-axis which implies that the grain rotates in a plane with ϕ = constant. When there are many particles, all such planes are equally likely, there is no favorite ϕ. The observer in figure 11.7 finds that the starlight is polarized. The net polarization is due mostly to grains with $\phi \simeq 0$, while those with $\phi \simeq \pi/2$ do not contribute. An observer who looks at the star from below in the direction of the velocity vector **v** in figure 11.7 would detect no polarization.

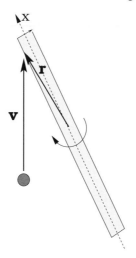

Figure 11.8. An impacting atom of momentum mv causes the cylinder to rotate in the plane of the drawing, indicated by the curved arrow. The same is true for a photon coming from the same direction as the atom; its momentum is hv/c. But the photon also transfers angular photon momentum $\hbar$ and this leads to rotation out of the plane of the drawing.

The thermal motion of the gas has a disordering effect as it introduces random velocities between gas and dust particles. But such disorientation may obviously be neglected when the streaming motion is supersonic.

11.4.2 Anisotropic illumination

When the grains absorb radiation exclusively from one side, as in the vicinity of a star, the situation is similar to gas streaming. Only now the impinging particles are photons and not gas atoms, and their momentum is hv/c instead of mv. So each transfers the angular momentum

$$dL_{\rm abs} = \frac{1}{2}l \cdot \frac{hv}{c}.$$

There is another effect related to photons that creates independent additional alignment [Har70]. Photons have, on top of their momentum hv/c, an intrinsic angular momentum $\hbar$ that is imparted to the grain in each absorption event. It is directed along the line of photon propagation, either in a positive or negative direction, and not perpendicular to it, as $dL_{\rm abs}$ before. The average (root mean square) angular momentum of the grain after the capture of N photons becomes $\sqrt{N}\hbar$. There are two alignment processes through photons, one transfers momentum hv/c, the other angular momentum $\hbar$. The latter dominates when $\hbar > dL_{\rm abs}$. This condition is fulfilled when the grain is smaller than the wavelength.

Each emitted photon carries off a quantum $\hbar$ in an arbitrary direction. In the end, the grain will acquire from this process an angular momentum L_{emis}. As the energy budget of heating and cooling is always balanced very well, a grain in the diffuse medium, which is heated by few UV and cooled by many IR photons, emits, very roughly, a hundred times more photons than it absorbs. So L_{emis} will be some $\sqrt{100} = 10$ times larger than L_{abs}. But the associated vector $\mathbf{L}_{emis}$ has no preferential orientation. Inside a molecular cloud, where only a few energetic photons fly around, the discrepancy between L_{emis} and L_{abs} is less severe.

Chapter 12

PAHs and spectral features of dust

Gas atoms or molecules emit and absorb narrow lines, from the UV to the microwave region, by which they can be uniquely identified. Because of their sharpness, the lines also contain kinematic information via the Doppler shift in their radial velocity. Observing several transitions simultaneously allows us to infer the conditions of excitation, mainly gas temperature and density.

The spectroscopic information from dust, however, is poor. Being a solid, each grain emits very many lines but they overlap to give an amorphous continuum. Moreover, the emission is restricted to infrared wavelengths. The smallest grains, PAHs, which show very characteristic bands are a fortunate exception. Resonances are also observed from silicate material and ice mantles. These features are the topic of the present chapter. Because of the strong disturbance of any oscillator in a solid, the transitions are broad and do not allow us to extract velocity information. Precise chemical allocation is also not possible but individual functional groups partaking in the transition can be identified and exciting conditions may occasionally be derived.

12.1 Thermodynamics of PAHs

12.1.1 What are PAHs?

These are planar molecules, made of a number of benzene rings, usually with hydrogen atoms bound to the edges. The smallest consist of barely more than a dozen atoms, so they are intermediate between molecules and grains and constitute a peculiar and interesting species of their own. The first observation of PAHs, without proper identification, dates back to the early 1970s when in the near and mid infrared a number of hitherto unknown features were found [Gil73]. Although their overall spectral signature was subsequently observed in very diverse objects (HII regions, reflection nebulae, planetary nebulae, cirrus clouds, evolved stars, star burst nuclei and other external galaxies), their proper identification took place only in the mid 1980s. PAHs is a short form for

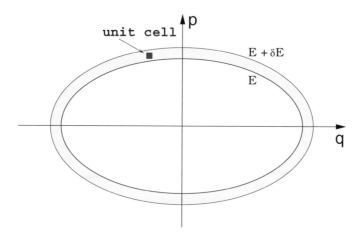

Figure 12.1. The phase space accessible to a one-dimensional oscillator with energy between E and $E + \delta E$ is, according to equation (8.2), given by the shaded annulus. The ellipses correspond to total (kinetic plus potential) energy E and $E + \delta E$, respectively. The unit phase cell has the size $\delta q \delta p = \hbar$.

Polycyclic Aromatic Hydrocarbons and they are called this way even when the hydrogen atoms are missing. In interstellar space, PAHs are excited by the absorption of a single ultraviolet photon upon which they emit their characteristic infrared lines.

12.1.2   Microcanonic emission of PAHs

Let us investigate the possible states of a PAH of total energy E. It is given in equation (8.2) as the sum of kinetic and potential energy of all atoms which are approximated by harmonic oscillators. The accessible states are in a narrow range from E to $E + \delta E$. Figure 12.1 illustrates the locus of the states in phase space when the system has only one degree of freedom ($f = 1$). There are then two coordinates, position q and momentum p, and all states lie in an annulus around an ellipse. The oscillating particle is most likely to be found near the greatest elongation, which is at the left and right edge of the ellipse. There the velocity is small and, for a given Δq, one finds the highest density of cells. For $f > 1$, the situation is multi-dimensional but otherwise analogous.

   The population of the cells in phase space is statistical and the probability of finding the ith oscillator of a PAH in quantum state v is, in view of (5.48),

$$P_{i,v} = \rho(E; i, v)/\rho(E).$$

$\rho(E)$ is the density of states at energy E and $\rho(E; i, v)$ is the density of states at energy E subject to the condition that oscillator i is in quantum state v. When one extends expression (5.42) for $\rho(E)$ to the case when the oscillators are not identical but each has its own frequency ν_i, and further refines the expression by adding to E the zero-point energy E_0 of all excited oscillators, one obtains [Whi63]

$$\rho(E) \simeq \frac{(E + aE_0)^{f-1}}{(f-1)! \prod_{j=1}^{f} h\nu_j} \qquad (12.1)$$

with

$$a \simeq 0.87 + 0.079 \ln(E/E_0) \qquad E_0 = \tfrac{1}{2} \sum_j h\nu_j.$$

The fudge factor a is of order unity. $\rho(E; i, v)$ is also calculated after (12.1) but now one has to exclude oscillator i. There is one degree of freedom less and E, as well as the zero-point energy E_0, have to be diminished accordingly. This gives

$$\rho(E; i, v) = \left[(E - vh\nu_i) + a(E_0 - \tfrac{1}{2}h\nu_i)\right]^{f-2} \left[(f-2)! \prod_{j \neq i} h\nu_j\right]^{-1}.$$

Hence the relative population of level v is

$$P_{i,v} = \frac{\rho(E; i, v)}{\rho(E)} = (f-1)\xi \left[1 - \left(v + \frac{a}{2}\right)\xi\right]^{f-2} \qquad \xi = \frac{h\nu_i}{E + aE_0}. \qquad (12.2)$$

The ith oscillator emits in the transition $v \rightarrow v - 1$ the power

$$\varepsilon_{\text{mic}}(E; i, v) = P_{i,v} h\nu_i A^i_{v,v-1}$$

and the total microcanonic emission from oscillator i is, therefore,

$$\varepsilon_{\text{mic}}(E; i) = h\nu_i A^i_{1,0} \sum_{v \geq 1} v P_{i,v}. \qquad (12.3)$$

12.1.3 The vibrational modes of anthracene

As an illustration, we consider the PAH anthracene, $C_{14}H_{10}$, which has a simple and fairly symmetric structure (see insert in figure 12.2) and is, therefore, easy to handle theoretically. The PAH is built up of three benzene rings; all bonds at its edges are saturated by hydrogen atoms. There are altogether $f = 3 \cdot (14 + 10) - 6 = 66$ degrees of freedom. The vibrational modes can be computed if one knows the force field (force constants) or the potential in the molecule. In the model on which figure 12.2 is based seven vibrational types are considered. Besides the four obvious ones,

- CC stretching,
- CH stretching,

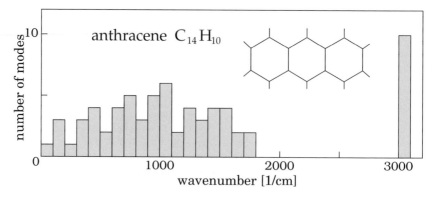

Figure 12.2. The polycyclic aromatic hydrocarbon anthracene has 66 degrees of freedom. They correspond to 66 oscillators (modes), each having its eigenfrequency. The wavenumbers are tabulated in [Whi78] and grouped here into intervals of 100 cm^{-1} width.

- CC out-of-plane bending and
- CH out-of-plane bending,

three mixed modes in which several atoms take part in the oscillation are also included. These are CCC and CCH bending as well as CCCC torsion. The frequencies of the 66 oscillators by which the molecule is replaced are found by introducing normal coordinates (section 8.4) and solving equation (8.25). Despite the simplicity of the model, the achieved accuracy is good, the theoretical frequencies agree well with laboratory data.

Some vibrational modes lie very close together because of the symmetry of the molecule. It is clear that atoms in similar force fields must swing in a similar manner. For instance, all CH stretches of the ten hydrogen atoms in anthracene are crowded between 3030 and 3040 cm^{-1}. For reasons of symmetry, at most four of these CH frequencies can be different but these four are really distinct because they correspond to sites where the surroundings of the H atoms have different geometries. The distribution of the number of modes over energy is displayed in figure 12.2. It contains several noteworthy features:

- The modes strongly cluster around 3035 cm^{-1}. This range is completely isolated and due to CH stretching. When κ is the force constant of CH binding and m the mass of an H atom, the frequency is roughly $\omega = \sqrt{\kappa/m}$ (see (12.8)). Hydrogen atoms are also involved in the modes around 1000 cm^{-1} which are due to a CH bend.
- The rest of the histogram is due to the skeleton of carbon atoms and surprisingly smooth.
- In section 8.4, we gave the Debye temperatures, $\theta_z = 950$ K and $\theta_{xy} = 2500$ K, for a graphite plane. Excluding the CH bend at 3000 cm^{-1}, the high frequency cutoff, which, by definition, is the Debye frequency, is at

1800 cm^{-1} and corresponds nicely to θ_{xy} (see (8.33)).

● At long wavelengths, the modes extend into the far infrared. The lowest energy is 73 cm^{-1} (137 μm) and again in qualitative agreement with the estimate for the minimum frequency ν_{min} from (8.45) if one uses $\theta_z = 950$ K.

12.1.4 Microcanonic versus thermal level population

We have learnt how to compute the microcanonic population of levels; this is always the statistically correct way. We now apply our skills to anthracene and check whether, for this PAH, the assumption of thermalization, which when fulfilled makes life so much easier, is permitted. Let the PAH absorb a photon of energy E. When the levels are thermalized, E is equal to the sum of the mean energies of all 66 oscillators (see (5.11)),

$$E = \sum_{i=1}^{66} \frac{h\nu_i}{e^{h\nu_i/kT} - 1}.$$

Because all ν_i are known, we can calculate from this equation the temperature of a thermalized anthracene molecule of energy E. Figure 12.3 demonstrates that if the PAH has absorbed a typical stellar photon of $h\nu \simeq 4$ eV, we may indeed use the Boltzmann formulae (dotted lines) for the level population, and thus also for line emission. For higher photon energies, things only improve.

We also learn from the microcanonic results of level populations that those oscillators in the PAH which correspond to the major bands (table 12.1) are mostly found in the lowest quantum numbers, $v \leq 1$. This implies that effects produced by the anharmonicity of the oscillator potential are, to first order, not important. However, if one looks closer, they have to be taken into account. Because of anharmonicity, the spacing between the upper levels decreases and the lines from higher v-states are shifted towards the red. The best example is presented by the C–H stretches at 3.28 μm ($v = 1 \rightarrow 0$, see table 12.1) and at 3.42 μm ($v = 2 \rightarrow 1$). The shift is sufficiently large so that the $v = 2 \rightarrow 1$ line can occasionally be seen well detached from the $v = 1 \rightarrow 0$ transition. The $v = 2 \rightarrow 1$ line is always weaker because of low level population but note that the Einstein A coefficient increases proportionally with v. If the shift due to anharmonicity is small, as for the C–H out-of-plane bend at 11.3 μm, the high-v lines blend with the ground transition producing an overall broadening of the band.

To estimate how big PAHs behave, we consider a fictitious molecule with 330 degrees of freedom but with the same relative mode distribution as in figure 12.2. There are now for each energy interval five times more modes. We see, from figure 12.4, that for such a big PAH, when it is excited by a photon of 4 eV, the thermal assumption is usually bad. It is acceptable only for low-frequency vibrations, whereas it grossly underestimates the population numbers at short wavelengths. If the photons are more energetic ($E > 10$ eV), the situation

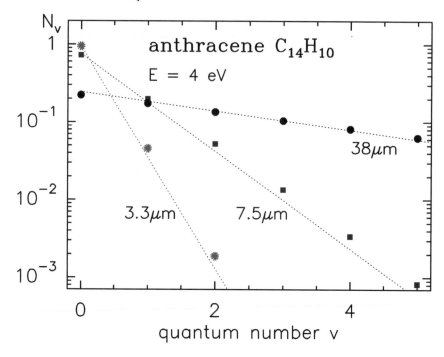

Figure 12.3. Level population N_v in anthracene for three oscillators (their wavelengths are indicated) after absorption of a photon with $E = 4$ eV from the ground state. Dots show the microcanonic fractional population as approximated by (12.2); always $\sum N_v = 1$. The dotted lines, which only have a meaning at integral quantum numbers v, give the thermal results at a temperature $T = 1325$ K; see also text.

improves towards thermalization. Note that while such a highly excited PAH cools, the assumption of thermal emission gets poorer.

Of course, if the size of the PAH is further increased, it must eventually be thermalized like a normal grain. As the PAH gets bigger, the threshold for collisional and photoexcitation, v_{min} after (8.45) and v_{cut} after (12.5), steadily declines.

12.1.5 Does an ensemble of PAHs have a temperature?

The concept of temperature is only valid if the system has many degrees of freedom. To an ordinary grain made of $N = 10^5 \dots 10^{10}$ atoms, one may certainly assign a temperature but it is doubtful whether one can do so in the case of a PAH, which may bear more affinity to a big molecule than to a solid body. For example, for a PAH with $f = 100$ degrees of freedom that is excited to an energy of $E = 10$ eV and emits IR photons of 0.1 eV (13 μm wavelength), the number of states whose energy does not exceed E is, from (5.41), roughly

Table 12.1. Approximate numbers for PAH resonances. The cross sections integrated over the band, $\sigma_{int} = \int \sigma(\lambda)\,d\lambda$, are per H atom for C–H vibrational modes, and per C atom for C–C modes. The space divides, somewhat artificially, major (top) from minor bands (bottom). See also figure 12.5.

Center wavelength λ_0 (μm)	Damping constant $\gamma\,(10^{12}\,\mathrm{s}^{-1})$	Integrated cross section σ_{int} (cm^2 cm)	vibrational type
3.3	20	12	C–H stretch
6.2	14	14	C–C stretch
7.7	22	51	C–C stretch
8.6	6	27	C–H bend in-plane
11.3	4	41	C–H bend out-of-plane
11.9	7	27	C–H bend out-of-plane
12.8	3.5	47	C–H bend out-of-plane
13.6	4	12	C–H bend out-of-plane
5.2	12	1.1	C–C vibration
7.0	5.9	12	?
14.3	5	2.5	C–C skeleton
15.1	3	1.2	C–C skeleton
15.7	2	1.2	C–C skeleton
16.4	3	1.7	C–C skeleton
18.2	3	1.0	C–C skeleton
21.1	3	2.0	C–C skeleton
23.1	3	2.0	C–C skeleton

$N_s \simeq (E/f\hbar\omega)^f$. This is not a fantastically large number, as it should be for a good temperature but one of order unity. So one has to check in a microcanonic treatment, as before, whether the PAH is approximately thermalized or not.

If one cannot assign a temperature to a single PAH, maybe one can to an ensemble of PAHs in the way one assigns a temperature to the level population of an ensemble of gas atoms? At least this is possible when the atoms are in a blackbody radiation field or when their density is high. The excitation then carries the signature either of the Planck function or of Maxwell's velocity distribution and it is completely described by the Boltzmann formula (5.8). However, when exciting events are rare, for example, at low gas density, the levels are not thermalized. The kinetic temperature of the gas does not then describe the population of the internal levels of the atoms.

For PAHs, excitation through energetic photons is much too rare an event and gas collisions are negligible altogether. Furthermore, prior to absorption of a photon, the PAH is usually in its ground state. If the photons were monoenergetic, so would the energies of the PAHs freshly excited from the

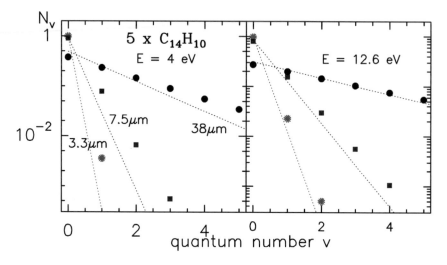

Figure 12.4. Level population N_v of a PAH with $f = 330$ degrees of freedom. It has the same relative mode distribution as anthracene (see figure 12.2). The energy of the absorbed photon equals 4 and 12.6 eV. The dotted lines are calculated under the assumption of thermalization at $T = 545$ K (left box) and $T = 1000$ K (right box). Otherwise as figure 12.3.

ground state be. Their distribution in energy would not show the spread typical of thermal systems but resemble a δ-function. Although the actual exciting photons are not monoenergetic but come from the interstellar radiation field and have a considerable spread, their frequencies do not mimic thermal conditions.

12.2 PAH emission

12.2.1 Photoexcitation of PAHs

In the diffuse interstellar medium, a neutral PAH molecule is almost all the time in its electronic ground state, a singlet S_0. Collisions with gas atoms are rare and the gas temperature is too low anyway to thermalize the PAH. The vibrational modes are, therefore, also in their lowest level. Emission of infrared lines from the PAH molecule is initiated by absorption of a UV photon of energy E_{UV}, say, once a day. Afterwards the molecule is highly excited and in a new electronic singlet state S_i. Then various and complicated things happen, very fast, some on time scales of 10^{-12} s. But by and large, the outcome is that the ultraviolet energy is thermalized, i.e. distributed over all available vibrational modes, and then degraded into infrared photons, mainly by emission in the fundamental modes (see table 12.1). The emission lasts typically for 1 s.

  Before the IR photons are emitted, immediately after the UV photon has been absorbed, the PAH may change from the electronic state S_i without radiative loss to another singlet state. Or it may undergo intersystem crossing to a triplet state. In this case, because the triplet system has a metastable ground level T_1, some of the energy E_{UV} will eventually go into phosphorescence from the triplet T_1 to S_0 and thus be lost for the IR features.

For a singly photo-ionized PAH, the excitation scheme is similar; only the electronic ground state is then a doublet because of the unpaired electron. A PAH may also be negative or bear multiple charges. The degree and kind of ionization depends on the radiation field, the gas temperature, the electron density and, of course, on the UV cross sections of the PAH; some of the basic ideas are outlined in section 9.2. The ionization details may not be decisive for evaluating the IR emission of PAHs but probably have an influence on the cutoff wavelength and thus on the conditions for excitation (see later).

12.2.2 Cutoff wavelength for electronic excitation

There is a minimum photon energy for electronic excitation of a PAH corresponding to a cutoff wavelength λ_{cut}. Light from wavelengths greater than λ_{cut} is simply not absorbed and thus cannot produce the infrared resonances. For PAHs of circular shape and size $\langle a \rangle$, one finds empiricallly, in the laboratory, the crude relation

$$\lambda_{cut} = 1630 + 370\sqrt{N_C} = 1630 + 450\langle a \rangle \qquad [\text{Å}]. \qquad (12.4)$$

Here $\langle a \rangle$ is measured in Å and N_C is the number of carbon atoms in the molecule. Small PAHs ($N_C < 50$) are only excited by UV radiation, larger ones also by optical photons. One can understand the existence of the threshold by considering a free electron in a linear metallic molecule of length L. The electron is trapped in the molecule and its energy levels are given by (6.52),

$$E_n = \frac{h^2 n^2}{8m_e L^2} \qquad n = 1, 2, \ldots .$$

The phase space available to electrons with energy $E \leq E_n$ is $hn/2$, so if there are s free electrons, all levels up to $n = s$ are populated; a denser population is forbidden by Pauli's exclusion principle. A transition with $\Delta n = 1$ has, therefore, the energy

$$h\nu_{cut} = \left. \frac{dE_n}{dn} \right|_{n=s} \Delta n = \frac{h^2 s}{4m_e L^2} = \frac{h^2}{4m_e(\frac{L}{s})L}.$$

If the atoms have a size of ~ 2 Å and if there is one optical electron per atom, then $L/s = 2$ Å and

$$\lambda_{cut} = \frac{4m_e(\frac{L}{s})Lc}{h} \sim 330L \qquad [\text{Å}]. \qquad (12.5)$$

This qualitative and simplified model of a linear metallic grain already yields the right order of magnitude for the cutoff.

12.2.3   Photo-destruction and ionization

Because the PAHs are so small, absorption of a UV photon can lead to dissociation. In a strong and hard radiation field, PAHs will not exist at all. Because of the anharmonicity of the potential of a real oscillator, an atom detaches itself from the PAH above a certain vibrational quantum number v_{dis}. The chance $P(E)$ of finding, in a PAH with f degrees of freedom and total energy E, the oscillator i in a state $v \geq v_{dis}$ is given by equation (12.2), and $P(E)$ is proportional to the dissociation rate $k_{dis}(E)$ of the molecule. In slight modification of (12.2), one, therefore, writes

$$k_{dis}(E) = v_0 \left(1 - \frac{E_0}{E}\right)^{f-1}. \tag{12.6}$$

v_0 has the dimension of a frequency, E_0 may be loosely interpreted as the bond dissociation energy. The two factors, v_0 and E_0, can be determined experimentally. For example, with respect to hydrogen loss, $E_0 = 2.8$ eV and $v_0 = 10^{16}$ s^{-1}; with respect to C_2H_2 loss, $E_0 = 2.9$ eV, $v_0 = 10^{15}$ s^{-1} [Joc94]. If hydrogen atoms detach, the aromatic cycles are unscathed, if however acetylene evaporates, the carbon rings are broken.

After UV absorption, the excited PAH is hot and has a certain chance to evaporate. When relaxation through emission of IR photons is faster than dissociation, the PAH is stable. The infrared emission rate of PAHs, which is of order 10^2 s^{-1}, can thus be considered as the critical dissociation rate, k_{cr}. When $k_{dis}(E) > k_{cr}$, evaporation will occur. k_{cr} is associated, from (12.6), with a critical internal energy E_{cr} and because $E_{cr} > E_0$, one has, approximately,

$$\ln k_{cr} \simeq \ln v_0 - (f-1)\frac{E_0}{E_{cr}}.$$

So there is a rough linear relationship between the degrees of freedom of the PAH and the critical energy. As k_{cr} enters logarithmically, an exact value is not required. The critical temperature T_{cr} can be found from E_{cr} via equation (5.11),

$$E_{cr} = \sum_{i=1}^{f} \frac{h v_i}{e^{h v_i / k T_{cr}} - 1}.$$

Ionization after absorption of an energetic UV photon is another reaction channel, besides dissociation and infrared emission, and it has the overall effect of stabilizing the molecule against destruction. Big PAHs are easier to ionize than small ones; the ionization potential of benzene is 9.24 eV and declines to ~6 eV for molecules with 100 carbon atoms.

For the astronomically relevant question, i.e. under what conditions do PAHs survive in a given, harsh radiation field, one has to balance their destruction rate with the rate at which they are built up by carbon accretion. In the interstellar radiation field (figure 8.2), we estimate that PAHs with only 30 carbon atoms can survive; near an O star (distance $r = 10^{16}$ cm, $L = 10^5 L_\odot$), N_C should be above 200.

As the physical processes connected with PAH destruction are most complicated, one may be tempted to use a purely phenomenolgical approach and simply assume that PAHs evaporate if the fraction of PAHs above some critical temperature T_{cr} exceeds a certain value f_{evap}, i.e. when

$$\int_{T_{cr}}^{\infty} P(T) \, dT \geq f_{evap}.$$

The probability function $P(T)$ of (8.46) has to be computed anyway for the evaluation of PAH emission. $f_{evap} \sim 10^{-8}$ and $T_{cr} \sim 2500$ K do not seem to be unreasonable numbers and one may use such an approach to decide whether PAHs evaporate or not.

12.2.4 Cross sections and line profiles of PAHs

A PAH has several hundred vibrational modes but one typically detects only half a dozen major resonances around the frequencies where the strong modes cluster. Table 12.1 compiles their vibrational type, center wavelength and integrated line strength but it also lists some of the weaker or minor resonances which are, however, more variable in intensity from source to source and their origin is less clear. The most prominent representative of the minor features is the band at 3.4 μm which was mentioned in section 12.1.4. It is not listed in table 12.1 because it is an overtone.

There are two classes of bands: one involves hydrogen atoms, which are all located at the periphery of the PAH; the other only carbon atoms. The strength of the emission is, therefore, proportional either to N_H or N_C, the number of hydrogen and carbon atoms, respectively. As the total amount of absorbed and emitted energy is determined solely by the carbon skeleton, a PAH without hydrogen atoms will have weaker C–C bands than one with hydrogen atoms (and the same N_C).

We approximate the line shapes by Lorentzian profiles (see (1.106)),

$$\frac{\gamma v^2}{\pi^2 [v^2 - v_0^2]^2 + [\gamma v/2]^2} \tag{12.7}$$

where $v_0 = c/\lambda_0$ is the center frequency, $\sigma_{int} = \int \sigma(\lambda) \, d\lambda$ the cross section of the particular band integrated over wavelength, and N the number of C or H atoms, respectively. The damping constant γ determines the band width and is probably the result of an overlap of many transitions with frequencies clustering around v_0.

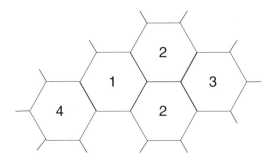

Figure 12.5. In a PAH, hydrogen atoms are attached to the edge of the carbon skeleton. The exact frequency of the C–H out-of-plane bend depends on whether the hydrogen atoms are isolated (mono positions) or whether there are two (duos), three (trios) or four (quartets) adjacent H atoms. Due to the coupling between the vibrating H atoms, the center wavelength increases from approximately 11.3 to 13.6 μm (see table 12.1). The number of neighboring H atoms is indicated for each (peripheral) carbon ring.

12.3 Big grains and ices

A few infrared resonances can also be observed in grains of normal size. They allow us to identify certain characteristic groups in molecules of which the grains are made, for example OH in water molecules when water has frozen out. The features are grain-size independent and serve as a tool to unravel the chemical composition of interstellar dust.

Let us consider the vibrations in a characteristic group that consists of just two molecules. In a simplified picture, the fundamental frequency ω between two atoms of mass m_1 and m_2 connected by a spring of force constant κ is given by

$$\omega = 2\pi\nu = \sqrt{\kappa/\mu} \tag{12.8}$$

where the reduced mass is

$$\mu = \frac{m_1 m_2}{m_1 + m_2}. \tag{12.9}$$

The frequency is high when the molecules are strongly coupled and the masses of the involved atoms small. For example, the stretching mode of the light CH molecule occurs at $\sim$3 μm, for the heavier CN, the resonance is at $\sim$10 μm. Likewise, in a single C–C bond, the stretching band lies at $\sim$10 μm, in a triple C$\equiv$C bond at $\sim$5 μm.

The resonance is broad in an amorphous material, and fairly sharp in a crystal where high structural order leads to long-range forces. Its width and center frequency depend weakly on temperature: when the solid cools, the force constant κ and thus the frequency ν in (12.8) increase because the lattice shrinks. At the same time, the feature becomes narrower because collisional damping with neighboring atoms is reduced.

12.3.1 The silicate features and the band at 3.4 μm

The most prominent and widely studied resonance is the one at 9.7 μm due to an Si–O stretching vibration in silicates. It is the strongest indicator of silicon in space. Towards star-forming regions, it is observed in absorption, towards oxygen-rich mass loss giants it is seen in absorption and occasionally in emission. At around 18 μm, there is an O–Si–O bending mode which is weaker and much more difficult to observe from the ground because of atmospheric attenuation; the absorption coefficient ratio $\kappa_{18}/\kappa_{10} \simeq 0.5$.

To derive the ratio of the visual to the 9.7 μm optical depth, $\tau_V/\tau_{9.7}$, one measures the extinction through a cloud that lies in front of a carbon star. In these stars, all oxygen is bound in CO and silicate grains are absent. The stars have shells of carbon dust that produce a smooth 10 μm emission background with a temperature of $\sim$1000 K and without the 9.7 μm feature. On average,

$$\frac{\tau_V}{\tau_{9.7}} \simeq 18 \tag{12.10}$$

with a scatter of about 10%. Knowing the ratio $\tau_V/\tau_{9.7}$, one can determine the visual extinction A_V from the 10 μm absorption, provided there are no hot grains along the line of sight that emit themselves at 9.7 μm. The method still works when stellar photometry is no longer possible. For example, from the depression of the 10 μm feature in figure 12.6 we estimate that the object IRS9 in NGC 7538 suffers about 35 mag of visual extinction corresponding to $\tau_{9.7} \simeq 2$.

Another feature that is widely observed in absorption in the diffuse medium but, quite interestingly, absent in molecular clouds is the one at 3.4 μm. It is usually attributed to an aliphatic (the carbon atoms are arranged in a chain, and not in a ring) C–H stretching mode, for example, in CH_3 but its exact identification has not yet been settled.

12.3.2 Icy grain mantles

Ice mantles form in cold and dense molecular clouds as a result of gas accretion onto refractory cores (section 9.1). To stay on the grain and not to become photodesorbed, they have to be well hidden from the UV radiation field. For instance, the 3.1 μm feature due to OH stretching seems to require shielding by more than 8 mag of visual extinction. The mantles consist of a mixture of molecules like H_2O, CO, CO_2, CH_3OH, NH_3, CH_4, ... in varying abundance ratios. An impressive example of absorption by ice mantles is presented in figure 12.6.

The vibrational spectrum of diatomic molecules in an ice mantle has several peculiarities. For free molecules, in a gas, there is generally an R- and a P-branch, each consisting of many rotational lines (see section 14.2.1 and figure 14.1). In a solid, the atoms are squeezed in by other molecules and cannot rotate freely, so these branches are supressed. Only hindered rotations may be possible in

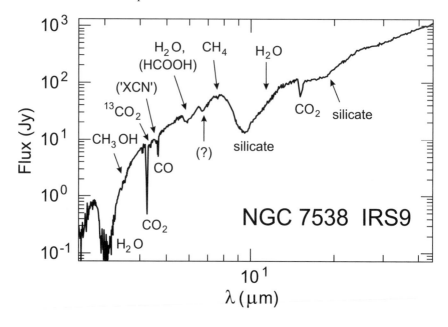

Figure 12.6. The infrared spectrum towards the infrared object IRS9 in the star-forming region NGC 7538. The resonances at 10 and 18 μm are due to refractory silicate cores. The other features come from molecules that have frozen out onto refractory grains and are now in ice mantles. Reproduced from [Whi96] with permission of EDP Sciences.

which the molecule switches from one position to another, which is energetically different, by rotating. The center frequency is also influenced by the composition of the matrix into which the molecule is embedded.

12.4   An overall dust model

We are now in a position to present a tentative but complete model of interstellar dust with respect to its interaction with light. It will be applied in chapter 16 to compute the spectral energy distribution of dusty astronomical objects. A dozen of rival dust models can be found in the literature. They are all capable of explaining the basic data on extinction, polarization, scattering and infrared resonances and, if employed in radiative transfer calculations, yield successful matches to observed spectra. Our model certainly does not produce superior fits, its advantage, if any, lies in its relative simplicity. It uses published sets of optical constants, without modifying them at certain wavelengths to improve fits. The grains have the simplest possible structure and shape: they are uncoated spheres. (Therefore, to account for the interstellar polarization, we have to turn some of the spheres into infinite cylinders, see chapter 10.) Altogether, only 12 types of

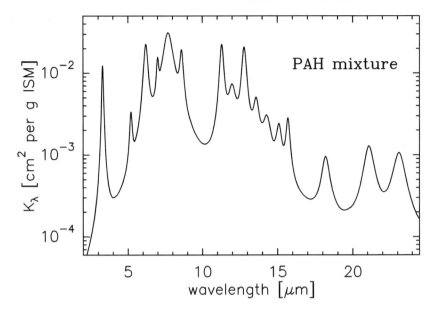

Figure 12.7. The mass absorption coefficient, K_λ, of a mixture of two kinds of PAHs, each with an abundance $Y_C^{PAH} = 3\%$, in the wavelength range where the infrared resonances are observed. The PAH parameters are described in the text and in table 12.1. K_λ refers to 1 g of interstellar matter. Below 3 μm, K_λ slumps but increases again when the wavelength falls below the cutoff for electronic excitation, λ_{cut}.

dust particles are needed, including PAHs, where one particle type differs from another either by its size or chemical composition.

All major dust models agree, in principle, on the solid state abundances, the approximate grain sizes and the minimum number of dust components: one needs PAH–ike particles, very small grains with temperature fluctuations and big grains consisting mainly of silicate and carbon. Such concensus on important issues does, however, not insure against common error. None of the models is certain because hard checks on their validity are missing. Interstellar grains are not within our reach and we cannot verify our assumptions in the laboratory.

Differences between models show up mostly in the details of the chemical composition (mixing of components, coating, degree of fluffiness, fraction of built in refractory organic compounds). All models can be refined by varying the shape, size and optical constants of the grains. This further enhances the agreement with observational data but does not necessarily bring new insight. A more promising attempt is to increase the internal consistency of the models. Steps in this direction have been undertaken with respect to the smallest particles (PAHs and vsg) by *computing* grain properties, such as enthalpy and vibrational modes, under the assumption of a likely grid structure for the atoms ([Dra01] and [Li01]).

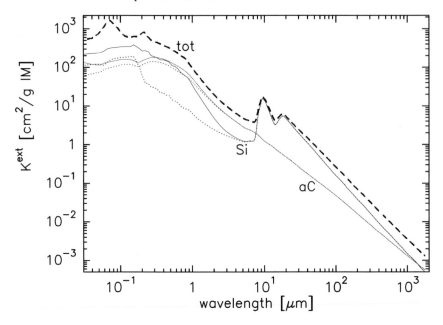

Figure 12.8.   The total extinction coefficient (MRN+vsg+PAH, broken curve) as a
function of wavelength for the dust model of the diffuse medium. It refers to 1 g of
interstellar matter adopting a dust-to-gas ratio $R_d = 140$. Also shown are the extinction
(full curve) and absorption (dotted curve) coefficients for big silicates (Si, $a_- = 300$ Å,
$a_+ = 2400$ Å) and big amorphous carbon grains (aC, $a_- = 150$ Å, $a_+ = 1200$ Å). One
can read the amount of scattering as the difference between extinction and absorption. The
bumps at 2200 and 700 Å are due to very small graphite particles ($a_- = 10$ Å, $a_+ = 40$ Å).

12.4.1   The three dust components

Our dust model for the diffuse interstellar medium consists of three components
with the following properties:

- *Big grains.* Most of the solid matter in the interstellar medium resides in big
  grains, also called MRN grains (section 7.5). They consist of amorphous
  carbon (abundance in solid [C]/[H] = 2.5×10^{-4}) and silicate particles
  ([Si]/[H] = 3×10^{-5}), both with bulk density ~ 2.5 g cm^{-3}. Their chemical
  structure has been discussed in section 7.4. They are spheres and their radii
  range from ~ 200 to 2400 Å with an $a^{-3.5}$ size distribution (section 7.5).
  Absorption and scattering cross sections are calculated from Mie theory
  employing optical constants from figure 7.19. Big grains attain a constant
  temperature in any radiation field.
- *Very small grains (vsg).* A few percent of the solid matter, we adopt 5–10%,
  form a population of *very small* grains (vsg) with radii less than 100 Å. They

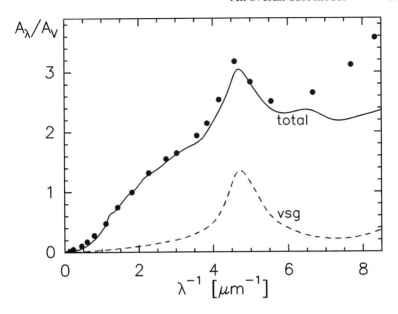

Figure 12.9. The interstellar extinction curve: observations (dots) and model (curves). The extinction coefficient of our dust model is displayed in figure 12.8. The fit could be substantially improved by modifying the abundance, sizes and UV properties of vsg and PAHs but it would then lose some of its simplicity.

may be regarded as the small-size extension of the big particles. Optical cross sections are still calculated from Mie theory, although the concept of a continuous solid medium may no longer be fully valid. We adopt, for them, a size spectrum $n(a) \propto a^{-4}$, which is somewhat steeper than for MRN grains (but this is an arguable minor point); and assume that the carbon component is graphitic, and not amorphous, so optical constants of graphite are appropriate. The main reason why one has to include vsg is their time-variable temperature, as demonstrated extensively in sections 8.5 and 8.6.

• *PAHs.* Finally, we add PAHs to the dust mixture. Although there must be a plethora of different species, we consider only two: small ones with $N_C = 40$ carbon atoms; and big ones with $N_C = 400$. Such a restriction, besides speeding up the computations, makes it easier to elucidate the influence of the PAH size on the spectrum. The temperature of the PAHs also fluctuates violently and has to be calculated according to the recipes of section 8.5.

In estimating PAH properties, we widely follow [Sch93]. For the hydrogenation parameter, $f_{H/C}$, which is the ratio of hydrogen-to-carbon atoms, we put, for $N_C > 30$,

$$f_{H/C} = 2.8/\sqrt{N_C}.$$ (12.11)

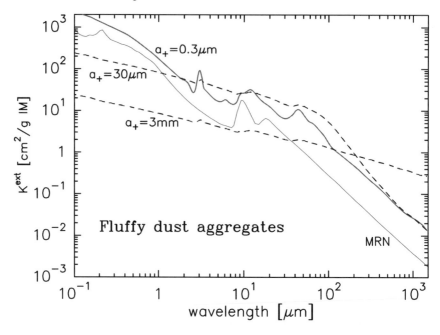

Figure 12.10. Mass extinction coefficient per gram of interstellar matter (IM) as a function of wavelength for composite grains which are fluffy aggregates of silicate, carbon, ice and vacuum. The lower limit a_- of the size distribution is in all curves 0.03 μm, the upper limit a_+ is indicated. The line labeled MRN shows the diffuse medium for reference; there the dust consists of pure (unmixed) silicate and amorphous carbon grains with $a_- = 0.02$ μm and $a_+ \sim 0.3$ μm. The MRN curve is more or less identical to the broken curve of figure 12.8, except in the far UV due to slight differences in the vsg abundance.

$f_{H/C}$ is thus proportional to the number of carbon atoms at the periphery of the PAH, where the H atoms are found, over the total number N_C. The formula gives $f_{H/C} = 0.44$ (0.14) for $N_C = 400$ (40). In reality, there is probably a large scatter in the hydrogenation parameter due to variations in the radiation field.

Each of the two PAH types has an abundance Y_C^{PAH} of 3–6% relative to the big grains. Y_C^{PAH} gives the mass fraction of all solid carbon that is in a certain kind of PAHs. For the absorption cross section of a single PAH, C^{PAH}, we disregard differences between neutral ond ionized PAH species, although this is a severe simplification. Ionized PAHs can be excited by less energetic photons and have greater infrared absorption coefficients than neutral ones [Li01]. At wavelengths shorter than λ_{cut}, which is usually in the UV (see (12.4)), we put $C^{PAH} = N_C \times 10^{-17}$ cm^2, so C^{PAH} does not depend on frequency and is proportional to the number of carbon atoms, irrespective of their geometrical arrangement. The cross sections in the resonances are taken from table 12.1. In the far infrared, at $\lambda > 25 \ \mu$m, one expects a continuum (see the smooth frequency distribution in the modes of even a small PAH in figure 12.2) and C_λ should fall off like λ^{-m} with m between one and two. However, this continuum is usually swamped by the emission from big grains and not observable. The absorption coefficient of PAHs per gram of interstellar matter is displayed in figure 12.7.

Assuming that helium accounts for 28% of the gas mass, these cited abundances for all three dust components (big grains, vsg, PAHs) imply a total dust-to-gas mass ratio in the diffuse medium of $R_d = 1 : 140$.

12.4.2 Extinction coefficient in the diffuse medium

Figures 12.8 and 12.9 further quantify our dust model. We use only four particles sizes for the big aC grains, four for the big Si grains and two for the very small grains. Nevertheless, the reproduction of the interstellar extinction curve is not bad. The deficit in the far UV is irrelevant for all models of infrared sources where the observer does not receive far UV radiation.

12.4.3 Extinction coefficient in protostellar cores

We extend our dust model to cold protostellar clouds. There grains are modified with respect to the diffuse interstellar medium by coagulation and by ice mantles. Consequently, their size increases and the dust mass doubles ($R_d = 1 : 70$), possibly even triples (see discussion on icy and fluffy grains in section 2.6). The effect of all three processes (fluffiness, frosting and grain growth) on the particle cross sections is investigated in section 2.6.4 and 2.6.5. In figure 12.10, we compute the mass extinction coefficient of dust that consists of such icy and porous grains. The particles are spheres and have a size distribution $n(a) \propto a^{-3.5}$ with lower limit $a_- = 0.03 \ \mu$m $= 300$ Å; the upper limit a_+ is varied. All grains are fluffy and composed of four components: silicate (Si), amorphous carbon (aC), ice and vacuum. Average optical constants are calculated from the Bruggeman mixing rule (section 2.6). In all grains, the ice mass is equal to the mass of the refractory components, Si plus aC, whereas the ice volume is 2.5 times bigger than the volume of Si plus aC. The vacuum fraction is always constant, $f^{\text{vac}} = 0.5$. Silicate has twice the volume of carbon, $f^{\text{Si}}/f^{\text{aC}} = 2$. The PAHs have disappeared and are dirty spots in the ice mantles. Without UV radiation, the vsg are irrelevant. The optical constants of silicate and amorphous carbon are from figure 7.19, for ice from the tables of [Ber69] and [Leg83]. Several ice resonances are visible, for instance, the one at 3.1 μm.

The curve marked $a_+ = 0.3 \ \mu$m is an educated guess for dust in protostellar clouds before it is warmed up by the star and the ices evaporate. When the grains are very big, as in stellar disks, the lines with $a_+ = 30 \ \mu$m or 3 mm might be appropriate.

Chapter 13

Radiative transport

This chapter deals with the transfer of radiation in a dusty medium. For example, when one observes a star-forming region, most photons created within it do not reach us on a straight path but are scattered, absorbed and re-emitted before they are collected in the telescope. To extract from the observational data the full information on the physical conditions in the star-forming region, one has to understand how the photons propagate within the source.

We summarize the elements of radiative transfer, discuss the general transfer equation and sketch how to derive, for a homogeneous medium, estimates on temperature, density and masses. We present a set of equations for spherical dust clouds that treats radiative cooling and heating in an energetically consistent way and outline how to solve it numerically. We append solution strategies of radiative transfer for two other astronomically relevant geometries: dusty accretion disks and a dust-filled star clusters, both of which can, under minor assumptions, be handled in one dimension. Finally, we also discuss the radiative transport in molecular lines because line observations are always complementary to dust observations when investigating stars in their making.

13.1 Basic transfer relations

13.1.1 Radiative intensity and flux

The intensity $I_\nu(\mathbf{x}, \mathbf{e})$ is the fundamental parameter describing the radiation field. It is defined in the following way (see figure 13.1): Consider a surface element $d\sigma$ at location $\mathbf{x}$ with normal unit vector $\mathbf{n}$. It receives, within the frequency interval $\nu \ldots \nu + d\nu$, from the solid angle $d\Omega$ in the direction $\mathbf{e}$, which is inclined by an angle θ against $\mathbf{n}$, the power

$$dW = I_\nu(\mathbf{x}, \mathbf{e}) \cos\theta \, d\sigma \, d\Omega \, d\nu. \qquad (13.1)$$

$I_\nu(\mathbf{x}, \mathbf{e})$ has the dimension erg s^{-1} cm^{-2} Hz^{-1} ster^{-1} and is a function of

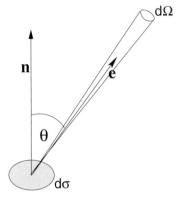

Figure 13.1. Defining the radiative intensity I_ν.

- place **x**,
- direction **e** and
- frequency ν.

Imagine an extended astronomical source in the sky of solid angle Ω_S. The direction **e** towards it can be expressed by two angles, say, θ and ϕ. Let $I_\nu(\theta, \phi)$ be the intensity distribution over the source. The total flux which we receive on the Earth is given by

$$F_\nu^{\text{obs}} = \int_{\Omega_S} I_\nu(\theta, \phi)\, d\Omega \qquad (13.2)$$

where

$$d\Omega = \sin\theta\, d\theta\, d\phi$$

is the element of solid angle. In practice, the integration may be performed by mapping the source with a telescope of high spatial resolution. If the intensity is constant over the source, $I_\nu(\theta, \phi)$ can be taken out of the integral and then

$$F_\nu^{\text{obs}} = I_\nu\, \Omega_S. \qquad (13.3)$$

Note that I_ν is independent of the distance D of the source because the solid angle Ω_S and the flux F_ν^{obs} are both proportional to D^{-2}. If we recede in a space ship from the Sun, we receive less light but the intensity in the direction of the Sun stays the same.

- The *mean intensity* J_ν is defined as the average of I_ν over the total solid angle 4π,

$$J_\nu = \frac{1}{4\pi} \int_{4\pi} I_\nu\, d\Omega = \frac{1}{4\pi} \int_0^\pi d\theta \int_0^{2\pi} d\phi\, I_\nu(\theta, \phi) \sin\theta. \qquad (13.4)$$

● The *net flux* F_ν through a unit surface follows from (13.1):

$$F_\nu = \int_0^\pi d\theta \int_0^{2\pi} d\phi I_\nu(\theta, \phi) \cos\theta \sin\theta. \tag{13.5}$$

13.1.2   The transfer equation and its formal solution

13.1.2.1   The source function

When a light ray propagates in the direction **e** along an axis s, its intensity I_ν changes. It is weakened by extinction and reinforced by emission and by light that is scattered from other directions into the direction **e**. This is written in the following way:

$$\frac{dI_\nu}{ds} = -K_\nu^{\text{ext}} I_\nu + \epsilon_\nu. \tag{13.6}$$

K_ν^{ext} is the coefficient for extinction, and ϵ_ν comprises emission and scattering. Note that the coefficient ϵ_ν defined in (8.1) stands for 'true' emission only and is thus different. We see from (13.6) that per unit length, the intensity I_ν decreases by an an amount $K_\nu^{\text{ext}} I_\nu$ and increases by ϵ_ν. Both K_ν^{ext} and ϵ_ν refer to unit volume and their dimensions are cm^{-1} and $\text{erg s}^{-1} \text{ cm}^{-3} \text{ Hz}^{-1} \text{ ster}^{-1}$, respectively.

A fundamental quantity in light propagation is the *optical depth* or *optical thickness*, τ_ν. It depends on frequency. When light travels from point S_1 to point S_2, its value is given by the integral

$$\tau_\nu = \int_{S_1}^{S_2} K_\nu^{\text{ext}} \, ds, \tag{13.7}$$

so along the line of sight

$$d\tau_\nu = K_\nu^{\text{ext}} \, ds.$$

An object of low optical depth ($\tau \ll 1$) is transparent and radiation reaches us from any internal point. An object of large optical depth ($\tau \gg 1$) is opaque and we can only look at its surface and do not receive photons from its interior. When we introduce the *source function*

$$S_\nu = \frac{\epsilon_\nu}{K_\nu^{\text{ext}}} \tag{13.8}$$

we can rewrite (13.6) as

$$\frac{dI_\nu}{d\tau_\nu} = -I_\nu + S_\nu. \tag{13.9}$$

This equation tells us that the intensity increases with optical depth as long as it is smaller than the source function ($I_\nu < S_\nu$) and decreases when it is stronger. The source function is thus the limiting value of the intensity for high optical thickness.

In a dusty medium, the intensity of a light ray decreases because of extinction, which is the sum of absorption and scattering, and increases because the grains radiate thermally and scatter light. If scattering is isotropic and the grains are at temperature T, the source function becomes

$$S_\nu = \frac{K_\nu^{\text{abs}} B_\nu(T) + K_\nu^{\text{sca}} J_\nu}{K_\nu^{\text{ext}}} \tag{13.10}$$

where J_ν is the mean intensity of the radiation field after (13.4).

13.1.2.2 *Analytical solutions to the transfer equation*

We solve equation (13.6) for three scenarios:

- *Pure extinction without emission* ($\epsilon_\nu = 0$). An example is starlight weakened on the way to us by an intervening dust cloud. The obvious solutions to (13.6) and (13.9) are:

$$I_\nu(s) = I_\nu(0) \exp\left[-\int_0^s K_\nu(s')\,ds' \right] \tag{13.11}$$

$$I_\nu(\tau) = I_\nu(0) \cdot e^{-\tau_\nu}. \tag{13.12}$$

  $I_\nu(0)$ is the intensity of the background star if it were not attenuated. The observed intensity diminishes exponentially with optical depth.
- *Pure emission without extinction* ($K_\nu^{\text{ext}} = 0$). This situation applies quite well to submillimeter radiation from an interstellar dust cloud, now

$$I_\nu(s) = I_\nu(0) + \int_0^s \epsilon_\nu(s')\,ds'. \tag{13.13}$$

  Without a background source ($I_\nu(0) = 0$), which is the standard scenario, and for constant emission coefficient ϵ_ν, the observed intensity grows linearly with the path length s.
- *Emission plus absorption.* In this most general case, we multiply (13.9) by e^τ and integrate by parts which leads to

$$I(\tau) = I(0)e^{-\tau} + \int_0^\tau S(\tau')e^{-(\tau-\tau')}\,d\tau'. \tag{13.14}$$

The index for the frequency has been dropped for convenience of notation. Although it appears as if we had mastered the radiative transfer problem, in reality, we have not. Equation (13.14) is only a formal solution to the transfer equation (13.9). The hard part is to find the source function $S_\nu(\tau)$ in formula (13.14). If $S_\nu(\tau)$ is constant, it may be taken out from under the integral yielding

$$I_\nu(\tau) = I_\nu(0)e^{-\tau} + S_\nu(\tau) \cdot [1 - e^{-\tau}]. \tag{13.15}$$

13.1.2.3   *A dust cloud of uniform temperature*

In a dust cloud, emission occurs only in the infrared. At these wavelengths, grain scattering is practically absent. For dust at temperature T, the source function is then equal to the Planck function at that temperature,

$$S_\nu = B_\nu(T).$$

Towards a uniform cloud of optical depth τ_ν, one, therefore, observes, according to (13.15), the intensity

$$I_\nu(\tau) = I_\nu(0)e^{-\tau_\nu} + B_\nu(T) \cdot [1 - e^{-\tau_\nu}]. \tag{13.16}$$

If the background is a point source of flux $F_\nu(0)$, one receives, out of a solid angle Ω, the flux

$$F_\nu(\tau) = F_\nu(0)e^{-\tau_\nu} + B_\nu(T) \cdot [1 - e^{-\tau_\nu}]\Omega. \tag{13.17}$$

When the background is absent or its spectrum known and multi-frequency observations are available, equation (13.17) yields an estimate of the source temperature and its optical depth, and thus its mass. The limiting solutions to (13.16) are:

$$I_\nu(\tau) = \begin{cases} B_\nu(T)\tau_\nu + I_\nu(0) & \text{for } \tau_\nu \ll 1 \\ B_\nu(T) & \text{for } \tau_\nu \gg 1. \end{cases} \tag{13.18}$$

They imply that without a background source and if the cloud is translucent ($\tau \ll 1$), the intensity is directly proportional to the optical depth: doubling the column density, doubles the strength of the received signal. In this case, the intensity I_ν carries information about the amount of dust but also about the spectral behavior of the grains via the frequency dependence of the optical depth. However, if the source is opaque ($\tau \gg 1$), all information about the grains is wiped out and the object appears as a blackbody of temperature T.

13.1.3   The brightness temperature

Observations at wavelengths greater than 100 μm fall in the domain of radioastronomy. There it is common to replace the intensity I_ν by the brightness temperature T_b which is defined through

$$I_\nu = B_\nu(T_b). \tag{13.19}$$

$B_\nu(T)$ is the Planck function. Because T_b expresses an intensity, it depends, of course, on wavelength, the only exception is a blackbody for which T_b is independent of ν. Although T_b is not a direct physical quantity, its use offers practical advantages:

- One now has for the intensity handy numbers in a simple unit (Kelvin), instead of weird numbers in awkward units.

- When the source is optically thick, T_b generally, but not always, indicates its physical (thermodynamic equilibrium) temperature.
- At low frequencies ($h\nu/kT_b \ll 1$), in the Rayleigh–Jeans approximation, the brightness temperature T_b is directly proportional to the observed intensity, $T_b = I_\nu c^2/2k\nu^2$. When we put in equation (13.15) $S_\nu(\tau) = B_\nu(T)$, the brightness temperature of a source of optical thickness τ becomes

$$T_b(\tau) = T_0\, e^{-\tau} + T(\tau) \cdot [1 - e^{-\tau}]. \qquad (13.20)$$

In the case of dust, T is the grain temperature T_d, in the case of line radiation, T is the excitation temperature T_{ex} (see (13.66)).

13.1.4 The main-beam-brightness temperature of a telescope

The quantity in which radioastronomers express the strength of a signal, especially in the case of line radiation, is the main-beam-brightness temperature T_{mb}. To understand it, we have to briefly discuss the standard measuring apparatus in radioastronomy. It consists of a parabolic antenna with a detector at its focus. For the moment, we replace the detector at the focus by an emitting device, as in a radar antenna. Let $P(\vartheta, \phi)$ denote the flux emitted by the antenna in the direction of the sky coordinates (ϑ, ϕ). The normalized function

$$P_n(\vartheta, \phi) = \frac{P(\vartheta, \phi)}{P_{max}}$$

is called the antenna pattern and determines the telescope beam. In the direction (ϑ_0, ϕ_0), which is parallel to the normal of the antenna surface, $P(\vartheta, \phi)$ attains its maximum value P_{max}; one may, of course, always use sky coordinates such that $(\vartheta_0, \phi_0) = (0, 0)$. The antenna pattern is the result of diffraction phenomena. Fortunately, most of the power is emitted into the main beam, in directions close to the normal of the antenna surface; however, some radiation is spilled into the side lobes. In a one-dimensional analog, the situation can be compared to the diffraction pattern on a screen behind an illuminated slit: the intensity has a central maximum and the range around it down to the first zero points on both sides corresponds to the main beam of a telescope.

The total solid angle of the antenna is defined by

$$\Omega_A = \int_{4\pi} P_n(\vartheta\phi)\, d\Omega \qquad (13.21)$$

with $d\Omega = \sin\vartheta\, d\vartheta\, d\phi$ (see (2.2)). Likewise one defines the solid angle of the main beam, Ω_{mb}, where the integral extends only over the main beam, and, therefore, $\Omega_{mb} < \Omega_A$.

Now we return to the common occupation of astronomers: they do not send messages into space but receive radiation from there. Note that, according to the reciprocity theorem of section 3.5.5, the spatial distribution of the electromagnetic field does not change when one switches from emission to reception. Let $B(\vartheta, \phi)$ be the intensity distribution on the sky and let us assume that the influence of the atmosphere of the Earth on $B(\vartheta, \phi)$ has been eliminated, either because the telescope is flying on a satellite or, if it stands on the ground, by way of calibration. The observed brightness B_{obs} is then the following mean over the main beam:

$$B_{obs} = \frac{\int_{\Omega_{mb}} B(\vartheta, \phi) P(\vartheta, \phi) \, d\Omega}{\int_{\Omega_{mb}} P(\vartheta, \phi) \, d\Omega}.$$

It is converted into the main beam brightness temperature, T_{mb}, by setting it equal to the Planck function,

$$B_{obs} = B_\nu(T_{mb})$$

or to its Rayleigh–Jeans approximation, $B_{obs} = T_{mb} 2k\nu^2/c^2$. It is often necessary to state explicitly which conversion is used.

The main-beam-brightness temperature T_{mb} is also called the antenna temperature, denoted T_A, although historically the definition of T_A was somewhat different. It is justified to say that the antenna temperature constitutes the high-tech counterpart to stellar magnitudes (section 7.3), equally fathomless in its subtleties. The major reason why T_A evades a precise determination lies in the fact that the telescope picks up a little bit of radiation from everywhere, even from the kitchen of the observatory. To account for it, any ambitious calibration attempt, therefore, requires the introduction of an efficiency η_{kitch}, on top of a dozen other ηs. In practice, astronomers do not worry about this and pretend to have all ηs under control. Nevertheless, one should be aware that the absolute calibration accuracy at 1 mm wavelength is, at best, 5% and in the submillimeter range, at best, 10%.

The main-beam-brightness temperature T_{mb} and the brightness temperature T_b are related, as they should be when names are so similar. When a source of constant brightness temperature T_b subtends a solid angle Ω_S much bigger than the solid angle Ω_A of the antenna, one measures $T_{mb} = T_b$. If, however the beam is much larger than the source ($\Omega_A \gg \Omega_S$),

$$T_{mb} = \frac{\Omega_S}{\Omega_{mb}} T_b$$

and then the antenna temperature increases when one uses a smaller beam, although the received physical power does not change with Ω_{mb}.

13.2   Spherical clouds

The dust emission of a cloud is observed at various wavelengths with the goal of deriving its internal structure. The observational data consist either of maps or of single point measurements towards different positions on the source obtained with different spatial resolution. When the cloud is heated from inside by a star, it is obviously inhomogeneous and estimates of the dust column density and temperature based on the assumption of a uniform cloud (see (13.17)) are no longer acceptable. A more refined model is needed.

Easy-to-model geometrical source configurations are infinite slabs, infinite cylinders and spheres. The structure of the medium then depends only on one spatial coordinate, the problem is one-dimensional. The radiative transfer may be solved in many ways and fast and efficient public computer codes exist. In practice, however, one may encounter difficulties using a public code because it usually has to be modified to meet the demands of a specific astronomical problem. It may then be better to write a code oneself, even at the expense of less professionality. We now present the equations and sketch the numerical strategy for such an endeavour in spherical symmetry.

13.2.1   Moment equations for spheres

In a spherical cloud, the dust density $\rho(r)$, the dust temperature $T(r)$, the mean radiative intensity $J_\nu(r)$ and the flux $F_\nu(r)$ are only functions of distance r to the cloud center. However, the basic quantity, the frequency-dependent intensity I, also depends on direction θ, where θ is the inclination angle to the radial vector, so altogether

$$I = I_\nu(r, \theta).$$

Putting $\mu = \cos\theta$, we obtain, from (13.4) and (13.5), for the mean intensity and the flux:

$$J_\nu = \tfrac{1}{2} \int_{-1}^{1} I_\nu(\mu)\,d\mu \tag{13.22}$$

$$F_\nu = 2\pi \int_{-1}^{1} I_\nu(\mu)\mu\,d\mu. \tag{13.23}$$

$(4\pi/c)J_\nu$ is the radiative energy density. J_ν and F_ν are the zeroth and first moment of the intensity. The second moment reads[1]

$$K_\nu = \tfrac{1}{2} \int_{-1}^{1} I_\nu(\mu)\mu^2\,d\mu. \tag{13.24}$$

In the case of an isotropic radiation field, $I_\nu = J_\nu$ and the net flux is zero, $F_\nu = 0$. Then the ratio

$$f_\nu = \frac{K_\nu}{J_\nu}$$

which is called the *Eddingon factor*, equals $\tfrac{1}{3}$ and the quantity $(4\pi/c)K_\nu$ gives the radiation pressure. It is a simple exercise to show that for spherical symmetry

---

[1] Note that volume coefficients for absorption, scattering and extinction are denoted by the slightly different symbol K_ν.

equation (13.6) becomes

$$\mu \frac{\partial I_\nu}{\partial r} + \frac{1 - \mu^2}{r} \frac{\partial I_\nu}{\partial \mu} = -K_\nu^{\text{ext}} I_\nu + \epsilon_\nu. \tag{13.25}$$

Multiplying (13.25) by $\mu^0 = 1$ and integrating over all directions gives

$$\frac{1}{r^2} \frac{d}{dr} (r^2 F_\nu) = -4\pi J_\nu K_\nu^{\text{ext}} + 4\pi \epsilon_\nu. \tag{13.26}$$

This equation looks much simpler than (13.25) because there is no dependence on the angle μ. However, we cannot solve it for $J_\nu(r)$ because it contains the new variable $F_\nu(r)$. One could try to find an additional equation by multiplying (13.25) with μ^1 and integrating over 4π yielding

$$\frac{dK_\nu}{dr} + \frac{3K_\nu - J_\nu}{r} = -\frac{F_\nu K_\nu^{\text{ext}}}{4\pi} \tag{13.27}$$

but the problem of having one more unknown than equations remains because now K_ν comes into play. The introduction of the next higher moment (multiplication of (13.25) by μ^2) will not remedy the situation, either. Instead, to be able to use formulae that do not depend explicitly on the direction angle μ, one needs physical guidance. For instance, for low optical depth, the radiation field is streaming radially outwards and $J_\nu = K_\nu = F_\nu/4\pi$. For high optical depth, the radiation field is isotropic and $f_\nu = \frac{1}{3}$.

13.2.2 Frequency averages

One can radically simplify the numerical problem by taking frequency averages. Mean values for the extinction coefficent K_ν in the case of low and high optical depth are found in the following way:

- In an optically thick configuration, where the mean intensity J_ν is close to the Planck function $B_\nu(T)$, one uses the *Rosseland mean* $K_R(T)$ defined through

$$\frac{1}{K_R(T)} = \frac{\int \frac{1}{K_\nu} \frac{\partial B_\nu(T)}{\partial T} d\nu}{\int \frac{\partial B_\nu(T)}{\partial T} d\nu}. \tag{13.28}$$

As $\int B_\nu(T) \, d\nu = \sigma T^4/\pi$, where σ is the radiation constant of (5.80), the denominator in (13.28) equals $4\sigma T^3/\pi$. The Rosseland mean gives the greatest weight to the lowest values of K_ν, as in a room darkened by thick curtains: it is not their thickness but the open slits between them that determine how much light enters from outside. If F_ν is the energy flux at frequency ν per unit area and time, the definition of K_R in equation (13.28) must be, and indeed is, although not evident at first glance, equivalent to

$$K_R \int F_\nu \, d\nu = \int K_\nu F_\nu \, d\nu.$$

For example, the equation for the radiative energy transport in the stellar interior may be approximated by using the average opacity K_R and assuming $J_\nu = B_\nu(T)$ and $f_\nu = \frac{1}{3}$. Then the total flux $F = \int F_\nu\, d\nu$ is, from formula (13.27), determined by the temperature gradient in the star:

$$F = -\frac{16\sigma T^3}{3K_R(T)}\frac{dT}{dr}.$$ (13.29)

We point out that because $P_r = aT^4/3$, with a from (5.84), is the radiation pressure; equation (13.29) corresponds to Fick's first law of diffusion, applied to a photon gas.

- In a tenuous cloud, one takes, as a frequency average, the *Planck mean K_P*. It is now not a question of which frequency penetrates deepest, they all do. Instead one has to take the average over the emitted energy,

$$K_P(T) = \frac{\int K_\nu B_\nu(T)\, d\nu}{\int B_\nu(T)\, d\nu}.$$ (13.30)

Using the approximation $K_\nu = K_0\nu^m$, which is fine for interstellar dust (section 8.2), one may compute the coefficients analytically. For $m = 2$, one gets a quadratic increase with temperature,

$$K_P = 8.14 \times 10^{21} K_0 T^2 \qquad K_R = 1.08 \times 10^{22} K_0 T^2.$$ (13.31)

The two values are astonishingly close. When we put for the absorption efficiency $Q_\nu = 10^{-23} a\nu^2$ after (8.10), where a is the grain radius in cm, and let K_R refer to 1 g of interstellar matter, $K_0 = 7.5 \times 10^{-24} R/\rho_{gr}$. A dust-to-gas ratio $R_d = 0.01$ and a matter density $\rho_{gr} = 2.5$ g cm^{-3} leads to

$$K_R \sim 3 \times 10^{-4} T^2 \text{ cm}^2 \text{ per gram of interstellar matter.}$$ (13.32)

The Rosseland mean opacity may be acceptable in protostellar cocoons or accretion disks, the Planck mean K_P may be used to assess the effective optical depth of clumps in a molecular cloud. As grains in early-stage protostellar condensations are probably fluffy and possess ice mantles (section 2.6), it may be better under such conditions to use the enhanced estimate

$$K_P \sim 10^{-3} T^2 \text{ cm}^2 \text{ per gram of IM.}$$ (13.33)

13.2.3 Differential equations for the intensity

We give here a set of equations for the intensity that retains the full information on frequency and direction. Our choice of coordinates is illustrated in figure 13.2. Light is propagating at constant impact parameter p parallel to the z-axis either to the left (I_ν^-) or to the right (I_ν^+). The coordinates p and z, which are used instead of r and μ, are connected to the radius by

$$r^2 = p^2 + z^2$$

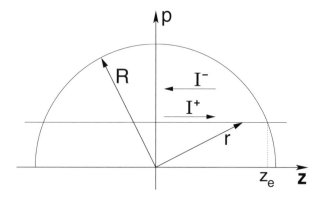

Figure 13.2. The coordinate system for the intensity in a spherical cloud of outer radius R. Light rays at constant impact parameter p are directed either to the left (I^-) or to the right (I^+). The ray I^+ enters the cloud at $(p, -z_e)$ and leaves at (p, z_e). The cloud is immersed into a radiation field of isotropic intensity I_0.

and the intensities obey the differential equations (see (13.9))

$$\frac{dI_\nu^+}{dz} = -K_\nu^{\text{ext}} \cdot [I_\nu^+ - S_\nu] \tag{13.34}$$

$$\frac{dI_\nu^-}{dz} = K_\nu^{\text{ext}} \cdot [I_\nu^- - S_\nu]. \tag{13.35}$$

If the dust particles scatter isotropically, the source function is given by (13.10). K_ν^{abs}, K_ν^{sca} and K_ν^{ext} are the volume coefficients for absorption, scattering and extinction referring to 1 cm^3 of interstellar space. In view of the symmetry of the spherical cloud, (13.34) and (13.35) have to be solved only in the right-hand quadrant of figure 13.2, i.e. for $z > 0$. The equations must be complemented by two boundary conditions:

- The first states that at the cloud's edge, where $z = z_e$, the incoming intensity I_ν^- is equal to the intensity of the external radiation field $I_{0\nu}$:

$$I_\nu^-(p, z_e) = I_{0\nu} \quad \text{with } z_e = \sqrt{R^2 - p^2}. \tag{13.36}$$

- The formulation of the second boundary condition depends on whether there is a central star or not. Without it or when there is a star but the impact parameter p is larger than the stellar radius R_*, one has, for reasons of symmetry,

$$I_\nu^+(p, z) = I_\nu^-(p, z) \quad \text{at } z = 0. \tag{13.37}$$

With a star of stellar surface flux $F_{*\nu}$ and when $p \leq R_*$ (see (5.74)),

$$I_\nu^+(p, z_*) - I_\nu^-(p, z_*) = \frac{F_{*\nu}}{\pi} \quad \text{with } z_* = \sqrt{R_*^2 - p^2}. \tag{13.38}$$

13.2.4   Integral equations for the intensity

Instead of solving the differential equations (13.34) and (13.35), it is, for reasons of numerical stability, advisable to use the corresponding integral equations (see (13.14))

$$I^+(\tau) = e^{-\tau}\left[I^+(0) + \int_0^\tau S(x)e^x \, dx\right] \tag{13.39}$$

$$I^-(t) = e^{-t}\left[I_0 + \int_0^t S(x)e^x \, dx\right]. \tag{13.40}$$

The boundary conditions stay the same. In (13.39), the optical depth τ increases from left to right, so $\tau(z = 0) = 0$ and $\tau(z > 0) > 0$. In (13.40), the optical depth t equals zero at the cloud surface $(t(z_e) = 0)$ and grows from right to left, so $t(z < z_e) > 0$.

Solving (13.39) and (13.40) is not eneough, one still has to determine the dust temperature. One arrives at a full solution to the radiative transfer problem iteratively. Suppose that after iteration cycle i we approximately know the grain temperature $T(r)$, the mean intensity of the radiation field $J_\nu(r)$, and thus after (13.10) the source function $S_\nu(r)$. We find improved values of $T(r)$, $J_\nu(r)$ and $S_\nu(r)$ in the following way: First, we determine for each impact parameter p the intensity I_ν^- from (13.40). Using the inner boundary condition, we then calculate, for all p, the intensity I_ν^+ from (13.39). From I_ν^+ and I_ν^- we form the quantities

$$u_\nu(p, z) = \tfrac{1}{2}\left[I_\nu^+(p, z) + I_\nu^-(p, z)\right] \tag{13.41}$$

$$v_\nu(p, z) = \tfrac{1}{2}\left[I_\nu^+(p, z) - I_\nu^-(p, z)\right]. \tag{13.42}$$

From the first, u_ν, one gets an updated mean intensity

$$J_\nu(r) = \frac{1}{r}\int_0^r \frac{p u_\nu(p, r)}{\sqrt{r^2 - p^2}} \, dp. \tag{13.43}$$

It is not difficult to see that this agrees with the definition of J_ν from from (13.22). The new $J_\nu(r)$ is then inserted into equation (8.8) or into

$$\int J_\nu(r)K_\nu(r) \, d\nu = \int B_\nu(T(r))K_\nu(r) \, d\nu$$

which yields new temperatures when solved for T. Updating the source function $S_\nu(r)$ (see (13.8)) finishes iteration cycle $i + 1$.

13.2.5   Practical hints

- As starting values for the first cycle $(i = 1)$, an initial mean intensity $J_\nu(r) = 0$ should be acceptable. For the temperature, one may put $T(r) = $ constant if there is no central star and the outer radiation field is not too hard and strong.

With a star, $T(r) \sim 20(L_*/r^2)^{1/6}$ should work when r and L_* are expressed in cgs units (see (8.17)).

- Use linear interpolation (trapezium rule) for evaluating integrals. The accuracy is then proportional to the square of the step size h. In Simpson's rule (quadratic interpolation), it is proportional to h^4 and that is worse when the grid is coarse.

- The iterations converge when the changes of the temperature $\Delta T(r)$ become smaller with each new cycle. As a check on the accuracy of the calculations, one can compute, for each radius r, the net flux

$$F_\nu(r) = \frac{4\pi}{r^2} \int_0^r p\, v_\nu(p, r)\, dp. \qquad (13.44)$$

This formula can be derived from (13.23) and (13.42). The integrated flux through a shell of radius r is

$$L(r) = 4\pi r^2 \int F_\nu \, d\nu.$$

If the cloud has a central star of luminosity L_*, the condition $L(r) = L_*$ should be numerically fulfilled everywhere to an accuracy of $\sim 1\%$.

- Without an embedded source, heating comes from an outer isotropic radiation field of intensity $I_{0\nu}$. The integrated flux $L(r)$ is then zero because what enters at the cloud surface must also leave. Nevertheless, the dust may be quite hot, especially in the outer envelope. Numerically, $|L(r)|$ should now be small compared to

$$4\pi R^2 \pi \int I_{0\nu} \, d\nu$$

which is the total flux from the external radiation field through the cloud surface.

- To select a sufficiently fine radial grid which is also economical in terms of computing time, the number of points and their spacings must be adapted during the iterations. For example, in a cloud of 50 mag of visual extinction with a central hot star, the dust column density between neighboring grid points must be small near the star because UV radiation is absorbed very effectively. At the cloud edge, however, there are only far infrared photons and the grid may be coarse. Generally speaking, as one recedes from the star, the radiation field softens and the mean free path of the photons increases.

At each radius r, there therefore exists a cutoff frequency with the property that only photons with $\nu < \nu_{\text{cut}}$ are energetically relevant, whereas those with $\nu > \nu_{\text{cut}}$, whose mean free path is, as a rule of thumb, shorter, are unimportant. ν_{cut} may be defined by and computed from

$$\int_{\nu_{\text{cut}}}^{\infty} J_\nu(r) \, d\nu = \alpha \int_0^{\infty} J_\nu(r) \, d\nu \qquad (\alpha \sim 0.2).$$

Choosing the radial step size in such a way that the optical depth at the cutoff frequency is of order 0.1 automatically ensures that the grid is fine only where necessary.

- Very close to the star, dust evaporates. The exact radius r_{evap} is part of the solution of the radiative transfer; it depends, of course, on the grain type. To account for dust evaporation, one may evaluate the grain temperature T_d in the local radiation field and add a flag at all radii to each grain type indicating whether T_d is above or below the evaporation threshold T_{evap}. The flag then controls whether the grain contributes in the next cycle to the extinction and emission or not. For heavily obscured sources ($A_V > 20$ mag), the exact value of the evaporation radius has little influence on the spectrum, if at all, only in the near infrared.

- For N radial grid points $r_1 = R_*, r_2, r_3, \ldots, r_N = R$, there are $N+1$ impact parameters $p_i = r_i$ with $i = 1, \ldots, N$ and $p_0 = 0$. Because to first order the intensity over the stellar disk does not change, one needs only two impact parameters ($p = 0$ and $p = R_*$) for the star. The second radial grid point r_2 should only be a tiny bit (0.1%) bigger than r_1.

- The dust temperature T_d is computed from (8.8). Let T_j be the value of the grain temperature in iteration cycle j. Writing (8.8) in the form

$$\int Q_\nu^{abs} J_\nu \, d\nu = \overline{Q}(T) \frac{\sigma}{\pi} T^4$$

where σ is the radiation constant of (5.80) and $\overline{Q}(T)$ is defined by

$$\overline{Q}(T) = \frac{\int Q_\nu^{abs} B_\nu(T) \, d\nu}{\int B_\nu(T) \, d\nu}$$

one gets an improved estimate T_{j+1} for the new iteration cycle $j + 1$ from

$$T_{j+1}^4 = T_j^4 \int Q_\nu^{abs} J_\nu \, d\nu / \overline{Q}(T_j).$$

The sequence $T_j, T_{j+1}, T_{j+2}, \ldots$ quickly converges towards T_d.

13.3 Passive disks

13.3.1 Radiative transfer in a plane parallel layer

All stars are surrounded during their early evolution by an accreting disk. A disk is passive when it is only heated by the star and internal dissipational losses are negligible. Imagine a flat and thin disk with density structure $\rho(r, z)$ given in cylindrical coordinates (r, z). Let it be illuminated by a central star of luminosity L_*, radius R_* and temperature T_* whose mid-point is at ($r = 0, z = 0$). To determine the disk emission, we consider, in figure 13.3, a small section at radius r from the star.

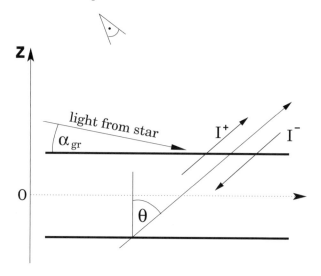

Figure 13.3. A small portion of a thin disk. The star is far away to the left at a distance r and its light falls on the disk under some small grazing angle α_{gr}. The basic transfer equation (13.9) is solved under all inclination angles θ for incoming (I^-) and outgoing (I^+) radiation. The situation is symmetric with respect to the mid-plane which is indicated by the dotted line at $z = 0$. The position of the observer is also shown.

The formalism of the radiative transfer developed for spheres can be carried over to the slab geometry. Again one starts by solving the pair of equations (13.39) and (13.40) for I^+ and I^-, which are now the intensities along straight lines under various angles θ with respect to the z-axis. Then one calculates from these intensities a new vertical temperature run $T(z)$; the procedure is iterated until $T(z)$ converges.

The starlight falls on the disk under a small effective grazing angle α_{gr}, which is evaluated in equation (13.49) for a flat disk and in (13.53) for an inflated disk. Viewed from the disk surface (figure 13.3), the upper hemisphere of the star subtends, at large distances, a solid angle

$$\Omega_* = \frac{1}{2}\pi \frac{R_*^2}{r^2}.$$

In order not to break the slab symmetry, the stellar hemisphere is replaced by a luminous ring of the same intensity and solid angle. This ring encircles the whole sky at an elevation α_{gr}, its width is $\Omega_*/2\pi \ll \alpha_{gr}$.

The first boundary condition of the transfer equations states that at the top of the disk the incoming intensity I^- is zero except towards the ring where it is equal to $B_\nu(T_*)$. The second boundary condition, $I^+ = I^-$ at $z = 0$, follows from symmetry requirements.

The radiative transfer at radial distance r_0 yields the temperature structure $T(r_0, z)$ and the intensities $I_\nu^+(r_0, z, \theta)$, $I_\nu^-(r_0, z, \theta)$ for all frequencies ν, heights z and angles θ. One thus also obtains I_ν^+ at the top of the disk, for all directions θ. To obtain the total flux that an observer receives when he views the disk under a certain angle θ_{obs}, one has to solve the radiative transfer at all radii r and then sum up the contributions. Furthermore, one has to include the direct emission from the star, and possibly apply a correction for foreground extinction.

13.3.1.1 Disks of high optical thickness

When the opacity in the vertical direction is large, as happens in young stellar disks where A_V is 10^4 mag or greater, one may encounter numerical difficulties. We then recommend the following strategy: Divide the disk in its vertical structure into a completely opaque mid layer sandwiched between two thin top layers (figure 13.4). The optical depth of the top layer in the z-direction is chosen to be sufficiently large for all starlight, including the scattered part, to become absorbed but small enough so that all re-emitted radiation, which has infrared wavelengths, can leave; a typical value would be $A_V \sim 5$ mag. Let f_ν be the monochromatic stellar flux that falls in z-direction on the disk (see (13.47)). The integrated incident flux in the z-direction is then

$$f_* = \int f_\nu \, d\nu.$$

Without scattering, all the flux f_* is absorbed in the top layer and then re-emitted. Half of it will penetrate into the mid layer, the other half will leave the top layer into space.

The mid layer is assumed to be isothermal at temperature T_{mid}, so that the radiative transfer inside it is very simple (see (13.16)). One can immediately write down the intensity that enters the top layer from below under the direction $\mu = \cos\theta$,

$$I_\nu^+(z = 0, \mu) = B_\nu(T_{mid}) \cdot \left[1 - e^{\tau_\nu(\mu)}\right]. \qquad (13.45)$$

Here $\tau_\nu(\mu)$ is the optical depth of the mid layer. The monochromatic flux received by the top layer from below is, therefore, (see (13.23)),

$$2\pi B_\nu(T_{mid}) \int_0^1 \left[1 - e^{\tau_\nu(\mu)}\right] \mu \, d\mu.$$

Defining the last integral as

$$E_\nu \equiv \int_0^1 \left[1 - e^{\tau_\nu(\mu)}\right] \mu \, d\mu$$

the temperature T_{mid} can be numerically evaluated from the condition

$$f_* = 4\pi \int B_\nu(T_{mid}) E_\nu \, d\nu. \qquad (13.46)$$

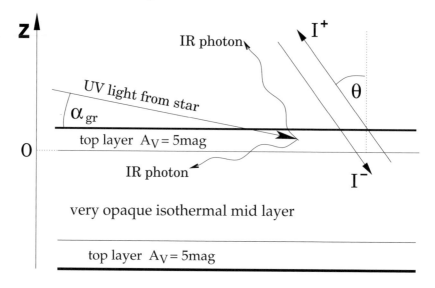

Figure 13.4. A small portion of a geometrically thin but optically very thick disk. The UV light from the star on the left is completely absorbed in the top layer. About half of the re-emitted IR photons heat the mid layer which is assumed to be isothermal at temperature T_{mid}. One calculates the radiative transfer for the top layer ($z \geq 0$) using the boundary condition (13.45) for I^+. Note that the zero point of the z-axis is different here from that in figure 13.3.

The assumption of isothermality of the optically thick mid layer is valid because a temperature gradient would imply in the diffusion approximation of (13.29) a net energy flux. However, there can be none because there are no internal sources and the disk is symmetric about its mid plane.

13.3.1.2 *The grazing angle in a flat disk*

We compute the monochromatic flux f_ν that falls in the z-direction on the disk and heats it. If the disk is flat and thin or if it has a constant opening angle, a surface element of unit area receives out of a solid angle $d\Omega$ pointing towards the star the flux $B_\nu(T_*) \, d\Omega \sin \theta \sin \phi$ (figure 13.5); $B_\nu(T_*)$ denotes the uniform stellar intensity. By the whole upper half of the star, the surface element is illuminated by the flux

$$f_\nu = B_\nu(T_*) \int_0^{\theta_{max}} \sin^2 \theta \, d\theta \int_0^\pi \sin \phi \, d\phi = B_\nu(T_*) \cdot \left[\theta_{max} - \tfrac{1}{2} \sin 2\theta_{max}\right]$$

$$= B_\nu(T_*) \left[\arcsin(R_*/r) - (R_*/r)\sqrt{1 - (R_*/r)^2}\right]. \tag{13.47}$$

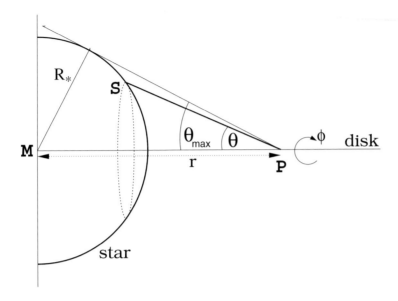

Figure 13.5. At position P on the disk, a unit area is illuminated by an infinitesimal surface element S of the star by the flux $B_\nu(T_*)\, d\Omega\, \sin\theta\, \sin\phi$. Seen from P, the surface element S subtends the solid angle $d\Omega = \sin\theta\, d\theta\, d\phi$, where θ is the angle between the lines $\overline{PM}$ and $\overline{PS}$. The factor $\sin\theta\, \sin\phi$ takes into account the inclination of the disk towards the incoming light. In the figure, S is drawn near $\phi = 90°$ and the dotted line near S is a circle with $\theta = $ constant.

We express f_ν through an effective grazing angle, α_{gr}, defined by

$$f_\nu = \Omega_* \cdot B_\nu(T_*) \cdot \alpha_{gr} \tag{13.48}$$

where Ω_* is the solid angle of the star. At large distances ($x \equiv R_*/r \ll 1$), $\arcsin x - x\sqrt{1-x^2} \simeq 2x^3/3$ and, therefore, $f_\nu = \frac{2}{3}x^3 B_\nu(T_*)$ and

$$\alpha_{gr} = \frac{4}{3\pi}\frac{R_*}{r}. \tag{13.49}$$

To obtain the total flux F_ν falling on an infinite disk, we integrate (13.47) over the entire plane outside the star,

$$F_\nu = \int_{R_*}^{\infty} 2\pi r f_\nu \, dr$$

$$= 2\pi B_\nu(T_*) \int_{R_*}^{\infty} \left[r \arcsin \frac{R_*}{r} - R_*\sqrt{1 - \frac{R_*^2}{r^2}} \right] dr = \tfrac{1}{2}\pi^2 R_*^2 B_\nu(T_*).$$

The integrals are solved using formulae (A.18) and (A.19). To evaluate the expressions at the upper bound $r = \infty$, we set $x = 1/r$, expand around $x = 0$ and let x go to zero.

Summing F_ν up over all frequencies, one finds that each side of an infinite disk intercepts exactly one-eighth of the stellar luminosity $L_* = 4\pi R_*^2 \sigma T_*^4$.

If the disk is a blackbody, it will have a surface temperature

$$T_{\text{disk}}^4(r) = \frac{T_*^4}{\pi} \left[\arcsin(R_*/r) - (R_*/r)\sqrt{1 - (R_*/r)^2} \right]. \tag{13.50}$$

For $r \gg R_*$, the radiative heating rate has a $r^{-3/4}$-dependence,

$$T_{\text{disk}}(r) = T_* \left(\frac{2}{3\pi} \right)^{1/4} \left(\frac{R_*}{r} \right)^{3/4}. \tag{13.51}$$

13.3.2   The grazing angle in an inflated disk

In a Kepler disk, when the gas is in hydrostatic equilibrium in the z-direction, the disk height H increases with the distance r from the star (subsection 15.4.3); the disk is said to be inflated or flared. Assuming that the gas and dust are well mixed, the dust disk is then not flat nor is the ratio H/r constant. To derive the grazing angle α_{gr} for this general case, we define the height H by the condition that the visual optical depth A_V from there towards the star is one. If $\rho(z)$ denotes the dust density distribution in vertical direction and K_V the visual absorption coeffcient,

$$1 = \frac{K_V}{\alpha_{\text{gr}}} \int_H^\infty \rho(z)\, dz. \tag{13.52}$$

The level $z = H$ represents the photosphere of the dusty disk and it is the approximate place where stellar photons are absorbed. The disk thickness is $2H$. A flared disk intercepts more stellar light than one with constant H/r. Indeed, the effective grazing angle α_{gr} of the incident radiation increases over the expression (13.49) by an amount

$$\beta - \beta' = \arctan \frac{dH}{dr} - \arctan \frac{H}{r} \simeq \frac{dH}{dr} - \frac{H}{r} = r\frac{d}{dr}\left(\frac{H}{r} \right)$$

so that now we have to use in (13.48)

$$\alpha_{\text{gr}} = \frac{4}{3\pi} \frac{R_*}{r} + r\frac{d}{dr}\left(\frac{H}{r} \right). \tag{13.53}$$

The second term in (13.53) is usually greater than the first, so it is desirable to have a good estimate for H and its derivative dH/dr. The surface temperature of a blackbody disk, whether it is inflated or not, has the form

$$T_{\text{disk}}^4(r) = \Omega_* \frac{\alpha_{\text{gr}}}{\pi} T_*^4. \tag{13.54}$$

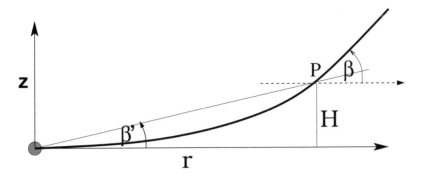

Figure 13.6. For a surface element P on a flared disk, the effective grazing angle α_{gr} of stellar light is enhanced with respect to the flat disk by $\beta - \beta'$. The bold line marks the top of the flared disk atmosphere; r and H are the radial distance and height above the mid-plane. The small shaded sphere represents the star.

13.4   Galactic nuclei

13.4.1   Hot spots in a spherical stellar cluster

Another class of objects whose infrared appearance one is eager to model and which possess a fair degree of spherical symmetry are galactic nuclei, like the center of the Milky Way. Many galactic nuclei have large infrared luminosities as a result of internal activity. Its origin may be either rapid star formation or accretion onto a massive black hole. In both cases, the hard primary radiation, from the star or the black hole, is converted into infrared photons by dust.

When the luminosity of the black hole dominates, one may, to first order, approximate it by a spherical central source with a power-law emission spectrum and use the formalism outlined in section 13.2 for single stars. However, when the luminosity comes from a star cluster, the configuration is intrinsically three-dimensional because the dust temperature varies not only on a large scale with galactic radius r, which is the distance to the center of the stellar cluster but also locally with the separation to the nearest star. This circumstance must be taken into account in some approximate manner. Loosely speaking, the temperature of a grain located in the immediate vicinity of a star is determined by the distance to and the properties of that star, whereas a grain at the same galactic distance but not very close to any star, absorbs only the mean interstellar radiation field $J_\nu^{\text{ISRF}}(r)$ and will thus be cooler. The spherical symmetry is broken on a small scale because the surroundings of a star constitute a hot spot (abbreviated HS) in the interstellar medium. Their presence significantly changes the spectral appearance of a galactic nucleus.

Let us assume that the hot spots are spheres, each with a star of monochromatic luminosity L_ν at the center. It turns out that their total volume fraction is small, so they are distributed over the galactic nucleus like raisins in a cake. The radius R^{HS} of a hot spot is determined by the condition that outside, a dust grain is mainly heated by the interstellar radiation field (ISRF), whereas inside, heating by the star dominates. If τ_ν is the optical depth from the star to the boundary of the hot spot, R^{HS} follows from

$$\int Q_\nu J_\nu^{\text{ISRF}} \, d\nu = \frac{1}{(4\pi R^{\text{HS}})^2} \int Q_\nu L_\nu e^{-\tau_\nu} \, d\nu. \qquad (13.55)$$

This equation can readily be generalized to a mixture of dust grains. The radius R^{HS} of a hot spot is a function of galactic radius r, but it depends, of course, also on the monochromatic stellar luminosity L_ν, the dust distribution within the hot spot, and on the grain type via the absorption efficiency Q_ν.

13.4.2 Low and high luminosity stars

It is useful to divide the stars in a galactic nucleus into two categories, each containing identical objects. Stars of the first class have small or moderate luminosity, a space density $n^*(r)$ and a monochromatic and bolometric luminosity L_ν^* and L^*, respectively, where

$$L^* = \int L_\nu^* \, d\nu.$$

They represent the population of old stars. Stars of the second class are of spectral type O and B. They are young and luminous and were formed in a starburst. The corresponding values are denoted $n^{\text{OB}}(r)$, L_ν^{OB} and L^{OB}.

- The low luminosity stars are very numerous, altogether typically 10^9, and the contribution of their hot spots to the overall spectrum may be neglected. To see why, consider a nucleus of integrated luminosity L_{nuc} containing N identical stars of luminosity L^*. When we fix $L_{\text{nuc}} = NL^*$ but increase N and thus lower L^*, the intensity of the interstellar radiation field in the galactic nucleus is, to first order, independent of N. In view of the definition for R^{HS} in (13.55), one finds that $R^{\text{HS}} \propto \sqrt{L^*} \propto N^{-1/2}$. Therefore, the total volume $N(R^{\text{HS}})^3$ of all hot spots decreases as $N^{-1/2}$. A large population of low-luminosity stars may be smeared out smoothly over the galactic nucleus and the structure of their hot spots need not be evaluated. To account for the radiation of these stars, one only has to introduce in the numerator of the source function (13.56) the volume emission coefficient

$$\Gamma_\nu^*(r) = n^*(r) \cdot L_\nu^*.$$

  If the stellar atmospheres are blackbodies of temperature T^*, one may put $L_\nu^* \propto B_\nu(T^*)$. Note that L_ν^* need not be specified, only the product $\Gamma_\nu^*(r) = n^*(r) \cdot L_\nu^*$.
- The OB stars, however, are very bright and not so numerous. There are rarely more than 10^6 in a galactic nucleus and their space density $n^{\text{OB}}(r)$

is moderate, typically one star per pc^3 or less in the starburst region. The emission of their hot spots is not negligible and has to be evaluated explicitly. Before solving the radiative transfer on a large scale in the galactic nucleus, one has to determine the luminosity $L_\nu^{HS}(r)$ emerging from a hot spot. The frequency integral $\int L_\nu^{HS} d\nu$ is, of course, equal to the luminosity $L^{OB} = \int L_\nu^{OB} d\nu$ of a single OB star but L_ν^{OB} and L_ν^{HS} are different because much of the hard stellar UV flux is converted by dust into infrared radiation.

To obtain $L_\nu^{HS}(r)$, we calculate for each galactic radial grid point the radiative transfer of a spherical cloud centrally heated by an OB star. The cloud radius R^{HS} is determined from (13.55). The hot spot is illuminated at its edge by an interstellar radiation field (ISRF), which fixes the outer boundary condition (13.36); at the inner boundary, we use (13.38) with the flux from the surface of the OB star. The volume emission coefficient due to the hot spots is

$$\Gamma_\nu^{HS}(r) = n^{OB}(r) \cdot L_\nu^{HS}.$$

Altogether, the source function in a spherical galactic nucleus with isotropic dust scattering becomes (see (13.10))

$$S_\nu = \frac{\Gamma_\nu^{HS}(r) + \Gamma_\nu^*(r) + K_\nu^{abs} B_\nu(T_d) + K_\nu^{sca} J_\nu^{ISRF}}{K_\nu^{ext}}. \qquad (13.56)$$

It is straightforward to generalize to a dust mixture with different temperatures T_d and to include small grains with temperature fluctuations. One then solves with this source function the integral equations (see (13.39) and (13.40))

$$I(\tau) = I(0)e^{-\tau} + \int_0^\tau S(x)e^{-(\tau-x)} dx \qquad (13.57)$$

for the spherical nucleus as a whole. This is done along lines of constant impact parameter for the intensities in positive and negative direction, I^+ and I^-. They yield the mean intensity $J_\nu^{ISRF}(r)$. The inner boundary condition is now always $I^+ = I^-$ and, at the outer edge, $I^- = 0$ (see (13.36) and (13.37)).

As the galactic nucleus in this model consists of two phases, a dusty medium interspersed with hot spots, equation (13.57) is not strictly correct. It gives, however, a good approximation as long as the volume fraction γ of the hot spots,

$$\gamma(r) = \frac{4\pi}{3} n^{OB}(r) \cdot \left[R^{HS}(r)\right]^3$$

is small; $\gamma(r)$ shrinks when $n^{OB}(r)$ becomes large. For typical space densities of OB stars in starburst nuclei, γ is of order 10^{-3}.

13.5   Line radiation

13.5.1   Absorption coefficient and absorption profile

In a gas of atoms or molecules, line radiation arises from transitions between discrete energy states. Consider the levels j and i with statistical weights g_j, g_i, energies $E_j > E_i$, and population numbers n_j, n_i. The center frequency ν_{ji} of the line is at

$$h\nu_{ji} = E_j - E_i.$$

The basic equation of radiative transfer (13.6),

$$\frac{dI_\nu}{ds} = -K_\nu I_\nu + \epsilon_\nu$$

is still valid but K_ν refers now only to absorption. Scattering of the kind that an atom absorbs a photon from level i to j and then returns to the same i is treated as two independent processes: one of absorption, the other of emission. The molecules absorb radiation not only at exactly the center frequency ν_{ji} but over some interval. How effectively they absorb away from the line center is described by the absorption profile $\Phi(\nu)$ (see section 6.3.1). This is a normalized function,

$$\int_{\text{line}} \Phi(\nu)\, d\nu = 1 \tag{13.58}$$

which for a Gaussian of half-width $2H$ has the form

$$\Phi(\nu) = \frac{\sqrt{\ln 2}}{H\sqrt{\pi}} \cdot \exp\left[-\frac{(\nu - \nu_{ij})^2}{H^2} \ln 2 \right]. \tag{13.59}$$

In our applications, the energy levels are rather sharp and the line broadening is due to the Doppler shift arising from the motion of the molecules (relative to their local standard of rest). In other environments, like stellar atmospheres, the levels themselves are broadened by collisional perturbations.

The emission and absorption coefficient can be expressed through the Einstein coefficients A_{ji} and B_{ji} for spontaneous and induced emission. Following the discussion in subsection 6.3.1, a beam of light of intensity I_ν, solid angle $d\Omega$ and frequencies in the interval $[\nu,\, d\nu + d\nu]$ induces, per unit length,

$$I_\nu B_{ij} N_i \Phi(\nu) \frac{d\Omega\, d\nu}{c}$$

upward transitions. The frequency-dependent absorption coefficient results from the difference between upward and downward induced transitions and can be written as

$$K_\nu = \left[n_i B_{ij} - n_j B_{ji} \right] \frac{h\nu}{c} \Phi(\nu). \tag{13.60}$$

The term $-n_j B_{ji}$ gives the correction for stimulated downward transitions and (without proof) is linked to the same Φ. The integrated line absorption coefficient is

$$K = \int K_v \, dv = \frac{hv}{c} \left[n_i B_{ij} - n_j B_{ji} \right] \qquad (13.61)$$

and the optical depth element

$$d\tau_v = K_v \, ds = K \Phi(v) \, ds. \qquad (13.62)$$

Likewise, the emission coefficient ϵ_v may be expressed as

$$\epsilon_v = n_j A_{ji} \Phi'(v) \frac{hv}{4\pi} \qquad (13.63)$$

with the profile function $\Phi'(v)$. Usually one may put $\Phi(v) = \Phi'(v)$. Note that spontaneous emission is isotropic, whereas the induced photons fly in the same direction as the photons in the incident beam.

13.5.2 The excitation temperature of a line

When the radiative transfer equation is written as

$$\frac{dI_v}{d\tau_v} = -I_v + S_v$$

it describes the variation of the intensity I_v with optical depth τ_v. The source function S_v becomes, in view of (13.60) and (13.63),

$$S_v = \frac{\epsilon_v}{K_v} = \frac{2hv^3}{c^2} \left[\frac{g_j n_i}{g_i n_j} - 1 \right]^{-1} \qquad (13.64)$$

and the profile function has canceled out. The ratio of the level populations defines an excitation temperature T_x,

$$\frac{n_j}{n_i} = \frac{g_j}{g_i} e^{(E_j - E_i)/kT_x}. \qquad (13.65)$$

When we insert T_x into the Planck function, we see from equation (13.64) that the source function is just the Planck function at temperature T_x,

$$B_v(T_x) = S_v. \qquad (13.66)$$

Different transitions of the same molecule have usually different excitation temperatures. Only in thermodynamic equilibrium does one have one value of T_x, equal to the kinetic temperature of the gas T, and then (13.65) is equivalent to the Boltzmann equation (5.9). When for a particular transition $T_x = T$, the corresponding levels are said to be thermalized. Employing the excitation temperature and setting

$$K_0 = \frac{g_j}{g_i} \frac{c^2}{8\pi v^2} A_{ji} \tag{13.67}$$

the integrated absorption coefficient of (13.61) becomes

$$K = n_i K_0 \left[1 - e^{-hv/kT_x} \right]. \tag{13.68}$$

The brackets contain the correction for stimulated emission. One uses the so called b_i-factors to indicate the deviation of the population of a level i from its LTE value denoted n_i^*,

$$b_i \equiv \frac{n_i}{n_i^*}. \tag{13.69}$$

Thus, if we replace in equation (13.68) the excitation temperature T_x by the kinetic temperature T,

$$\frac{b_j}{b_i} e^{-hv/kT} = e^{-hv/kT_x}. \tag{13.70}$$

Because the brightness temperature T_b is defined by (13.19) from the intensity,

$$I_v = B_v(T_b) \tag{13.71}$$

the transfer equation (13.9) can be written as

$$\frac{dB_v(T_b)}{d\tau_v} = -B_v(T_b) + B_v(T_x). \tag{13.72}$$

The Rayleigh–Jeans approximation,

$$\frac{dT_b}{d\tau} = -T_b + T_x \tag{13.73}$$

then has, in a uniform cloud, the solution

$$T_b = T_x(1 - e^{-\tau}) + T_0 e^{-\tau}. \tag{13.74}$$

The term $T_0 e^{-\tau}$ representing a background source of strength T_0 has been added to stress the analogy to (13.20).

13.5.3 Radiative transfer in lines

For line radiation, the absorption coefficient K in (13.61) depends on the statistical level population n_j. Generally, one cannot assume LTE conditions but has to determine the abundance n_j of each level j from the balance between processes that populate and depopulate it. These are, first, radiative transitions due to spontaneous and induced emission and to induced absorption (section 6.3), all subject to certain selection rules; second, there are collisional transitions.

13.5.3.1 *Level population of carbon monoxide*

Let us consider the excitation of rotational levels in a simple diatomic molecule, for example, CS or CO (see section 14.2.1). Radiative transitions coupling to rotational level j are then restricted by the selection rules to levels $j \pm 1$. Collisions, however, can go from j to any $k \neq j$ and back, the main collisional partners being helium atoms and ortho- and para-hydrogen. The collision rate C_{jk} from $j \to k$ is a function of temperature only. The number of such transitions per s and cm^3 is given by $n_j C_{jk}$. The inverse rate C_{kj} follows from

$$g_j C_{jk} = g_k C_{kj} e^{-(E_k - E_j)/kT}.$$

Let there be $N + 1$ levels altogether and let us put

$$i = j - 1 \qquad k = j + 1 \qquad C_{jj} = 0.$$

The set of equations for the equilibrium population n_j then reads

$$n_k \left[A_{kj} + B_{kj} u_{kj} \right] + n_i B_{ij} u_{ij} + n(\text{H}_2) \sum_{m=0}^{N} n_m C_{mj}$$

$$= n_j \left[A_{ji} + B_{ji} u_{ji} + B_{jk} u_{jk} + n(\text{H}_2) \sum_{m=0}^{N} C_{jm} \right] \qquad (13.75)$$

for $j = 0, 1, 2, \ldots, N - 1$. The left-hand side populates, the right-hand side depopulates level j. The radiative energy densities u_{ij} in the lines are taken with respect to transitions between levels i and j and are defined in (6.38). For $j = 0$, terms connecting to $i = j - 1$ disappear because level $i = -1$ does not exist. The total number of CO molecules is fixed by the additional condition

$$\sum_{j=0}^{N} n_j = n(\text{CO}). \qquad (13.76)$$

13.5.3.2 *LTE population*

We point out limiting cases of (13.75):

- When the density is high, radiative transitions are negligible and collisions dominate. Because

$$C_{ji} = \frac{g_i}{g_j} C_{ij} e^{-(E_i - E_j)/kT}$$

we are led to

$$\sum_i C_{ij} \left[n_i - n_j \frac{g_i}{g_j} e^{(E_j - E_i)/kT} \right] = 0.$$

These equations are fulfilled if the brackets vanish which is true when all levels are in LTE.

• In the case of a blackbody radiation field, $u_\nu = B(\nu, T)$. When collisions are negligible,

$$\left[n_j \frac{8\pi h}{c^3} \nu_{ji} + \left(n_j - n_i \frac{g_j}{g_i} \right) B(\nu_{ji}, T) \right] B_{ji} = 0$$

which also implies thermalization of the levels at temperature T.

13.5.3.3   The microturbulent approximation

The equations (13.75) and (13.76) for the abundances n_j contain the radiative energy densities u_{ij}. They must, therefore, be solved together with the radiative transfer. For the radiative transfer we again use the integral form (13.39) and (13.40). However, we now have to follow the intensity not only in the right quadrant of figure 13.2, from $z = 0$ to $z = z_e$ as for dust but from $-z_e$ to z_e because the line-of-sight velocity has, in spherical symmetry, opposite signs in the near and far quadrant. This changes the boundary conditions in an obvious fashion.

According to (13.60) and (13.63), the local line profile $\Phi(\nu)$ appears in the expression for the absorption and emission coefficient, K_ν and ϵ_ν. The function $\Phi(\nu)$ is rarely of purely thermal origin. For example, purely thermal and optically thin molecular lines from a cold interstellar cloud of 10 K would have a width of typically 0.1 km s^{-1} (see (5.13)), however, observed lines are usually a few times wider. Broadening is, therefore, dominated by other, non-thermal motions, their precise origin being unclear. The observed profiles are, nevertheless, usually of Gaussian shape which indicates that the motions have a random, turbulent character. In the microturbulent approximation it is assumed that the velocity distribution of the non-thermal motions follows a Gauss curve and that the mean free photon path is much bigger than the scale length of turbulence. In the line profile (13.59), one, therefore, puts

$$H^2 = \left(\frac{2kT}{m} + \xi_{\text{turb}}^2 \right) \ln 2$$

where m is the molecular mass and ξ_{turb} the most likely turbulent velocity. One solves the radiative transfer equations (13.39) and (13.40) for all lines and afterwards the abundance equations (13.75) and (13.76), and repeats this sequence until convergence is reached.

13.5.3.4   The Sobolev approximation

When the velocity field **v** in a cloud has a large gradient, the radiative energy densities in the lines, u_{ij}, may be computed locally, without knowledge of the radiation field in other parts of the cloud [Sob60, Ryb70]. This significantly simplifies the procedure. Atoms do then not interact with one another radiatively over large distances because their emission is Doppler-shifted by comfortably more than the linewidth. The standard example is the outward flow in a stellar wind. For a radial inflow, there are, however, two interacting points along the line of sight (figure 16.8). The mean intensity J (or the radiative energy density u_{ij}) in the case of large velocity gradients is derived from a photon escape probability β by assuming that J is proportional to the source function S,

$$J = (1 - \beta)S.$$

If $\beta = 1$, photons leave unimpeded and there are no induced transitions in (13.75). If $\beta = 0$, photons cannot escape and J equals the source function of (13.64). To determine β, consider a point $\mathbf{x}$ in a cloud. Let τ be the optical depth from $\mathbf{x}$ towards the edge of the cloud along some axis s, with $s = 0$ at point $\mathbf{x}$. Let $\beta_\mathbf{e}$ be the corresponding escape probability given by

$$\beta_\mathbf{e} = e^{-\tau}.$$

Here $\mathbf{e}$ denotes the unit vector in the direction of the s-axis. The escape probability $\beta_\mathbf{e}$ depends on the optical depth τ but not on the radiation field. Because the velocity gradient is large, all relevant contributions to the optical depth τ come from the immediate vicinity of point $\mathbf{x}$. As there is also emission along the line for $s > 0$, the mean $\beta_\mathbf{e}$ is given by an average over $e^{-\tau}$,

$$\langle \beta_\mathbf{e} \rangle = \langle e^{-\tau} \rangle = \frac{1}{\tau} \int_0^\tau e^{-x} \, dx = \frac{1 - e^{-\tau}}{\tau}.$$

If τ is large, $\langle \beta_\mathbf{e} \rangle = 1/\tau$. If τ is small, $\langle \beta_\mathbf{e} \rangle = 1$, which is all reasonable. To find the effective β, one must further average $\langle \beta_\mathbf{e} \rangle$ over all directions $\mathbf{e}$. For a spherical cloud of radius r, where θ is the inclination angle to the radial vector and $\mu = \cos\theta$, we get

$$\beta = \frac{1}{2} \int_{-1}^1 \frac{1 - e^{-\tau}}{\tau} \, d\mu. \tag{13.77}$$

To evaluate this integral, one has to determine τ as a function of frequency v, radius r and direction μ. If K is the integrated line absorption coefficient of (13.61) and v_0 the frequency at the line center, the optical depth becomes

$$\tau_v = K \int_0^\infty \Phi\left(v - v_0 + \frac{v_0 s}{c}\frac{dv}{ds}\right) ds = \frac{Kc\,ds}{v_0\,dv} \int_{v-v_0}^\infty \Phi(x)\,dx = \frac{Kc\,ds}{v_0\,dv}$$

and we see that it is independent of the frequency v. The velocity gradient along the s-axis is given in spherical symmetry by

$$\frac{dv}{ds} = \mu^2 \frac{dv}{dr} + (1 - \mu^2)\frac{v}{r}.$$

This expression must be positive. If it is not, one takes $|dv/ds|$ which corresponds to flipping the direction vector of the s-axis. The intensity which one detects along a line sight at a certain frequency v is only due to the emission from a small region of size ℓ, where $\ell|dv/ds|$ is about equal to the local linewidth, and may, therefore, be found from (see (13.15) and (13.60)):

$$I_v = S_v\left[1 - e^{-K_v\ell}\right].$$

Integration over all lines of sight and all frequencies yields the line profile which one would observe towards such a cloud.

Chapter 14

Diffuse matter in the Milky Way

14.1   Overview of the Milky Way

To put interstellar dust into a broader astronomical context, we give in the remaining chapters an overview of the Milky Way, talk about the interstellar medium in general and about star formation in particular because it is there that the study of interstellar dust finds its most important application.

14.1.1   Global parameters

The Milky Way is a rotating stellar system of some 10^{11} stars that has been created about 15×10^9 yr ago. After an initial and violent phase it has settled into a quasi-stationary state where global parameters, like luminosity, spectral appearance, gas mass or supernova rate change only gradually, on a time scale of $\sim 10^{10}$ yr. It is a flat disk about 30 kpc across and several hundred parsec thick with a global star formation rate of ~ 5 $M_\odot$ per year. The center of the Milky Way harbours a huge stellar cluster, the nuclear bulge, with a diameter of order 2 kpc which contains $\sim 1\%$ of the total stellar mass of the Milky Way. The flat disk is enveloped by a halo of some 200 globular clusters.

   The total power radiated by the Milky Way is estimated to 4×10^{10} $L_\odot$ and has its origin in the luminosity of the stars. About two-thirds of the starlight leave the Milky Way unimpeded, one-third is absorbed by dust grains and transferred to infrared radiation ($\gtrsim 10$ μm).

   The space between the stars is not empty but filled with gas, dust grains, relativistic particles (moving close to the speed of light), magnetic fields and photons. Together these components constitute the interstellar medium. It exists in several phases which differ in all basic parameters like temperature, density, chemical composition or radiation field. In each of them, the dust is modified by the specific environment. Usually, one classifies the interstellar medium according to the state of hydrogen, the most abundant element:

- In molecular clouds, hydrogen is mostly molecular (H_2). Besides H_2, one finds in them all kinds of molecules.
- In the diffuse or HI gas, hydrogen is mostly atomic (H).
- In HII regions, hydrogen is ionized (H^+) and forms a hot plasma.

Roughly speaking, half of interstellar the mass is in molecular clouds, half in the diffuse gas, and of order 1% in HII regions. Together, the three phases account for 99% of the mass of the interstellar medium or about 5% of the stellar mass of the Milky Way. The total gas mass is $M_{gas} \sim 5 \times 10^9$ $M_\odot$. The dust is just a small admixture. Assuming a dust-to-gas mass ratio of 1:150, its mass in the Milky Way amounts to $M_{dust} \sim 3 \times 10^7$ $M_\odot$.

As a result of its rotation, the Milky Way has formed a spiral pattern which can be admired in face-on external galaxies (figure 11.4). About every 10^8 yr a density wave sweeps over a parcel of gas in the Milky Way and causes a drastic change in its physical state. The dust then changes its environment leaving, for example, the warm and tenuous medium and entering a cold and dense phase. This affects the composition and structure of the grains and happens repeatedly during their lifetime. The passage of spirals wave initiates star formation.

14.1.2 The relevance of dust

If there is so relatively little dust, why is it so important? There are many reasons: practical, astronomically fundamental and philosophical. Some are listed here, without aiming at completeness. Dust is so important:

- because it weakens the visual light and thus limits our optical view in the Galaxy;
- because it converts the radiation from stars into the infrared; in some galaxies up to 99%, in the Milky Way about 30%;
- because it determines the spectral appearance of stars during their formation; in this phase, it is also essential for the dynamics of the gas;
- because it strongly influences the physical and chemical conditions of the interstellar medium (thermal balance, molecular abundances, radiation field);
- because it is the basis for the formation of planets; and
- because man is made of interstellar dust. Although not an argument of natural sciences but of humanities, the fact that there is a lineage between us and dust is probably the ultimate reason for our interest in it.

When did astronomers realize the relevance of interstellar dust? As a legendary beginning one may count the year 1784 when *Friedrich Wilhelm Herschel* (the father of *John Frederick William*) looked at the dust cloud ρ Ophiuchus and exclaimed: *Hier ist ein Loch im Himmel* (There is a hole in the sky). The scientific investigation of dust started with the Sky photography of *E Barnard* and *M Wolf* in the 1920s which revealed the existence of dark clouds. Their concentration towards the galactic plane could explain the *zone of avoidance* for globular clusters and galaxies. *J H Oort* argued in 1932 that the absorbing particles cannot be larger than the wavelength of visible light as otherwise they would produce noticeable effects on the dynamics of stars perpendicular to the galactic plane. In 1937 *J S Hall* established the extinction law in the visible and in 1949 *Hall* and *W A Hiltner* discovered interstellar polarization. But the full relevance of dust for astronomy was grasped only with the advent of infrared detectors in the 1960s. The breakthrough came when it was realized that the sky was littered with bright infrared sources. In the 1980s and 1990s, satellites mapped the total sky at infrared wavelengths not accessible from the ground; the origin of the infrared emission is usually dust.

14.2   Molecular clouds

Roughly half of the mass of the interstellar medium resides in molecular clouds. There, as the name suggests, one finds many molecular species and almost all hydrogen is in the form of H_2. The prevalence of molecules is due to the fact that the dust shields them from destructive ultraviolet radiation. Molecular clouds block the starlight and, therefore, on photographic plates, the near ones appear as holes in the sky that is otherwise littered by myriads of stars. This has prompted their second name: dark clouds.

The biggest molecular clouds have masses greater than 10^5 $M_\odot$ and are labeled *giant molecular clouds* (GMC). These GMCs are further structured into clumps which are the birth place of stars. The density contrast between clumps and their environment is about one order of magnitude. Molecular clouds trace the spiral arms of the Milky Way but are also found in the interarm medium. Examples of well-studied nearby molecular clouds are the complexes in Taurus (distance $D = 140$ pc, mass $M \sim 10^4$ $M_\odot$, extension $L = 20$ pc), in Ophiuchus (similar parameters but containing much denser concentrations), and the Orion GMC.

Although on maps in the CO line (see later) most molecular clouds have a sharp edge, the transition of the gas density to the intercloud medium is not abrupt. There is always an intermediate layer with a dust optical depth $A_V \sim 1$ mag where CO is photo-dissociated. Therefore, molecular clouds are enveloped by a halo of atomic hydrogen which, by mass, is not always negligible with respect to the molecular cloud proper.

It is not clear how molecular clouds are made, whether through coagulation of smaller units or by large-scale instabilities but the shocks in the spiral arms are probably an important ingredient in their creation. The likely destruction mechanism for GMCs is the internal formation of massive stars which produce strong winds and eventually supernova explosions. The average lifetime of a giant molecular cloud, τ_{GMC}, depends then, in some loose way, on the star-formation efficiency; estimates for τ_{GMC} are several 10^7 yr.

Table 14.1. Average parameters of molecular clouds.

Average gas density	$n = 100$ cm^{-3}
Average temperatures	$T_{gas} = T_{dust} = 20$ K
Mass fraction of interstellar medium	$f \simeq 50\%$
Volume fraction in disk of Milky Way	$v = $ a few percent
Scale height (in z-direction)	H = 80 pc near Sun
Observable mainly through	Molecular lines and IR dust emission
Mass spectrum of GMCs in Milky Way	$N(M) \propto M^{-1.5}$
Mass spectrum of clumps within GMC	$N(m) \propto m^{-1.5}$

14.2.1 The CO molecule

14.2.1.1 *Why CO is the most frequently observed molecule*

Over 100 molecular species are known in space today and about ten of them are regularly used as probes of the interstellar medium; without being too precise, they are H$_2$, OH, SiO, CO, CS, H$_2$O, HCN, NH$_3$, H$_2$CO and HCO$^+$. These molecules have very diverse structures and trace, in various lines, different physical conditions. We only discuss carbon monoxide (CO) which is the most widely observed molecule in the study of galactic and extragalactic molecular clouds because it possesses a number of favourable properties:

- It is very abundant. Where it exists, its number density relative to molecular hydrogen is [CO]/[H$_2$] $\sim 10^{-4}$. All other molecules are at least one order of magnitude less frequent.
- It is very stable with a dissociation energy of 11.09 eV.
- Its lowest rotational levels are easily populated by collisions. The level $j = 1$ is only 5 K above the ground state and requires for its excitation a density of only $n(H_2) > 100$ cm^3.
- It has several observable isotopes: CO, ^{13}CO, C^{18}O and C^{17}O spanning three orders of magnitude in abundance and likewise in column density and optical thickness.

We, therefore, exemplify the discussion of interstellar molecules by CO.

14.2.1.2 *Nomenclature of levels*

CO possesses rotational, vibrational and electronic levels which require progressively higher energies for their excitation,

$$E_{el} \gg E_{vib} \gg E_{rot}.$$

- Rotational states have quantum numbers $j = 0, 1, 2, 3, \ldots$. The lowest transition $j = 1 \to 0$ has a wavelength $\lambda = 2.6$ mm.

- Vibrational states are labeled $v = 0, 1, 2, 3, \ldots$. The lowest $v = 1 \to 0$ band is at 4.67 μm.
- Electronic states bear the names $X, A, B, C, E, \ldots$. The transition from the ground state to the next higher electronic state, $X\,^1\Sigma^+ \to A\,^1\Pi$, is at 1544 Å.

The lowest CO state is denoted $X\,^1\Sigma^+ v = 0\; j = 0$. The letter X stands for the electronic ground state, the Greek Σ indicates that the total orbital angular momentum of the electrons vanishes, so the quantum number $\Lambda = 0$; for $\Lambda = 1$, the state is denoted Π. The superscript 1 at the letter Σ shows the multiplicity. The total eigenfunction ψ of the molecule is approximately the product of the electronic, vibrational and rotational wavefunctions,

$$\psi = \psi_{\mathrm{el}} \cdot \psi_{\mathrm{vib}} \cdot \psi_{\mathrm{rot}}.$$

The plus superscript in $X\,^1\Sigma^+$ state means that the electronic state ψ_{el} does not change when all particles of the molecule are reflected at the origin (inversion); for a Σ^-, it would change sign.

The change in rotational quantum number Δj in a radiative transition is subject to the selection rule

$$\Delta j = j' - j = 0, \pm 1$$

with the further restriction that transitions between $j' = 0$ and $j = 0$ are forbidden as well as transitions between Σ-states with $\Delta j = 0$. Vibrational transitions split into branches depending on the changes in the rotational quantum number j:

- P-branch, $\Delta j = -1$
- Q-branch, $\Delta j = 0$ and
- R-branch, $\Delta j = 1$.

The P- and R-branches are displayed in figure 14.1 for $\Delta v = 1$ transitions under LTE at $T = 1000$ and 2000 K, the Q-branch is absent. Such highly excited CO is found in accretion disks of young stars.

14.2.1.3   *Frequencies, energies and A-values of pure rotational transitions*

The rotational eigenstates for $v = 0$ are found from the Schrödinger equation of a *rigid rotator*. If I is the moment of inertia of the molecule about its rotation axis, the energy levels are

$$E_j = \frac{\hbar^2 j(j+1)}{2I}$$

and the angular momenta

$$L_j = \hbar\sqrt{j(j+1)}.$$

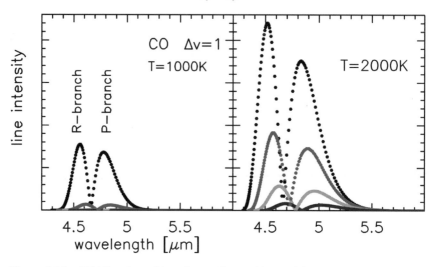

Figure 14.1. LTE line intensities of rot–vib transitions with $\Delta v = 1$. The upmost sequences of dots show the $v = 1 \rightarrow 0$ bands, the lower curves refer to $v = 2 \rightarrow 1$, $3 \rightarrow 2$ and $4 \rightarrow 3$. The intensity scale is linear, equal in both boxes, and goes from zero to 4.6×10^{-14} erg s^{-1}. It gives the average emission per molecule over 4π.

The eigenfunctions are similar to those of the hydrogen atom except that the function with the radial dependence, which reflects the potential energy of the atom, is missing. The rotational levels are degenerate with statistical weight

$$g_j = 2j + 1. \tag{14.1}$$

Defining the rotational constant B by

$$B = \frac{h}{8\pi^2 c I} \tag{14.2}$$

it follows that

$$\frac{E_j}{k} = \frac{hB}{k} j(j+1) \tag{14.3}$$

and the frequency of the transition $j \rightarrow j - 1$ is

$$\nu = 2Bj.$$

For ^{12}CO, the rotational constant $B = 5.763 \times 10^{10}$ s^{-1} and the energy levels are $E_j/k \simeq 2.77 j(j+1)$ in the unit Kelvin. Inserting the eigenfunctions of the rigid rotator into the Einstein coefficient (6.41), one finds for the $j \rightarrow j - 1$ transition

$$A_{j,j-1} = \frac{512\pi^4 B^3 \mu^2}{3hc^3} \frac{j^4}{2j+1} \tag{14.4}$$

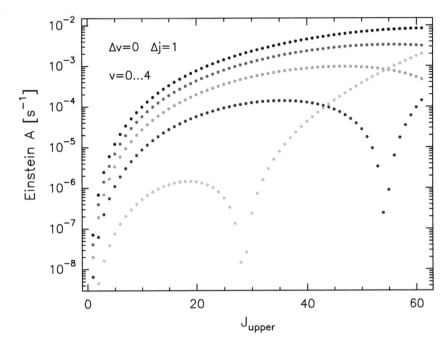

Figure 14.2. Radiative transition probabilities of the CO molecule for pure rotational lines ($\Delta v = 0$) in various vibrational levels v. The upmost sequence of points refers to $v = 0$.

where the dipole moment $\mu = 0.112$ Debye $= 1.12 \times 10^{-19}$ cgs. The A-values increase like $\sim j^3$; for the lowest transition $A_{1,0} = 7.4 \times 10^{-8}$ s^{-1}. With higher j-number, the internuclear distance increases and the rigid rotator model, which assumes a bar of fixed length between the nuclei, is not adequate. Figure 14.2 plots the A-coefficients of rotational lines in various vibrational levels and figure 14.3 the energies $E_{v,j}$ of the vib–rot levels, all for the electronic ground state.

14.2.2 Population of levels in CO

14.2.2.1 LTE population and critical density

In local thermodynamic equilibrium at temperature T, the population n_i of any level i is determined by (5.8),

$$\frac{n_i}{n} = \frac{g_i e^{-E_i/kT}}{Z(T)}.$$

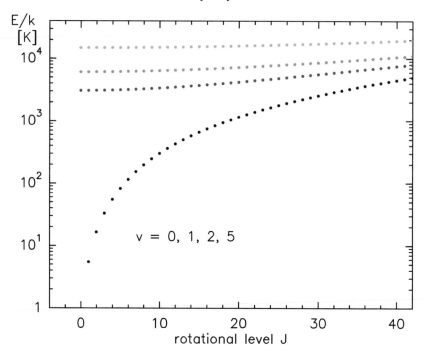

Figure 14.3. Energies above ground of the CO molecule in the electronic ground state for vibrational quantum numbers $v = 0, 1, 2, 5$ as a function of the rotational level j. The lowest sequence of dots is for $v = 0$. Energies E/k are expressed in Kelvin.

n is the total number of CO molecules per cm^3 and $Z(T)$ the partition function of (5.7). $Z(T)$ can be split into its rotational, vibrational and electronic part,

$$Z(T) = Z_{rot} \cdot Z_{vib} \cdot Z_{el}.$$

For example, in LTE at $T = 10, 20$ and 30 K, the most densely populated levels are $j = 1, 2, 3$, respectively. At these temperatures, the molecule is in its vibrational and electronic ground state ($Z_{vib} = Z_{el} = 1$) and Z_{rot} equals 3.96, 7.56 and 11.17, respectively.

When level populations are close to LTE, the molecule is said to be thermalized. This is always true in the limit of high densities. The temperature T computed from the Boltzmann equation (5.8) is then the gas kinetic temperature T_{kin}. If the molecule is not in LTE and the levels are not thermalized, the ratio of level population n_j/n_i defines the excitation temperature T_x according to (13.65). In molecular clouds, usually only the lowest rotational levels of CO are close to thermalization, the higher ones are not.

To estimate the minimum or critical gas density n_{cr} necessary to excite the lowest rotational line $j = 1 \rightarrow 0$, we neglect photons so that n_{cr} depends only on the collision rates C_{01} and C_{10} and on the Einstein A coefficient A_{10} (see equation (13.75)). For para-H_2, $C_{10} \simeq 3 \times 10^{-11}$ cm^3 s^{-1} when T is in the range from 10 to 50 K, typical of molecular clouds. The balance of level population requires

$$n(H_2)n_0 C_{01} = n_1[A_{10} + n(H_2)C_{10}].$$

With $C_{01} \sim 3 \times C_{10}$, one finds that already 10% of the molecules are excited ($n_1/n_0 \sim 0.1$) when $n_{cr} = n(H_2) \sim A_{10}/C_{01} \sim 100$ cm^{-3}. This rough estimate indicates that the $j = 1 \to 0$ line of CO is always seen in molecular clouds. Because $A_{j,j-1} \propto j^3$, the upper transitions need higher densities and, of course, also higher gas temperatures. Because $A_{j,j-1} \propto \mu^2$, molecules with a larger dipole moment (for example, CS with $\mu = 1.98$ Debye) require denser gas for their excitation.

14.2.2.2 *Population of rotational levels including radiative excitation*

Abundances of rotational levels for representative values of density, temperature and velocity gradient of molecular clouds are shown in figure 14.4. The values have been calculated including radiative excitation under the Sobolev approximation (section 13.5.3).

14.2.2.3 *CO abundance and isotopic ratios*

From many pieces of indirect evidence, the average abundance ratio [CO]/[H_2] $\sim$ 10^{-4}. The value is much smaller in diffuse clouds because of photo-destruction, and possibly also in dense cold clouds because CO freezes out onto dust grains. Direct evidence for [CO]/[H_2] can, in principle, be gained from simultaneous measurements of rot–vib absorption bands in CO and H_2 towards strong background infrared continuum sources but this has not yet been done for molecular clouds. Only in the diffuse medium have CO, H and H_2 absorption lines been observed in the far UV towards weakly reddened hot stars stars ($A_V \le 1$ mag).

The solar value of the isotopic ratio [C]/[^{13}C] is 89. By and large, in other places of the Galaxy it tends to be smaller, somewhere between 40 and 80, and in the Galactic Center it is noticeably lower ([CO]/[^{13}CO] $\sim$ 20). The ratio is influenced by the following processes:

- *Self-shielding in the dissociating lines* which favours CO.
- *Isotopic fractionation* which favours ^{13}CO. At its root is the charge exchange reaction

$$CO + {}^{13}C^+ \quad \Leftrightarrow \quad {}^{13}CO + C^+. \tag{14.5}$$

The vibrations of the molecules have, like all harmonic oscillators, a non-zero energy $E_0 = \frac{1}{2}h\nu$ in the ground level $\nu = 0$. The frequency ν decreases with the molecular mass, $\nu \propto m^{-1/2}$. Because ^{13}CO is heavier than CO,

$$E_0({}^{13}CO) \quad < \quad E_0(CO)$$

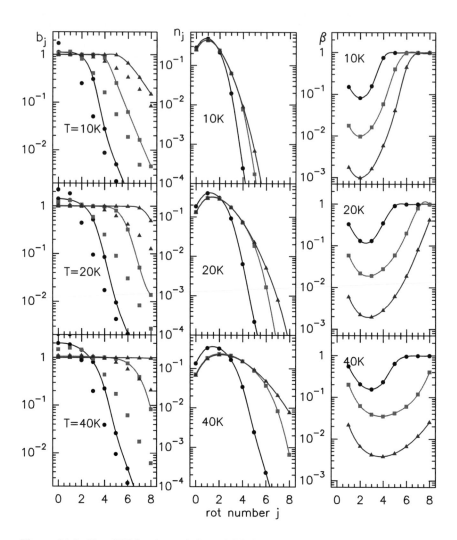

Figure 14.4. Non-LTE level population of CO for $T = 10$, 20 and 40 K and for H_2 densities 10^3 cm^{-3} (circles), 10^4 cm^{-3} (squares) and 10^5 cm^{-3} (triangles). The CO abundance is [CO]/[H$_2$] $= 10^{-4}$. The deviation from LTE, b_j, is defined in (13.69), the number densities n_j are normalized so that their sum equals one. The escape probability β is computed in the Sobolev approximation from (13.77) for a velocity gradient of $dv/dr = 1$ km s^{-1} pc^{-1}. In the three vertical frames on the left, those dots which are *not* connected by a line are calculated without a radiation field and thus without an escape probability. We see that for $j \geq 3$, radiative excitation is more important than collisional excitation.

with an energy difference $\Delta E_0/k = 35$ K. Therefore, the reaction (14.5) is temperature-dependent and goes preferentially from left to right in cold clouds ($T < 35$ K).

- *Mass loss from stars heavier than* 1 $M_\odot$, which, because of the specific preceding nucleosynthesis, enriches the interstellar medium more efficiently with ^{13}C than with ^{12}C.

The standard isotopic ratio of $C^{18}O$ and $C^{17}O$ are 500 and 1800, respectively, so these molecules are some five and 20 times less abundant than ^{13}CO.

14.2.3 Molecular hydrogen

Molecular hydrogen, by far the most abundant molecule, is hardly excited under normal conditions of the interstellar medium and, therefore, generally inconspicuous, it just supplies the mass for the interstellar gas. However, it is well observable in photo-dissociation regions and shocks and there its emission gives information about the ultraviolet radiation field, gas temperatures and densities or shock velocities. Wherever there is high-mass star formation, with jets and copious UV radiation, rot–vib lines of H_2 are seen, even in external galaxies.

There are two forms of molecular hydrogen: ortho-H_2 with parallel and para-H_2 with opposite proton spins. When the temperature is high and many rotational levels are excited, the ortho form is three times as abundant as the para form, corresponding to the ratio of the statistical weights. Radiative transitions between ortho and para hydrogen are forbidden; however, spin exchange can occur by collisions with molecules and ions or on grain surfaces but the rates are low. The existence of the ortho and para forms is a consequence of the exclusion principle; the isotope HD, where the nuclei are not identical, does not show this property.

The electronic ground state of H_2 is X $^1\Sigma_g^+$ (see nomenclature in section 14.2.1). The g (*gerade*) signifies that the the homonuclear molecule is symmetric with respect to swapping the two nuclei, whereas u (*ungerade*) would imply asymmetry. Being homonuclear, only quadrupole rot–vib transitions are allowed (section 1.4.4) subject to the selection rules $\Delta j = \pm 2$ or 0 (excluding $0 \to 0$ transitions). The spacing of rotational levels is given by (14.3). The lowest excited level of para-H_2 (X $^1\Sigma_g^+$ $v = 0$ $j = 2$) lies 510 K above the ground ($j = 0$), too high to be populated at normal temperatures. The $j = 2 \to 0$ emission line from this level appears at 28 μm and, as a quadrupole transition, has a small Einstein coefficient ($A = 3 \times 10^{-11}$ s^{-1}). The lowest ortho-line ($j = 3 \to 1$, $A = 4.8 \times 10^{-10}$ s^{-1}) is at 17 μm and even harder to excite.

14.2.4 Formation of molecular hydrogen on dust surfaces

The major destruction route of H_2 in the diffuse gas, outside HII regions, is photo-dissociation. Absorption of photons with wavelengths between 912 and 1108 Å leads to electronic transitions X $^1\Sigma_g^+$ $\to$ B $^1\Sigma_u$ (Lyman band) and X $^1\Sigma_g^+$ $\to$ C $^1\Pi_u$ (Werner band). They are immediately followed by electric dipole decay ($A \sim 10^8$ s^{-1}) to the electronic ground state. If the final vibrational quantum number v is greater than 14, this happens in about one out of seven cases, the state is unbound and the molecule dissociates.

Molecular hydrogen is observed in the diffuse medium if the ultraviolet radiation field is not too strong. Its photo-dissociation can then be calculated and must be balanced by H_2 formation. Radiative association $H + H \to H_2 + h\nu$ is strongly forbidden, three-body reactions are excluded as they require densities $n \geq 10^{12}$ cm^{-3}, other reactions are negligible. The only remaining alternative to create H_2 is on grain surfaces. The reaction rate R (per cm^3 and second) is obviously proportional to the number of H atoms impinging on a grain ($n(HI)\, \sigma_{gr}\langle v \rangle$) times the grain density n_{gr},

$$R = \eta\, n(HI)\, \sigma_{gr}\, \langle v \rangle\, n_{gr}.$$

The efficiency η includes the probability for sticking, finding another hydrogen atom by tunneling over the grain surface, molecule formation and desorption. $n(HI)$ is the density of hydrogen atoms, $\langle v \rangle$ from (5.16) their mean velocity and $\sigma_{gr} = \pi a^2$ the geometrical cross section of spherical grains of radius a. With a gas–to–dust of ratio 150 and a bulk density of 2.5 g cm^{-3},

$$n_{gr} \simeq 1.5 \times 10^{-27}\frac{n_H}{a^3}$$

where $n_H = n(HI) + n(H_2)$ is the total hydrogen density. Therefore

$$R \simeq 7 \times 10^{-23}\, \eta\, n(HI)\, n_H \frac{T^{1/2}}{a}. \tag{14.6}$$

In the diffuse medium, $n(HI) \simeq n_H$. The smallest grains should get the highest weight (see (7.29)), so $a \sim 10^{-6}$ cm seems appropriate. If $\eta \sim 0.2$,

$$R \sim 10^{-17} n_H^2 T^{1/2}\ \text{cm}^{-3}\ \text{s}^{-1}$$

and this should indicate the right order of magnitude. On leaving the grain surface, it is unclear how much of the formation energy of 4.5 eV is transmitted to the dust particle, which has *many* modes of energy disposal, and how much is carried away by the H_2 molecule in the form of excited rot–vib levels.

14.2.4.1   *A few remarks on H_2 and CO chemistry*

In the envelope of a cloud illuminated by ultraviolet light with wavelengths $\lambda > 912$ Å, the chemistry of the major gas species is determined by photo-dissociation. As the radiation penetrates into the cloud, it becomes absorbed (mainly by dust) and the UV flux weakens. This leads to the formation of a sequence of chemically distinct layers. If the zero point of the visual optical depth, τ_V, is taken at the cloud surface, then, roughly speaking, molecular hydrogen is dissociated for $\tau_V \leq 1$, carbon is ionized for $\tau_V \leq 2$, and most gas-phase carbon is locked in CO for $\tau_V \geq 3$. So at the transition between atomic and molecular hydrogen ($\tau_V \simeq 1$), carbon is in the form of C^+ and CO is absent.

Although molecular hydrogen is chemically rather inert, a small fraction becomes ionized by cosmic rays (CR) and is transformed into H_3^+,

$$\begin{aligned} H_2 + CR &\longrightarrow H_2^+ \\ H_2^+ + H_2 &\longrightarrow H_3^+ + H. \end{aligned}$$

The new species, H_3^+, is reactive and relevant for the chemistry of the interstellar medium as it can initiate the formation of important molecules like H_2O,

$$\begin{aligned} H_3^+ + O &\longrightarrow OH^+ + H_2 \\ OH^+ + 2H_2 &\longrightarrow H_3O^+ + 2H \\ H_3O^+ + e &\longrightarrow H_2O + H \end{aligned}$$

or remove carbon monoxide,

$$H_3^+ + CO \longrightarrow HCO^+ + H_2. \tag{14.7}$$

When the UV shielding by dust is low ($\tau_V < 3$), CO formation is promoted by ionized carbon, for example, via

$$C^+ + OH \longrightarrow CO + H^+.$$

CO is then destroyed by line absorption ($\lambda < 1118$ Å) from the ground state X $^1\Sigma^+$ to a bound level of a high electronic state (E, F, ...) that couples radiationless to a dissociating state. The line absorption leads to the effect of self-shielding. For an isotope like ^{13}CO, the dissociating lines are slightly shifted in frequency. Because the ^{13}CO column density is much smaller, ^{13}CO is destroyed in a larger volume than ^{12}CO. When UV shielding is high, CO is formed by neutral–neutral reactions like CH + O → CO + H.

14.2.4.2 *The early universe*

In the early universe, at a redshift $z \lesssim 1000$ when hydrogen was already neutral but before the appearance of the first stars, there were no heavy elements and thus no dust, only hydrogen and helium. H_2 could not then form on grains, however, it could be created via

$$\begin{aligned} H + H^+ &\longrightarrow H_2^+ + h\nu \\ H_2^+ + H &\longrightarrow H_2 + H^+ \end{aligned}$$

or

$$\begin{aligned} H + e &\longrightarrow H^- + h\nu \\ H^- + H &\longrightarrow H_2 + e \end{aligned}$$

Table 14.2. Typical parameters of HI clouds. They consist mainly of atomic hydrogen.

Temperature	$T_{gas} = 50 \ldots 100$ K
	$T_{dust} \sim 20$ K
Density	$n(HI) \sim 1 \ldots 10$ cm^{-3}
Cloud diameter	$D \sim 10$ pc
Column density	$N(HI) \sim 10^{20}$ cm^{-2}
Distance between clouds	$l \sim 3$ pc
Mass fraction	50% of the HI gas and 25% of all gas
Volume fraction	a few percent of galactic disk
21 cm linewidth	$\Delta v \sim 30$ km s^{-1}
21 cm line optical thickness	$0.2 < \tau < 1$

and be destroyed radiatively or by collisions with H atoms. Although the reactions were not efficient, they led to a small fraction of molecular hydrogen which was important for the formation of the first stars. The critical mass for gravitational instability of a gas cloud depends on temperature as $T^{3/2}$ (see (15.48)) and contraction is possible only if the release of potential energy can be radiated away. Here molecular hydrogen could have acted as a coolant down to 100 K by emission of its lowest rotational lines.

14.3   Clouds of atomic hydrogen

14.3.1   General properties of the diffuse gas

About half of the mass of the interstellar medium is in diffuse clouds. They contain only a few molecular species. As hydrogen is mainly atomic there, this component is called a HI gas. Its properties, to a large degree, are derived from observations of the famous 21 cm line but also from Hα and atomic fine structure lines, infrared dust emission and pulsar dispersion. Although the HI gas is very inhomogeneous, it is useful to define a standard HI cloud with parameters as in table 14.2. Such clouds are seen in absorption and emission and make up what is called the *cold neutral medium* (CNM) of the diffuse gas because the hydrogen there is mostly neutral and relatively cold. The scale height of the clouds at the locus of the Sun is about 250 pc, which is a few times bigger than for CO and increases outside the solar circle.

There are other components which are less dense and completely penetrated by UV radiation. They are, therefore, more strongly ionized and much warmer. In the *warm ionized medium* (WIM), $n = 0.1 \ldots 1$ cm^{-3} and gas temperatures are several thousand K. This component is only seen in emission, it is much more voluminous, has a larger scale height and a very low optical depth in the 21 cm line, $\tau < 0.01$. There also exists some *hot ionized medium* (HIM) with

$T_{\text{gas}} \sim 10^5 \ldots 10^6$ K. It is, of course, very tenuous ($n = 0.01$ cm^{-3}) and detected either by soft X-ray emission or through UV absorption lines of highly excited ions (for instance, O VI at 1030 Å) towards hot stars. It occupies a lot of space, probably most of the galactic disk but its mass fraction is negligible.

14.3.2 The 21 cm line of atomic hydrogen

In its electronic ground state, a hydrogen atom has two hyperfine sublevels because the spin of the electron couples to the spin of the proton. There is level splitting for the same reason as in the Zeeman effect, only here the magnetic field is supplied by the nuclear magnetic dipole. Let F denote the quantum number of total angular momentum which is the sum of electron spin S and nuclear spin I,

$$F = S + I.$$

In the upper state, electron and proton spin are parallel, $F = 1$ and the statistical weight is $g_1 = 3$; it is a triplet. In the lower state, the spins are anti-parallel, $F = 0$ and $g_0 = 1$; it is a singlet. The transition from an upper to the lower level occurs at $\nu = 1420$ MHz or $\lambda = 21.11$ cm and one, therefore, speaks of the 21 cm line. As its propagation is not impeded by dust, in the 1950s it played a dominant role in elucidating the structure of the interstellar medium of the Milky Way. Indeed, most of what we know about the diffuse component, its temperature, density, kinematic behavior, total mass and overall distribution, is due to the study of the 21 cm line.

14.3.2.1 *Frequency and transition probability of the line*

One obtains an estimate of the energy difference ΔE between triplet and singlet from the classical formula (11.16) which gives the potential energy of a magnetic dipole in a magnetic field. The dipole moment of the proton is given by

$$\mu_{\text{p}} = 2.793 \cdot \mu_{\text{N}}$$

and is thus 2.793 times greater than a nuclear magneton μ_{N} (compare with (11.10) for the Bohr magneton),

$$\mu_{\text{N}} = \frac{e\hbar}{2m_{\text{p}}c}.$$

The magnetic field in which the proton aligns is supplied by the electron. It is a dipole field (see the electric analog in (3.10)) of moment μ_{B} and decays with distance like r^{-3}. When one evaluates its strength at one Bohr radius $a_0 = 5.29 \times 10^{-9}$ cm, (11.16) yields a frequency $\nu = \Delta E / h \simeq 530$ MHz. This is the right order of magnitude but an underestimate because a_0 is the mean distance and the positions where the electron is closer to the nucleus have a higher weight. A very good number is obtained when one averages over the radius and takes the

1s wavefunction as the weight. Quantum mechanically, the frequency has been determined to extremely high precision, it is a number with 11 significant digits.

The Einstein A-coefficient of the line follows from the general formula (6.41) but this time μ is a *magnetic* dipole moment given by the sum $\mu_p + \mu_B$. The value of A_{10} comes out very small ($A_{10} = 2.869 \times 10^{-15}$ s^{-1}): first, because it is not an electric but a magnetic dipole transition which reduces A_{10} by a factor $\sim 10^6$; and second, because the frequency is low ($A \propto \nu^3$ after (6.41)). The excited state has an extremely long lifetime, $\tau = 1/A \sim 10^7$ yr, compared to optical transitions where $\tau \sim 10^{-8}$ s. Nevertheless, there is enough neutral hydrogen around to make the line easily observable.

14.3.3 How the hyperfine levels of atomic hydrogen are excited

14.3.3.1 The excitation or spin temperature

In determining the population of the hyperfine levels, we include, for generality, a radiation field u_ν. It may come, for example, from a supernova remnant or the 3 K microwave background. The atomic levels change because of collisions as well as spontaneous and radiatively induced transitions. As there are only two levels involved, 0 and 1, we have, in standard notation,

$$nn_1 C_{10} + n_1 A_{10} + n_1 u_\nu B_{10} = nn_0 C_{01} + n_0 u_\nu B_{01}. \tag{14.8}$$

The left-hand side populates state 0, the right-hand side depopulates it. Here $n = n_1 + n_0$ is the total number of hydrogen atoms per cm^3 and C_{10} is the collision rate for de-excitation ($F = 1 \rightarrow 0$). In equilibrium at kinetic temperature T,

$$g_0 C_{01} = g_1 C_{10} e^{-\alpha} \qquad \alpha \equiv \frac{h\nu}{kT}.$$

The excitation temperature T_x, which is often called spin temperature, and the radiation temperature T_R are defined by

$$\frac{n_1}{n_0} = \frac{g_1}{g_0} e^{-h\nu/kT_x}$$

and

$$u_\nu = \frac{8\pi \nu^2}{c^3} k T_R.$$

With the relations (6.39) and (6.40) between the Einstein A and B coefficients, and

$$\alpha_x \equiv \frac{h\nu}{kT_x}$$

one finds

$$e^{-\alpha_x} \left[nC_{10} + A_{10} + \frac{kT_R}{h\nu} A_{10} \right] = nC_{10} e^{-\alpha} + \frac{kT_R}{h\nu} A_{10}. \tag{14.9}$$

This is a general result for a two-level sysytem. The statistical weights g_i are gone. For the 21 cm line, the exponents α and α_x are much smaller than one. After a linear expansion of the e-function, the spin temperature T_x can be directly expressed by the kinetic gas temperature T and the radiation temperature T_R,

$$T_x = \frac{ynT + T_R}{1 + yn} \quad \text{with } y \equiv \frac{C_{10}}{A_{10}}\frac{h\nu}{kT}. \tag{14.10}$$

14.3.3.2 The critical density for excitation

Under most circumstances, the radiaton field is negligible. Then the levels are thermalized ($T_x \simeq T$) above the critical density:

$$n_{cr} = y^{-1} = \frac{A_{10}}{C_{10}}\frac{kT}{h\nu}.$$

One often uses for the critical density the expression $n_{cr} \sim A_{10}/C_{01}$. We did so, for example, in estimating n_{cr} for the CO rotational levels in section 14.2.1. Indeed, when there are only two levels and no radiation, equation (14.9) implies $T_x \sim T$ when $n = A_{10}/C_{01}$, provided that $\alpha \geq 1$. However, at very low frequencies ($h\nu \ll kT$), thermalization requires a density that is $kT/h\nu$ times larger. Because the temperature difference between the levels of the 21 cm line is only $h\nu/k = 0.068$ K and the gas temperature $T \sim 100$ K, the usual formula $n_{cr} \simeq A_{10}/C_{01}$ is wrong by a factor ~ 1000.

The cross section for collisional deexcitation ($F = 1 \rightarrow 0$) in a gas of 100 K is $\sigma_{10} \sim 10^{-15}$ cm^2, corresponding to a collision rate $C_{10} = \langle u\sigma_{10}\rangle \sim 2 \times 10^{-10}$ cm^3 s^{-1}. The brackets denote an average over the Maxwellian velocity distribution and u is the velocity. In a standard HI cloud, which has $T = 100$ K and $n \geq 1$ cm^{-3}, the sublevels are, in view of the small Einstein coefficient, thermalized, so the population ratio

$$\frac{n_1}{n_0} \simeq 3$$

and the total hydrogen density n is close to $4n_0$. However, in a tenuous and hot cloud ($n = 0.05$ cm^{-3}, and $T = 1000$ K implying $y \simeq 7$ in equation (14.10)), there is, rather surprisingly, a substantial excursion of the hyperfine levels from LTE and the ratio T_x/T is only one-quarter.

14.3.3.3 Spin exchange by collisions among H atoms

The physical process that determines the cross section, σ_{10}, is spin exchange. Suppose a hydrogen atom labeled (1) with $F = 1$ and electron spin up is hit by another hydrogen atom. In the collision, the electron spin of atom (1) can flip so that afterwards it is in the state $F = 0$; the proton of atom (1) is shielded by the electron cloud and its spin is unaffected. The electron spin of the hitting atom is random and thus some linear combination of up and down; its proton spin is irrelevant.

During the impact, the two atoms form a molecular complex. If the electron spin of the hitting atom is also up, the complex is in the repulsive electronic *t*riplet state ($^3\Sigma$, potential V_t); if it is down, the complex is in the attractive electronic *s*inglet state ($^1\Sigma$, potential V_s), as in stable H_2 molecules. At large separations, the two potentials are equal. In quantum mechanics, it is not possible to specify the state of the atoms separately. Instead, the wavefunction of the complex is written as a sum of a symmetric and an anti-symmetric function with respect to the interchanging particles. The symmetric part is associated with the phase factor $\exp(-iV_t t/\hbar)$, because when the electrons are swapped, the wavefunction stays the same, the anti-symmetric part is associated with the phase factor $\exp(-iV_s t/\hbar)$. So in a time interval dt, the phase between triplet and singlet changes by $(V_t - V_s)dt/\hbar$. Therefore the atoms exchange electron spins at a rate $(V_t - V_s)/\hbar$. As the potentials V_t and V_s are known, one can semi–classically follow the collision over the time of the impact and thus calculate the probability that afterwards atom (1) is in state $F = 0$ [Pur56]. This yields σ_{10}. Because the interaction potential is steep, σ_{10} changes only weakly with temperature ($\propto T^{-1/4}$).

14.3.3.4   Excitation by electrons and *L*α radiation

Spin flip is also possible as a result of collisions with free electrons. Here the cross sections are about half as large but electrons fly faster, so the collision rates are higher. Electron impact becomes important when the fractional ionization of hydrogen exceeds a few percent. In principle, spin flip can also occur due to the interaction between the magnetic moments of the electrons but this process is several orders of magnitude weaker.

Another interesting way of exciting the hyperfine levels of hydrogen atoms is through *L*α photons [Wou52]. They connect the $n = 1$ and $n = 2$ states and are created in HI clouds by cosmic-ray ionization and subsequent proton–electron recombination. The state $n = 2$ consists of three sublevels,

$$2P_{1/2} \qquad 2S_{1/2} \qquad 2P_{3/2}$$

each being double because of the hyperfine splitting. The $2S_{1/2}$ and $2P_{1/2}$ states, which coincide in Dirac theory, are actually separate due to the Lamb shift. Only the 2P levels are allowed to combine with the ground state $1S_{1/2}$. As a resonance line, the *L*α optical depth is immense and each *L*α photon is scattered *many* times, i.e. undergoes many collisions with H atoms at the gas kinetic temperature T. The radiation field in the *L*α line therefore acquires the same shape (spectral slope) as the Planck function at that temperature. Because in the absence of collisions and radiatively induced transitions, the population of the hyperfine levels is determined by the *L*α radiation field, the spin temperature T_x is also forced towards the kinetic gas temperature T.

14.3.4   Gas density and temperature from the 21 cm line

14.3.4.1   Optical depth and column density

The optical depth element in the HI line is, from (13.62),

$$d\tau_\nu = K\Phi(\nu)\,ds$$

where $\Phi(\nu)$ is the normalized profile function and the absorption coefficient K is from formula (13.68). Because $h\nu \ll kT_x$, one can expand the e-function in (13.68) and obtain

$$K \simeq n_0 K_0 \frac{h\nu}{kT_x}.$$

Here n_0 is the number of atoms in the lower singlet state and K_0 is after (13.67) proportional to the Einstein coefficient A_{10}. The factor $h\nu/kT_x$ is due to stimulated emission. Without it, one would have $K \simeq n_0 K_0$. So stimulated emission, which is usually negligible at optical wavelengths, reduces the cross section in the 21 cm line by a factor of order $10^3(!)$ at typical excitation temperatures $T_x \sim 100$ K. If $N(\text{HI})$ denotes the total column density of neutral hydrogen, the optical thickness becomes

$$\tau_\nu = \frac{N(\text{HI})}{4} \cdot K_0 \cdot \frac{h\nu}{kT_x} \cdot \Phi(\nu). \tag{14.11}$$

Because of stimulated emission, the hotter the gas, the lower the optical depth. But note that the intensity of optically thin line emission, which is proportional to $\tau_\nu T_x$, is independent of temperature.

It is often convenient to refer the profile function to the Doppler velocity v, and not to the frequency ν. It is then denoted by $P(v)$, again with $\int P(v)\,dv = 1$. At the line center, $v = 0$. Because $P(v)\,dv = \Phi(\nu)\,d\nu$,

$$\phi(\nu) = \frac{c}{\nu} P(v)$$

and, analogous to (14.11), the optical depth at velocity v is

$$\tau_v = C^{-1} \frac{N(\text{HI})P(v)}{T_x}. \tag{14.12}$$

When v, $N(\text{HI})$ and T_x are measured in cm s^{-1}, cm^{-2} and K, respectively,

$$C = 1.82 \times 10^{13}.$$

The brightness temperature T_b at velocity v observed towards a cloud in front of a background source of intensity T_c is, from (13.74),

$$T_b(v) = T_0 \cdot e^{-\tau_v} + T_x(1 - e^{-\tau_v}). \tag{14.13}$$

Without the background, one obtains, from (14.12),

$$N(\text{HI}) = \int N(\text{HI}) P(v)\,dv = C \int \frac{T_b(v)\tau_v}{1 - e^{-\tau_v}}\,dv. \qquad (14.14)$$

Therefore, when the optical depth is small everywhere in the line, as is often the case for the 21 cm transition, integration over the observed line profile directly yields the HI column density:

$$N(\text{HI}) = C \int T_b(v)\,dv. \qquad (14.15)$$

14.3.4.2   How to measure the optical depth and the spin temperature

With the help of a strong background source, say, a quasar, one can determine the optical depth τ_v as well as the spin temperature T_x of an HI cloud. First, one performs a measurement towards the quasar at a frequency outside the line where the HI cloud is transparent ($\tau_v = 0$). This provides the brightness T_0 of the quasar after (14.13). Then one observes the quasar at various velocities in the line. The actual frequency changes are very small, so T_0 is practically constant. If the quasar is much brighter than the foreground HI gas, it will be seen in absorption and equation (14.13) reduces to

$$T_b(v) = T_0\,e^{-\tau_v}.$$

This gives τ_v. One can now evaluate the column density $N(\text{HI})$ after (14.14); the optical thickness need not be small. In the line core, at a position immediately adjacent to the quasar,

$$T_b = T_x(1 - e^{-\tau_v}).$$

As τ_v is known, such a measurement yields the spin temperature T_x. Of course, one has to assume that T_x is constant along the line of sight and that the cloud parameters do not change as one switches from the quasar to a nearby off-position.

14.3.5   The deuterium hyperfine line

Deuterium, the hydrogen isotope, also shows hyperfine splitting in its electronic ground state.   The importance of this line rests in the fact that it could improve the estimate of the deuterium abundance of the interstellar medium, with cosmological consequences because deuterium has been created during the Big Bang. So far all attempts to detect the transition have failed because the isotope is so rare ($[D]/[H] \sim 10^{-5}$).

In deuteron, the spins of proton and neutron are parallel. So the nuclear spin quantum number I of deuterium equals 1 and the spin flip of the electron corresponds to an $F = 3/2 \to 1/2$ transition. The magnetic moment of the neutron is

$$\mu_n = -1.913\,\mu_N$$

and directed opposite to the magnetic moment of the proton. The magnetic moment of deuteron is quite close to the sum of the two,

$$\mu_D \simeq \mu_n + \mu_p \simeq 0.857 \, \mu_N.$$

When one aims at greater precision for μ_D, one must take into account that the deuteron nucleus possesses some *orbital* angular momentum (quantum number l). This is witnessed by its electric quadrupole moment, which also gives rise to a magnetic dipole moment. In its ground state, deuteron is a mixture consisting to ~95% of an S ($l = 0$) wavefunction and to ~5% of a D ($l = 2$) wavefunction.

To find the frequency of the deuterium hyperfine line relative to the hydrogen 21 cm line, we cannot simply scale with the ratio μ_D/μ_p. Instead, we have to translate the classical formula (11.16) for the potential energy of a magnetic dipole **m** in an outer field into an interaction operator W according to the rules of quantum mechanics. But for a constant factor, we can write $\mathbf{m}_{nuc} = g_{nuc}\mathbf{I}$ for the nucleus and $\mathbf{m}_{el} = g_{el}\mathbf{S}$ for the electron, where g is the Landé factor, so

$$W = \mathbf{m}_{nuc} \cdot \mathbf{m}_{el} = g_{nuc}g_{el}\,\mathbf{I} \cdot \mathbf{S}.$$

For deuterium, $I_D = 1$ and $g_D = 0.857$; for hydrogen, $I_H = \frac{1}{2}$ and $g_H = 2 \cdot 2.793$, for the electron $S = \frac{1}{2}$ and $g_{el} = 2$. The square of the angular momentum vector $\mathbf{F} = \mathbf{I} + \mathbf{S}$ is obviously

$$\mathbf{F} \cdot \mathbf{F} = (\mathbf{I} + \mathbf{S}) \cdot (\mathbf{I} + \mathbf{S}) = \mathbf{I} \cdot \mathbf{I} + \mathbf{S} \cdot \mathbf{S} + 2\mathbf{I} \cdot \mathbf{S}$$

so that

$$\mathbf{I} \cdot \mathbf{S} = \tfrac{1}{2}\big[\mathbf{F} \cdot \mathbf{F} - \mathbf{I} \cdot \mathbf{I} - \mathbf{S} \cdot \mathbf{S}\big].$$

When $\mathbf{I} \cdot \mathbf{S}$ is turned into a quantum mechanical operator, it therefore has the expectation value

$$\langle \mathbf{I}, \mathbf{S} \rangle = \tfrac{1}{2}\big[F(F + 1) - I(I + 1) - S(S + 1)\big]\hbar^2.$$

The difference Δ in these averages between upper and lower state is $\Delta_D = 3$ for deuterium ($S = \frac{1}{2}$; $I = 1$; $F = \frac{3}{2}, \frac{1}{2}$) and $\Delta_H = 2$ for hydrogen ($S = \frac{1}{2}$; $I = \frac{1}{2}$; $F = 1, 0$). Therefore, the deuterium hyperfine line has a frequency that is a factor $g_H\Delta_H/g_D\Delta_D \simeq 4.34$ smaller corresponding to a wavelength of 91.6 cm.

Another hyperfine line of cosmological relevance is emitted by ^{3}He$^+$ at 8.665 GHz ($F = 0 \rightarrow 1$), again very hard to observe. Of course, one can only see it where helium is ionized. This is in the inner parts of HII regions where hydrogen is ionized too. As the ionization zones of H and He are not co-extensive (section 14.4.2), the derivation of the abundance ratio $[^3$He$]/[$H$]$ along the line of sight is problematic.

14.3.6   Electron density and magnetic field in the diffuse gas

14.3.6.1   The dispersion measure

One can use the pulses emitted from rotating neutron stars to determine the electron density n_e in the diffuse interstellar medium. A pulse, which is a composite of many frequencies, approaches the Earth with a group velocity v_g given by (1.131). As the observing frequency ω is always much greater than the plasma frequency $\omega_p = \sqrt{4\pi e^2 n_e/m_e}$ (the latter is of order 1 kHz), we can write, approximately,

$$\frac{1}{v_g} = \frac{1}{c}\left(1 + \frac{\omega_p^2}{2\omega^2}\right).$$

A signal from a pulsar at a distance L reaches us after a travel time

$$\tau = \int_0^L \frac{ds}{v_g} = \frac{L}{c} + \frac{e^2}{2\pi m_e c}\frac{1}{v^2}\int_0^L n_e(s)\,ds.$$

The travel time depends on the observing frequency v: long waves are slower than short waves. The quantity

$$\mathrm{DM} = \int_0^L n_e(s)\,ds \tag{14.16}$$

is called the *dispersion measure* and has the unit pc cm^{-3}. When one measures the arrival time of a pulse at two frequencies of which $v_2 > v_1$, there is a delay $\Delta\tau$ between the two,

$$\Delta\tau = \frac{e^2}{2\pi m_e c}\left(\frac{1}{v_1^2} - \frac{1}{v_2^2}\right)\mathrm{DM}. \tag{14.17}$$

If the distance L is known, one thus finds, via the dispersion measure DM, the mean electron density, a fundamental quantity of the interstellar medium. Typical values for the diffuse medium, outside HII regions, are $n_e \sim 0.01$ cm^{-3}.

14.3.6.2   The rotation measure

A linearly polarized wave can be decomposed into two waves of opposite circular polarization. When they are traveling parallel to a magnetic field B, equation (1.126) yields for their difference in phase velocity

$$\Delta v_{\mathrm{ph}} = \Delta\left(\frac{c}{n_\pm}\right) = \frac{n_e e^2}{2\pi^2 m_e}\frac{eB}{m_e}\frac{1}{v^3}. \tag{14.18}$$

The relative phase changes per wavelength λ by

$$2\pi\frac{\Delta v_{\mathrm{ph}}}{v_{\mathrm{ph}}} \simeq 2\pi\frac{\Delta v_{\mathrm{ph}}}{c}$$

and the rotation angle Ψ of the *linearly* polarized wave, therefore, by half this amount. The total change on the way from the source at distance L to the observer is

$$\Psi = \frac{\pi}{\lambda} \int_0^L \frac{\Delta v_{ph}}{c} \, ds = RM\lambda^2 \tag{14.19}$$

with the *rotation measure* RM defined by

$$RM = \frac{e^3}{2\pi m_e^2 c^4} \int_0^L n_e B \, ds. \tag{14.20}$$

Because the dispersion measure DM is equal to $\int n_e \, ds$ and the rotation measure RM proportional to $\int B n_e \, ds$, their ratio yields the mean magnetic field strength

$$\langle B \rangle \propto \frac{RM}{DM}.$$

From systematic observations in the Milky Way, one finds $\langle B \rangle \sim 3 \; \mu G$. The **B** vector lies preferentially parallel to the galactic plane. Although the rotation angle Ψ is usually smaller than $90°$, in principle, it may be greater than 2π. To determine its value, not only modulo 2π, one needs measurements at more than two wavelengths.

14.3.6.3 The Zeeman effect in HI

The strength of the interstellar magnetic field may be estimated from the Zeeman effect in neutral hydrogen. In the presence of a magnetic field **B**, the degeneracy of the upper hyperfine state (the triplet with $F = 1$) is removed. When **B** is perpendicular to the line-of-sight vector **e**, one observes a central π-component and two symmetric σ-components, all with different polarizations (section 6.3.2). This is summarized here, together with the case when **B** is parallel to **e**.

(1) $\mathbf{e} \perp \mathbf{B}$

- σ is linearly polarized perpendicular to **B**
- π is linearly polarized parallel to **B**, no frequency shift
- σ is linearly polarized perpendicular to **B**, but opposite to other σ-component

(2) $\mathbf{e} \parallel \mathbf{B}$

- σ is right–hand circularly polarized
- π is missing
- σ is left–hand circularly polarized.

The frequency shift between the components of opposite circular polarization is, obviously,

$$\Delta v = \frac{e B_\parallel}{2\pi m_e c} = 2.80 \frac{B_\parallel}{\mu G} \quad \text{[Hz]}.$$

It is much smaller than the linewidth (~ 20 kHz) and, therefore, not easy to observe. Measured values for $B_\parallel$ also cluster around $3 \; \mu G$.

14.4   HII regions

14.4.1   Ionization and recombination

14.4.1.1   *The Lyman continuum flux*

The hottest stars on the main sequence are of spectral type O and B. As their effective surface temperature $T_{\rm eff}$ exceeds $20\,000$ K, they ionize the surrounding hydrogen. The resulting nebula of hot plasma is called the HII region. The gas temperature is fairly constant there ($T_{\rm gas} \sim 8000$ K), the exact number depends foremost on the abundance of metals which produce cooling lines. Dust that survives the harsh conditions in the nebula is very warm ($T_{\rm dust} \sim 50$ K), much warmer than the average dust in HI or molecular clouds. Most HII regions are large and diffuse with number densities $n \sim 10$ cm^{-3} but some are very compact ($n > 10^5$ cm^{-3}) and these are always embedded in molecular clouds. HII regions can be observed in many ways: via free–free radiation (bremsstrahlung), radio and optical recombination lines, fine structure lines, infrared emission of dust. By order of magnitude, HII regions fill about 1% of the volume of the galactic disk and their mass fraction of the interstellar medium is of the same order.

The ionization of hydrogen requires Lyman continuum (abbreviated Lyc) photons. They have energies greater than the ionization potential of hydrogen, $h\nu \geq h\nu_{\rm L} = \chi({\rm H}) = 13.56$ eV, or wavelengths

$$\lambda \leq \lambda_{\rm L} = 912 \,\text{\AA}$$

(see table 14.4 for a list of symbols). The number of Lyc photons emitted each second from a star of radius R_* and effective temperature $T_{\rm eff}$ is given by

$$N_{\rm L} = 4\pi R_*^2 \int_{\nu_{\rm L}}^{\infty} \frac{B_\nu(T_{\rm eff})}{h\nu}\, d\nu. \tag{14.21}$$

This is a simplistic formula. In reality, the atmospheres of OB stars are not blackbodies but are extended and non-LTE effects play an important role. One often wishes to relate the Lyc flux $N_{\rm L}$ of a main-sequence star to its mass m. If m is in solar units, we suggest from various sources of the literature the approximation

$$\log N_{\rm L} = -5.13(\log m)^2 + 19.75 \log m + 30.82 \quad \text{s}^{-1}. \tag{14.22}$$

Table 14.3 lists $N_{\rm L}$, the spectral type and some other stellar parameters. The numbers are not precise but sufficient for many purposes.

14.4.1.2   *Relation between ionization and recombination coefficient*

In thermodynamic equilibrium, a situation which does *not* prevail in an HII region, there is a detailed balance so that each process has its exact counterpart running in the opposite direction. Hence, the number of electrons with velocities

Table 14.3. Spectral type, effective temperature, bolometric luminosity, Lyman continuum flux, mass and radius for main sequence stars of various spectral type. The radius and mass of the HII region, r_S and M_{HII}, are calculated for an electron density $n_e = 1000$ cm^{-3}.

SpT	T_{eff} [K]	L [$L_\odot$]	N_L [s^{-1}]	M_* [$M_\odot$]	R_* [$R_\odot$]	r_S [cm]	M_{HII} [$M_\odot$]
O4	50000	1.3×10^6	8.0×10^{49}	60	15.0	4.2×10^{18}	260
O6	42000	2.5×10^5	1.2×10^{49}	30	10.0	2.2×10^{18}	40
O9	35000	4.6×10^4	1.2×10^{48}	19	6.0	1.0×10^{18}	4
B3	18000	1.0×10^3	5.0×10^{43}	7.6	3.4	3.6×10^{16}	
A0	10000	54		2.9	2.4		

Table 14.4. List of symbols used in this section.

ν_L	= Threshhold frequency for ionizing hydrogen
λ_L	= $c/\nu_L = 912$ Å, threshhold wavelength
N_L	= Number of Lyc photons emitted by star [s^{-1}]
$N_c(r)$	= Number of Lyc photons passing through shell of radius r [s^{-1}]
$h\nu_c$	= Mean energy of Lyc photons
n_e, n_p, n_H	= Number density of electrons, protons and hydrogen atoms [cm^{-3}]
N_n	= Number density of neutral hydrogen atoms in level n [cm^{-3}]
α_n	= Recombination coefficient into level n
$\alpha^{(i)}$	= Recombination coefficient into all levels $i \geq n$,
	at 8000 K, $\alpha^{(1)} = 4.6 \times 10^{-13}$, $\alpha^{(2)} = 3.1 \times 10^{-13}$ cm^{-3} s^{-1}
β_n	= Cross section for electron capture into level n
$K_{n,\nu}$	= Cross section for ionization from level n [cm^2]
K_ν	= $K_{1,\nu}$, cross section for ionization from the ground state [cm^2]
K	= Frequency average of K_ν
χ	= Ionization potential from ground state
χ_n	= Ionization potential from level n
$g_{n,\nu}$	= Gaunt factor ($\sim$1)
J_ν	= Mean radiation intensity in the nebula
L_ν	= Emission of star [erg s^{-1} Hz^{-1}]
ϵ_ν	= Emission coefficient of gas [erg s^{-1} Hz^{-1} ster^{-1}]

in the interval $[\nu, \nu + d\nu]$ captured by protons into level n equals the number of ionizations from level n by photons with frequencies $[\nu, \nu + d\nu]$,

$$n_e n_p \beta_n(\nu) f(\nu) \nu \, d\nu = N_n K_{n,\nu} \left(1 - e^{-h\nu/kT}\right) \frac{c u_\nu}{h\nu} d\nu. \qquad (14.23)$$

If χ_n is the energy necessary to ionize an atom in state n, frequencies and velocities are related through

$$h\nu = \chi_n + \tfrac{1}{2}m_e v^2.$$

In equation (14.23), $f(v)$ is the Maxwellian velocity distribution (5.13) for electrons, u_ν the blackbody radiation density of photons after (5.71), both at temperature T. The brackets on the right contain the correction factor for induced emission, $\beta_n(v)$ denotes the cross section for electron capture and $K_{n,\nu}$ for photo-ionization. The latter (without proof) is

$$K_{n,\nu} = \frac{2^6 \pi^4}{3^{3/2}} \frac{m_e e^{10}}{ch^6} \frac{g_{n,\nu}}{n^5 \nu^3} \simeq 2.82 \times 10^{29} \frac{g_{n,\nu}}{n^5 \nu^3} \qquad (h\nu > \chi_n) \qquad (14.24)$$

where the Gaunt factor $g_{n,\nu}$ is close to one. When one expresses in (14.23) $N_n/n_e n_p$ by the Saha equation (5.40) with $E_n = -\chi_n$ and substitutes $K_{n,\nu}$ from (14.24), one finds after integration over all velocities and frequencies, respectively, the total recombination coefficient

$$\alpha_n(T) \equiv \int_0^\infty \beta_n(v) f(v) v\, dv \simeq \frac{3.22 \times 10^{-6}}{T^{3/2} n^3} e^x E_1(x). \qquad (14.25)$$

$E_1(x)$ is defined in (A.20), and $x = \chi_n/kT$. Using the Saha equation, (14.23) yields the *Milne* relation

$$\beta_n(v) = \frac{2n^2 h^2 \nu^2}{c^2 m_e^2 v^2} K_{n,\nu}. \qquad (14.26)$$

It must also hold when the medium is not in thermodynamic equilibrium because $\beta_n(v)$ and $K_{n,\nu}$ are atomic quantities and independent of the radiation density u_ν and the particle velocity distribution $f(v)$. There is thus a remarkable link between the ionization coefficient $K_{n,\nu}$ from principal quantum number n and the cross section for electron capture $\beta_n(v)$ into that level.

14.4.2   Dust–free HII regions

14.4.2.1   Ionization balance

In each cm^3 of a stationary and dust-free HII region *of pure hydrogen*, the rate of ionization of neutral hydrogen atoms by the radiation field of mean intensity J_ν equals the rate of recombinations of electrons and protons to all quantum levels $n \geq 1$. If the corresponding recombination coefficient is denoted $\alpha^{(1)}$, then

$$n_H \int_{\nu_L}^\infty \frac{4\pi J_\nu}{h\nu} K_\nu(H)\, d\nu = n_e n_p \alpha^{(1)}. \qquad (14.27)$$

The radiation field has two components: one from the *star* (subscript s) which is directed radially outwards, the other arising from recombinations within the HII region. The latter is *diffuse* (subscript d) and travels in all directions, altogether $J_\nu = J_{s\nu} + J_{d\nu}$. The stellar flux $F_{d\nu}$ declines with distance r from the star because it is geometrically diluted and weakened by absorption. In spherical symmetry,

$$F_{s\nu}(r) = \frac{R_*^2}{r^2} F_{s\nu}(R_*) e^{-\tau(\nu)}. \tag{14.28}$$

The optical depth

$$\tau(\nu) = \int_{R_*}^{r} n_H K_\nu(H) \, ds$$

is due to absorption of hydrogen atoms in their ground state 1^2S. Almost all neutral hydrogen atoms reside there because the probability for downward transitions from upper principal quantum numbers n is very high (A-coefficients $10^4 \ldots 10^8$ s^{-1} for $\Delta L \pm 1$). Equation (14.28) can now be written as (see (13.26))

$$\frac{1}{r^2} \frac{d}{dr} (r^2 F_{s\nu}) = -4\pi J_{s\nu} n_H K_\nu(H). \tag{14.29}$$

The expression for the net flux of outwardly directed diffuse radiation $F_{d\nu}$ is similar but contains additionally as a source term the emission coefficient $\epsilon_{d\nu}$ arising from recombinations to the ground level $n = 1$,

$$\frac{1}{r^2} \frac{d}{dr} (r^2 F_{d\nu}) = -4\pi J_{d\nu} n_H K_\nu(H) + 4\pi \epsilon_{d\nu}. \tag{14.30}$$

Adding (14.29) and (14.30), dividing by $h\nu$ and integrating over ν, one obtains

$$\frac{d}{dr} \left[4\pi r^2 \int \frac{F_{s\nu} + F_{d\nu}}{h\nu} \, d\nu \right]$$
$$= -4\pi r^2 \left[\int n_H K_\nu(H) \frac{4\pi J_{s\nu} + 4\pi J_{d\nu}}{h\nu} \, d\nu - \int \frac{4\pi \epsilon_{d\nu}}{h\nu} \, d\nu \right]. \tag{14.31}$$

The integrals extend from ν_L to infinity. The bracket on the left gives the number of Lyc photons that pass per s through a shell of radius r; we denote this quantity by $N_c(r)$ (unit s^{-1}); by definition,

$$N_c(r = R_*) = N_L.$$

In the bracket on the right-hand side of (14.31), the first term equals $\alpha^{(1)} n_e n_p$, which is the number of *all* recombinations per second and cm^3, and the second term equals $\alpha_1 n_e n_p$, which is the number of recombinations to the ground level only. Here it is assumed that every diffuse Lyc photon arises from a recombination to $n = 1$ and is absorbed *on the spot* by a neutral H atom. The difference between both terms is $\alpha^{(2)} n_e n_p$, where the coefficient $\alpha^{(2)} = \alpha^{(1)} - \alpha_1$ stands for all recombinations to levels $n \geq 2$. We thus obtain

$$\frac{dN_c(r)}{dr} = -4\pi r^2 \alpha^{(2)} n_e n_p. \qquad (14.32)$$

In an HII region, hydrogen is almost fully ionized, $n_H \ll n_e = n_p$. Therefore, we may directly integrate equation (14.32) from the stellar surface ($r = R_*$) to the edge of the HII region. The radius there is denoted r_S and called *Strömgren radius*. Because $r_S \gg R_*$, in the case of *uniform plasma density*

$$N_L = N_L(r{=}R_*) = \frac{4\pi}{3} r_S^3 n_e^2 \alpha^{(2)}.$$

14.4.2.2   *Strömgren radius, ionization structure, recombination time*

- For the Strömgren radius, one immediately finds, from the last equation, that

$$r_S = \left[\frac{3 N_L}{4\pi n_e^2 \alpha^{(2)}} \right]^{1/3}. \qquad (14.33)$$

- Although $n_H \ll n_e$ is a very good approximation, it does not tell us the density of hydrogen atoms, n_H. If $n \equiv n_p + n_H$, the fraction $y = n_H/n$ of hydrogen atoms follows from the quadratic equation

$$N_c K(H) y = 4\pi r^2 \alpha^{(2)} n (1 - y)^2. \qquad (14.34)$$

y is, of course, a function of radius r. Note that the degree of ionization, $1 - y = n_e/n_H$, is not determined by the Saha formula (5.38).
- The transition from the HII region to the neutral gas is sharp and called ionization front. Its width d_{IF} is determined by the condition that the optical depth τ with respect to Lyc radiation is of order unity,

$$d_{IF} = \frac{1}{n_H K(H)}. \qquad (14.35)$$

For densities 10^3 cm^{-3} and $K(H) \sim 10^{-17}$ cm^2 (table 14.4), the width d_{IF} is only $\sim 10^{14}$ cm, whereas the diameter of the HII region may be greater than 10^{18} cm.
- The mass of an HII region depends not only on the spectral type of the exciting star but also on the environment. It is inversely proportional to the electron density,

$$M_{HII} = \frac{N_L m_H}{n_e \alpha^{(2)}}. \qquad (14.36)$$

- Per unit volume, the recombination rate $n_e^2 \alpha^{(1)}$ equals the rate of ionization. An individual electron will, therefore, recombine after an interval

$$\tau_{ion} = \frac{1}{n_e \alpha^{(1)}}. \qquad (14.37)$$

This is the characteristic time for recombination and ionization and also the time an HII region would go on shining after the star were switched off. For a plasma of moderate density ($n_e = 10$ cm^{-3}), equation (14.37) implies $\tau_{ion} \sim 10^4$ yr.

- When helium is present in the gas (typical galactic abundance by number [He]/[H] $\simeq 0.1$), it competes with hydrogen for photon capture. Its Lα radiation from the singlet system of para–helium at $\lambda = 584$ Å is also able to ionize hydrogen. Because helium ionization requires more energetic photons (χ (He) $= 24.6$ eV $> \chi$ (H) $= 13.6$ eV), the HeII region is never greater than the HII region, at best co-extensive. If $N_{L,He}$ is the number of Lyc photons with $h\nu > \chi$ (He), the Strömgren radius r_S (He) for helium becomes

$$r_S(\text{He}) = \left[\frac{3N_{L,He}}{4\pi n_e n(\text{He}^+)\alpha^{(2)}(\text{He})} \right]^{1/3}. \tag{14.38}$$

The flux $N_{L,He}$ depends sensitively on the stellar effective temperature: For B0 stars, the HeII zone is practically absent, for stars of spectral type O6 or earlier, r_S (He) $= r_S$ (H).

14.4.3 Dusty HII regions

14.4.3.1 Ionization balance of a dusty HII regions

The presence of dust reduces the number of Lyc photons available for ionization of the gas and, therefore, the size of an HII region. Let τ be the optical depth in the Lyman continuum with respect to dust only, and K_d the corresponding dust cross section per H atom. So $d\tau = n K_d dr$, where $n = n_e + n_H$ and helium has been negelected. As the degree of ionization within an HII region is close to one, we get from (14.32)

$$\frac{1}{N_L} \frac{dN_c}{d\tau} = -\frac{4\pi \alpha^{(2)} n r^2}{N_L K_d} - \frac{N_c}{N_L}.$$

Inserting r_S from (14.33) leads to the ordinary differential equation

$$\frac{1}{N_L} \frac{dN_c}{d\tau} = -\frac{3}{n^3 K_d^3 r_S^3} \tau^2 - \frac{N_c}{N_L} \tag{14.39}$$

with the boundary condition $N_c(\tau=0) = N_L$. The equation may be solved numerically but if the density n is constant, it offers the analytical solution [Pet72]

$$\frac{N_c}{N_L} = e^{-\tau} - \frac{3\tau^2 - 6\tau + 6(1 - e^{-\tau})}{n K_d r_S} \tag{14.40}$$

which gives $N_c(\tau)$ directly as a function of the dust optical depth $\tau = r n K_d$. The dusty HII region extends to the dusty Strömgren radius r_{dS}, where all Lyman continuum photons have been absorbed either by gas or by dust. One finds r_{dS} by putting (14.40) to zero and solving for τ. This yields the optical thickness $\tau_{dS} = r_{dS} n K$ of a dusty HII region and thus r_{dS}. For example, a region with optical depth $\tau_{dS} = 1$ has a 56% smaller volume than one without dust where $\tau_{dS} = 0$.

There are no observations of the extinction curve in the Lyman continuum, the dust cross section K_d is, therefore, quite uncertain for $\lambda < 912$ Å. Of course, we have some guidance from the dielectric permeability ε of figure 7.19. According to it, when applying Mie theory, the cross section of an *individual grain* is at $\lambda = 1000$ Å three times greater than in the visual and continues to rise towards shorter wavelengths up to a resonance at ~ 700 Å.

For a compact HII region ($n = 10^4$ cm^{-3}) around an O6 star ($N_L = 10^{49}$ s^{-1}) the dust-free Strömgren radius $r_S = 4.5 \times 10^{17}$ cm. If the dust in the HII region had the same abundance and properties as in the diffuse interstellar medium, its cross section per hydrogen atom at 1000 Å would be $K_d \simeq 1.5 \times 10^{-21}$ cm^2, and in the Lyman continuum greater still. This would imply a huge optical depth ($\tau = r n K_d \sim 10$) so that the HII region would almost vanish. However, compact HII regions are being observed. Consequently, the dust must be modified, most likely it is depleted, partially destroyed by the strong and hard far UV field. In HII regions, dust absorption coefficients per gram of interstellar matter are probably reduced by a factor of four with respect to neutral gas clouds.

The dust in an HII region is heated radiatively and its temperature T_d is decoupled from the gas temperature. The radiation field J_ν which the grains absorb includes stellar light, Lα photons and optical cooling lines. T_d then follows from (8.7). In a compact HII region, more than two-thirds of the radiation is emitted by dust (disregarding the surrounding shell of neutral gas).

14.4.3.2   *Lyα photons of hydrogen*

Each Lyman continuum photon ionizing a hydrogen atom leads inevitably to a proton–electron recombination. No matter how the newly formed hydrogen atom cascades down from its excited state, eventually it produces a Lyα photon of energy $h\nu_\alpha = 10.2$ eV. Like all photons connecting to the ground state, they are trapped inside the HII region. Therefore their radiation density is very high. But it cannot build up indefinitely because of the following three destruction routes:

- Absorption of Lyα photons on grains. This is by far the most important process and works independently of the amount of dust in the HII region. If there is very little dust because most has been destroyed, the few remaining grains receive all the Lyα flux and will be very hot. They absorb the fraction $N_L h\nu_\alpha / L_*$ of the stellar luminosity L_*, typically 25%. This energy then escapes as infrared radiation. If the HII region is not almost dust-free and the dust optical depth τ_{dS} not small, the infrared luminosity (without cooling fine structure lines) will be even greater than $\sim 25\%$. In addition, the grains then also directly absorb stellar photons of all wavelengths, as well as diffuse photons of the cooling lines.

- Decay of the metastable upper level $2\,^2S$ ($n = 2, l = 0$, lifetime 0.12 s) into the ground state $1\,^2S$ by emission of *two* photons (see end of section 6.3.4.3) of total energy $h\nu_\alpha$. Averaged over many transitions, the decay forms a continuum.
- Diffusion of Lα photons into the optically thin line wings from where they can escape the HII region.

14.4.4 Bremsstrahlung

In a hydrogen plasma of temperature T and electron density n_e, the electrons have a mean velocity proportional to $T^{1/2}$. The protons are much heavier and, therefore, almost at rest. As an electron passes a proton, it is accelerated and changes its direction, usually only a little. The acceleration of the electron leads after the dipole formula (1.97) to electromagnetic radiation.

- The monochromatic emission coefficient is (without proof)

$$\epsilon_\nu = \frac{8}{3}\sqrt{\frac{2\pi m_e}{3kT}}\,\frac{e^6}{m_e^2 c^3}\,g_{\mathrm{ff}}\,n_e^2\,e^{-h\nu/kT}.\tag{14.41}$$

The emission spectrum is at low frequencies flat,

$$\epsilon_\nu \simeq 5.44 \times 10^{-39} n_e^2 T^{-1/2}\ \mathrm{erg\,cm^{-3}s^{-1}Hz^{-1}ster^{-1}}$$

because $\exp(-h\nu/kT) \simeq 1$ and the Gaunt factor $g_{\mathrm{ff}} \simeq 1$. In a plasma at 8000 K, the Boltzmann factor $e^{-h\nu/kT}$ becomes important only at wavelengths $\lambda < 10\ \mu$m.

- Denoting by $\bar{g}_{\mathrm{ff}}$ the frequency average of the Gaunt factor, the total bremsstrahlung emission is

$$4\pi \int \epsilon_\nu\,d\nu = \sqrt{\frac{2\pi kT}{m_e}}\,\frac{32\pi e^6}{3h m_e c^3}\,\bar{g}_{\mathrm{ff}}\,n_e n_p$$

$$\simeq 1.42 \times 10^{-27}\,\bar{g}_{\mathrm{ff}}\,n_e^2\sqrt{T}\quad \mathrm{erg\,cm^{-3}s^{-1}}.\tag{14.42}$$

- The continuum absorption coefficient of the plasma follows from Kirchhoff's law. At radio wavelengths

$$K_\nu = \frac{B_\nu(T)}{\epsilon_\nu} \simeq 0.173\left(1 + 0.13\log\frac{T^{3/2}}{\nu}\right)\frac{n_e n_p}{\nu^2}\quad \mathrm{cm^{-1}}\tag{14.43}$$

implying a continuum optical depth $\tau_c \propto \nu^{-2}$. Defining the *emission measure* by

$$\mathrm{EM} = \int n_e^2\,ds\tag{14.44}$$

τ_c can be approximated by [Alt60]

$$\tau_c = 0.082 a(v, T) T^{-1.35} \left(\frac{v}{\text{GHz}}\right)^{-2.1} \text{EM} \qquad (14.45)$$

where the factor $a(v, T) \simeq 1$ and the emission measure is in the unit pc cm^{-6}.

- Below a turnover frequency v_{turn}, the HII region is opaque, above transparent. When it is opaque, one receives a flux S_v that is proportional to the Planck function, $S_v \propto B_v \propto v^2$; when it is transparent, $S_v \propto \tau_v B_v \propto v^{-0.1}$. Bright and compact HII regions have typical emission measures 10^7 pc cm^{-6}. At $T = 8000$ K, they become optically thin at frequencies above $v_{\text{turn}} \sim 2$ GHz.
- Measuring the integrated radio flux S_v of an optically thin source allows to determine the number of stellar ionizing photons N_c,

$$N_c = 7.54 \times 10^{46} \left(\frac{v}{\text{GHz}}\right)^{-0.1} \left(\frac{T}{10^4 \text{ K}}\right)^{-0.45} \left(\frac{D}{\text{kpc}}\right)^2 \left(\frac{S_v}{\text{Jy}}\right) \text{ s}^{-1}. \quad (14.46)$$

One can then infer the spectral type of the exciting star (table 14.3). In case of a compact source, one has to make allowance for the absorption of Lyc photons by dust. Because the integrated radio flux is

$$S_v \propto \int n_e n_p \, dV$$

one can also derive from S_v the mean electron density if the volume V is estimated from the linear size of the HII region.

- A young HII region appears, first of all, as an infrared source whose spectral energy distribution is dominated by dust and peaks typically at 60 μm. With the help of table 14.3 or equation (14.22), one can predict from (14.46) the radio flux S_v that a star of certain stellar type or mass will produce. Because the dust emission drops with frequency like v^4, whereas the radio spectrum is flat, bremsstrahlung exceeds dust emission at long long wavelengths. The cross-over is usually at $\lambda \sim 3$ mm.

14.4.5 Recombination lines

Important information on HII regions is provided by recombination lines. They arise when electrons recombine with protons, or other ions, and cascade downwards. For hydrogen-like atoms, the frequencies follow from the *Rydberg* formula $(n > m)$

$$v = cRZ^2 \left(\frac{1}{m^2} - \frac{1}{n^2}\right).$$

Z is the electric charge of the atomic nucleus $(Z = 1$ for H$)$ and R the Rydberg constant (for infinite nuclear mass, $R = 109\,737.3$ cm^{-1}). Transitions

$n \to n - 1$ are called α-lines, $n \to n - 2$ are β-lines, and so forth. Because the recombination and Einstein coefficients are known, the strength of the lines yields directly the number of ionizing photons in an HII region.

Optical lines (Balmer lines, like Hα) can be very bright but in order to be interpreted they have to be corrected for dust extinction. De-reddening by means of an extinction curve (figure 7.8) requires observations of several lines at different wavelengths. The procedure is often ambiguous because one does not know how much dust resides within the HII region and how much is in the foreground. The problem is less severe in the near infrared (Brackett and higher-order lines).

Interstellar dust extinction is irrelevant for transitions between high quantum states. Such radio recombination lines can be detected from very obscured and optically not accessible places. When combining the radial velocity with the rotation curve of the Milky Way, one obtains the (kinematic) distance of the source. The lines thus reveal the spiral structure of the Milky Way as traced by bright star-forming regions. We add three other interesting features of radio recombination lines:

- As the Einstein coefficients are small, the levels have a long lifetime and they are reshuffled by collisions leading to a population that is close to thermodynamic equilibrium, at the temperature of the ionized gas (see (5.40)).
- Because electrons with high quantum numbers are far from the atomic nucleus, nuclei of the same charge but of different mass produce only slightly shifted frequencies. For example, the relative difference $\Delta v / v$ between H und He radio recombination lines is 4.1×10^{-4} which allows us to observe both lines in one measurement, thus enhancing the accuracy of line and abundance ratios.
- The highest transitions observed are around $n \sim 700$. The atomic radius r increases with quantum number like $r = a_0 n^2$ (Bohr radius $a_0 = 0.529$ Å) and such highly excited atoms have macroscopic sizes (> 1 μm).

14.5   Mass estimates of interstellar clouds

The amount of interstellar gas, M_{gas}, that an astronomical object contains, be it a galaxy, a molecular cloud or a stellar disk, is one of its prime parameters, like luminosity or size. We summarize a few methods for estimating M_{gas}. None of them is very precise, so one employs as many methods as possible and takes an average.

14.5.1   From optically thin CO lines

Consider rotational CO transitions $j + 1 \to j$ in a homogeneous cloud of temperature T and column density N_j of molecules in level j. If $\Phi(v)$ is the

Gaussian line profile from (13.59), the optical thickness τ_ν becomes

$$\tau_\nu = N_j \frac{g_{j+1}}{g_j} \frac{c^2}{8\pi\nu^2} A_{j+1,j} \left(1 - \frac{b_{j+1}}{b_j} e^{-\frac{h\nu}{kT}}\right) \Phi(\nu) \qquad (14.47)$$

where $b_j = N_j/N_j^*$ denotes the deviation of N_j from the LTE value N_j^* (see (13.69)). Let $\Delta\nu = 2H$ be the half-width in frequency of the absorption profile and τ_c the optical depth in the line center. One then finds that in LTE ($b_j = b_{j+1} = 1$), the column density of the upper level is given by

$$N_{j+1} = \frac{\tau_c \Delta\upsilon}{e^{h\nu/kT} - 1} \frac{3h}{16\pi^3\mu^2} \sqrt{\frac{\pi}{\ln 2}} \frac{2j+3}{j+1}.$$

μ is the dipole moment of the molecule and $\Delta\upsilon = c\Delta\nu/\nu$. The optical depth in the center of millimeter CO lines is always high ($\tau_c \gg 1$) but for the isotope $C^{18}O$, τ_c is often below one. Then $\Delta\upsilon$ is also the observed linewidth and τ_c follows from the line temperature $T_L = \tau_c T$. If $\Delta\upsilon$ is given in cm s^{-1}

$$N_{j+1} = 6.81 \times 10^9 \frac{T_L}{T} \frac{\Delta\upsilon}{e^{h\nu/kT} - 1} \frac{2j+3}{j+1} \quad \text{cm}^{-2}.$$

The column density of the upper level, N_{j+1}, is roughly proportional to the line area $T_L\Delta\upsilon$. It depends only weakly on the kinetic gas temperature T and can be converted into the total CO column density $N(CO)$ with the help of the Boltzmann distribution (5.8) and from there to the molecular hydrogen column density, $N(H_2)$, assuming the standard ratio $[H_2]/[CO] = 10^4$.

14.5.2 From the CO luminosity

Most molecular clouds do not collapse under their own gravity but seem to be virialized. Their typical internal systematic velocity $\Delta\upsilon$ is then determined by the cloud mass M (section 15.2.1). If R denotes the cloud radius,

$$\Delta\upsilon \simeq \sqrt{GM/R}.$$

The CO luminosity is defined by

$$L_{CO} = A_{\text{beam}} \int_{\text{line}} T_{\text{mb}}(\upsilon) \, d\upsilon \qquad (14.48)$$

where A_{beam} is the area of the beam at the distance of the source and T_{mb} is the observed main-beam-brightness temperature (section 13.1.4) of a CO rotational line, usually the (1–0) or (2–1) transition. It is customary to measure L_{CO} in pc^2 K km s^{-1}. The integral in (14.48) extends over the whole line. As long as the source is smaller than the beam, the product $A_{\text{beam}}T_{\text{mb}}$ and thus L_{CO} is constant, independent of the beam size. If the mass fraction of stars within the cloud is negligible such that $M \simeq M_{\text{gas}}$, then

$$L_{\text{CO}} \propto \frac{T}{\sqrt{n}} M_{\text{gas}} \tag{14.49}$$

where T is the cloud temperature and n the number density of the gas. If one assumes that the ratio $T/\sqrt{n}$ is constant for all clouds, the CO luminosity is directly proportional to the cloud mass, although the CO line may be optically very thick. Of course, the relation (14.49) must be calibrated. The assumed constancy of $T/\sqrt{n}$ implies that the gas temperature goes up when a cloud becomes more compact. By and large, this may be so, possibly as a result of increased star formation in dense regions but it is certainly not a rule. Nevertheless, the method seems to work, at least, better than we understand.

The CO luminosity is also used to derive gas masses of galaxies. Gas motions there are large, of order 200 km s^{-1}, comparable to the rotational velocity of the Milky Way. The gas kinematics are entirely determined by the gravitational potential of the stars, and not by the gas. Nevertheless, equation (14.49) is applicable. When one makes an observation towards the center of a galaxy, the CO line will have a width of the same order, say, 200 km s^{-1}. As the beam usually covers a linear scale of a few hundred pc or more, at the distance of the galaxy, it samples many giant molecular clouds. Each GMC produces a relatively narrow line, say, of 10 km s^{-1} width. The observed line is then a superposition of many sublines that are well separated, either spatially because they come from different locations in the galaxy or in frequency space because they have different Doppler velocities. The contributions from the GMCs thus add up linearly to give the total gas mass of the galaxy.

In the CO (2–1) line, for example, an average over a large sample of spiral galaxies yields $L_{\text{CO}} \sim 4M_{\text{gas}}$ when M_{gas} is measured in $M_{\odot}$ and L_{CO} in pc^2 K km s^{-1}. Of course, the masses from the CO luminosity refer only to the molecular gas. How to estimate column densities of the HI gas was discussed in section 14.3.4. Masses from dust emission (see later) include both components, HI and H_2.

14.5.3 From dust emission

Because the dust absorption coefficient K_λ drops quickly with wavelength (figure 12.10), almost all interstellar clouds are transparent in the submillimeter and millimeter region. The observed flux S_λ from the dust is then directly proportional to the dust mass M_{d}. If the dust temperature T_{d} is uniform over the source,

$$S_\lambda = \frac{K_\lambda B_\lambda(T_{\text{d}}) M_{\text{d}}}{D^2}. \tag{14.50}$$

The distance D is usually known, T_{d} has to be derived as a color temperature from the fluxes at, at least, two wavelengths (see (8.20) and discussion thereafter) or from molecular line data. At long wavelengths, when $hc/\lambda kT \ll 1$, the Planck function $B_\lambda(T_d)$ is not very sensitive to the exact value of T_d. A favourite observing wavelength is $\lambda = 1.3$ mm because the atmosphere there has a window of good transmission.

The critical quantity in (14.50) is the absorption coefficient per gram of interstellar dust, K_λ. At $\lambda = 1.3$ mm, one is in the Rayleigh limit (grain radius $a \ll \lambda$), and the grain size distribution is fortunately irrelevant. One may, therefore, use the absorption efficiency (3.3),

$$Q_\lambda^{abs} = \frac{8\pi a}{\lambda} \frac{6nk}{(n^2 - k^2 + 2)^2 + 4n^2k^2}.$$

In our dust model, whose optical constants are displayed in figure 7.19 and which is detailed in section 12.4, bare silicates have at this wavelength $(n, k) \simeq (3.4, 0.04)$ and amorphous carbon (aC) has $(n, k) \simeq (12, 4)$. These numbers imply

$$Q_{1.3mm} = \begin{cases} 0.86a & \text{for silicate dust} \\ 2.13a & \text{for amorphous carbon dust} \end{cases} \tag{14.51}$$

when a is measured in cm. The mass absorption coefficient per gram of silicate or carbon dust becomes

$$K_{1.3mm} = \frac{3\pi a^2 Q_{1.3mm}}{4\pi a^3 \rho} = \begin{cases} 0.26 \text{ cm}^2 & \text{per g of Si} \\ 0.64 \text{ cm}^2 & \text{per g of aC .} \end{cases} \tag{14.52}$$

For the material density of the dust, we put for both components $\rho = 2.5$ g cm^{-3}. With a dust-to-gas mass ratio in the diffuse medium of $R_d = 1 : 140$, one obtains

$$K_{1.3mm} = \begin{cases} 0.0026 \text{ cm}^2 & \text{per g of interstellar matter} \\ 0.41 \text{ cm}^2 & \text{per g of dust .} \end{cases} \tag{14.53}$$

Equation (14.53) may be extrapolated to far infrared wavelengths by assuming $Q_\lambda \propto \lambda^2$,

$$K_\lambda = 4.4 \times 10^{-5} \left(\frac{\lambda}{\text{cm}}\right)^{-2} \text{ cm}^2 \text{ per g of interstellar matter.} \tag{14.54}$$

Fluffiness of the grains as well as ice mantles tend to increase K_λ, so that in dense and cold clouds (see figure 12.10) it may be five to ten times bigger than that given in (14.54).

Chapter 15

Stars and their formation

To get an idea of how the elements that make up interstellar dust are created, section 15.1 deals with nulear burning, in particular, with the pp chain and the 3α process. We then present approximate formulae for the luminosity and lifetime of individual stars, discuss the *initial mass function* (IMF) which specifies how masses are distributed when stars are born, and show averages for star clusters displaying an IMF. In section 15.2, we delineate the structure of cloud clumps at the verge of gravitational instability and present *Jeans's* criterion for collapse to actually occur. In section 15.3, we numerically analyse the equations for spherical, isothermal collapse and derive, for the earliest stages of protostellar evolution, the approximate density and velocity structure of the protostellar clump. Section 15.4 treats disks which inevitably form as the result of cloud rotation and which have a strong effect on the appearance and dynamical evolution of protostars, to say nothing about planet formation. The results of section 15.3 and 15.4 serve as the basis for models of dust and molecular line emission in chapter 16.

15.1 Stars on and beyond the main sequence

15.1.1 Nuclear burning and the creation of elements

Stars form the most obvious component of the Milky Way. They are, by and large, the ultimate source of luminosity in galaxies. The energy is generated in the stellar cores by fusion of atomic nuclei. During the main-sequence phase, hydrogen is burnt into helium. In the post-main-sequence phase, when hydrogen has been exhausted, elements of atomic mass number greater than four are produced by compounding helium nuclei. The energy yield per gram of matter is then one order of magnitude smaller. Fusion ends with the generation of iron which has the lowest binding energy per nucleon; heavier nuclei release energy not in fusion but in fission. An example is uranium ^{235}U where fission, here on Earth, has very diverse technical applications: to supply electricity to our homes, or to remove 10 000 homes in one blow.

Table 15.1. Standard solar abuncance of selected elements and their origin from [And89] and [Tri89].

Element	Relative abundance	Origin
H	1	Big bang
He	0.10	Big bang, H-burning
C	3.6×10^{-4}	He-burning
N	1.1×10^{-4}	From carbon during H-burning
O	8.5×10^{-4}	He-burning
Ne	1.2×10^{-4}	C-burning
Mg	3.9×10^{-5}	Ne-burning
Al	3.0×10^{-6}	Ne-burning
Si	3.6×10^{-5}	O-burning
S	1.9×10^{-5}	O-burning
Fe	3.2×10^{-5}	Pre-supernovae

Nuclear burning not only supplies energy but also creates elements. We are mostly interested in the origin and abundance of carbon, nitrogen, oxygen, magnesium, silicon and iron as they are, together with hydrogen, the building blocks of interstellar dust and constitute the most frequently observed molecules. They are synthesized during the major burning phases and that qualitatively explains why they are so common. Estimates of their abundances are summarized in table 15.1, the burning phases are described later.

Nuclear astrophysics is the branch where the synthesis of elements is studied. According to standard cosmological theories, none of the atoms heavier than helium can have been created in the Big Bang in any relevant amount. So they were made afterwards. But obviously, not all elements can have been synthesized by α-particles in nuclear burning. Nor does adding a proton to a helium core and then another and another usually work in building up heavier nuclei. For example, helium has more binding energy per nucleon than its heavier neighbors in the Periodic Table: lithium, beryllium and boron. So the latter three cannot be fused in this way. Instead, the light elements are probably made in cosmic-ray spallations.

Most other elements (besides C, N, O, Ne, Mg, Si) and the plethora of isotopes are built by neutron capture. Neutrons are made in various side-branch reactions. They do not feel the Coulomb repulsion and easily penetrate an atomic nucleus. The resulting new atom can be a stable isotope or be unstable to β-decay.

- When the rate of neutron production and capture is *slow* compared to β-decay, as during the quiescent post-main-sequence burning stages, one speaks of s–processes.
- When it is *rapid*, as in supernova explosions and some late stellar phases,

more than one neutron is captured before the nulceus β-decays. In this case, one speaks of r–processes.

15.1.2   The binding energy of an atomic nucleus

To understand nuclear burning and the resulting element abundances, we must know how tightly atomic nuclei are bound. The approximate energy of a nucleus containing N neutrons of mass m_n and Z protons of mass m_p, altogether

$$A = N + Z$$

particles, can be derived from the liquid drop model where one considers the binding of the nucleons in analogy to the chemical binding of molecules in a water drop (section 9.5). According to this model, the density in the nucleus is constant and there is mutual attraction only between neighboring particles. The total energy of the atomic nucleus consists of the energy equivalent of the mass of the *free* particles minus the binding energy $E_b > 0$,

$$W = Zm_p c^2 + Nm_n c^2 - E_b. \tag{15.1}$$

For $A > 4$, approximately

$$E_b = \alpha A - \beta A^{\frac{2}{3}} - \gamma \frac{e^2 Z(Z-1)}{A^{1/3}} - \eta \frac{(N-Z)^2}{A} \pm \frac{\delta}{A}. \tag{15.2}$$

The binding energy per nucleon, E_b/A, is plotted in figure 15.1 as a function of atomic number Z. We discuss the individual terms in (15.2):

- The first is due to the strong force and its value per nucleon is constant, equal to α.
- The second term is a correction to the first because some nucleons, their number being proportional to $A^{2/3}$, are on the surface of the nucleus and have fewer neighbors. It produces in figure 15.1 the decline at small Z. Term 1 and 2 correspond to equation (9.58) which gives the energy of an ordinary liquid drop.
- Term 3 comes from the Coulomb repulsion which acts not only between neighboring particles but everywhere and here over a mean distance proportional to the radius of the nucleus, or to $A^{1/3}$. This also brings E_b/A down at large Z so that the function E_b/A obtains a maximum at intermediate Z.
- Term 4 follows from the Pauli principle and very much improves the approximation. When it is included, E_b/A peaks near iron ($A = 56$) at about 8.8 MeV. The term is zero if there is an equal number of neutrons and protons, $N = Z$. To derive it, one assumes that the energy levels available to the nucleons have a constant separation ϵ and that the same level can be occupied by a neutron–proton pair but not by two neutrons. It is then easy to

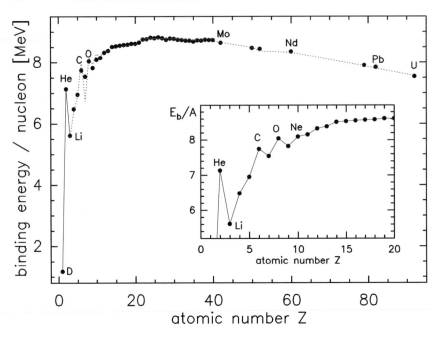

Figure 15.1. Binding energy E_b/A per nucleon of the most abundant isotope (but including deuterium) of each element. Filled circles are derived from mass excess as tabulated by [Lan80] and are connected by a full line. The dotted curve is a fit after equation (15.1). It is quite satisfactory all the way from uranium down to lithium, except for nitrogen. Parameter values are given in the text. The insert is merely a magnification.

show that the binding energy decreases proportionally to $\epsilon(N - Z)^2$ when $N > Z$.

- The final term is relatively small and takes into account the effect of pairing among nucleons. It is positive when N and Z are both even, negative when both are odd, and zero otherwise. It introduces the zigzag and the local peaks when an atom consists of a multiple of α-particles (^{4}He, ^{12}C, ^{16}O, ^{20}Ne, ^{24}Mg, ^{28}Si, ...).

The coefficients in equation (15.2) are determined by a combination of theory and experiment and are listed in the special literature. We used, in figure 15.1, $\alpha = 15.7$, $\beta = 17.8$, $\gamma = 0.71$, $\eta = 23.6$ and $\delta = 132$ MeV, with no attempt for optimization.

15.1.3 Hydrogen burning

15.1.3.1 Energy yield in hydrogen burning

When stars are on the main sequence, they burn hydrogen by melting four protons into helium,

$$4\,^1\mathrm{H} \longrightarrow \,^4\mathrm{He}.$$

A helium nucleus (mass $m_{\mathrm{He}^{2+}}$) is a bit lighter than four protons. The mass difference

$$\Delta m = m_{\mathrm{He}^{2+}} - 4m_\mathrm{p}$$

is such that $\Delta m/4m_\mathrm{p} = 7.12 \times 10^{-3}$. In other words, $\sim 0.7\%$ of the mass is converted into energy. The fusion of one helium nucleus thus liberates

$$\Delta mc^2 = 4.282 \times 10^{-5}\,\mathrm{erg} = 26.73\,\mathrm{MeV}. \tag{15.3}$$

From burning 1 g of hydrogen into helium one receives 6.40×10^{18} erg and this provides the greatest energy reservoir of a star, excluding gravitational collapse into a black hole. The 26.73 MeV released in reaction (15.3) appear in the form of γ-rays, neutrinos, positrons and as the kinetic energy of the reaction products. The neutrinos have an extremely small cross section and escape from the burning core without interacting with the stellar material. The energy of 0.263 MeV per fused He nucleus they carry away is lost. The positrons unite with electrons and are annihilated producing γ-rays. All γ-photons collide with other particles and deposit their energy in the form of heat.

15.1.3.2 The proton–proton chain

In all low-mass stars, up to masses a little more than 1 $M_\odot$, helium is created in the proton–proton or pp chain. Because this process is at work in the Sun, we discuss a few details. The chain consists of three major steps:

In the first step, deuterium is generated according to

$$^1\mathrm{H} + \,^1\mathrm{H} \longrightarrow \,^2\mathrm{D} + \mathrm{e}^+ + \nu_\mathrm{e} + 1.442\,\mathrm{MeV}. \tag{15.4}$$

The 1.442 MeV include the annihilation energy of the positron e^+ and the energy of the electron–neutrino ν_e. Usually the Coulomb force prevents protons from coming close together but at very short distances the *strong nuclear force* sets in and overcomes electric repulsion. The strong force has a range of order $\sim 10^{-13}$ cm, it binds the nucleons but is not felt outside the nucleus. The strong force is insensitive to the electric charge of the nucleons, so it does not discern between protons and neutrons.

The potential barrier between two protons separated by a distance $r = 10^{-13}$ cm is $E_\mathrm{b} = e^2/r \sim 1.5$ MeV. In the Sun, which has a central temperature of 1.5×10^7 K, the average kinetic energy of atoms is only 2 keV or a thousand times smaller. The chance, $P_\mathrm{Maxw}(v)$, of finding in the tail of the Maxwellian velocity distribution atoms with a kinetic energy of 1.5 MeV is negligible (see (5.13)). The only way protons can unite is by tunneling through the Coulomb potential barrier.

We can estimate the tunneling probability $P_{tun}(v)$ for protons of velocity v from (6.57), although this formula applies strictly only to square potentials in one dimension. The distance L at which tunneling sets in follows from $\frac{1}{2}m_p v^2 = e^2/L$. The penetration probability, P_{pen}, is given by

$$P_{pen} = P_{Maxw}(v) \cdot P_{tun}(v).$$

As the protons get faster, $P_{tun}(v)$ increases, while $P_{Maxw}(v)$ decreases. When we evaluate the velocity where $P_{pen}(v)$ has its maximum, using calculus, we find that tunneling already occurs at a distance $L \sim 5 \times 10^{-12}$ cm, far beyond the range of the strong force.

After the tunneling, there is an ephemeral pp complex but it is not stable: ^{2}He does not exist. The reason is that the potential of the strong force depends on the spin direction. When the spins of the two nucleons are anti-parallel, the system is unbound, when they are parallel, the complex is bound, however, this state is excluded because the two protons are identical and as fermions their quantum numbers must not coincide.

It is, therefore, not enough to have the two protons approach each other via tunneling to a nuclear distance. One of the protons must also be converted into a neutron (see reaction (15.4)). Only then does one have a stable system, a deuteron, with a binding energy of 2.23 MeV. Because a proton consists of two up and one down quark, a neutron of one up and two down quarks, there is a transformation

$$p = (up, up, down) \quad \longrightarrow \quad n = (up, down, down).$$

It is mediated by the *weak nuclear force* which has an extremely short range ($\sim 10^{-15}$ cm) and is 10^{12} times weaker than the *electromagnetic force*. Because the conversion of a proton into a neutron must take place in the short time while the proton traverses a distance of $\sim 10^{-13}$ cm, the reaction (15.4) proceeds very slowly. A proton in the Sun has to wait $\sim 10^{10}$ yr before it is incorporated into a deuterium atom.

The next step,

$$^2D + {}^1H \quad \longrightarrow \quad {}^3He + \gamma + 5.493 \, \text{MeV} \tag{15.5}$$

follows only seconds later. It is so much faster because it requires only the electromagnetic and not the weak force. The appearance of a γ-quantum is necessary for the conservation of energy and momentum. The photon energy is accounted for in the 5.493 MeV.

In the final step the build-up is completed by

$$^3He + {}^3He \quad \longrightarrow \quad {}^4He + 2\,{}^1H + 12.859 \, \text{MeV}. \tag{15.6}$$

For this reaction to proceed, the preceding two must have occurred twice. The ^{3}He nuclei are doubly charged, so the Coulomb repulsion is strong and as they are also rare, the reaction is fairly slow with a time scale of order 10^6 yr. In the sun there are other channels besides (15.6) involving the elements Li, Be, B but they are far less common. The seemingly simple reaction ^{3}He + p → ^{4}He is irrelevant because it involves the weak force.

15.1.3.3 The CNO cycle

In all massive stars, there is an alternative mode of hydrogen burning: the CNO cycle. It may be started by adding a proton to a ^{12}C atom, and then another and another. In this way, one builds up nuclei of subsequently higher mass number but eventually ends up again with a ^{12}C atom plus an α-particle, symbolically

$$^{12}\text{C} \xrightarrow{\text{P}} {}^{13}\text{N} \xrightarrow{\beta} {}^{13}\text{C} \xrightarrow{\text{P}} {}^{14}\text{N} \xrightarrow{\text{P}} {}^{15}\text{O} \xrightarrow{\beta} {}^{15}\text{N} \xrightarrow{\text{P}} {}^{12}\text{C} + {}^4\text{He}. \quad (15.7)$$

The letter p above an arrow indicates proton capture, β stands for β-decay. The elements C, N and O that give the burning its name take part in the reactions only as catalysts: in equilibrium, their abundance does not change. The net result of the reaction chain is again the conversion of hydrogen into helium.

The CNO cycle requires higher temperatures than the pp cycle because the Coulomb barriers are higher. It is also more sensitive to temperature: the energy production rate is at 18×10^6 K roughly proportional to T^{18} versus T^4 in the pp chain. The CNO cycle dominates when the central stellar temperature is higer than $\sim$18 million K or, equivalently, when the stellar mass is slightly above 1 M$_\odot$.

The β-decay reactions in (15.7) proceed quickly ($10^{2...3}$ s), on about the same time scale that a free neutron disintegrates into a proton, electron and anti–neutrino. The proton captures, however, are much slower, by far the slowest step being the formation of ^{15}O from ^{14}N which takes about 300 million years. Therefore, even if ones starts out with no nitrogen at all, ^{14}N will build up until an equilibrium abundance is reached. Indeed, the CNO cycle is the major source for nitrogen in the universe. Of course, the cycle can only operate if initially one of the catalysts is present.

15.1.4 The 3α process

The next element, after helium, synthesized in stars by burning is carbon. It is made in red giants by fusing three α-particles. Because of the strong Coulomb repulsion between α-particles, the temperatures have to be above 10^8 K. The burning process consists of two steps. First, two α-particles unite to ^{8}Be, then another α-particle is added to transform beryllium into carbon. The beryllium atom resulting from the first step,

$$\alpha + \alpha \quad \longrightarrow \quad {}^8\text{Be} + \gamma - 92 \text{ keV}$$

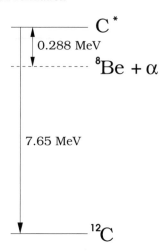

Figure 15.2. The formation of ^{12}C from the unstable compound (^{8}Be + α) proceeds through the resonant state C*.

is, however, unstable and α-decays. This reaction is endothermic requiring $Q = 92$ keV; it goes in both ways, so we may write

$$2\alpha \quad \longleftrightarrow \quad {}^8\text{Be}. \tag{15.8}$$

When the temperature in the stellar core reaches 1.2×10^8 K, the average kinetic energy per particle is $\frac{3}{2}kT \simeq 15$ keV and in thermal equilibrium about 1 out of 10^{10} atoms will be on the right-hand side of (15.8), in the form of ^{8}Be. This follows from the Saha formula when one makes some obvious modifications in (5.38). The small fraction of beryllium nuclei reacts further to form carbon,

$$^8\text{Be} + \alpha \quad \longrightarrow \quad {}^{12}\text{C} + \gamma + 7.37 \text{ MeV}.$$

Now it is important that carbon possesses an excited state, C*, which lies $E^* = 7.65$ MeV above the ground level but only $E_R = 0.288$ Me above the complex ^{8}Be + α (figure 15.2). For those atoms whose kinetic energy is close to E_R, the cross section for creating C* has a sharp maximum (resonant reaction). Once formed, C* may then de-excite to the ground state ^{12}C (α-decay is, however, more frequent).

The existence of the state C* speeds up the reaction ^{8}Be + $\alpha \rightarrow {}^{12}$C by a factor 10^7. It allows the triple-α process to operate already at a temperature of 1.2×10^8 K. Without this, the stellar interior would have to be as hot as 2×10^8 K for helium to burn into carbon. This makes a tremendous difference considering the extreme temperature sensitivity of the 3α process. At $\sim 10^8$ K, the energy production rate is proportional to T^{30}, and a temperature rise of only 5% makes the reaction ten times faster. The energy yield in the 3α process per gram of matter is ten times lower than in hydrogen burning.

Once carbon is present, oxygen, neon, magnesium and so forth can be generated through the capture of α-particles. Here are some reactions with their energy yield:

$$^{12}C + \alpha \longrightarrow {}^{16}O + \gamma \ \ + 7.16 \, \text{MeV}$$

$$^{16}O + \alpha \longrightarrow {}^{20}Ne + \gamma \ + 4.73 \, \text{MeV}$$

$$^{20}Ne + \alpha \longrightarrow {}^{24}Mg + \gamma + 9.32 \, \text{MeV}$$

$$^{24}Mg + \alpha \longrightarrow {}^{28}Si + \gamma \ \ + 9.98 \, \text{MeV}.$$

When taken per gram of stellar matter, the yield decreases as heavier elements are synthesized.

15.1.5 Lifetime and luminosity of stars

Stars are born with masses, m, between 0.08 and 100 $M_\odot$, the upper limit being rather uncertain. Their lifetime τ and luminosity L determine, to a large degree, the stellar population and the optical and infrared appearance of the Milky Way. To first order,

$$\tau \propto \frac{m}{L}.$$

The stellar mass fixes the amount of fuel, and the luminosity the speed of its consumption. For example, assuming that 20% of the mass of the Sun is in its hydrogen burning core and that the energy yield is $0.007c^2$ erg per gram of burnt matter (see (15.3)), one gets a lifetime of 20 billion years, not bad as a first estimate.

Most of this time, a star spends on or near the main sequence. The subsequent evolution is very quick: first, because the star has become a giant and is generally much more luminous, so the consumption rate is higher; and second, because 90% of the fuel reserve has already been used up. Therefore, the total lifetime is not much longer than the main sequence lifetime, τ_{ms}. However, because giants are so bright, the luminosity averaged over the total lifetime of a star, L_{av}, can be several times greater than the main sequence luminosity L_{ms}, despite the fact that the post-main-sequence phase is short.

Stars are best classified by their line spectrum. The analysis yields the spectral type (O, B, A, F, G, K, M) or effective temperature T_{eff}, and the luminosity class (roman numbers from I to VI) or surface gravity. The stellar mass m is not an observational quantity and much less certain. It can be directly determined only in binary systems.

Formulae (15.9)–(15.13) and figure 15.1 are approximations of the parameters τ_{ms}, L_{ms} and L_{av}, as a function of the stellar mass m. They are based on computations of stellar evolution and adapted from [Mas88]. The initial chemical composition of the star as well as its mass loss rate have some influence on the results but it is not important for our purposes.

When lifetimes are expressed in years, stellar masses in $M_\odot$ and luminosities in $L_\odot$, one gets the following results

- *Main sequence lifetime:*

$$\tau_{ms} = \begin{cases} 3 \times 10^6 + 1.2 \times 10^9 m^{-1.85} & m > 7.293 \\ 10^7 + 5 \times 10^9 m^{-2.7} & 2.594 \le m \le 7.293 \\ 1.1 \times 10^{10} m^{-3.5} & m < 2.594 \,. \end{cases} \quad (15.9)$$

- *Main sequence luminosity:* For $m \ge 1 \, M_\odot$,

$$\log L_{ms} = 4.989 \times 10^{-3} + 4.616 \log m - 0.6557 \log^2 m - 0.06313 \log^3 m. \quad (15.10)$$

For $m < 1 \, M_\odot$,

$$L_{ms} = \begin{cases} 1.012 m^{4.34} & 0.6 \le m \le 1 \\ 0.51 m^3 & m \le 0.6. \end{cases} \quad (15.11)$$

- *Average luminosity* (mean over lifetime of star including giant phase):

$$L_{av} = 9.77 \times 10^{10} \max\left(\frac{m_{rem}}{\tau_{ms}}, L_{ms}\right). \quad (15.12)$$

L_{av} is expressed through the mass m_{rem} of the stellar remnant. It is a neutron star or a black hole when the progenitor explodes as a supernova, i.e. when the initial mass $m > m_{SN} = 6 \, M_\odot$. It is a white dwarf when the initial mass $0.65 \le m \le m_{SN}$:

$$m_{rem} = \begin{cases} m & m \le 0.65 \\ 0.65 & 0.65 \le m \le 2.2 \\ 0.105(m - 2.2)^{1.4} + 0.65 & 2.2 \le m \le m_{SN} \\ 0.1 m^{1.4} & m > m_{SN}. \end{cases} \quad (15.13)$$

The constant 9.77×10^{10} in (15.12) assumes an energy yield of 6×10^{18} erg per gram of burnt remnant matter with a mass fraction of 0.7 of the heavy elements (the ashes).

15.1.6   The initial mass function

Stars generally form in clusters, almost simultaneously. An outstanding example is the very compact cluster in Orion around the Trapezium which contains over 2000 members with a stellar density of more than 10^4 pc^{-3}. A fundamental problem concerns the distribution of masses of young stars, either born in clusters, as in Orion, or elsewhere. If $dN(m)$ stars are born in the mass range $m \ldots m+dm$, then

$$dN(m) \propto \xi(m)\, dm \quad (15.14)$$

defines the *initial mass function* $\xi(m)$, abbreviated IMF. One way to derive the IMF observationally is to count the number of stars in the solar neighborhood per interval of apparent visual magnitude m_V. Then one determines their distances to obtain the luminosity function $\phi(M_V)$. It gives the number of stars per unit volume and interval of absolute visual magnitude. When one restricts the counting to main sequence stars, M_V can be uniquely translated into a stellar mass m (see figure 15.3). This leads to a function $\psi(m)$ such that $\psi(m)\,dm$ is the present stellar density for the mass interval $m \ldots m + dm$. One arrives at the initial mass function $\xi(m)$ by deconvolving $\psi(m)$ with the stellar lifetime $\tau(m)$ (see again figure 15.3) and assuming that the star-formation rate has been constant over the past.

A much more direct approach to find $\xi(m)$ is to count in a young stellar cluster, where no member has yet evolved off the main sequence, the number of stars per mass interval. Such a method is, however, prone to observational bias by the detection limit as young clusters are always heavily obscured by dust. Also, the result from any one cluster is not representative for the Milky Way as a whole. One may alternatively try to determine $\xi(m)$ theoretically by making certain assumptions about the processes during star birth. Two analytical expressions for the IMF are in wide use [Sal55, Mil78],

$$\xi(m) = \begin{cases} m^{-2.35} & \text{Salpeter IMF} \\ \dfrac{1}{m} \exp(-0.966 \ln m - 0.206 \ln^2 m) & \text{Miller–Scalo IMF.} \end{cases}$$

(15.15)

The functions are not normalized; they may, for example, refer to 1 $M_\odot$ of gas that has transformed into stars. For the lower and upper mass limit of stars, m_l and m_u, one often assumes 0.1 $M_\odot$ and 50 $M_\odot$, respectively but these numbers are debatable and depend on the environment. Figure 15.4 illustrates how stars of different masses contribute to the total luminosity and thermal radio emission of a cluster with mass distributions as given in (15.15).

15.2 Clouds near gravitational equilibrium

15.2.1 Virialized clouds

Consider an interstellar cloud of volume V consisting of N atoms, all of equal mass m and with coordinates x_j. An outer pressure P_0 is exerted on the cloud surface. The virial theorem asserts a relation between the various forms of energy when the cloud is in equilibrium so that all time derivatives with respect to the cloud as a whole but not with respect to an individual particle, vanish. The theorem follows immediately from the equation of motion,

$$\mathbf{F}_j = m\ddot{\mathbf{x}}_j \qquad i = 1, \ldots, N$$

(15.16)

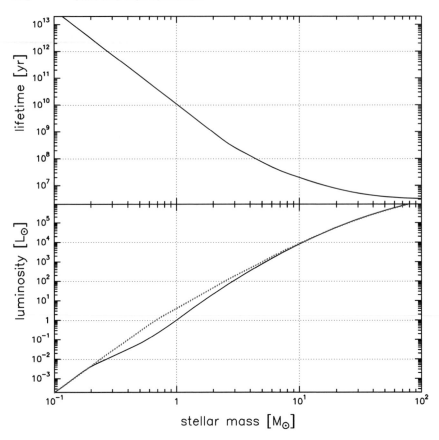

Figure 15.3. The top panel shows the lifetime of stars according to equations (15.9), The bottom one their luminosity after (15.10) to (15.13). The full curve gives main sequence luminosity L_{ms}, dotted line shows L_{av} which is an average over the whole stellar life including the giant phase. For stars of very high and very low mass, $L_{\mathrm{ms}} \simeq L_{\mathrm{av}}$. The former move after the exhaustion of hydrogen in the Hertzsprung–Russell diagram horizontally to the right, without increasing their luminosity. The latter never turn into giants. They also live much longer than 10^{10} yr, so their post-main-sequence evolution does not bother us.

when one multiplies by $\mathbf{x}_j$ and sums over all particles. Because

$$\mathbf{x} \cdot \ddot{\mathbf{x}} \;=\; \frac{d}{dt}(\mathbf{x} \cdot \dot{\mathbf{x}}) - \dot{\mathbf{x}}^2$$

one finds for the *time averages* (we do not mark them explicitly)

$$\sum_j \mathbf{F}_j \mathbf{x}_j \;=\; -\sum_j m \dot{\mathbf{x}}_j^2. \tag{15.17}$$

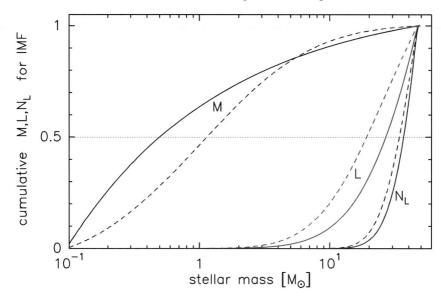

Figure 15.4. The cumulative mass M, luminosity L and Lyman continuum flux N_L for the Salpeter (full curve) and Miller–Scalo (broken curve) IMF with lower and upper mass limit $m_l = 0.1$ $M_\odot$ and $m_u = 50$ $M_\odot$. For example, in a Salpeter IMF, stars with masses smaller than 10 $M_\odot$ account for only 9% of the luminosity, although they comprise 90% of the stellar mass. In other words, one-tenth of the stars by mass accounts for almost all the luminosity (precisely, for 91%). The dependence of the luminosity on mass, $L(m)$, is from (15.10) and (15.11); the Lyc flux $N_L(m)$ is from (14.22).

Without systematic motion, all kinetic energy is thermal and $\sum m \dot{\mathbf{x}}_j^2 = 3NkT$. The other sum $\sum \mathbf{F}_j \mathbf{x}_j$ can be split into a term containing only those forces $\mathbf{F}_j$ which act from outside, the other containing the inner forces. The outer forces are due to the external pressure P_0 which results in a force $P_0\,df$ that is perpendicular to each surface element df and directed inwards. To evaluate this sum, one has to integrate over the cloud surface and transform the expression into a volume integral according to Gauss' theorem. The inner forces come from mutual gravitational attraction of the atoms. The end result links the thermal energy $E_{\text{therm}} = \frac{3}{2}NkT$, the gravitational (potential) energy E_{grav}, and $P_0 V$,

$$P_0 V = NkT + \tfrac{1}{3} E_{\text{grav}} \tag{15.18}$$

with

$$E_{\text{grav}} = -\sum_{i<j} \frac{Gm^2}{|\mathbf{x_i} - \mathbf{x}_j|}.$$

An interstellar cloud fulfilling (15.18) is said to be virialized. Virialization is a special cloud property. For a spherical cloud of uniform density, radius R and mass M,

$$E_{grav} = -\frac{3GM^2}{5R}. \tag{15.19}$$

When the outer pressure can be neglected with respect to the central pressure, (15.18) yields

$$-E_{grav} = 2E_{therm} = \frac{3kTM}{m}. \tag{15.20}$$

For example, a low-mass protostar in its late stage of evolution moves towards the main sequence along the Hayashi track; its surface temperature is then approximately constant, whereas the stellar radius and the luminosity shrink. When accretion has ended and nuclear burning not yet started, the luminosity L is entirely due to gravitational contraction. The star is always in equilibrium and exactly half of the liberated gravitational energy goes according to the virial theorem into heating the gas; the other half is radiated away.

When the star is on the main sequence and, in a thought experiment, one switches off nuclear burning, the luminosity would be sustained over the Kelvin–Helmholtz time scale

$$t_{KH} = \frac{E_{therm}}{L} \simeq \frac{3kTM}{2mL} \tag{15.21}$$

where T is an average by mass over the star and m, as before, is the mean particle mass. For the Sun, t_{KH} equals 30 million years; for O stars, t_{KH} is a thousand times smaller.

Equation (15.20) also holds for planets and other objects whirling around the Sun if one replaces E_{therm} by the average kinetic energy E_{kin} over a revolution. It is not valid for comets that fly on a parabolic orbit of infinite extension.

15.2.2   Isothermal cloud in pressure equilibrium

In a gas sphere, when pressure and gravitational forces balance exactly everywhere,

$$\frac{dP}{dr} = -\frac{GM_r}{r^2}\rho \tag{15.22}$$

$$\frac{dM_r}{dr} = 4\pi r^2 \rho. \tag{15.23}$$

M_r denotes the mass contained in a sphere of radius r. Formulae (15.22) and (15.23) can be combined into one, which is of second order and called the Emden equation [Emd07],

$$\frac{1}{r^2}\frac{d}{dr}\left(\frac{r^2}{\rho}\frac{dP}{dr}\right) = -4\pi G\rho. \tag{15.24}$$

We solve (15.22) and (15.23) subject to the boundary conditions

$$\rho(0) = \rho_c \qquad \left.\frac{d\rho}{dr}\right|_{r=0} = 0 \tag{15.25}$$

i.e. we prescribe the central density ρ_c and its gradient, which has to be flat. One still has to add the equation of state,

$$P = \frac{kT}{m}\rho = v_s^2\rho \tag{15.26}$$

where m denotes the mean mass of the gas particles and

$$v_s = \sqrt{\frac{\partial P}{\partial \rho}} = \sqrt{\frac{kT}{m}} \tag{15.27}$$

the isothermal sound speed. For polytropes, the pressure P can be expressed solely by ρ,

$$P = K\rho^{1+\frac{1}{n}} \tag{15.28}$$

where K is some constant. The temperature does not enter but T is nevertheless not uniform over the cloud. General solutions of polytropes are discussed in [Cha39]. Here we are interested in the structure and stability of an isothermal sphere which is described by a polytrope of index $n = \infty$.

The assumption of isothermality correctly reflects the conditions in protostellar clouds before they collapse. Heating of such a cloud, beneath its surface layer, happens either through absorption of far infrared radiation by grains or by interaction of cosmic rays with the gas. In either case, the heating rate per gram of interstellar matter is fairly uniform over the cloud and does not depend on density. Cooling comes mainly from grains (although some is due to molecular lines, especially of H_2O and CO). As the cloud temperatures are low, the maximum dust emission lies at submillimeter wavelengths and such photons can freely escape. At the prevailing high densities, dust and gas are thermally coupled by collisions between gas molecules and grains.

When the cloud collapses, the PdV work of the gravitational compression is transferred from the gas to the dust and radiated away. As long as the cloud is optically thin to dust radiation, which is true during the early protostellar phases, the collapse does not enhance the temperature.

15.2.3 Structure and stability of Ebert–Bonnor spheres

Equations (15.22) and (15.23) are integrated from the center outwards. At the edge, where $r = R$, the cloud is confined by an outer pressure P_0 which is equal to the pressure just below the cloud surface, so $P_0 = v_s^2\rho(R)$. For every mass M, there is a family of solutions, the central density ρ_c being the parameter. By introducing the variables

$$y = \rho/\rho_c \tag{15.29}$$

$$x = r\sqrt{4\pi Gm\rho_c/kT} \tag{15.30}$$

equation (15.24) can be made dimensionless and brought into the form

$$y'' - \frac{y'^2}{y} + \frac{2y'}{x} + y^2 = 0 \qquad (15.31)$$

with the boundary conditions $y(0) = 1$ and $y'(0) = 0$. Its solution describes the structure of isothermal gas balls in equilibrium; they are called Ebert–Bonnor spheres. If we further put $y = e^u$, (15.31) simplifies to

$$u'' + \frac{2u'}{x} + e^u = 0 \qquad (15.32)$$

with $u(0) = 0$ and $u'(0) = 0$. To solve this second-order differential equation by a Runge–Kutta method, we still have to know how u'/x in (15.32) behaves near zero. Because obviously $u'/x \to u''(0)$ for $x \to 0$, it follows that $(u'/x)_{x=0} = -\frac{1}{3}$. The result of the numerical integration is shown in figure 15.5. The density is flat in the core, and ρ falls off in the envelope approximately like r^{-2}. The mass of an Ebert–Bonnor sphere is computed from

$$M = 4\pi \int_0^R r^2 \rho \, dr = \frac{1}{\sqrt{4\pi \rho_c}} \left[\frac{kT}{Gm} \right]^{3/2} \int_0^{x_u} y x^2 \, dx \qquad (15.33)$$

where $x_u = R \cdot \sqrt{4\pi G \rho_c / v_s^2}$. With the help of figure 15.6, which displays the integral

$$I(x_u) = \int_0^{x_u} y(x) \, x^2 \, dx \qquad (15.34)$$

one can determine to which value of x_u (or radius R) a sphere of given mass M, central density ρ_c and temperature T extends.

Figure 15.7 depicts how the radius R of an Ebert–Bonnor sphere varies with the boundary pressure P_0. The function $P_0 = f(R)$ and the ensuing stability criterion can be studied analytically but the treatment is laborious [Ebe55, Bon56]. In figure 15.7, clouds from A over B to infinite radii are stable because $\partial P_0 / \partial R < 0$. When such a cloud is perturbed, it reverts to the configuration whence it started from. However, between A and C, which is the point of minimum cloud radius, $\partial P_0 / \partial R > 0$. When somewhere on this part of the curve the radius is decreased by some inevitable fluctuation, the cloud arrives at a new configuration where the actual outer pressure P_0 surpasses the value required for equilibrium. Thus the cloud is further compressed and finally collapses. As one increases the central density and moves on the curve beyond the point C towards the eye of the spiral, the instability continues. Although $\partial P_0 / \partial R$ may again be negative for the cloud as a whole, interior parts are unstable.

A cloud in a low-pressure environment settles into an equilibrium configuration, somewhere to the right of point A on the curve in figure 15.7. Its environment may vary but the cloud is stable and adjusts to a changing outside pressure P_0. Collapse starts only when the Ebert–Bonnnor sphere becomes

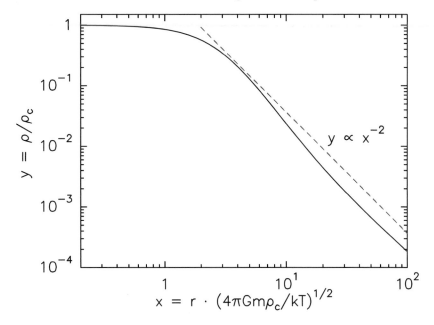

Figure 15.5. The structure of an (isothermal) Ebert–Bonnor sphere after (15.31). The density y is in units of the central density, $y = \rho/\rho_c$. The radial coordinate x is proportional to the cloud radius, $x = r(4\pi G\rho_c/v_s^2)^{1/2}$. In this logarithmic plot, the broken curve shows for comparison the slope of a density distribution falling off like $1/r^2$.

critical. To determine the critical values, we write the outside pressure P_0 and the cloud radius R as a function of the dimensionless variable x,

$$P_0(x) = v_s^2 \rho_c y(x) = \left(\frac{kT}{m}\right)^4 \frac{1}{G^3 M^2} \frac{I^2(x)y(x)}{4\pi} \tag{15.35}$$

$$R = x \left(\frac{kT}{4\pi Gm\rho_c}\right)^{1/2} = \frac{Gm}{kT} M \frac{x}{I(x)}. \tag{15.36}$$

The expressions $I^2 y$ and x/I are readily evaluated numerically. For a fixed cloud mass M and temperature T, P_0 attains its maximum at $x = 6.450$, where $I^2 y/4\pi = 1.398$ and $x/I = 0.4106$. Therefore, the cloud is stable only if the outer pressure is smaller than the critical value $P_{\rm crit}$ and the radius greater than $R_{\rm crit}$,

$$P_{\rm crit} = 1.40 \frac{k^4}{G^3 m^4} \frac{T^4}{M^2} \tag{15.37}$$

$$R_{\rm crit} = 0.411 \frac{Gm}{kT} M. \tag{15.38}$$

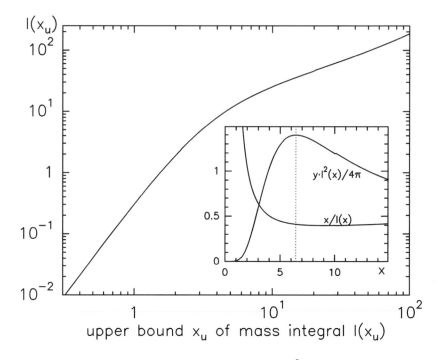

Figure 15.6.   The mass integral $I(x_u) = \int yx^2\,dx$ of equation (15.34); $x_u = R(4\pi G\rho_c v_s^2)^{1/2}$ marks the cloud boundary. For large values, roughly $I(x_u) \propto x_u$, for small ones $I(x_u) \propto x_u^3$. *Insert:* The functions $x/I(x)$ and $y(x)\,I^2(x)/4\pi$ near the critical value $x_{cr} = 6.45$.

Assuming some reasonable pressure to prevail in the interstellar medium, say $P_0 = nkT \sim 10^{-12}$ dyn cm^{-2}, as in a tenuous HII region, the largest possible stable mass for a cold H_2 cloud of 10 K is thus only $\sim$6 $M_\odot$.

A cloud in pressure equilibrium is also virialized and the virial theorem in the form (15.18) follows directly from the equation for hydrostatic equilibrium (15.22). Indeed, multiplication of (15.22) with $4\pi r^3$ and integration by parts over the cloud radius gives

$$\int_0^R 4\pi r^3 \frac{dP}{dr}\,dr = 3P_0V - 3NkT = -\int_0^R \frac{GM_r}{r}4\pi r^2\rho(r)\,dr = E_{grav}.$$

When the cloud has a large volume, its density is almost constant and the relations between P_0 and R are practically identical for an Ebert–Bonnor sphere and a virialized homogeneous sphere. However, when the cloud is compact, pressure gradients, which are taken into account by the virial theorem in (15.18) only in an integral form, become important and the curves diverge.

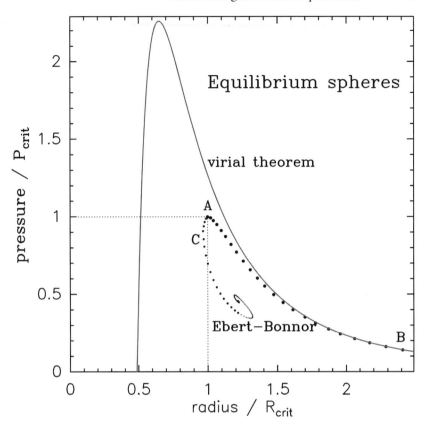

Figure 15.7. The sequence of points shows the general relation between the radius R and the outer pressure P_0 for an isothermal gas ball in hydrostatic equilibrium after (15.24) when R and P_0 are normalized to the values of the critical sphere. The central density ρ_c increases by a constant factor of 1.2 from one point to the next as one follows the spiral of dots from right to left. Stable configurations are only possible from A over B to infinity. For an H_2 cloud of 1 $M_\odot$ and $T = 10$ K, at the critical point A, $\rho_c = 1.19 \times 10^{-18}$ g cm^{-3}, $R = 1.33 \times 10^{17}$ cm and $P_0 = 3.41 \times 10^{-11}$ dyn cm^{-2}; at point B, $\rho_c = 2.15 \times 10^{-20}$ g cm^{-3}. For comparison, the full curve is from the virial relation (15.52) for unmagnetized spheres of constant density.

15.2.4 Free-fall of a gas ball

The ultimate simplification in the study of cloud collapse is to treat the contraction of a constant density sphere under the influence of gravity alone, thus neglecting all other forces that arise from pressure gradients, magnetic fields or rotation. Consider a gas sphere of total mass M which at time $t = 0$ is at rest and has an outer radius R. Let $r(m, t)$ denote the radius of an internal sphere that contains the mass $m \leq M$ at time t. The collapse is then described by

$$\ddot{r} = -\frac{Gm}{r^2} \tag{15.39}$$

together with the conditions that, at time $t = 0$, the cloud is at rest, $\dot{r}(m, 0) = 0$. When $r_0 \equiv r(m, 0)$ and ρ_0 denotes the initial constant density, then obviously $4\pi\rho_0 r_0^3/3 = m$. Integration of (15.39) yields

$$\frac{\dot{r}}{r_0} = -\sqrt{\frac{8\pi G\rho_0}{3}\left(\frac{r_0}{r} - 1\right)}.$$

Putting $r/r_0 = \cos^2\beta$, one finds

$$\beta + \tfrac{1}{2}\sin 2\beta = t\sqrt{\frac{8\pi G\rho_0}{3}}. \tag{15.40}$$

β and thus also r/r_0 do not depend on the mass m and the collapse therefore proceeds homologously. It is completed when the cloud has contracted to a point, i.e. when $r = 0$. This happens for $\beta = \pi/2$, from which follows the free-fall time

$$t_{\mathrm{ff}} = \sqrt{\frac{3\pi}{32G\rho_0}}. \tag{15.41}$$

t_{ff} depends only on the initial density ρ_0 and is inversely proportional to its square root. The free-fall time gives the right order of magnitude for the duration of gravitational collapse but rather as a lower limit because all retarding effects, like rotation or pressure gradients, are excluded. It is often useful to compare t_{ff} with the time scale of other processes, such as fragmentation of the cloud into subunits, gas cooling, frosting of molecules onto dust particles, grain coagulation and so forth.

15.2.5   The critical mass for gravitational instability

Although in any interstellar cloud there is always some gravitational pull which tries to compress it, clouds are generally not in the stage of collapse. If they were, the star-formation rate would be orders of magnitude higher than what one observes. *J Jeans* was the first to give a quantitative answer to the question under what conditions a cloud collapses and the stability criterion bears his name. It is valid only when rotation, magnetic fields, turbulence or outer forces are absent.

In a medium of constant temperature T, the basic gas–dynamical equations of momentum and mass continuity for the time-dependent variables velocity $\mathbf{v}$ and density ρ are

$$\frac{\partial \mathbf{v}}{\partial t} + (\mathbf{v} \cdot \mathrm{grad})\mathbf{v} = -\frac{1}{\rho}\,\mathrm{grad}\,P - \mathrm{grad}\,\phi \tag{15.42}$$

$$\frac{\partial \rho}{\partial t} + \mathrm{div}(\rho\mathbf{v}) = 0. \tag{15.43}$$

The gravitational potential ϕ is given by the Poisson equation

$$\Delta\phi = 4\pi G\rho \tag{15.44}$$

and the gas pressure P by the equation of state (15.26), $P = k\rho T/m = \rho v_s^2$, where m is the mean particle mass and v_s the sound velocity of (15.27). Let the medium intitially be uniform and at rest. We introduce a small perturbation, denoted by the subscript 1, and examine whether it grows or not. If the initial values of density and velocity are $\rho_0 = 0$ and $v_0 = $ constant, at some later time

$$\rho = \rho_0 + \rho_1 \qquad \mathbf{v} = \mathbf{v_0} + \mathbf{v_1}.$$

Writing the density change ρ_1 in the form of a plane wave,

$$\rho_1 \propto e^{i(kz+\omega t)}$$

one obtains

$$\omega^2 = k^2 v_s^2 - 4\pi G\rho_0. \tag{15.45}$$

If the perturbation ρ_1 increases exponentially, the cloud is unstable. This occurs when ω is imaginary ($\omega^2 < 0$), i.e. when the wavenumber k is below some critical value k_J (the letter J stands for Jeans) given by

$$k_J = \frac{\pi}{L_J} = \sqrt{\frac{4\pi Gm\rho_0}{kT}}.$$

Do not confuse the symbol k for the wavenumber with Boltzmann's constant k. So a cloud of density ρ and temperature T is unstable to collapse if it has a size greater than

$$L_J = \sqrt{\frac{\pi kT}{4Gm\rho}}. \tag{15.46}$$

For a cloud of molecular hydrogen ($m = 2m_H$),

$$L_J \approx 1.2 \times 10^{19}\sqrt{T/n} \text{ cm} \tag{15.47}$$

where n is the number density of hydrogen molecules per cm^3. Therefore, the maximum mass of an interstellar cloud of given density and temperature that does not collapse is

$$M_J \approx L_J^3\rho = \left(\frac{\pi k}{4Gm}\right)^{3/2}\sqrt{\frac{T^3}{\rho}}. \tag{15.48}$$

It is called the Jeans mass. For convenient comparison with observations, (15.48) may be written as

$$\frac{M_J}{M_\odot} \approx 2.4 \times 10^{-19}\left(\frac{L}{\text{cm}}\right)T. \tag{15.49}$$

A gravitationally unstable clump of one solar mass has, therefore, a size of ~0.1 pc. This derivation of the critical mass M_J from perturbation theory is logically not clean. A homogeneous medium at rest cannot be in equilibrium. If it were, $\nabla\phi$ and thus $\Delta\phi$ would have to vanish but that is impossible because of the Poisson equation (15.44). All constants in the preceding equations (15.46)–(15.49) should, therefore, not be taken too literally. However, the Jeans criterion is qualitatively correct and very useful. For example, by measuring the temperature T, linear size L and mass M of a suspected protostellar condensation, formula (15.49) helps decide whether it should collapse or not.

15.2.6 Implications of the Jeans criterion

15.2.6.1 The role of dust in gravitational stability

Because an isothermal collapse of a molecular cloud requires radiation losses, the dust optical depth τ, from the cloud center to the surface, must be smaller than unity at wavelengths where dust emission is efficient, so

$$\tau = \tfrac{1}{2}L_J K_d \rho < 1.$$

When the optical depth is transformed into

$$\tau = \left(\frac{\pi k}{4Gm}\right)^2 \frac{K_d T^2}{\tfrac{1}{2} M_J}$$

one sees that clumps of very low mass cannot freely radiate the gravitational energy released in the collapse. Using for K_d the Planck mean absorption coefficent K_P of equation (13.33), the minimum fragment mass that can undergo free collapse is

$$\min(M_J) = 2.4 \times 10^{-7} T^4 \ M_\odot. \tag{15.50}$$

For $T = 20$ K, $\min(M_J) \sim 0.04$ $M_\odot$. Of course, these numbers should not be taken at their face value but the remarkably strong dependence of M_J on the cloud temperature may explain the observational fact that in starburst galactic nuclei, where the dust is warm ($T_d \geq 30$ K), the IMF is heavily biased towards massive stars and low-mass stars do not seem to form at all.

15.2.6.2 Fragmentation and thermodynamic instability

During cloud contraction, gravitational energy is converted into heat. Initially, the collapse is isothermal because dust is a very effective coolant; gas cooling is relatively inefficient. Continuing our crude analysis of cloud collapse, we thus notice from (15.48) that a gravitationally unstable and contracting cloud has the tendency to break up into fragments because the critical mass M_J decreases proportionally with T^3/ρ as the average cloud density ρ rises. Smaller subunits of the mother cloud become unstable as long as the temperature does not rise faster than $\rho^{1/3}$. This process, called fragmentation, is fundamental.

The isothermal phase ends when the cloud becomes opaque to the cooling radiation. In a homogeneously contracting cloud, for which $R^3\rho = $ constant, the dust optical depth τ grows proportionally to R^{-2}; for inhomologous contraction, the rise is even steeper. So as the cloud shrinks, τ must at some moment exceed unity. Compression then leads to heating and pressure gradients build up which stop the collapse: a hydrostatic core forms. When the gravitational energy is not radiated away, the gas behaves adiabatically, $\delta Q = dU + P\,dV = 0$, and

$$T \propto \rho^{\gamma-1}$$

where $\gamma = C_p/C_v$ is the ratio of specific heat at constant pressure to that at constant volume. Isothermality implies $\gamma = 1$. According to (15.48), there is gravitational instability if γ falls below

$$\gamma_{cr} = \tfrac{4}{3}. \tag{15.51}$$

Besides fragmentation, the value γ_{cr} also figures in protostellar evolution (section 15.3.2).

15.2.6.3 Further remarks on the Jeans criterion

- The Jeans instability criterion is also valid when the gas pressure is due to supersonic turbulence (because it just compares kinetic with gravitational energy). One must then replace, in the corresponding formulae, the sound velocity v_s by the turbulent velocity, and the thermal energy by the energy of the turbulent motions.
- Suppose in some place of a cloud the gas is contracting. The information about the contraction travels through the cloud in a time t_{trav}. If t_{trav} is shorter than the free-fall time t_{ff} of (15.41), the cloud will be stable. Otherwise, it cannot respond quickly enough to the density perturbation and collapse will ensue. The condition

$$t_{trav} = \frac{R}{v_s} = \sqrt{\frac{3\pi}{32G\rho}} = t_{ff}$$

together with (15.27) for the sound velocity, also yields the Jeans mass or Jeans length.
- When one applies the virial theorem in the form of (15.20), where the outer pressure is absent and the gravitational energy equals twice the thermal energy, to a gas sphere of mass M, radius R and with M/m gas particles, one gets

$$\frac{3GM^2}{5R} = 2\frac{M}{m}\cdot\frac{3}{2}kT.$$

Within a factor of order unity, one immediately recovers the formulae for L_J and M_J, (15.46) and (15.48). The radius of the critical Ebert–Bonnor sphere also has the same form as the Jeans length (see (15.38) and (15.46)) and the two expressions are also practically equivalent.

- Any spherical cloud supported against its gravity by an internal pressure P becomes unstable when $P \propto \rho^{4/3}$, independently of the physical origin of P. So the critical value γ_{cr} also holds for magnetic pressure P_{mag}. We can thus extract from the Jeans criterion (15.48) the critical mass M_{mag} in the presence of a magnetic field (see section 15.2.7). Indeed, when we rewrite equation (15.48) as $M_J = (\pi/4G)^{3/2} P^{3/2}/\rho^2$ and insert, for P, the magnetic pressure $P_{mag} = B^2/8\pi$, we are led to M_{mag} in (15.53), again but for a factor of order unity.

15.2.7 Magnetic fields and ambipolar diffusion

15.2.7.1 Critical mass for magnetized clumps

When a cloud is pervaded by a magnetic field, its net effect is to lend further support against collapse. One can approximately extend the virial equation (15.18) so that it incorporates magnetic forces. As the terms in (15.18) represent global cloud energies, one expects the new term for the magnetic field to be equal to the total magnetic field energy (see (1.56)). Indeed, for a spherical cloud threaded by a uniform field of strength B, equation (15.18) is generalized to

$$4\pi R^3 P_0 = 3NkT - \frac{3GM^2}{5R} + \frac{1}{3}R^3 B^2. \tag{15.52}$$

If the field is frozen into the cloud, the magnetic flux is conserved and the relation $BR^2 = $ constant holds no matter whether the cloud is contracting or expanding. The gravitational and magnetic term in (15.52) have then the same $1/R$-dependence and if one form of energy, gravitational or magnetic, dominates at the beginning, it will do so throughout the subsequent evolution. The cloud can, therefore, only collapse if $3GM^2/5R > R^3 B^2/3$. This condition leads to a minimum mass for instability in the presence of a magnetic field, M_{mag}, which can obviously be written in two ways,

$$M_{mag} = \frac{5^{3/2}}{48\pi^2 G^{3/2}} \frac{B^3}{\rho^2} = \frac{5^{1/2}}{\pi G^{1/2}} \Phi. \tag{15.53}$$

The second expression for M_{mag} contains the flux $\Phi = \pi R^2 B$ through the clump. When the analysis is extended and cloud flattening taken into account, the proportionality factor drops, from $\sqrt{5}/3\pi = 0.24$ to 0.13; but such refinements are academic as estimates based on the virial theorem are rough anyway. We conclude that for $R = 0.1$ pc and $B = 10 \, \mu$G, the critical mass for collapse is $\sim 1 \, M_\odot$.

It is not easy to determine the strength of the magnetic field in molecular cloud cores but if one extrapolates from the lower density environment, where measurements are easier and B is at least 1 μG, to clump densities, one ends up with a field strength of $\sim$10 μG or more. A 1 $M_\odot$ cloud can, therefore, only contract if its radius is smaller than $\sim$0.1 pc. For the extrapolation, one assumes a relation $B \propto \rho^n$ between gas density and field strength. For homogeneous contraction ($R^3 \propto \rho^{-1}$), flux freezing ($BR^2 = $ consant) implies $n = \frac{2}{3}$; in the case of equipartition between thermal energy of the gas and magnetic energy, $n = \frac{1}{2}$; a more elaborate analysis yields $\frac{1}{3} \leq n \leq \frac{1}{2}$ [Mou91].

15.2.7.2   Time scale of ambipolar diffusion

Of course, all this is correct only if the magnetic field is coupled to the gas, otherwise there is no interaction. Although the interstellar matter is mainly neutral, cosmic rays alone can probably keep one out of 10^7 gas particles ionized, even in molecular cloud cores, and this should be sufficient to effectively freeze in the magnetic field. The strength of the coupling between gas and field is expressed through a diffusion timescale t_d, which we now define for a spherical cloud of mass M and radius R.

Let the cloud be close to equilibrium so that the gravitational force is balanced by the magnetic field. The ions are then kept at their position while the neutrals drift slowly across the field lines under the influence of gravity. The process is called ambipolar diffusion. Assuming that the density of the neutrals is much larger than the ion density, $n \gg n_i$, and that both kinds of particles have the same mass m, each ion imparts to the neutrals per second the momentum $n\langle \sigma u \rangle \cdot m w_d$, where w_d is the drift speed between the two components and $\langle \sigma u \rangle$ an average over thermal velocity times cross section. The force balance for 1 cm^3 is thus

$$\frac{GM}{R^2}nm = n_i n \langle \sigma u \rangle m w_d$$

from which follows the ambipolar diffusion time

$$t_{ad} = \frac{R}{w_d} = \frac{n_i}{n}\frac{3}{\pi Gm}\langle \sigma u \rangle \simeq 7 \times 10^{13}\frac{n_i}{n} \quad \text{yr.} \tag{15.54}$$

Because the slippage occurs between ions and neutrals, the cross section for momentum transfer is large, $\langle \sigma u \rangle \sim 2 \times 10^{-9}$ cm^3 s^{-1}; for neutral–neutral collisions it would be a thousand times smaller. With $n_i/n = 10^{-7}$, the diffusion time t_{ad} is several million years. All this suggests that a cloud must first get rid of its magnetic field through ambipolar diffusion before it can collapse. As ambipolar diffusion is a slow process compared to dynamical contraction, it limits the pace at which stars can form. Should the degree of ionization be much above 10^{-7}, for example, because the medium is clumpy and stray UV photons manage to penetrate into the cloud interior, the diffusion time t_{ad} would exceed the cloud lifetime and collapse could not occur.

15.3   Gravitational collapse

15.3.1   The presolar nebula

15.3.1.1   Gravitational and thermal energy

To get a feeling for the parameters of a protostellar cloud, let us consider the presolar nebula. It must have had a mass of 1 $M_\odot$, possibly a bit more as some matter was dispersed and did not fall onto the star. Evidence from molecular lines and dust radiation from low-mass protostellar clumps indicates a temperature $T \sim$ 10 K. Equation (15.38) then yields a critical cloud radius of about $R = 0.05$ pc or 1.54×10^{17} cm and a density $\rho \sim 1.3 \times 10^{-19}$ g cm^{-3}, corresponding to a molecular hydrogen density somewhat above 3×10^4 cm^{-3}. The free-fall time from (15.41) then becomes $t_{ff} \simeq 1.9 \times 10^5$ yr, about equal to the travel time of a sound wave through the cloud.

Today, the Sun has a radius $R_\odot = 7 \times 10^{10}$ cm and a mean density $\bar{\rho}_\odot \simeq 1.4$ g cm^{-3}. So on the way from the protostellar cloud to the main sequence, there are vast changes: the radius shrinks by a factor $\sim 10^7$, the density increases by 20 powers of ten, and the cold neutral gas turns into a hot plasma of over 10^7 K with thermonuclear reactions.

The initial gravitational and thermal energy of the protosolar nebula are:

$$E_{\text{grav}} \simeq \frac{GM^2}{R} \simeq 2 \times 10^{42} \text{ erg} \tag{15.55}$$

$$E_{\text{therm}} = \frac{3kTM}{2m} \simeq 1 \times 10^{42} \text{ erg.} \tag{15.56}$$

By the time the accretion process is over, the cloud radius has shrunk to $\sim 2R_\odot$. We know this from the position of T Tau stars in the Hertzsprung–Russell diagram. Half of the released gravitational energy, namely $GM^2/4R_\odot \sim 10^{48}$ erg, has gone into heating. As a result, the gas has acquired an average temperature of $\sim 3 \times 10^6$ K, a reasonable mean value for the Sun. The other half was radiated away. Over one free-fall time, this gives a mean luminosity of ~ 40 $L_\odot$. So very elementary considerations suggest that the Sun was, at least, during some of its protostellar evolution more luminous than it is today.

15.3.1.2   Rotational and magnetic energy

A protostellar cloud rotates. This is what the Milky Way does and what all other celestial objects do. The galactic disk does not rotate rigidly but differentially and the angular frequency ω decreases with galactic radius R as described by Oort's constants A and B:

$$A = -\frac{1}{2}R_0\frac{d\omega_0}{dR_0} \simeq 15 \text{ km s}^{-1} \text{ kpc}^{-1} \tag{15.57}$$

$$-B = \frac{1}{2}R_0\frac{d\omega_0}{dR_0} + \omega_0 \simeq 10 \text{ km s}^{-1} \text{ kpc}^{-1}. \tag{15.58}$$

They are determined observationally and refer to the locus of the Sun where $\omega = \omega_0$ and $R = R_0$. Rigid rotation of the disk would imply $A = 0$ and $B = -\omega_0$. From (15.57) and (15.58), one finds $\omega_0 = A - B \simeq 8^{-16}$ s^{-1} corresponding to a rotation period of 250 million years. In the differentially rotating galactic disk, the orbital velocity $v = \omega_0 R$ decreases towards the periphery like

$$\frac{dv}{dR_0} = R_0 \frac{d\omega_0}{dR_0} + \omega_0 = -A - B \simeq -5 \text{ km s}^{-1} \text{ kpc}^{-1}. \qquad (15.59)$$

Here we have used $dv = \omega_0 dR_0 + R_0 d\omega_0$. Because of the differential rotation of the galactic disk, the orbital speed of an interstellar cloud is higher on the near side facing the galactic center than on the far side. We may envisage a large cloud to revolve around its own axis at an angular frequency $\omega = |dv/dR_0| \simeq 1.6 \times 10^{-16}$ s^{-1}, a few times smaller than ω_0. As protostellar clouds have undergone some contraction by which they were spun up, $\omega \sim 10^{-15}$ s^{-1} seems a fair guess for them. Note: the previously cited values of A and $-B$ are not up-to-date but easy to remember. Their uncertainty is about 20%. In more recent analyses, the numbers for both A and $-B$ tend to converge around 12.5 km s^{-1} kpc^{-1}; this would reduce ω accordingly [Oll98].

A constant density sphere has a moment of inertia $I = \frac{2}{5}MR^2$. For a presolar cloud, the initial rotational energy becomes

$$E_{\rm rot} = \tfrac{1}{2}I\omega^2 \sim 10^{37} \text{ erg.}$$

Although this is small compared to $E_{\rm grav}$ or $E_{\rm therm}$, conservation of angular momentum $I\omega$ implies $E_{\rm rot} \propto R^{-2}$, so the rotational energy rises steeply as the radius shrinks. This raises a fundamental problem in the theory of star formation: How do we get rid of the angular momentum?

When the protosolar nebula is pervaded by a magnetic field of strength $B = 10 \ \mu$G, the magnetic energy is

$$E_{\rm mag} = \frac{B^2}{8\pi} \frac{4\pi}{3} R^3 \simeq 6 \times 10^{40} \text{ erg.}$$

Flux conservation implies for the contracting cloud $BR^2 = $ constant and thus $E_{\rm mag} \propto R^{-1}$. The magnetic field is initially relatively weak but grows with R^{-2} and should thus be 10^{13} times stronger in the final star. This, however, is not what one observes. The main mechanism by which a protostellar cloud rids itself of the magnetic field is probably ambipolar diffusion.

15.3.2 Hydrodynamic collapse simulations

Modern hydrodynamic calculations of protostellar evolution are multi-dimensional, with two or three spatial coordinates. They are based on sphisticated codes and carried out on computers as powerful as available at the time. Because of the complexity of the numerical results, they usually cannot be adequately presented in plots but have to be visualized in motion pictures. Inevitably, a naive understanding of what is happening gets lost. Here and in section 15.3.3, we, therefore, sketch simplified scenarios which nevertheless catch important aspects of the collapse stages.

The dynamical equations of a spherical cloud can be written as (cf. (15.42) and (15.43))

$$\frac{\partial u}{\partial t} + u\frac{\partial u}{\partial r} + \frac{1}{\rho}\frac{\partial P}{\partial r} + \frac{GM_r}{r^2} = 0 \tag{15.60}$$

$$\frac{\partial \rho}{\partial t} + \frac{1}{r^2}\frac{\partial}{\partial r}(r^2\rho u) = 0. \tag{15.61}$$

The geometry is here one-dimensional: density ρ, pressure $P = v_s^2\rho$ and velocity u depend only on radius r, and on time t. The gas is driven by the gradient of the pressure P and by the gravitational attraction. M_r is the mass of a sphere of radius r, so

$$\frac{\partial M_r}{\partial r} - 4\pi r^2\rho = 0. \tag{15.62}$$

During the early phases, the collapse is isothermal. The sound velocity v_s is then constant and in (15.60) the derivative $\partial P/\partial r = v_s^2\partial\rho/\partial r$. In this case, and as long as there are no shocks, it is straightforward to transform (15.60) to (15.62) into difference equations and solve them numerically; an example is presented in figure 15.8. For a fixed cloud volume, the non-trivial outer boundary condition reads as $u = 0$.

When the cloud becomes optically thick (see section 15.2.6), the assumption of isothermality has to be abandoned. One must then add to (15.60)–(15.62) an equation for the conservation of energy (first law of thermodynamics) and for the transport of radiation. For the latter, one often uses a frequency–averaged (grey) approximation in the spirit of (13.29). The numerical solution of the full set of equations [Lar69] leads to a $\rho \propto r^{-2}$ density distribution even when starting at constant density (see figure 15.8). The core becomes opaque when about 0.01 $M_\odot$ of gas have been accumulated within a radius $r \sim 10^{14}$ cm. The submillimeter dust radiation is then trapped, the core warms up and settles into hydrostatic equilibrium. When the gas temperature reaches $\sim$2000 K, molecular hydrogen dissociates. The released gravitational energy is then no longer used to increase T but to separate the hydrogen atoms. Initially, the cloud is made up mostly of H_2 for which $\gamma = 7/5 > \gamma_{cr}$ of (15.51). But when hydrogen dissociates, γ falls below the critical value of 4/3 and the hydrostatic core begins to collapse. This second collapse is halted only when γ rises again above 4/3, which happens at a central density of $\sim$$10^{-3}$ g cm^{-3}.

As one can read from figure 15.8 or 15.9, flow velocities become quickly supersonic because the sound speed is only $\sim$1 km s^{-1} in a cloud of 10 K. When a hydrostatic core has formed, the supersonically falling gas is stopped at the core

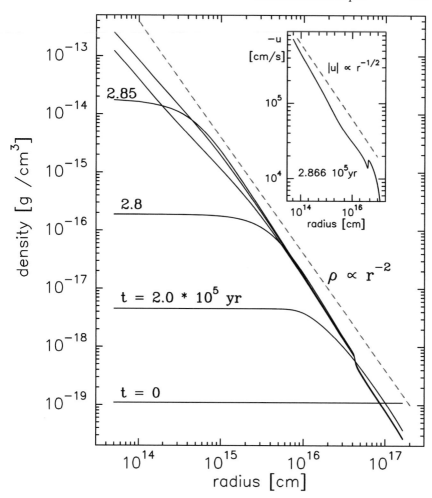

Figure 15.8. The distribution of density ρ and velocity u during the isothermal contraction of a spherical and initially homogeneous H_2 cloud of 1 $M_\odot$. The cloud is at temperature $T = 10$ K, its radius is fixed ($u = 0$ at the surface), and the free-fall time $t_{ff} = 1.60 \times 10^5$ yr. The collapse is very inhomologous. After 2.86×10^5 yr, the core ($r < 10^{14}$ cm) turns opaque to its own far infrared radiation and the assumption of isothermality is no longer valid. If the density at that moment is approximated by a power law, the exponent equals -2.3 in the envelope and -1.6 in the inner regions. For an Ebert–Bonnor sphere (slightly beyond the brink of stability), the evolution would be similar.

surface giving rise to a shock. The width of the shock is determined by the mean free path of the gas particles, so practically it is a discontinuity described by the Rankine–Hugoniot jump conditions and this poses a numerical challenge. In the

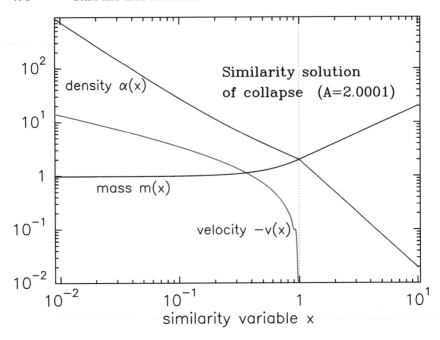

Figure 15.9. The solution of the ordinary coupled differential equations (15.66) and (15.67) describing spherical isothermal collapse. The time evolution goes along the lines from right to left. The starting values at $x = 10$ are from (15.68) with $A = 2.0001$. One can determine, from this figure, the cloud structure in terms of r, u, ρ of an initially ($t = 0$) hydrostatic cloud of density $\rho = v_s^2/2\pi Gr^2$ at any later time $t_0 > 0$. To get the radius r, one multiplies the abscissa x by $v_s t_0$; to get the velocity u, one multiplies v by v_s; and to obtain the density ρ, one multiplies α by $(4\pi Gt_0^2)^{-1}$ (see (15.64)).

shock front, the gas kinetic energy is converted into radiation. If R_* and M_* are the radius and mass of the hydrostatic core, the protostellar luminosity, prior to nuclear burning and neglecting the contraction of the protostellar core, becomes

$$L_{shock} = \frac{GM_*\dot{M}}{R_*}. \tag{15.63}$$

The accretion rate $\dot{M}$, the size and radius of the hydrostatic core are therefore the parameters that determine the protostellar luminosity. To evaluate it correctly, a proper treatment of shocks is obviously necessary.

When the numerical effort to compute protostellar collapse is much enhanced and the physical processes are refined, new problems arise concerning the stability of the core. For example, when the collapse calculations are carried out in two dimensions and cloud rotation is included, the core has the tendency to split up. To avoid fission of the core, one may, for instance, invoke strong turbulent friction (of speculative origin). It would guarantee an efficient outward transport of angular momentum and thus enable the formation of a single star [Tsc87].

15.3.3 Similarity solutions of collapse

15.3.3.1 Equations for spherical collapse

Similar solutions to the one-dimensional hydrodynamical equations for an isothermal cloud also exist [Shu77]. By definition, the spatial distribution of $u(r, t)$ and $\rho(r, t)$ at a fixed moment t then differs from the spatial distribution at any other moment t' only by scale factors that depend only on time. This restricted class of solutions does not add anything new to the one-dimensional numerical calculations of [Lar69] but because of its simplicity, it is very instructive. The similarity solutions can be found by introducing a dimensionless variable which expresses some relation between r and t, let us choose

$$x = \frac{r}{v_s t}.$$

Because of similarity, the dependent variables velocity, density and mass are the product of a function of t times a function of x. A fruitful ansatz is

$$u(r, t) = v_s v(x) \qquad \rho(r, t) = \frac{\alpha(x)}{4\pi G t^2} \qquad M(r, t) = \frac{v_s^3 t}{G} m(x). \qquad (15.64)$$

Inserting these expression for u, ρ and M into (15.62) and into

$$\frac{\partial M}{\partial t} + u \frac{\partial M}{\partial r} = 0 \qquad (15.65)$$

which is an equivalent form of the continuity equation (15.61), yields

$$m = m'(x - v) \qquad m' = x^2 \alpha \qquad \Longrightarrow \qquad m = \alpha x^2 (x - v).$$

Now (15.60) and (15.61) can be written as two ordinary differential equations:

$$\left[(x - v)^2 - 1 \right] v' = \left[\alpha(x - v) - \frac{2}{x} \right] (x - v) \qquad (15.66)$$

$$\left[(x - v)^2 - 1 \right] \frac{\alpha'}{\alpha} = \left[\alpha - \frac{2}{x}(x - v) \right] (x - v). \qquad (15.67)$$

Instead of two independent variables, r and t, one has only one, x, and there are no more partial derivatives, only derivatives with respect to x, which are denoted by a prime. Two solutions are obvious. The first, $v = 0$ and $\alpha = 2x^{-2}$, corresponds to a static sphere with a diverging central density. The second, $x - v = 1$ and $\alpha = 2/x$, can be discarded as unphysical because it implies a static mass distribution $M(r, t) = 2a^2 r/G$, although the velocity u is not zero.

15.3.3.2   Initial cloud structure

Let us consider an infall scenario where initially ($t = 0$), the cloud is at rest. When $t \to 0 \Longrightarrow x \to \infty$. Then v goes to zero and (15.66) and (15.67) become

$$v' = \alpha - 2x^{-2} \qquad \alpha' = \alpha(\alpha - 2)x^{-1}.$$

They yield, for $x \to \infty$, the asymptotic solutions

$$\alpha \to Ax^{-2} \qquad v \to -\frac{A-2}{x} \qquad m \to Ax. \tag{15.68}$$

The constant A has to be greater than 2 for v to be negative but only a tiny bit so that velocities are small. When $A \simeq 2$, the cloud is close to hydrostatic equilibrium and has a structure similar to the envelope of an Ebert–Bonnor sphere. As one can sketch the global aspects of its evolution with elementary mathematics, the singular isothermal sphere with $A \simeq 2$ is often used to outline qualitatively protostellar collapse. Of course, one may be skeptical whether such an artificial configuration is realized in nature and the results given here can in no way replace hydrodynamic calculations.

The initial structure ($t \to 0$) of a singular isothermal sphere is, therefore,

$$\rho(r,t) = \frac{v_s^2 A}{4\pi G r^2} \qquad u(r,t) = 0 \qquad M(r,t) = \frac{A v_s^2}{G} r. \tag{15.69}$$

The density falls off like r^{-2}. One can integrate (15.66) and (15.67) from some large value of x by a Runge–Kutta method down to $x = 0$ (corresponding to $t \to \infty$), if one takes the initial parameters from (15.68). The solution for $A = 2.0001$ starting with $x = 10$ (that is large enough) is plotted in figure 15.9; for other values of A slightly above two, the solutions are similar.

15.3.3.3   Final cloud configuration

There are also analytical expressions for the configuration towards which this cloud evolves as $x \to 0$, corresponding to $t \to \infty$. We can read from figure 15.9 that for $x \to 0$, the absolute value of the dimensionless velocity $|v| \gg 1$ and $\alpha x|v| \gg 1$. From (15.66) one then finds $v' = \alpha$ and from (15.67) $v'' + v'^2/v + 2v'/x = 0$. This differential equation is satisfied by

$$\alpha \to \sqrt{m_0/2x^3} \qquad v \to -\sqrt{2m_0/x} \qquad \text{for } x \to 0$$

where $m_0 = -(x^2\alpha v)_{x=0}$ is the limiting value of $m(x)$; for $A \to 2$, $m_0 = 0.975$. The final cloud structure for $t \to \infty$ is, therefore,

$$\rho(r,t) = \sqrt{\frac{m_0 v_s^3}{32\pi^2 G^2}}\, t^{-\frac{1}{2}} r^{-\frac{3}{2}} \tag{15.70}$$

$$u(r,t) = -\sqrt{2m_0 v_s^3}\, t^{\frac{1}{2}} r^{-\frac{1}{2}}. \tag{15.71}$$

At a fixed time, the density now changes like $r^{-3/2}$ and the velocity

$$u(r, t) = -\sqrt{2GM_*(t)/r}$$

is the free-fall velocity towards the central mass

$$M_*(t) = \frac{m_0 v_s^3}{G} t.$$

The accretion rate $\dot{M}$, which is the quantity that determines how long the collapse will last, is time-independent and given in the self–similarity analysis by

$$\dot{M} = 4\pi r^2 \rho u = \frac{m_0 v_s^3}{G}. \tag{15.72}$$

15.3.3.4 Remarks on the cloud evolution

To solve the full equations (15.60) and (15.61), one needs a starting model and boundary conditions at the edge and center of the cloud. In the similarity solution, only the parameters α and v are required for some value of the variable $x \propto r/v_s t$. Inner and outer cloud boundaries are disregarded and should not influence the evolution of the flow.

With the help of figure 15.9, we can delineate the evolution of an initially hydrostatic cloud of density distribution $\rho = v_s^2/2\pi Gr^2$. To follow the density and velocity at some fixed radius in time, we note that as time proceeds the corresponding point of the similarity variable $x = r/v_s t$ moves along the curves $\alpha(x)$ and $v(x)$ from right to left. Because the initial density declines like r^{-2} and the free-fall time t_{ff} of (15.41) decreases as $\rho^{-1/2}$, the collapse begins at the center working its way outward.

- In the envelope, for $x > 1$, the gas is still at rest, it does not yet know that the inner regions are collapsing. The density has not changed, ρ is still proportional to r^{-2}.
- In the core, for $x < 1$, velocities become supersonic and approach the free-fall velocity towards a central object of mass $m_0 v_s^3 t/G$ that is accreting at a rate $m_0 v_s^3/G$. The density decreases with time and radius, from (15.71), proportionally to $t^{-1/2} r^{-3/2}$.
- At the interface between free-falling core and static envelope, $x \simeq 1$ and $r = v_s t$, so this transition region is moving outwards with the speed of sound.

For example, if the total cloud mass is 1 $M_\odot$ and $v_s = 0.2$ km s^{-1}, the cloud stretches out to $R = 1.67 \times 10^{17}$ cm. After $t = 5 \times 10^{12}$ s, or 1.6×10^5 yr, all gas interior to $r = 10^{17}$ cm is collapsing and the mass accreted at the center equals ~ 0.3 $M_\odot$. At $r = 10^{14}$ cm, the density is $\rho \simeq 1.0 \times 10^{-15}$ g cm^{-3} and the velocity $u \simeq -9 \times 10^5$ cm s^{-1}.

15.4 Disks

A rotating collapsing cloud inevitably flattens because the gravitational pull perpendicular to the rotation axis will, at some radius r_{cenfug}, be balanced by centrifugal forces, whereas parallel to the rotation axis, these forces are absent. Eventually, a disk forms. Disks are widely observed in young stars and their presence is predicted by theory. Take, for example, an initially rigidly rotating spherical cloud of angular velocity ω_0. A gas parcel at radial distance r_0 and under an angle θ_0 to the rotation axis has the specific angular momentum $j = \omega_0 r_0^2 \sin^2 \theta_0$. In the gravitational field of a dominant central mass M_*, the parcel can approach the center only to a distance

$$r_{cenfug} = \frac{j^2}{GM_*} \tag{15.73}$$

which depends on r_0 and θ_0. Because the gas parcels have different initial specific angular momenta, the final equilibrium configuration in which they collect is a disk.

15.4.1 Viscous laminar flows

The evolution and structure of an accretion disk is largely determined by internal friction. We, therefore, begin with an elementary experiment illustrating Newton's friction law (figure 15.10). A flat piece of arbitrary material and total surface area $2A$ is dipped into a vessel filled with a viscous fluid of density ρ and kinematic viscosity coefficient ν. To pull it out at a constant velocity u, one needs, after subtraction of gravity and buoyancy, a force F. The molecules of the liquid are always at rest on the walls of the vessel and on both sides of the plate. The flow in between is plane parallel at velocity $\mathbf{v} = (v_x, v_y, v_z) = (0, v_y, 0)$ but it has a gradient $\partial v_y / \partial x$ in the x-direction. When the width $2b$ of the vessel is small, the velocity increases linearly from the walls towards the sheet and

$$F = \nu\rho \frac{u}{b} 2A = \nu\rho \frac{\partial v_y}{\partial x} 2A. \tag{15.74}$$

The force F does not depend on the velocity itself but on its gradient $\partial v_y / \partial x$. The gradient implies that the fluid particles change their relative positions, so tangential or shear forces appear and there is friction which causes the plate to resist being pulled out. The friction also generates heat. The total power dissipated in the vessel is Fu, so each cm^3 of the liquid receives, per second, a heat input

$$D = \frac{Fu}{2Ab} = \frac{F}{2A} \frac{\partial v_y}{\partial x} \geq 0. \tag{15.75}$$

It is positive, as it should be, because of the term $(\partial v_y / \partial x)^2$. One can get a feeling for the strength of the force and the dissipation rate by using numbers

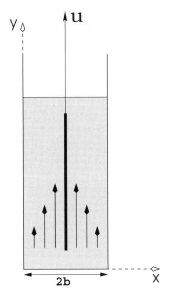

Figure 15.10. To move a sheet at a constant speed through a viscous fluid, one needs a force F after (15.74).

from everyday's life. For example, at 20 °C, water has $\nu = 10^{-2}$ cm^2 s^{-1} and $\rho = 1$ g cm^{-3}; for air, the kinematic viscosity ν is 15 times larger but the density ρ is 830 times smaller.

A rotational analog of this system consists of three coaxial cylinders of height h (figure 15.11). The inner and outer cylinder are corotating, the middle one has a different angular velocity $\omega \neq \omega_0$. The thin layers between the cylinders are filled with liquid or gas. Again the molecules are at rest at the cylinder walls and there is a shear flow in between. A torque τ then has to be applied to the middle cylinder in order to sustain the motion. It equals force per unit area multiplied by area and by radius,

$$\tau = \nu \rho r \frac{d\omega}{dr} 4\pi r h r.$$

Let us denote the derivative with respect to r by a prime and with respect to time by a dot. Because the work expended per second equals $\tau \omega$ after (11.4), we get, in complete analogy to (15.75), the dissipation rate per cm^3 of gas, $D = \tau\omega/4\pi rhb$ or

$$D(r) = \nu\rho(r\omega')^2 \geq 0. \qquad (15.76)$$

In the marginally modified experiment of figure 15.12, the angular velocities of the cylinders C_1, C_2, C_3 decrease with radius. C_3 exerts on C_2 the torque

$$\tau(r) = \nu \rho r \frac{d\omega}{dr} 2\pi r h r \qquad (15.77)$$

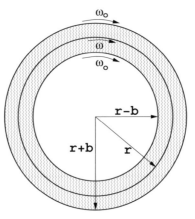

Figure 15.11. A viscous fluid is trapped between three closely spaced cylinders; the inner and outer one are corotating at angular velocity ω_0, the one in the middle has $\omega \neq \omega_0$.

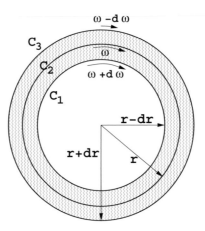

Figure 15.12. Like figure 15.11 but now the angular velocity ω decreases monotonically outward, as in a disk with Keplerian rotation.

trying to retard it. But for a change of sign, this is equal to the torque of C_2 on C_3 which tries to accelerate C_3. The same considerations apply to C_1 and C_2. Therefore, if the middle cylinder C_2 has a thickness Δr, it feels the *net* torque $\tau' \Delta r$. When we remove C_2 in the experiment, the gas that fills its place is subjected to exactly the same torque. The net torque on a gas ring of 1 cm width is $d\tau/dr$.

15.4.2   Dynamical equations of the thin accretion disk

We next explore the evolution of a thin viscous disk of height h and surface density $\Sigma = \rho h$. We use cylindrical coordinates (r, φ, z) and assume that the disk parameters do not depend on z or φ. The equation of mass continuity, $\dot{\rho} + \operatorname{div} \rho \mathbf{v} = 0$ (see (15.43)), becomes

$$\frac{\partial \Sigma}{\partial t} + \frac{1}{r}\frac{\partial}{\partial r}(r v_r \Sigma) = 0 \tag{15.78}$$

where the velocity has the components

$$\mathbf{v} = (v_r, v_\varphi = \omega r, v_z = 0)$$

v_φ is the Kepler velocity. The momentum equation (15.42) must be generalized to include the viscous force per cm³, $\mathbf{F}_{\text{visc}}$. Such a force is not conservative and cannot be derived from a potential. For a viscous fluid, the correct equation of motion is

$$\frac{\partial \mathbf{v}}{\partial t} + (\mathbf{v} \cdot \operatorname{grad})\mathbf{v} = -\frac{1}{\rho}\operatorname{grad} P - \operatorname{grad} \phi + \frac{\mathbf{F}_{\text{visc}}}{\rho}.$$

We now write it in a way that it directly expresses the conservation of angular momentum $\Sigma r^2 \omega$. It is then formally similar to (15.78) but for the creation term $d\tau/dr$ due to the torque. When we divide this term by $2\pi r$ so that it refers to the unit length of the circumference, we get

$$\frac{\partial}{\partial t}(\Sigma r^2 \omega) + \frac{1}{r}\frac{\partial}{\partial r}(r v_r \Sigma r^2 \omega) = \frac{1}{2\pi r}\frac{d\tau}{dr}. \tag{15.79}$$

The torque $\tau(r)$ is from (15.77). Combining (15.78) and (15.79) and assuming that ω is time-independent yields $r v_r \Sigma \cdot (r^2\omega)' = \tau'/2\pi$ from which it follows, again with the help of (15.78), that

$$\frac{\partial \Sigma}{\partial t} = -\frac{1}{2\pi r}\frac{\partial}{\partial r}\left[\frac{\tau'}{(r^2\omega)'}\right]. \tag{15.80}$$

This is the general formula for the evolution of the surface density. It contains a time derivative on the left-hand side and a second-order space derivative on the right; it is, therefore, a diffusion equation. We can immediately work out the radial gas velocity

$$v_r = \frac{\tau'}{2\pi r \Sigma (r^2\omega)'}$$

and the radial mass flow

$$\dot{M}(r) = -2\pi r \Sigma v_r.$$

The latter is positive when matter streams inwards. Significant changes in the structure of the disk occur over a diffusion timescale t_{dif}. If we put in a dimensional analysis $\dot{X} \sim X/t$ and $X' \sim X/r$, where X denotes some variable, we find, from the earlier expression for v_r or from equation (15.80), that

$$t_{\text{dif}} = \frac{r}{v_r} \sim \frac{r^2}{v}.$$

Without viscosity, the surface density would stay constant forever.

15.4.3 The Kepler disk

In a thin disk gravitationally dominated by a central mass M_*, when radial pressure gradients are absent, matter moves in Keplerian orbits. The angular velocity ω decreases outwards while the specific angular momentum j increases,

$$\omega(r) = \sqrt{GM_*} r^{-3/2}$$
$$j(r) = \sqrt{GM_*} r^{1/2} = r^2 \omega.$$

One finds $(r^2\omega)' = \frac{1}{2} r\omega$ and $\tau'/(r^2\omega)' = -3\pi[\nu\Sigma + 2r(\nu\Sigma)']$. With these two expressions, the diffusion equation (15.80) in a Kepler disk becomes

$$\frac{\partial \Sigma}{\partial t} = \frac{3}{r} \frac{\partial}{\partial r} \left[r^{1/2} \frac{\partial}{\partial r} \left(\nu \Sigma r^{1/2} \right) \right] \tag{15.81}$$

and the radial velocity

$$v_r = -\frac{3}{r^{1/2}\Sigma} \frac{\partial}{\partial r} (\nu \Sigma r^{1/2}). \tag{15.82}$$

The disk structure in the z-direction can be determined assuming that the gas is isothermal and in hydrostatic pressure equilibrium. The vertical component of the gravitational acceleration, g_z, is then balanced by the pressure gradient $\partial P/\partial z$,

$$g_z = -\frac{z}{r} \frac{GM_*}{r^2} = \frac{1}{\rho} \frac{\partial P}{\partial z}. \tag{15.83}$$

Because $P = \rho kT/m = \rho v_{\text{s}}^2$, one finds

$$\frac{\partial \rho}{\partial z} = -\rho z \frac{\omega^2}{v_{\text{s}}^2}$$

and this differential equation is satisfied by a vertical Gaussian density profile,

$$\rho = \rho_0 \exp\left[-\frac{z^2\omega^2}{2v_{\text{s}}^2} \right] = \rho_0 \, e^{-z^2/2h^2} \tag{15.84}$$

with

$$h^2 = \frac{v_{\text{s}}^2 r^2}{v_\varphi^2} = \frac{kT r^3}{GM_* m}. \tag{15.85}$$

h measures the disk thickness and is independent of the surface density. Here we have used the Kepler velocity $v_\varphi = \sqrt{GM_*/r}$. The temperature and isothermal sound speed v_s are, of course, functions of the distance r. Equation (15.85) has the interesting consequence that when the disk is thin, $h \ll r$, the Kepler velocity must be highly supersonic,

$$v_\varphi = \sqrt{GM_*/r} = v_s \frac{\sqrt{2}r}{h} \gg v_s. \qquad (15.86)$$

15.4.4 Why a star accretes from a disk

Let us follow the time evolution of a viscous Kepler disk according to the diffusion equation (15.81). We rewrite it as

$$\dot{\Sigma} = 3(\nu\Sigma'' + 2\nu'\Sigma' + \Sigma\nu'') + 9(\nu\Sigma' + \Sigma\nu')/2r$$

translate it into a difference equation and solve it numerically. In a radial mesh $r_1, r_2, \ldots, r_N$ with spacings $\Delta r = r_i - r_{i-1}$, one needs (at least) three grid points to express a second-order derivative and two for a first-order derivative. Let Σ_i denote the known value of the surface density at grid point i and time t, and Σ_i^{new} the new value a small time step τ later. We then make the following replacements (assuming constant spacings $r_i - r_{i-1}$)

$$\left.\frac{\partial\Sigma}{\partial t}\right|_i \simeq \frac{\Sigma_i^{new} - \Sigma_i}{\tau} \qquad \left.\frac{\partial\Sigma}{\partial r}\right|_i \simeq \frac{\Sigma_i - \Sigma_{i-1}}{\Delta r}$$

$$\left.\frac{\partial^2\Sigma}{\partial r^2}\right|_i \simeq \frac{\Sigma_{i+1} - 2\Sigma_i + \Sigma_{i-1}}{(\Delta r)^2}.$$

The derivatives of ν are written accordingly. All variables in the resulting difference equation for (15.81) refer to the old model, except for the term of the time derivative $\dot{\Sigma}$. Such an explicit scheme is very easy to solve but the time step τ must be fairly small for reasons of stability (Courant–Friedrichs–Lewy condition).

Let the surface density Σ of the starting model ($t = 0$) have a bell-shaped spike centered at radius $r_0 = 1$ of width $\sigma = 0.1$. So initially,

$$\Sigma(r) = \exp\left[-(r - r_0)^2/\sigma^2\right]$$

and $\Sigma(r_0) = 1$. At later times, the spike will broaden and we want to see in detail how. For the viscosity coefficient, we choose $\nu = r$ (units are not important in this section). For comparison, the radial dependence of ν in an α-disk in which the temperature $T(r)$ is proportional to $r^{-3/4}$ would be $\nu \propto r^{3/4}$ (see (15.92)). At the disk center ($r = 0$), the surface density is flat or $\partial\Sigma/\partial r = 0$ (see the analog (15.25) for a sphere). At the disk edge ($r = R$), the radial velocity v_r is zero, or $\nu\Sigma r^{1/2} = $ constant (see (15.82)). This corresponds to a solid confinement but

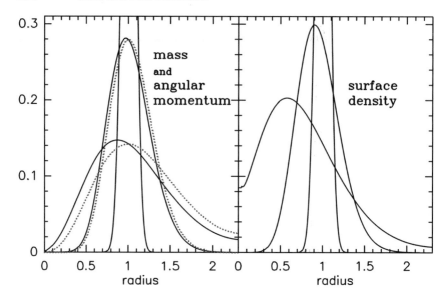

Figure 15.13. The spreading of a mass spike in a Kepler disk due to internal friction after equation (15.81). Curves are shown for time $t = 0, 0.01, 0.04$. The left panel gives the mass $2\pi r \Sigma$ (full curve) and the angular momentum $2\pi r \Sigma j$ (dots) of a ring of unit width where j is the angular momentum per unit mass. The right panel displays the surface density Σ. At time $t = 0$, the maximum values of the central spikes are all one, so their tops are cut off. Note how mass and angular momentum separate during the evolution.

as we assume the edge to be far out ($R = 5$), it has no influence on the early disk evolution. The inner and outer boundary condition are, therefore, $\Sigma_1 = \Sigma_2$ and $\Sigma_N = \Sigma_{N-1}(v_{N-1}/v_N)(r_{N-1}/r_N)^{1/2}$, corresponding to $\Sigma' = 0$ and $v_r = 0$, respectively.

The right box of figure 15.13 displays the numerical results for the evolution of the surface density. Because of the viscous forces, the spike flattens and broadens. There is mass influx towards the central gravitating mass for $r < r_0$, whereas farther out, the gas streams away from the center. The radial velocity v_r of (15.82) (not shown) is roughly proportional to $(r - r_0)$.

The full curves in the left frame are plots of $2\pi r \Sigma$ indicating how much mass is contained in a ring of unit width. The dotted curves give the angular momentum, $2\pi r \Sigma j$, in these rings. For $t = 0$, dots and solid line practically coincide. The essence of the diffusion mechanism in an accretion disk is not the spreading of the gas but the segregation of mass and angular momentum. The angular momentum dissipates outwards and the mass diffuses inwards while the total mass and angular momentum stay constant. Without this separation into low and high angular momentum gas, a star or a black hole could not accrete matter from a disk. In the end, not shown in the figure, almost all mass has accreted on the star and almost all angular momentum has been transported to large radii.

When the viscosity is a function of radius, equation (15.81) offers an analytical solution [Lyn74]. But because the mathematics are involved, it loses much of its instructiveness. However, the analytical solution is superior to the numerical results in showing the asymptotic behavior of the disk for $t \to \infty$.

15.4.5   The stationary accretion disk

When the gas flow is steady, one can readily derive the structure and luminosity of the disk. Setting in the equation of mass conservation (15.78) the time derivative $\partial / \partial t$ to zero implies that the accretion flow $\dot{M} = -2\pi r \Sigma v_r$ does not depend on space or time. Integration of the momentum equation (15.79) gives

$$r^3 \Sigma \omega v_r = r^3 v \Sigma \omega' + C. \tag{15.87}$$

For Kepler orbits around a stellar mass M_*, the angular velocity ω increases inwards like $\sqrt{GM_*/r^3}$. However, the increase cannot continue up to the stellar surface because the star rotates quite slowly, at least, T Tauri stars do. We can determine the constant C in (15.87) if we assume that ω is abruptly reduced in a thin layer immediately above the stellar surface and that in this layer, at the point where the first derivative ω' vanishes, ω can be approximated by $\sqrt{GM_*/R_*^3}$ where R_* is the stellar radius. At the position where $\omega' = 0$, one finds $C = \dot{M}\sqrt{GM_*R_*}/2\pi$ and, therefore, equation (15.87) becomes

$$\Sigma = v^{-1} \frac{\dot{M}}{3\pi} \left[1 - \sqrt{\frac{R_*}{r}} \right]. \tag{15.88}$$

The mass inflow rate $\dot{M}$ together with the viscosity v determine the surface density $\Sigma(r)$. It also follows that at distances comfortably greater than the stellar radius, the viscosity must be inversely proportional to the surface density,

$$v \propto \Sigma^{-1} \qquad \text{for } R_* \ll r.$$

15.4.6   The α-disk

15.4.6.1   *The viscosity coefficient*

So far, we have described internal friction only phenomenologically. We did not discuss the physical processes involved or how the kinematic viscosity coefficent v depends on the parameters of the gas. The reason for viscosity in a gas flow with a velocity gradient is the exchange of momentum perpendicular to the flow direction. In an ideal gas at temperature T, where the atoms have a mean velocity $\langle u \rangle$ and free path ℓ, one works out that

$$v = \tfrac{1}{3} \langle u \rangle \ell. \tag{15.89}$$

Because in a Maxwellian velocity distribution $\langle u \rangle \propto \sqrt{T}$ and because ℓ grows only weakly with T, one expects for an ideal gas something like $v \propto \sqrt{T}$. We will not prove formula (15.89) but it is qualitatively understandable. Looking at the flow depicted in figure 15.10, one realizes that the deeper the atoms penetrate from one layer into the next and the faster they move, the more momentum they transfer and the stronger the friction. If n is the number density of the particles and σ their collisional cross section, the free path length is

$$\ell = (n\sigma)^{-1}. \tag{15.90}$$

The product $v\rho$, which appears in the basic equation (15.81) of a Kepler disk, is thus independent of the density ρ (the surface density Σ equals ρ multiplied by the disk height h). For hydrogen molecules, σ is of order 10^{-15} cm^2 and the ensuing value $v\rho$ is far too small to produce any dynamical effect on the disk evolution.

But our previous analysis on internal friction applies to laminar flows. When the gas is turbulent, the viscosity is greatly enhanced. Turbulence means that superimposed on the average flow there are violent chaotic motions *on all scales* capable of very effectively exchanging momentum. The largest eddies determine the character of the flow. This becomes evident from the Reynolds number (15.93) which contains the linear dimension of the object. It is reasonable to use, for the size of the largest eddies, the disk height h and for their velocity the sound speed v_s; supersonic motions would be quickly damped in shocks. In analogy to (15.89), one writes the viscosity coefficient

$$v = \alpha v_s h \qquad (\alpha \leq 1) \tag{15.91}$$

where α is a fudge factor introduced to express the inherent uncertainty. Proposing $\alpha \sim 0.01$ would not lead to a riot among astronomers. Disks with v in the form of (15.91) are called α-disks [Sha73]. If they are in hydrostatic equilibrium in the vertical direction, one can substitute h from (15.85) so that for a turbulent Kepler disk the viscosity is

$$v = \alpha v_s^2 \sqrt{r^3/GM_*}. \tag{15.92}$$

Although this expression is too nice to be true, it gives some handle on the surface density Σ. With $v_s^2 \propto T \propto r^{-3/4}$ and equation (15.88), one finds for Σ the same power law decline as for the temperature, $\Sigma(r) \propto r^{-3/4}$.

15.4.6.2 The Reynolds number

To answer the question whether there is turbulence or not, one turns to the Reynolds number defined as

$$\text{Re} = \frac{Lv}{v}. \tag{15.93}$$

L is the dimension of the system and v the speed of the fluid. In laboratory experiments, the flow around objects is laminar when Re < 10, turbulent when Re $> 10^5$, and the critical range is Re ~ 3000. These numbers give some rough general guidance. To guarantee an effective outward transport of angular momentum in an accretion disk, very large Reynolds numbers seem to be required. Possible ways to achieve them are convection in the disk or, if the disk is permeated by a magnetic field, twisting of the field lines in the rotating plasma.

The Reynolds number (15.93) increases with velocity (a strong wind whirls up more dust), and with the size of the object (a big body disturbs the flow more effectively), and also when the viscosity decreases (water is more likely to be turbulent than honey). With respect to the last point, it is, however, puzzling that a small viscosity, which should mean little momentum exchange between neighboring stream lines yields a high Reynolds number and thus implies turbulence. The answer to this apparent contradiction follows from the analysis of the flow equation of an incompressible fluid. There, in the term containing the viscosity, the smallness of v is more than compensated by a large second-order gradient in the vorticity rot $\mathbf{v}$.

15.4.7 Disk heating by viscosity

In a steady state, the heat generated in the optically thick disk by internal friction is lost radiatively from its surface. Let $F(r)$ denote the outward flux at radius r on one side of the disk surface. Local balance with the frictional dissipation rate of (15.76) using (15.88) yields

$$F(r) = \tfrac{1}{2} v \Sigma (r\omega')^2 = \frac{3}{8\pi} \frac{GM_* \dot{M}}{r^3} \left[1 - \sqrt{\frac{R_*}{r}} \right]. \qquad (15.94)$$

To obtain the total disk luminosity L_{disk}, we must integrate over both sides of the disk,

$$L_{\text{disk}} = 2 \int_{R_*}^{\infty} 2\pi r \, F(r) \, dr = \frac{GM_* \dot{M}}{2R_*}. \qquad (15.95)$$

L_{disk} depends on the mass and radius of the star and on the accretion rate $\dot{M}$. The viscosity does not appear but is hidden in $\dot{M}$. For a blackbody surface, the effective temperature T_{visc} follows from

$$F(r) = \pi \int B_v(T_{\text{visc}}) \, dv = \sigma T_{\text{visc}}^4(r). \qquad (15.96)$$

Here σ is the radiation constant of (5.80). At large distances, $F(r) = 3GM_* \dot{M} / 8\pi r^3$ and

$$T_{\text{visc}} \propto r^{-3/4}.$$

A disk with a finite radius R_{out} has the total monochromatic luminosity

$$L_v = 4\pi^2 \int_{R_*}^{R_{\text{out}}} r B_v(T_{\text{visc}}(r)) \, dr \qquad (15.97)$$

with $T^4_{\text{visc}}(r) = 3GM_*\dot{M}[1 - \sqrt{R_*/r}]/8\pi\sigma r^3$. An observer sees, of course, only one side of the disk. If he views it face-on from a distance D, he receives the flux $L_\nu/2\pi D^2$. when the disk is inclined, one has to apply an obvious cosine correction.

15.4.7.1   The UV excess of T Tauri stars

L_{disk} in (15.95) corresponds to only half the potential energy that was available when matter was at infinity. The other half is stored in rotational energy. If the disk extends to the star, the power

$$\tfrac{1}{2}\dot{M}v_\varphi^2 = \frac{GM_*\dot{M}}{2R_*} \tag{15.98}$$

is liberated in a thin layer at the stellar surface where rapid braking takes place from the Kepler velocity v_φ to the usually much smaller rotational velocity of the star. As the energy is released over an area A small compared to the stellar surface, it will lead to hot UV continuum radiation of luminosity L_{UV}. An estimate of the sprectrum is obtained by assuming blackbody emission of temperature T_{UV} and putting $L_{\text{UV}} = A\sigma T^4_{\text{UV}}$.

Alternatively, the disk may be prevented from reaching the star by magnetic forces of the stellar magnetic field. At a distance of a few stellar radii, where magnetic and kinetic energy are about equal, the accreting gas is channelled from the disk along the magnetic field lines upward in the z-direction and then falls in a loop onto the star at high stellar latitudes. The result is again a hot continuum. Such a model can better explain the observational fact that the emission lines of T Tauri stars often have a width of several hundred km s^{-1}. In the case where the gas rotates in a disk all the way to the star, it is difficult to produce such large *radial* velocities, i.e. velocities along the line of sight. Examples of disk spectra are presented in chapter 16.

Chapter 16

Emission from young stars

We have learnt to compute how dust grains absorb and scatter light and how they emit. In this final chapter, we apply our faculties to astronomical examples relating to star formation. Cosmic dust is not now the object of our study but the agent by which we want to disclose the structure of an observed source. Its infrared continuum spectrum yields not only, in an obvious fashion, the infrared luminosity or an estimate of the dust temperature. An analysis of the spectrum by means of radiative transfer calculations often allows us to get an idea of the three-dimensional distribution of the dust with respect to the embedded star(s). For this purpose, one constructs models that are supposed to resemble the source and varies the model parameters until agreement between theory and observation is achieved. Fitting an extensive observational data set is laborious and requires occasional intuition. To solve such a task, a senior scientist needs a good student (the reverse is not true). Most of the spectra here are merely sketches, for the purpose of illustrating the magnitude of the emission.

16.1 The earliest stages of star formation

16.1.1 Globules

Usually, star formation is a contagious process. When a star is born in a molecular cloud, it disturbs the interstellar medium in its vicinity through its wind, jet or radiation pressure, and then neighboring clumps are pushed beyond the brink of gravitational stability, too. The isolated birth of a star is a rare event but it can happen in globules. These are cloudlets that have been stripped off from larger complexes. They have masses of $\sim 10\ M_\odot$ and appear as dark spots on photographic plates. Globules are bathed in a radiation field which heats them from outside and we want to evaluate their infrared emission in the absence of an internal source.

Let us consider a spherical globule of 2.6 $M_\odot$ at a distance of 200 pc with diameter of $2r = 1.2 \times 10^{17}$ cm. The hydrogen density n_H equals 4×10^6 cm^{-3}

in the inner part ($r \leq 3 \times 10^{16}$ cm) and declines linearly to 2×10^6 cm^{-3} at the surface. Using the standard dust model described in section 12.4, the overall visual optical thickness through the cloud center amounts to 282 mag, so the globule is very opaque.

We compute the radiative transfer for an external radiation field J_ν^{ext} that is, when integrated over frequency, four times stronger than the ISRF of figure 8.2. Such excursions in the radiative intensity can produce remarkable effects and occur when the globule is near a star (for example, at a distance of 1.5 pc from a B3V star). We assume that J_ν^{ext} consists of two components: a hard one described by a diluted blackbody of $T = 18\,000$ K that is meant to represent the UV flux from star(s), and a very soft one of spectral shape $\nu^2 B_\nu (T = 20$ K) that mimics the ubiquitous far infrared radiation. The relative strength of these radiation fields is chosen such that the grain temperatures lie between 17 and 27 K at the surface of the globule where the UV photons are responsible for heating and around 7 K in the cloud center.

The globule in figure 16.1 appears observationally first of all as a very cold submillimeter source. Its emission has a broad maximum around 250 μm. The millimeter spectrum comes from the coldest dust, it is fit by a modified Planck curve (section 5.4.3) of only 7 K. There is a skin layer of thickness $A_V \simeq 1$ mag where the big grains are much warmer ($\sim$20 K) and these grains account for the flux between 50 and 200 μm.

The presence of very small grains (vsg) and PAHs in the skin layer has the remarkable effect that the source becomes visible in the mid infrared. As discussed in section 8.5, such tiny grains with heat capacities of merely a few eV make a big upward leap every time they absorb an ultraviolet photon and this produces an emission spectrum entirely different from big grains. Although the tiny grains account for only a few percent of the total dust mass, they absorb some 20–30% of the UV light and their excitation therefore radically changes the infrared appearance of the source. Most noticeable are the PAH features (section 12.2). Without vsg and PAHs, the globule would only be detectable in the far infrared and beyond.

The presence of the PAHs has the further interesting consequence that it leads to important heating of the interior by mid infrared photons. Without them, the core would be extremely cold ($T_d \simeq 5$ K). A rise in dust temperature from 5 to 7 K implies an increase in the heating rate by a factor $(7/5)^6 \simeq 8$.

16.1.2   Isothermal gravitationally-bound clumps

A homogenous spherical cloud of mass M and temperature T has, at the brink of gravitational stability, from (15.19) and (15.20) a radius

$$R \simeq \frac{GMm}{5kT}.$$

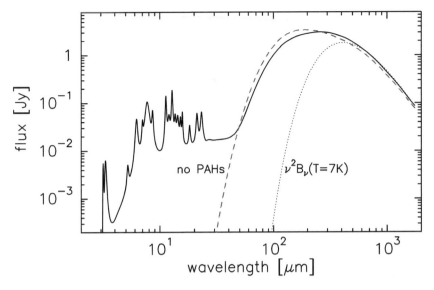

Figure 16.1. Dust emission of a globule irradiated by a UV field that is four times stronger than the ISRF of figure 8.2. The plotted spectra show the difference in the signal towards the source and in the off–beam where one still sees the external radiation field. The broken curve gives the flux when PAHs and vsg are absent, the dotted curve the emission from dust at 7 K. See text for further explanation.

A typical optical thickness of the clump is

$$\tau_\nu \simeq 2\rho K_\nu R$$

where ρ is the gas density and K_ν the absorption coefficient at frequency ν per gram of interstellar matter. The optical depth becomes

$$\tau_\lambda = \frac{3K_\lambda}{2\pi}\left(\frac{5k}{Gm}\right)^2\frac{T^2}{M} \simeq 10^{26}\frac{T^2}{M\lambda^2}$$

where we used cgs units for the clump mass and the wavelength to derive the factor 10^{26}. At a distance D, the clump subtends a solid angle

$$\Omega = \frac{\pi R^2}{D^2}$$

and an observer receives the flux

$$S_\nu = \Omega B_\nu(T) \cdot [1 - e^{-\tau_\nu}]. \tag{16.1}$$

- If the clump is tranparent ($\tau_\nu \to 0$),

$$S_\nu = \frac{MK_\nu B_\nu(T)}{D^2}$$

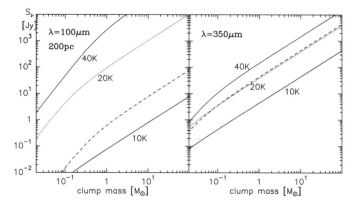

Figure 16.2. The flux S_ν from (16.1) emitted by dust from a gravitationally bound, homogeneous and isothermal clump at 100 and 350 μm. S_ν is shown as a function of cloud mass and for three dust temperatures (full curves). The broken curve in each box refers to a clump at 10 K with a tenfold the absorption coefficient as a result of grain coagulation and accretion of ice mantles (see figure 12.10). The source is at a distance of 200 pc.

- if it is opaque ($\tau_\nu \to \infty$),

$$S_\nu = \frac{\pi R^2 \cdot B_\nu(T)}{D^2}.$$

Cold clumps are best detected by their submillimeter dust emission. Their flux, S_ν, is given by (16.1). It is plotted in figure 16.2 for various temperatures and wavelengths as a function of clump mass M assuming a source distance $D = 200$ pc, a mean molecular mass $m = 2.34m_H$, corresponding to a normal mixture of H_2 molecules and helium atoms, and a far infrared dust absorption coefficient according to (14.54).

When the optical depth is below one, which is always true at 850 μm or longer wavelengths, the flux is simply proportional to the cloud mass. However, when τ_ν is large, which happens at $\lambda = 100$ μm for small masses, $S_\nu \propto M^2$. The break in the slope of the flux S_ν where the dependence on mass M changes from linear to quadratic occurs at $\tau_\nu \sim 1$.

16.2 The collapse phase

16.2.1 The density structure of a protostar

In sections 15.3.2 and 15.3.3, we derived the density distribution of a protostellar cloud under the assumption of spherical symmetry. We now discuss a protostar where it is possible to verify the theoretical predictions by direct measurement of the dust absorption [Sie00]. It is named HH108MMS and its surroundings

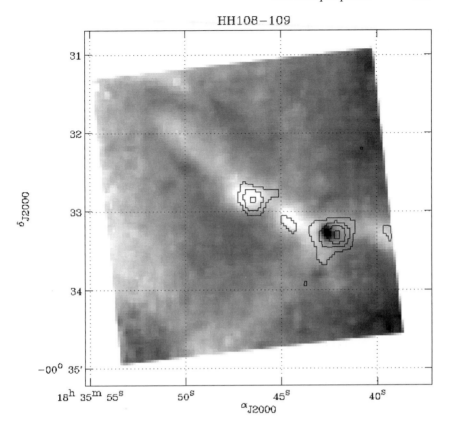

Figure 16.3. A 14 μm image around the protostar HH108MMS. The contour lines show the dust emission at 1.3 mm [Chi97]. The protostar is the white blob in the center of the figure. It coincides with a millimeter source and is seen at 14 μm in *absorption* against a diffuse background. The millimeter source on the lower right is a young stellar object (IRAS18331–0035) that *emits* in the mid infrared (black blob). Reproduced from [Sie00] with permission of EDP Sciences.

are depicted in the mid infrared map of figure 16.3. One sees a triplet of lined-up sources, one dark and two white spots, which can also be followed by their 1.3 mm dust emission (contour lines). The protostar, at a distance of 310 pc, is at the map center. The millimeter source to its lower right is a young stellar object of ∼3 L$_\odot$, still heavily obscured at optical but sticking out prominently at mid-infrared wavelengths as witnessed by the black spot in the map. It drives two Herbig–Haro objects but they lie outside the field of view. There is a third weak millimeter source in between the protostar (map center) and the young stellar object (black spot) that appears as a white blob in the 14 μm map. It is also a protostar but it will not be discussed any further.

No far infrared luminosity is detectable from HH108MMS, an embryo star has therefore not yet formed and the accretion phase not yet begun. HH108MMS appears at 14 μm in *absorption* against the diffuse background and this offers the opportunity to derive, from the variation of the surface brightness over the source, the optical depth and the density profile in the source. Alternative methods to obtain a density profile, by observing millimeter dust *emission* or certain molecular lines, are less direct and, therefore, less reliable.

The geometrical configuration for our analysis is drawn in figure 16.4. HH108MMS is embedded in a molecular cloud and viewed from Earth through the veil of zodiacal light of the solar system. Four regions contribute to the observed 14 μm flux:

- The background of intensity I_0. Our globular model in section 16.1.1 suggests that this radiation is due to emission by PAHs and vsg heated in the surface layer on the *backside* of the molecular cloud.
- The far and near side of the molecular cloud with temperatures T_1 and T_3 and optical depths τ_1 and τ_3, respectively. We put $\tau_{cl} = \tau_1 + \tau_3$.
- The absorbing core (protostar) of optical depth τ_2 and temperature T_2.
- The interplanetary dust of the solar system. It emits zodiacal light of intensity I_z and is, in fact, by far the strongest contributor to the 14 μm flux.

The molecular cloud and the absorbing core are certainly cold ($T < 20$ K) and have no mid infrared emission of their own ($B_{14\mu m}(T_i) \simeq 0, i = 1, 2, 3$). The intensity in the direction of the absorbing core (*source*) can, therefore, be written as $I_s = I_z + I_0 e^{-(\tau_{cl}+\tau_2)}$, and towards the *background*, at a position next to the absorbing core, $I_b = I_z + I_0 e^{-\tau_{cl}}$. As the zodiacal light is constant on scales of a few arcsec, one gets

$$I_s - I_b = I_0 e^{-\tau_{cl}} \cdot [e^{-\tau_2} - 1]. \tag{16.2}$$

I_s and I_b are known from observations. To find the optical depth τ_2 across the protostar, let us denote the maximum value of τ_2 by $\tau_{2,\max}$ and the corresponding intensity by $I_{s,\max}$. The 1.3 mm flux towards that position yields for the central beam an average optical depth $\tau_{1.3mm} \sim 10^{-2}$ (see (8.19)). The dust opacity appropriate for cold clumps is depicted by the upper full curve of figure 12.10; this gives $K_{14\,\mu m}/K_{1.3mm} \sim 1000$. The optical depth $\tau_{2,\max}$ is, therefore, of order 10 and well above unity. Towards the position of maximum obscuration, equation (16.2) can, therefore, be approximated by

$$I_0 e^{-\tau_{cl}} = I_b - I_{s,\max}.$$

Inserting $I_0 e^{-\tau_{cl}}$ into (16.2) yields the desired expression for the optical depth in the core as a function of position,

$$\tau_2 = -\ln\left[1 - (I_b - I_s)/(I_b - I_{s,\max})\right]. \tag{16.3}$$

τ_2 is plotted in figure 16.5 for two perpendicular cuts. From τ_2, one can derive the density structure in the protostellar core. If it is a sphere of radius R

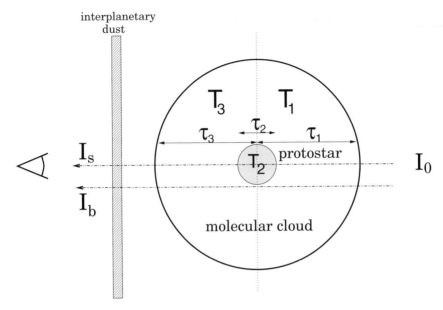

Figure 16.4. A sketch of how HH108MMS is embedded into a larger molecular cloud. Temperature and optical depth in the various regions are indicated. The intensity towards the *source* (protostar HH108MMS) and the *background* is denoted I_s and I_b, respectively. I_0 is the intensity behind the Serpens cloud. Reproduced from [Sie00] with permission of EDP Sciences.

and the dust inside has a power–law density $\rho(r) = \rho_0 r^{-\alpha}$, the optical depth $\tau_2(x)$ through the cloud at an offset $x \leq R$ from its center is

$$\tau_2(x) = K\rho_0 \int_0^X (x^2 + s^2)^{-\alpha/2}\, ds$$

where K is the dust mass absorption coefficient and $X = \sqrt{R^2 - x^2}$. For finding the exponent α, the product $K\rho_0$ is irrelevant. The variation of the optical depth and the best fit density distributions are shown in figure 16.6. Good fits are obtained for $\rho(r) \propto r^{-1.8\pm0.1}$.

The self–similarity solution of gravitational collapse in section 15.3.3 predicts for the early evolution, in the absence of rotation and magnetic fields, a density profile $\rho \propto r^{-2}$ in the outer envelope, and a less steep distribution $\rho \propto r^{-3/2}$ in the inner part. The dynamical calculation for a low-mass protostar in figure 15.8 gives similar results. In more sophisticated models, when magnetic fields and rotation are included, the density distribution in the envelope follows a power law with an exponent between -1.5 and -1.8 [Bas95]. So the observational results for HH108MMS confirm the theoretical predictions, at least, up to an optical thickness at 14 μm of 4 (see figure 16.6) corresponding to a visual extinction of 200 mag.

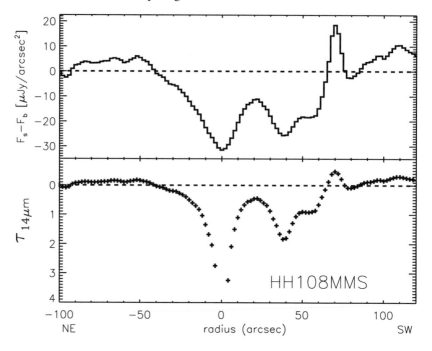

Figure 16.5. The top frame gives the difference between the flux towards the protostar HH1108MMS and the background in a cut along the major axis. The protostar is at offset zero, IRAS18331–0035 at offset $+70''$. The bottom frame shows the optical depth derived from (16.3). Reproduced from [Sie00] with permission of EDP Sciences.

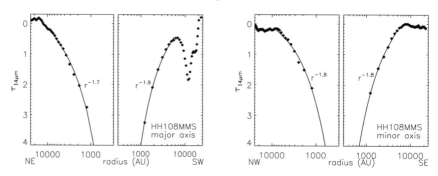

Figure 16.6. Optical depth profiles (diamonds) of HH108MMS. For the major axis, as in figure 16.5 but now on a logarithmic scale. The full curves show the theoretical variation of the optical depth in a spherical cloud with a power-law density distribution, $\rho \propto r^{-\alpha}$. The exponent α is close to 1.8; a deviation by only 0.2 from this value would imply an unacceptable fit. Reproduced from [Sie00] with permission of EDP Sciences.

16.2.2 Dust emission from a solar-type protostar

We compute the dust emission from a single low-mass protostar, a protosun. The total cloud mass, including the embryo star, is about one solar mass. In the early phase of collapse, the luminosity comes from the release of kinetic energy of the free-falling envelope in a shock front at the surface of the hydrostatic core. Its value, L_{shock}, is given by (15.63) and (15.72) but one needs hydrodynamic calculations to relate M_* to R_*. They suggest that for a clump of initially one solar mass, L_{shock} is 30 $L_\odot$ and fairly constant over one free-fall time [Win80]. Only near the end of the accretion phase when most of the matter in the envelope has rained onto the star does the shock luminosity decline and the energy generation from contraction (section 15.2.1) and later from hydrogen burning take over. In the Hertzsprung–Russell diagram, the protostar approaches the main sequence from above, at roughly constant color B − V.

In the example of figure 16.7, the outer cloud radius $R_{out} = 1.3 \times 10^{17}$ cm is determined by the critical radius (15.38) of an Ebert–Bonnor sphere. The density ρ changes in the protostellar envelope like $\rho \propto r^{-3/2}$ (see (15.70)). The total cloud optical depth thus depends foremost on the inner cloud radius, R_{in}. There are various ways to fix R_{in} in the model. One may equal it to the evaporation temperature of dust, which would imply a value less than 1 AU. Or one may use the centrifugal radius r_{cenfug} of (15.73) assuming that inside this radius gas and dust are strongly flattened into a disk. The disk emission would then have to be evaluated separately (section 16.3) and added, in a consistent fashion, to the flux from the spherical cloud. As the numbers extracted from equation (15.73) for r_{cenfug} are very rough, one may set R_{in} equal to the radius of the solar planetary system (40 AU). We opt for a value in between, we put $R_{in} = 2 \times 10^{13}$ cm and neglect the disk.

In the particular presentation of figure 16.7, we plot νF_ν instead of the observed flux F_ν. This has the advantage that for any two logarithmic frequency intervals $[\nu_1, \nu_2]$ and $[\nu_1', \nu_2']$ of the same length, i.e. $\nu_1/\nu_2 = \nu_1'/\nu_2'$, equal areas under the curve νF_ν imply equal spectral luminosities over these frequency intervals. One can thus directly compare the energy emission from different spectral regions.

Figure 16.7 shows models at five evolutionary stages, for a source distance of 150 pc. The curves labeled 1 to 5 follow a time sequence. During the collapse, the visual optical depth from the center to the cloud edge as well as the clump mass (without the embryo star) decline, the corresponding values are $A_V = 560, 560, 57, 11.4, 1.1$ mag and $M_{cloud} = 0.93, 0.93, 0.093, 0.019, 0.0018$ M$_\odot$. At stage 1, the accretion phase starts and the luminosity is still low ($L = 1$ L$_\odot$). During the main accretion period (stage 2 and 3), the luminosity is much higher ($L = 30$ L$_\odot$). It drops later to 3 L$_\odot$ (stage 4) and 2 L$_\odot$ (stage 5) in the T Tauri phase. The spectral index α between 2.2 and 20 μm, which is used in the classification of protostars (section 7.3.4) has, for curve 1–5, the values $12, 12, 1.7, 0.2, -0.2$.

The spectra have the following qualitative features as they change in time.

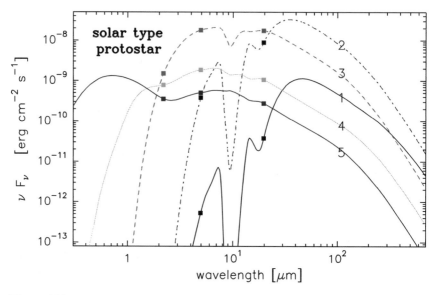

Figure 16.7. The spectral energy distribution of a protostellar cloud of 1 M$_\odot$ at five consecutive evolutionary stages, numbered 1 to 5. The curves are marked by squares at 2.2, 5.0 and 20 μm which are the wavelengths used for computing the spectral index α introduced in section 7.3.4. The models are described in the text.

The protostar appears first as a submillimeter source. In the course of the early protostellar evolution, the maximum of the emission shifts towards shorter but still far infrared wavelengths. There is a strong 10 μm silicate absorption feature but a near infrared counterpart of the embryo star is not yet detectable. In the advanced phases, the emission increases rapidly in the near infrared, the silicate absorption disappears and the star becomes visible optically.

Although the computer was busy for a few minutes to obtain the spectral energy distributions of figure 16.7, the output must not be taken too literally. First, because of the simplicity of the one-dimensional code, in state-of-the-art models, the radiative transfer is calculated consistently together with the multi-dimensional hydrodynamic evolution. Interestingly, the dust properties are not crucial, all dust models give similar curves, except for the outer cloud envelope where the grains may be icy and fluffy. Second, the protostar is always embedded in a larger cloud complex which distorts the spectrum of the protostar because it leads to additional extinction in the near and even mid infrared and to an enhancement of the submillimeter emission. Third, a real cloud is more complicated, it is asymmetric due to rotation and the presence of magnetic fields which makes the observed spectrum dependent on the viewing angle. Finally, one expects an accretion disk around the embryo star whose structure and emission have not been clarified.

16.2.3 Kinematics of protostellar collapse

The search for a protostar has occupied observing astronomers for decades. It may continue to do so for many more to come because there is no exact definition what a protostar is. The term is applied loosely to anything at the verge of gravitational stability to the stage when it becomes optically visible. The ultimate proof that one has found a collapsing object has to come from the velocity field of the interstellar clump: without a kinematic signature, the collapse cannot be identified with certainty. As interstellar dust does not supply us with velocity information, we have to turn to atomic or molecular lines. This is a healthy reminder that even dust is not everything.

Ideally, in the study of a protostellar clump, one has for each physical domain, characterized by a certain interval of temperature and density, a specific transition of a particular molecule with which one can probe the conditions in the gas. The source is mapped in various transitions of several such molecules to cover the full physical range encountered in the cloud. The observations are carried out with high spatial resolution so that one can resolve the cloud structure. When the observed lines are analysed in a radiative transfer model, infall betrays itself by distinctive features in the profiles.

The structure of a spherically symmetric, collapsing protostellar cloud was sketched in section 15.3: the gas density decreases with radius r like a $r^{-3/2}$ and the infall velocity v goes with $r^{-1/2}$. The isovelocity curves with respect to the Doppler shift toward the Sun are displayed in figure 16.8. Along a line of sight, emission from position 1 (3) interacts with the gas at position 2 (4) because the radial velocities at the two points are equal. After an embryo star has formed, the temperature will decrease away from the star because the cloud has an internal energy source. Point 2 (3) will, therefore, be hotter than point 1 (4). If the line is optically thick, one receives red-shifted emission only from the cold position 4 because emission from the warm gas at 3 is absorbed, and likewise blue-shifted emission is detected only from the warm position 2. This leads to the signature of infall: an asymmetric profile where the blue peak is stronger than the red one. If the line, however, is optically thin, as for rare isotopic species, the profiles will be symmetric because of the symmetric configuration in the near and far side of the cloud.

To illustrate what the lines might look like, we present in figure 16.9 and 16.10 model calculations for the early phase of protostellar collapse, before the release of gravitational energy in a shock front. The cloud is then isothermal. We choose a kinetic gas temperature $T_{\rm kin} = 17$ K, a $\rho \propto r^{-3/2}$ density dependence and an infall velocity $u = 5.5 \times 10^{12} r^{-1/2}$ cm s^{-1} (which implies an embryo star mass M of 0.11 M$_\odot$ if one puts $u^2 = 2GM/r$). The clump has a radius $R = 10^{17}$ cm and its total gas mass (without the embryo star) equals 0.65 M$_\odot$. The computations are performed in the microturbulent approximation described in section 13.5.3, assuming a constant turbulent linewidth of 0.2 km s^{-1}.

It is important to note that although $T_{\rm kin}$ is constant in our model, the

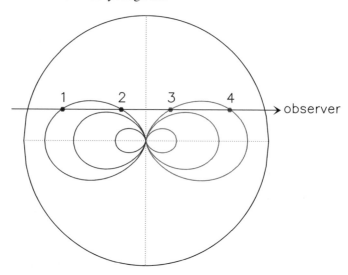

Figure 16.8. Kinematics in a spherically symmetric collapsing cloud. The infall velocity v equals 1 (in arbitrary units) at the circumference of the circle and increases with radius r like $r^{-1/2}$ (see (15.71)). The egg-shaped curves are lines of constant radial velocity v_{rad}, which is the component in the direction to the observer. Along a line of sight (arrow), at points 3 and 4, v_{rad} is positive (red-shifted gas) and has the value v_{rad}^{+}. At points 1 and 2, it has the value $v_{rad}^{-} = -v_{rad}^{+}$ (blue-shifted gas). On the egg-shaped curves, which have been computed and are drawn to scale, $v_{rad} = \pm 1.1, \pm 1.3$ and ± 2, respectively.

excitation temperature T_{ex}, which is the relevant quantity for line emission (see (13.65) and (13.66)), will decrease in the envelope if the level population of the particular molecule is sensitive to the gas density. A good tracer for the densities prevailing in a protostar are the rotational transitions of CS. The molecule is fairly abundant ($[CS]/[H_2] \sim 2 \times 10^{-9}$) and in its physics similar to CO (section 14.2.1). CS has, however, a much larger dipole moment ($\mu = 1.98$ Debye) than CO and thus requires higher densities for excitation (section 14.2.2). This, ideally, has the pleasant side effect that all tenuous foreground and background material, not related to the protostar, becomes invisible. Because the Einstein coefficients change roughly with j^3 (see formula (14.4)), one penetrates progressively deeper into the cloud as one climbs the j–ladder. So observations of the higher $j \rightarrow j-1$ transitions refer to the core region where densities and velocities are larger.

    The spectra in figures 16.9 and 16.10 show the CS lines of upper level $j_{up} = 2, 4, 5$ for different angular resolution of the telescope. Those in figure 16.9 refer to a large beam and their profiles are almost Gaussian. But when looking at the source at high resolution, the lines become very asymmetric. They then display a double peak as a result of self-absorption. The excitation temperature T_{ex} drops from 17 K in the center where the levels are thermalized (LTE population) to 4.6 K in the (5–4) line and 5.3 K in the (2–1) line at the cloud edge. There are other more

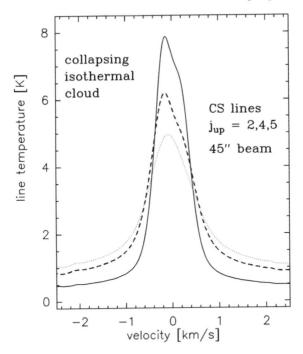

Figure 16.9. Selected rotational lines of CS as observed towards a collapsing isothermal clump: full curve, (2–1); broken curve, (4–3); and dotted curve, (5–4). The model is described in the text. The object is at a distance of 150 pc and observed with a beam half its angular size. The profiles are only slightly asymmetric as emission comes mainly from the cloud envelope.

subtle features in figure 16.10 that concern the change of the line shape with j-number. As j grows, the intensity ratio between red and blue peak increases and the velocity of the red peak as well as of the central dip (local minimum of the profile) shift by about one local linewidth to the red. Altogether, the red peak shows stronger variations with j-number than the blue one.

In the real world, outside our models, things are less simple. First, a protostar is always embedded into a larger gas complex and the latter, although of lower density is usually not transparent and modifies the observed line profiles. For example, when we increase the radius of our model cloud to $R > 10^{17}$ cm while retaining the density distribution $\rho(r)$, so when we just add a low-density outer envelope, the profiles change, not dramatically but noticeably. The line profiles are further influenced by all kinds of systematic (non-radial) or turbulent motions. Our microturbulent assumption that the mean free photon path is much bigger than the scale length of turbulence (section 13.5.3) may not be valid. Finally, observations support the theoretical expectation that in cold clouds the molecules

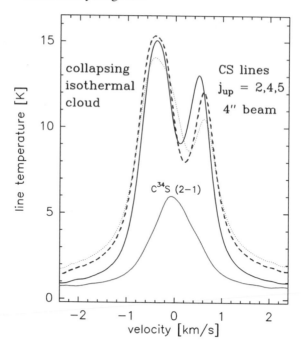

Figure 16.10. The same model as in figure 16.9 but here observed with a narrow beam, directed toward the cloud center. Now the core region dominates. The CS lines are double-peaked, the maximum on the blue side being stronger. Their wings are due to the high central infall velocities. The lines of the rare isotope ($[CS]/[C^{34}S] \simeq 20$) are optically thin and quite symmetric.

are frozen out (section 12.3). In such a scenario, one would not see collapsing gas at all, dust would be our only tracer. So to conclude with confidence that certain line profiles arise from collapsing gas in a protostar is, at present, not easy.

16.3 Accretion disks

16.3.1 A flat blackbody disk

We model the emission of a geometrically flat but not necessarily optically thin accretion disk of finite extension around a T Tauri star. If the disk is passive, its only source of energy comes from the illumination by the star. In a first approximation, let the disk be a blackbody. Its temperature profile in radial direction is given by (13.50) and plotted in figure 16.11. There are four contributors to the spectrum displayed in figure 16.12:

(i) The star. In our example, it has a luminosity $L_* = 2\,L_\odot$ and an effective temperature $T_* = 4070$ K.

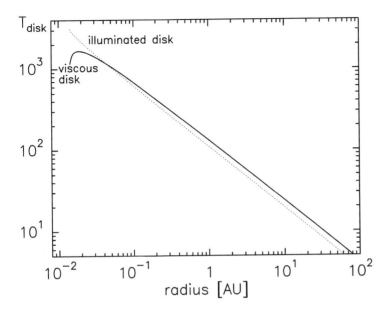

Figure 16.11. The temperature distribution in a blackbody disk. T_{illu} and T_{visc} refer to radiative and viscous heating (see (13.50) and (15.96)). When both processes operate, the effective temperature is $T = (T_{\text{visc}} + T_{\text{illu}})^{1/4}$. Model parameters are given in figure 16.12. T_{visc} is computed assuming an accretion rate $\dot{M} = 6 \times 10^{-8}\ M_\odot\ \text{yr}^{-1}$.

(ii) The hot spot on the stellar surface. It is the place where the kinetic energy of the accreted disk material is thermalized and radiated in the ultraviolet. We assume, somewhat arbitrarily, blackbody emission of 15 000 K. The associated luminosity in all directions (see (15.98)) equals $G\dot{M}M_*/2R_* = 0.33\ L_\odot$, for $\dot{M} = 6 \times 10^{-8}\ M_\odot\ \text{yr}^{-1}$, $M_* = 1\ M_\odot$ and $R_* = 2 \times 10^{11}$ cm.

(iii) The disk. It is illuminated by the star and its overall luminosity equals $\frac{1}{4}L_* = 0.5\ L_\odot$ when the disk stretches out to infinity. The observed flux is computed from

$$S_\nu = \frac{\cos i}{D^2} \int_{R_*}^{R_{\text{out}}} 2\pi r\, B_\nu(T)\, dr \qquad (16.4)$$

where D is the distance and T is taken from (13.50). The viewing angle i entails a reduction factor $\cos i$. When one looks face-on, $\cos i = 1$.

(iv) A potential further energy source, not included in figure 16.12, is viscous heating. The associated luminosity into all directions would be $G\dot{M}M_*/2R_* = 0.33\ L_\odot$, the same as for the UV hot spot (see (15.95)).

The spectrum is dominated by the ridge between 2 and 200 μm. Its slope is constant as a result of the power-law distribution of the temperature. Indeed, from

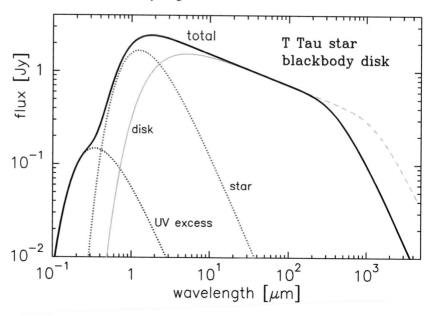

Figure 16.12. The spectrum of a T Tauri star with a flat blackbody disk viewed pole–on. The disk is heated by the star, there is no frictional dissipation. The stellar parameters are: distance $D = 200$ pc, $M_* = 1$ $M_\odot$, $L_* = 2$ $L_\odot$, $T_* = 4070$ K, $R_* = 2 \times 10^{11}$ cm. The full bold curve is the sum of three components: star, UV hot spot, plus disk. Inner and outer disk radius: $R_{in} = R_*$, $R_{out} = 20$ AU. For the broken curve beyond 200 μm, $R_{out} = 100$ AU.

$T(r) \propto r^{-q}$ it follows that

$$S_\nu \propto \int 2\pi r B_\nu(T)\, dr \propto \nu^{3-\frac{2}{q}} \int \frac{x^{\frac{2}{q}-1}}{e^x - 1}\, dx \qquad (16.5)$$

so that in our case where $q = 3/4$ (see (13.51)),

$$S_\nu \propto \nu^{1/3}.$$

In the submillimeter region, $S_\nu \propto \nu^2$. The flux has the same slope as the Planck function in the Rayleigh–Jeans part but only because of the finite disk radius which sets a lower boundary to the temperature. If the disk were infinite, T would go to zero as $r \to \infty$ and the wavelength range where S_ν is proportional to ν^2 would extend to infinity. So the radial cutoff, although irrelevant for the energy budget, is important for the shape of the long wavelength spectrum.

16.3.2 A flat non-blackbody disk

16.3.2.1 *Isothermality in the z-direction*

A real disk is not a blackbody, even if it is opaque at all wavelengths. The vertical optical thickness $\tau_\nu^\perp$ in the disk is the product of the surface density Σ multiplied by the mass absorption coefficient per gram of circumstellar matter, K_ν,

$$\tau_\nu^\perp(r) = \Sigma(r)K_\nu.$$

For a disk of arbitrary optical depth that is isothermal in the z-direction, the observed flux is given by

$$S_\nu = \frac{\cos i}{D^2} \int_{R_*}^{R_\text{out}} 2\pi r\, B_\nu(T(r)) \cdot [1 - e^{-\tau_\nu}]\, dr \tag{16.6}$$

with $\tau_\nu = \tau_\nu^\perp / \cos i$. The formula is similar to (16.4) but contains the additional term $[1 - e^{-\tau_\nu}]$. If τ_ν is large, (16.4) and (16.6) coincide. At wavelengths where the emission is optically thin, the factor $\cos i$ due to the disk inclination drops out and the flux S_ν is directly proportional to the disk mass M_disk. The problem with using (16.6) is to find the right radial temperature variation $T(r)$. As the disk is not a blackbody, equation (13.50) can only be an approximation.

16.3.2.2 *The influence of grain size*

Nevertheless, to get an idea how the grain size influences the spectrum, we apply (16.6) to the disk around a T Tau star with the same radial temperature profile $T(r)$ as in figure 16.12 and with an outer radius $R_\text{out} = 20$ AU. For the surface density, we assume

$$\Sigma(r) = \Sigma_0 r^{-\alpha}$$

with $\alpha = 1$. In the example of figure 16.13, $M_\text{disk} \simeq 0.01\ \text{M}_\odot$.

Before we can numerically evaluate (16.6), we have to specify the wavelength–dependent mass absorption coefficient, K_λ, of dust. As we are dealing with a disk, the grains may be quite big as a result of coagulation, after all, even planets might form. A power-law approximation $K_\lambda \propto \lambda^{-2}$ according to (8.10) is then not warranted and K_λ must be calculated from Mie theory (section 2.6.5). Figure 16.13 illustrates the influence of grain size on the spectrum for fixed surface density $\Sigma(r)$. If λ_0 denotes the wavelength where the outer, more tenuous parts of the disk become transparent, the total disk emission falls below the blackbody case (depicted in figure 16.13 by the full curve) for $\lambda > \lambda_0$.

- If the particles have radii of 10 μm or less, $\lambda_0 \simeq 500\ \mu$m. For $\lambda > \lambda_0$, the emission S_ν is proportional to ν^4, and not to ν^2, as in the Rayleigh limit of the Planck function.
- If the particles are bigger than 10 μm, λ_0 first increases with grain radius a (the line labeled A refers to $a = 100\ \mu$m), then decreases. For very big

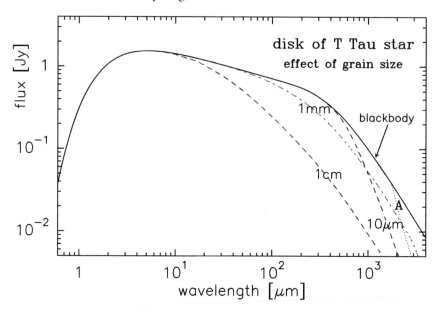

Figure 16.13. The spectra of a disk with the same radial temperature profile $T(r)$ used in figure 16.12. The flux is computed from (16.6), the full curve shows, for comparison, the blackbody of figure 16.12. The surface density of the gas $\Sigma(r) = 10^{16}r^{-1}$ (cgs units), the gas-to-dust ratio equals 150. The dust consists of identical silicate spheres, the grain radius, a, is indicated; the label A stands for $a = 100\ \mu$m.

grains ($a = 1$ cm), the spectrum steepens very gradually as the frequency decreases.

16.3.3 Radiative transfer in an inflated disk

We refine the disk model by discarding the assumption of isothermality in the z-direction. The temperature $T(z)$ is now determined from the radiative transfer outlined in section 13.3. As an example, we choose a main sequence star of type B9V ($L_* = 100\ L_\odot$, $M_* = 3\ M_\odot$, $T_* = 12\,000$ K, $R_* = 2.76\ R_\odot$). The disk is inflated, the grazing angle $\alpha_{\rm gr}$ is taken from (13.53). The surface density $\Sigma(r) = \Sigma_0 r^{-3/2}$ with $\Sigma_0 = 4.1 \times 10^{22}$ cgs. This implies that there are $0.01 M_\odot$ of gas within 100 AU and that at one astronomical unit from the star, the visual optical thickness of the disk due to dust is $\tau_V \simeq 2.1 \times 10^5$. The dust model is from section 12.4, except that the MRN grains have an upper size limit $a_+ = 1\ \mu$m. They are thus, on average, a bit bigger because of likely coagulation in the disk.

We first look at the emission from a narrow annulus thus avoiding the complexity that arises from the radial variations in temperature and surface density. The flux from an annulus of radius $r = 20$ AU, where the grazing

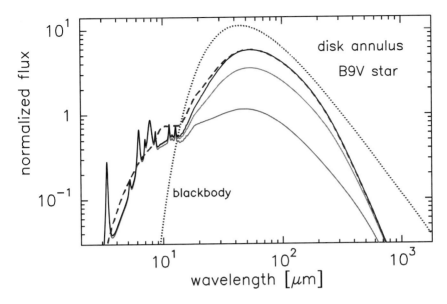

Figure 16.14. The flux, in arbitrary units, from a narrow annulus of a disk around a B9V star. The radius of the annulus is 3×10^{14} cm. The model and its parameters are expounded in the text and in figure 16.15. The dots show a blackbody surface. The broken curve is for dust without PAHs or vsg. The full curves refer to inclination angles $i = 90°$ (face-on), $78°$ and $52°$, respectively.

angle $\alpha_{gr} = 3.32°$, is displayed in figure 16.14. The temperature in the midplane ($z = 0$) is for all grain types equal, $T_{mid} = 91$ K. At the disk surface, the dust temperatures vary between 210 and 510 K. A blackbody would after (13.54) acquire a temperature of 115 K.

The full curves in figure 16.14 result from a full radiative transfer in the z-direction and include PAHs and vsg. The viewing angle i for the upper curve is $90°$, for the two lower ones, $i = 78°$ and $52°$, respectively. In the far infrared, the disk annulus is opaque up to $\sim$150 μm. The flux is then reduced because of geometrical contraction if one looks at the disk from the side ($i < 90°$). In the submillimeter region, the disk becomes translucent and the curves begin to converge. At 1 mm (not shown), the dependence on the position of the observer has disappeared and the slope approaches $S_\nu \propto \nu^4$. Note that an increase in surface density Σ would raise the submillimeter emission without changing much in the spectrum below 50 μm. For $\lambda \leq 15$ μm, the full curves coincide. At these wavelengths, one sees only the hot dust from a thin and transparent surface layer. The broken curve, again for an inclination angle $i = 90°$, is calculated without very small particles. It merges with the upper full curve for $\lambda > 30$ μm. The dotted curve shows, for comparison, the flux from a blackbody annulus $T = 115$ K viewed face-on. The integrated flux is the same as in the other curves

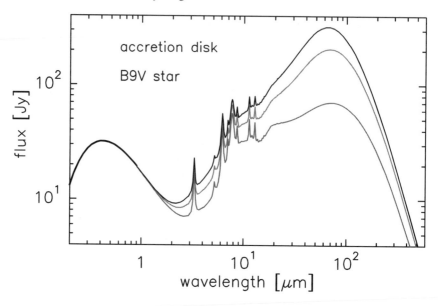

Figure 16.15. The flux from the whole disk, including the B9V star, at inclination angles $i = 90°$ (face-on), 78° and 52°. Stellar distance $D = 100$ pc, $R_{in} = 0.1$ AU, $R_{out} = 100$ AU, surface density $\Sigma = \Sigma_0/r^{3/2}$ with $\Sigma_0 = 4.1 \times 10^{22}$ cgs so that at 1 AU, $\Sigma = 700$ g cm^{-2} and $\tau_V \simeq 2.1 \times 10^5$.

with $i = 90°$.

Figure 16.15 shows the emission from a full disk of outer radius 100 AU. The inner radius $R_{in} = 0.1$ AU $= 9.2R_*$ is about equal to the evaporation temperature of dust. Note that a flat disk absorbs 92% (!) of the stellar flux within $r \leq 9.2R_*$, thus only 8% at larger radii. This explains the trough at ~ 3 μm with respect to the spectral energy distribution in figure 16.12. So there is a hole in our disk between 1 and ~ 9 stellar radii. There might be even much larger holes if planetesimals or planets have formed. The rise in the far infrared in figure 16.15 is due to the large grazing angle. It is the result of inflation. Should only the gas disk be inflated but not the dust disk because the grains have sedimented, the calculations would yield quite different numbers. Such uncertainties about the basic structure of the disk are more grave than the shortcomings of the simple one-dimensional radiative transfer code.

16.4 Reflection nebulae

Reflection nebulae are illuminated by stars, typically of type B3V, that are hot but not hot enough to create an HII region. Nevertheless, the radiation field within them is hard and the grains are in a harsh environment. This leads to ample excitation of very small grains and PAHs. We demonstrate in the models here the tremendous effect which they have on the spectrum of a reflection nebula.

The visual optical depth of a reflection nebula is never large ($\lesssim 1$) and we may, therefore, assume that the grains are heated directly by the star, without any foreground attenuation. Computing the radiative transfer is therefore not really necessary. The infrared emission from the dust itself is, of course, also optically thin. To evaluate it, we adopt the dust model outlined in section 12.4 but slightly simplify it to make the role of the PAHs more transparent. We use only one kind of PAHs with $N_C = 50$ carbon atoms, hydrogenation parameter $f_{H/C} = 0.3$ and cutoff wavelength $\lambda_{cut} = 0.42$ μm, and lump the four C–H out-of-plane bends (table 12.1) into one resonance, at 11.3 μm. The mass fraction of the PAHs is 6%. The very small grains (vsg), which are made of graphite and silicate, contain 10% of the mass of the big particles and have uniform radii of 10 Å. The big (MRN) grains are exactly as described in section 12.4.

The dust emission at various positions in the nebula is shown in figure 16.16. The source parameters are listed in the figure caption, absolute fluxes can be read from figure 16.18. Far from the star (right frame, distance $D = 5 \times 10^{18}$ cm), the big grains are cool, their mid infrared emission is negligible and the maximum occurs beyond 100 μm. The spectrum below 50 μm is then entirely due to PAHs. Closer to the star, the big grains become hotter. But even at $D = 5 \times 10^{16}$ cm, where the maximum in the spectrum has shifted to ~ 25 μm, the PAHs dominate below 15 μm.

The *relative* strengths of the PAH features, the band ratios, are not sensitive to the stellar distance, only the absolute flux weakens if the dust is far out. This becomes evident from figure 16.17. A similar situation is encountered in figure 8.14 with respect to the very small grains where the colors (flux ratios) are almost constant over the nebula. In the present example, however, the vsg play only a minor role.

To exemplify the influence of PAH size, the computations are repeated for a distance $D = 5 \times 10^{17}$ cm but now the number of carbon atoms in a PAH, N_C, is varied from 30 to 500 (figure 16.19). The hydrogenation parameter, $f_{H/C}$, changes according to (12.11). The big PAHs are, on average, cooler than the small ones and it is, therefore, harder to excite their high-frequency bands. This is best seen by comparing the C–H resonances at 3.3 and 11.3 μm. The emission ratio, $\epsilon_{3.3}/\epsilon_{11.3}$, is much larger for small (hot) than for big (cool) PAHs.

It is also instructive to see how the temperature of the big (MRN) particles changes with distance, grain size and chemical composition. Summarizing, the silicates are always colder than the amorphous carbon grains, and big particles always colder than small ones of the same chemistry. At a distance $D = 5 \times 10^{18}$ cm, the temperature of the silicates, $T(Si)$, ranges from 16 to 20 K, that of the carbonaceous particles, $T(aC)$, from 20 to 25 K. Corresponding numbers at distance $D = 5 \times 10^{17}$ cm are: $T(Si) = 33 \ldots 43$ K and $T(aC) = 46 \ldots 58$ K. At $D = 5 \times 10^{16}$ cm: $T(Si) = 71 \ldots 93$ K and $T(aC) = 107 \ldots 137$ K.

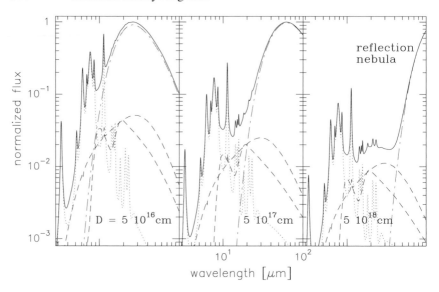

Figure 16.16. The emission by dust in a reflection nebula near a B3V star with $L = 10^3 \, L_\odot$ and $T_{\text{eff}} = 18\,000$ K at various distances to the star. The spectra are normalized at their maxima. Full curves: sum of all three dust components; dash-dotted lines: big (MRN) grains; broken lines: very small graphite and silicate grains (vsg) of 10 Å radius (the silicates have a hump at 10 μm); dots: PAHs.

16.5   Cold and warm dust in galaxies

Typically, one-third of the total flux of a galaxy is emitted at infrared wavelengths and, in some objects, this fraction exceeds 90%. The infrared luminosity results from the absorption and re–emission of stellar light by interstellar dust grains. The spectra appear at first glance rather similar, at least, they always peak around 100 μm. From the spectral energy distribution in the infrared, one can infer:

- *The total infrared luminosity* L_{IR}. This is done simply by integrating over wavelength (usually for $\lambda > 10 \, \mu$m). One only needs a few spectral points to obtain a reasonable estimate of L_{IR}. The ratio $L_{\text{IR}}/L_{\text{tot}}$ indicates how dusty the galaxy is and serves as a crude measure for its activity.
- *The temperature distribution of the dust.* Naturally, the infrared spectrum originates from grains in all kinds of environment but there are dominating components. Of special import is the coldest because it refers to the bulk of the dust, and thus of the gas in view of mixing. To determine its temperature T_{d}, one needs measurements at several (at least, two) well-separated wavelengths between, say, 200 μm and 1.3 mm. The T_{d} derived from such photometric observations is a color temperature. It represents the physical grain temperature only when one uses the correct slope $dK_\lambda/d\lambda$ for

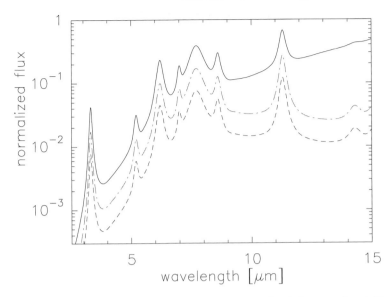

Figure 16.17. A blowup of figure 16.16 showing the PAH bands between 3 and 15 μm. The top line refers to the case when the dust is nearest to star star ($D = 5 \times 10^{16}$ cm), the bottom line when it is farthest away ($D = 5 \times 10^{18}$ cm).

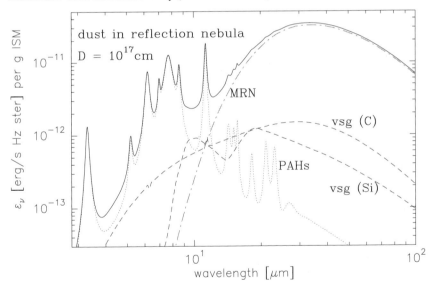

Figure 16.18. Very similar to figure 16.16, same kind of PAHs ($N_C = 50$, $f_{H/C} = 0.3$) and vsg but now with *absolute* fluxes per gram of interstellar matter and a distance $D = 10^{17}$ cm.

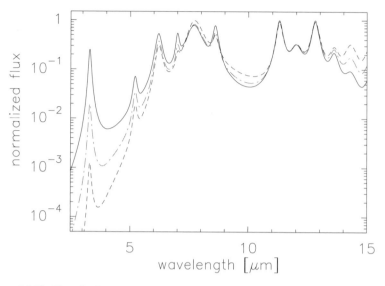

Figure 16.19. How the PAH emission varies with PAH size. The grains are at a distance $D = 5 \times 10^{17}$ cm from the star, again with $L = 10^3 \, L_\odot$ and $T_{\text{eff}} = 18\,000$ K. The broken curves are big ones, $N_C = 500$, $f_{H/C} = 0.13$; Dashes and dots, medium sizes, $N_C = 120$, $f_{H/C} = 0.26$; Full curves, small ones, $N_C = 30$, $f_{H/C} = 0.5$. Arbitrary but identical flux units are used for all three curves.

the wavelength dependence of the dust absorption coefficient (section 8.2.6). The emission shortward of the peak, usually below 100 μm, comes from dust that is warmer and its mass is always negligible compared to the total dust mass in the galaxy.

- *The total dust mass M_d.* This is, from equation (14.50), directly proportional to the detected flux S_λ but one has to use wavelengths greater than 350 μm in order to avoid the sensitivity of M_d with respect to T_d. When one adopts a reasonable dust-to-gas ratio (for example, $R_d \sim 0.007$), one gets the total gas mass M_{gas}. This is certainly one of the fundamental parameters of a galaxy because the gas provides the reservoir for star formation. Other fundamental galactic parameters are luminosity, size, total mass, morphology (spiral, elliptical, irregular) and activity class.

We exemplify a few conclusions which can be drawn from a decomposition of the far infrared dust spectrum ($\lambda > 40$ μm) in the case of the spiral galaxies Mkn799 and UGC2982 [Sie99]. Their optical images are displayed in figures 16.20 and 16.21 and some of their parameters are listed in the captions. We point out that when galaxies, like these two, have distances greater than $\sim$20 Mpc, their far infrared emission cannot be spatially resolved. The observations must be carried out from above the atmosphere, and space telescopes, to save weight, have a small diameter ($d \leq 1$ m). Their resolving power, given by $1.2\lambda/d$ in arcsec,

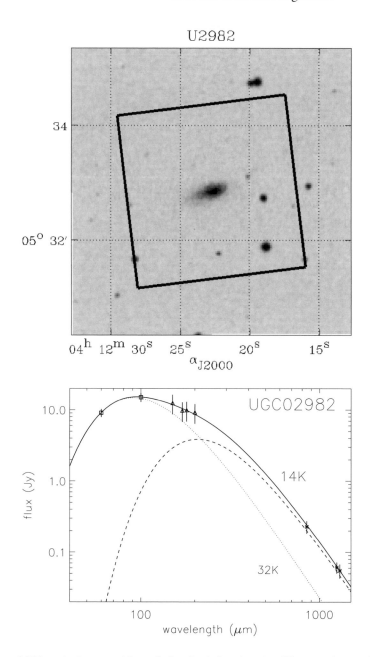

Figure 16.20.   A decomposition of the far infrared and millimeter dust emission of the spiral galaxy UGC2982, distance $D = 70$ Mpc (Hubble constant $H_0 = 75$ km s^{-1} Mpc^{-1}), total IR luminosity $L_{IR} = 1.4 \times 10^{11}$ L$_{\odot}$.

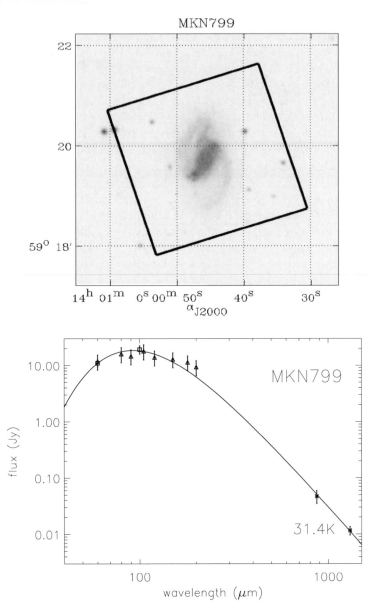

Figure 16.21. A decomposition of the far infrared and millimeter dust emission of the spiral galaxy Mkn799, $D = 38$ Mpc, $L_{IR} = 5 \times 10^{10}$ L$_\odot$.

is only of order 1 arcmin at $\lambda = 100$ μm. The far infrared spectrum, therefore, refers to the galaxy as a whole. The big squares in the upper boxes of figure 16.20 and 16.21 roughly indicate the beam at 200 μm.

In Mkn799, the flux S_λ between 50 μm and 1 mm can be approximated by *one* temperature component. Putting $S_\nu \propto \nu^2 B_\nu(T_d)$, a least–square fit yields $T_d = 31$ K. For UGC2982, however, we need *two* components, one at about the same temperature as in Mkn799 ($T_d = 32$ K), the other at 14 K, thus much colder and of lower integrated luminosity. Nevertheless, the cold component contains 90% of the mass. Because the total dust emission is proportional to T_d^6 (see (8.14)), the heating rate for most of the dust in Mkn799 is a factor $(31/14)^6 \simeq 120$ greater than in UGC2982 and the average stellar luminosity per unit volume correspondingly higher.

Closely related to the disparate temperatures of the coldest component (14 K versus 31 K) is the discrepancy in the ratio of infrared luminosity over gas mass, L_{IR}/M_{gas}. Expressing both quantities in solar units, $L_{IR}/M_{gas} \sim 5$ for UGC2982, whereas this value is almost 20 times larger in Mkn799. A high ratio L_{IR}/M_{gas} is often interpreted to imply a high efficiency for extracting luminosity from the gas via star formation. Exactly how much depends, of course, on the initial mass function (see figure 15.4). L_{IR}/M_{gas} is thus regarded as an indicator of star-formation activity. In this scenario, Mkn799 is active whereas UGC2982 is quiescent. Periods of extreme star-formation efficiency (starbursts) are usually restricted to the nucleus of a galaxy (see section 16.6). Due to the strong concentration of both dust and luminosity, grains with temperatures below 30 K are absent in the center of a starburst galaxy, which would explain $T_d = 31$ K for the coldest component in Mkn799.

16.6   Starburst nuclei

16.6.1   Repetitive bursts of star formation

Galactic nuclei are extended stellar clusters that form the nuclear bulge. In roughly one out of ten, one finds at their center a high concentration of OB stars that must have been created in an episode of violent star formation, a so called starburst. The rate of astration during the burst cannot be sustained over a period much longer than 10^7 yr, otherwise the nucleus would run out of interstellar matter. The burst takes place in a region only a few hundred parsec across, often considerably less. The starburst nucleus is very bright, up to $10^{12} L_\odot$, much brighter than the rest of the galaxy. The high luminosities advocate an initial mass function (IMF) heavily biased towards massive stars, with a lower limit m_l much above the value of the solar neighborhood ($\sim 0.1 M_\odot$, section 15.1.6).

Before we model the dust emission from a galactic nucleus, we sketch how the collective star formation possibly operates. Such a tentative explanation is always useful. It will guide us in the same way that a notion about the dynamics of protostellar collapse (section 15.3) guided us in calculating the protostellar spectra.

In a galactic nucleus, the gas is gravitationally trapped in the potential well created by the bulge stars. Initially, before the burst, it is supported against collapse by circular as well as turbulent non-circular motions. As the turbulent energy is inevitably dissipated in cloud–cloud collisions, the gas sinks in the potential well and its density increases. Above a certain threshold, violent star formation sets in. A few million years later, the most massive stars explode as supernovae and their ejecta replenish the turbulent energy of the gas. The gas is then dispelled and star formation ends. The process is likely to be repetitive, as argued later, and this would explain why bursts are fairly common. A sequence of bright bursts may be initialized by a strong inflow of gas from the disk into the nucleus, possibly as a result of tidal interaction of the host galaxy with a neighbor.

16.6.1.1 *Dynamics of the gas in a galactic nucleus*

To estimate the strength of a burst, its duration and spatial extent, and to show the likely cyclic character of the phenomenon, we treat the interstellar matter in a galactic nucleus as one homogeneous gas cloud of mass M, radius r and turbulent energy E_T [Tut95]. The following equations of a one–zone model describe the evolution of the cloud:

$$\dot{r} = u \tag{16.7}$$

$$\dot{u} = \frac{2}{3}\frac{E_T}{Mr} - \frac{G[M_*(r) + M]}{r^2} \tag{16.8}$$

$$\dot{E}_T = -\frac{E_T}{\tau_{dis}} + (\dot{M} - \mu)\frac{E_T}{M} + \nu_{SN} E_{SN} - \frac{2u E_T}{r} \tag{16.9}$$

$$\dot{M} = \mu - \begin{cases} M/\tau_{SF} & \text{if } M \geq f M_* \text{ and } \rho > \rho_{cr} \\ 0 & \text{else.} \end{cases} \tag{16.10}$$

There are four time-dependent variables: r, M, E_T and the velocity u. In (16.10), μ denotes a constant mass inflow, from the disk or from the wind of old stars in the nuclear bulge. The quantity f is the volume filling factor of the gas, $f < 1$ implies clumping. $M_*(r)$ specifies the mass of the bulge stars within radius r; their lifetime is very long ($\sim 10^{10}$ yr) and their total mass much greater than the mass of the gas or the newly–born stars. The time scales for star formation and dissipation of turbulent energy are τ_{SF} and τ_{dis}, respectively, ν_{SN} denotes the supernova rate and E_{SN} the turbulent energy input into the gas per supernova explosion.

It is easy to show that a protostellar clump can collapse only when the mean gas density exceeds the mean stellar density or when $M \geq f M_*$ in a galactic nucleus, otherwise tidal forces will disrupt the clump. The condition applies where strong gradients in the gravitational potential are present. In the present scenario, it is more stringent than the Jeans criterion (section 15.2.5). We also

Table 16.1. Typical parameters for a galactic nucleus with a starburst, like M82. See section 12.2 for nomenclature of PAHs and very small grains (vsg). Infrared spectra are displayed in figures 16.23 and 16.24.

OB stars	integrated luminosity	$3 \times 10^{10} \, L_\odot$
	radius of stellar cluster	$R^{OB} = 200$ pc
	stellar density	$n^{OB}(r) \propto r^{-1/2}$
	effective surface temperature	$T^{OB} = 30\,000$ K
	luminosity of one OB star	$L^{OB} = 10^5 \, L_\odot$
Low luminosity	integrated luminosity	$10^{10} \, L_\odot$
stars	radius of stellar cluster	$R^* = 700$ pc
	stelllar density	$n^*(r) \propto r^{-1.8}$
	effective surface temperature	$T^* = 4000$ K
Dust	radius of dust cloud	$R_{dust} = 800$ pc
	density $\rho_{dust}(r)$	= constant for $r \leq 230$ pc
	$\propto r^{-1/2}$ for $r > 230$ pc	
	A_V from cloud edge to center	29 mag
	A_V from $r = 0$ to $r = 230$ pc	10 mag
PAHs	small ones ($N_C = 50$)	$Y_C^{PAH} = 0.05$, $f_{H/C} = 0.4$
	big ones ($N_C = 300$)	$Y_C^{PAH} = 0.05$, $f_{H/C} = 0.16$
Very small grains	only graphites	$Y_C^{vsg} = 0.05$, $a = 10$ Å

stipulate in (16.10) that stars can only form when the gas density is above some critical value ρ_{cr}.

Equations (16.7)–(16.10) are solved numerically. One has to follow the history of newly–born stars in all stellar mass intervals: how bright they are, how long they shine and whether they explode as SN or die quietly. Figure 16.22 presents an example for a galactic nucleus that is meant to resemble M82, the archetype nearby starburst galaxy. The fundamental feature of the bursts, their periodicity, is not very sensitive to the intial conditions or to the parameters that appear in the equations (except for the lower stellar mass limit m_l, see later). In particular, the model of figure 16.22 assumes the following.

At time $t = 0$, the gas cloud has a mass $M = 2 \times 10^8 \, M_\odot$ and a radius $r = 200$ pc. The constant mass input from old stars or the disk is $\mu = 0.2 \, M_\odot \, \text{yr}^{-1}$. The gas is clumped with volume filling factor $f = 0.2$. For the dissipation time scale of turbulent energy, we choose $\tau_{dis} = 3 \times 10^7 \text{yr}$, for the critical gas density, $\rho_{cr} = 10^{-19}$ g cm^{-3}. When condition (16.10) for star formation is fulfilled, the gas is consumed in an astration time $\tau_{SF} = 3 \times 10^6$ yr.

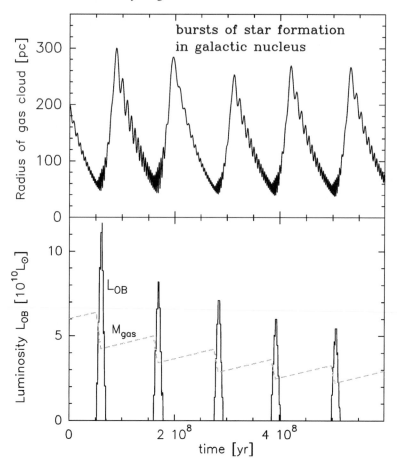

Figure 16.22. Evolution of a galactic nucleus similar to M82. Top: Radius r of the gas cloud. It first shrinks because energy is dissipated on a time scale τ_{dis} and then grows when supernovae explode. The ripples in r reflect quasi–adiabatic oscillations (the free-fall time $t_{\mathrm{ff}} \ll \tau_{\mathrm{dis}} = 3 \times 10^7\,\mathrm{yr}$). Bottom, dashes: Gas mass M on a linear scale starting from zero; at time $t = 0$, $M = 2 \times 10^8\,\mathrm{M_\odot}$. For $t \leq 2 \times 10^7\,\mathrm{yr}$, M increases because of mass inflow. In a burst, M decreases. Bottom, full curve: Total luminosity L_{OB} of OB stars. They are created in bursts of $\sim 10^7$ yr duration, the time interval between them is $\sim 10^8$ yr. Note that L_{OB} may be greater than the luminosity of the rest of the galaxy.

Stars are born in a burst with a Salpeter IMF, $\xi(m) \propto m^{-2.35}$ with upper and lower boundary $m_{\mathrm{u}} = 50\mathrm{M_\odot}$ and $m_{\mathrm{l}} = 3\mathrm{M_\odot}$. Stars of masses greater than $m_{\mathrm{SN}} = 8\mathrm{M_\odot}$ end as supernovae (section 15.1.5). The main sequence lifetime of the SN progenitors can be read from figure 15.3. Each supernova inputs $E_{\mathrm{SN}} = 10^{50}$ erg of turbulent energy into the gas. The mass distribution of the

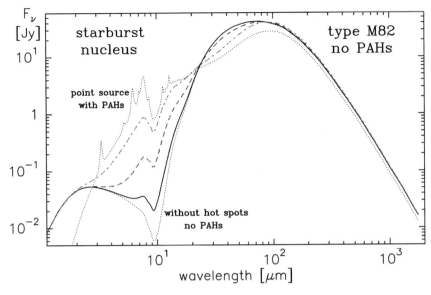

Figure 16.23. Radiative transfer models of a starburst nucleus similar to M82. Distance $D = 20$ Mpc, dust properties from section 12.2, further parameters in table 16.1. The two dotted curves are benchmarks for comparison with figure 16.24: The lower one is computed without hot spots, the upper depicts a point source; it is the only curve in this figure where PAHs are included. The full curves are models with hot spots of gas density $n(H) = 10^2$ cm^{-3} (solid), 10^3 cm^{-3} (dashed) and 10^4 cm^{-3} (dash–dot) and a standard dust-to-gas ratio.

bulge stars follows the power law $M_*(r) \propto r^{1.2}$, the same as for the center of the Milky Way. Within $r = 500$ pc, there are $M_*(r) = 10^9$ M$_\odot$ of stars.

16.6.2 Dust emission from starburst nuclei

The nucleus of a galaxy is usually deeply embedded in dust and can, therefore, only be probed by infrared observations. One way to derive the structure of the nucleus, to determine how dust and stars are geometrically arranged, what the total extinction and the gas mass are, or to which spectral type the stars belong, is to model the dust emission spectrum. For this end, one has to compute the radiative transfer in a dust-filled stellar cluster.

A starburst nucleus contains many OB stars. We first convince ourselves that it is indeed necessary to treat the OB stars and their surroundings separately, as hot spots, following the strategy in section 13.4. If one does not treat them properly, one grossly underrates the mid–infrared flux. This becomes evident from figure 16.23 where spectra are displayed calculated with and without hot spots. In the latter case, the luminosity of the OB stars is taken into account but

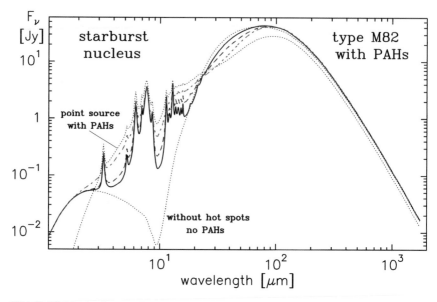

Figure 16.24. As figure 16.23 and with the same benchmark models (upper and lower dotted curves) but now with hots spots *and* PAHs and vsg.

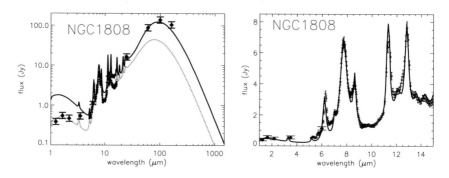

Figure 16.25. Left: Dust emission from the nucleus of the galaxy NGC 1808. The dotted (full) line shows the model flux received in a $25''$ ($100''$) aperture. The galaxy is at a distance of 11 Mpc where $1''$ corresponds to 53 pc. Circles represent observational data. Right: Blowup of the near and mid infrared data and of the $25''$ aperture model. The model fit is very good. Reproduced from Siebenmorgen, Krügel and Laureijs 2001 *Astron. Astrophys.* **377** 735 with permission of EDP Sciences.

their emission is smeared out smoothly over the cluster.

Except for the point source model, none of the stellar cluster spectra in figure 16.23 includes PAHs or very small grains (vsg). To get closer to reality, one has, of course, to incorporate them. This has been done in figure 16.24.

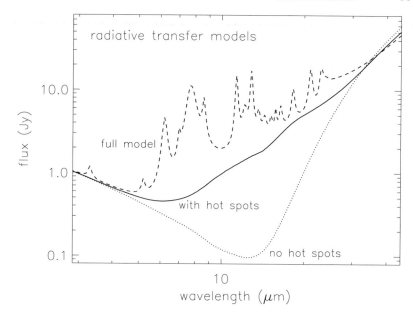

Figure 16.26. The influence of the hot spots and the PAHs on the spectrum of the galaxy NGC 1808. The line labeled full model includes PAHs and hot spots. Reproduced from Siebenmorgen, Krügel and Laureijs 2001 *Astron. Astrophys.* **377** 735 with permission from EDP Sciences.

The PAHs and vsg further enhance the mid infrared flux over the already strong level produced by the hot spots and totally dominate this wavelength region. Interestingly, it is now not obvious at all how to derive, from the 10 μm silicate feature which lies between PAH bands, an absorption optical depth. We add that in view of the strong and hard radiation fields prevalent in a galactic nucleus with OB stars, the radiative transfer code must allow for the possibility of photo-destruction of PAHs and vsg (see section 12.2.3).

A point source model of the same luminosity and spectral appearance as the exciting stars but where the stars are squeezed into a tiny volume at the center of the galactic nucleus is also of relevance and it is shown by the upper dotted curve in figures 16.23 and 16.24. It too yields a very strong mid-infrared flux but differs from the models with extended emission by its small angular size. Mid–infrared cross cuts of high spatial resolution over the galactic nucleus allow us to discriminate between a stellar cluster and a point source.

A massive black hole with an accretion disk would appear to an observer as a point source. But it is one with a considerably harder emission spectrum than OB stars. Usually one assumes for a black hole a power law $S_\nu \propto \nu^\alpha$ with $\alpha \simeq -0.5$. If the dust is exposed to the flux from such an object, PAHs will evaporate in its vicinity (they can only survive farther out) and the ratio of mid infrared over total luminosity is reduced. This can be checked observationally. However, spectral lines from multiply ionized atoms, which require for their excitation more energetic photons than are found in HII regions, are generally a better diagnostic for the presence of a powerful non-thermal source.

Hot spots and PAHs determine the mid infrared emission not only of starburst nuclei. Figure 16.25 shows the spectrum of the galaxy NGC 1808 where the star-formation activity in the center is very mild and only 10% of the total luminosity ($L_{tot} \simeq 5 \times 10^{10}\,L_{\odot}$) comes from OB stars. The nucleus of NGC 1808 is also only weakly obscured ($A_V \sim 5$ mag). Nevertheless, even here model calculations with and without hot spots or PAHs (figure 16.26) yield completely different results [Sie01].

Appendix A

Mathematical formulae

Notations

- Vectors are written boldface, for example, torque $\boldsymbol{\tau}$, velocity $\mathbf{v}$, electric field $\mathbf{E}$.
- The absolute values of these vectors are τ, v and E, respectively. As an exception to these customs, the position vector in Cartesian coordinates, (x, y, z), is sometimes denoted by $\mathbf{x}$, and then, of course, $|\mathbf{x}| = \sqrt{x^2 + y^2 + z^2}$ and not $|\mathbf{x}| = x$.
- A dot over a letter means a time derivative. When a function ϕ depends on the space coordinates (x, y, z) and on time t, its full differential is

$$d\phi = \frac{\partial \phi}{\partial t}\, dt + \frac{\partial \phi}{\partial x}\, dx + \frac{\partial \phi}{\partial y}\, dy + \frac{\partial \phi}{\partial z}\, dz.$$

In our notation, $\dot{\phi} = \partial\phi/\partial t$, which, of course, is different from $d\phi/dt$. A prime denotes space derivative: $f'(x) = df/dx$ where x is a space coordinate.

- For the vector operators we usually use the notations div, rot and grad. Other common symbols are

$$\text{rot} = \text{curl} = \nabla\times = \nabla\wedge$$
$$\text{div} = \nabla\cdot \qquad \text{grad} = \nabla \qquad \Delta = \nabla^2 = \text{div grad}.$$

The *Nabla* or *Hamilton* operator ∇ is mathematically more convenient, whereas div, grad and rot are physically more suggestive. Δ is the *Laplace* operator.

- The symbol $\simeq$ indicates that two quantities are quite similar, $\sim$ means that the agreement is only rough.

Hermitian polynomials

The Hermite polynomials are defined through

$$H_n(y) = (-1)^n e^{y^2} \frac{d^n}{dy^n} e^{-y^2} \qquad (n = 0, 1, 2, \ldots). \tag{A.1}$$

They may be generated from the recurrence relation ($n \geq 1$)

$$H_{n+1}(y) - 2y H_n(y) + 2n H_{n-1}(y) = 0. \tag{A.2}$$

The first four Hermite polynomials are

$$H_0(y) = 1$$
$$H_1(y) = 2y$$
$$H_2(y) = 4y^2 - 2$$
$$H_3(y) = 8y^3 - 12y.$$

The Hermite polynomials $H_n(y)$ are orthogonal with respect to the weight function e^{-y^2},

$$\int_{-\infty}^{\infty} H_n^2(y) e^{-y^2}\, dy = 2^n n! \sqrt{\pi}.$$

Exploiting (A.2) gives

$$\int_{-\infty}^{\infty} H_i(y) y H_j(y) e^{-y^2}\, dy = \int_{-\infty}^{\infty} H_i \left[H_{j+1} + 2j H_{j-1} \right] e^{-y^2}\, dy$$
$$= 2^{i-1} i! \sqrt{\pi} \left\{ \delta_{i,j+1} + 2j \delta_{i,j-1} \right\}. \tag{A.3}$$

Vector analysis

- Let $f(x, y, z)$ be a scalar function and $\mathbf{A}(x, y, z)$, $\mathbf{B}(x, y, z)$ vector functions. Then

$$\text{div}(f\mathbf{A}) = \mathbf{A} \cdot \text{grad}\, f + f\, \text{div}\, \mathbf{A} \tag{A.4}$$

$$\text{div}(\mathbf{A} \times \mathbf{B}) = \mathbf{B} \cdot \text{rot}\, \mathbf{A} - \mathbf{A} \cdot \text{rot}\, \mathbf{B} \tag{A.5}$$

$$\text{rot}\, \text{rot}\, \mathbf{A} = \text{grad}\, \text{div}\, \mathbf{A} - \Delta \mathbf{A} \tag{A.6}$$

$$\text{grad}(\mathbf{A} \times \mathbf{B}) = (\mathbf{B}\, \text{grad})\mathbf{A} + (\mathbf{A}\, \text{grad})\mathbf{B} + \mathbf{B} \times \text{rot}\, \mathbf{A} + \mathbf{A} \times \text{rot}\, \mathbf{B} \tag{A.7}$$

$$\text{rot}(\mathbf{A} \times \mathbf{B}) = (\mathbf{B}\, \text{grad})\mathbf{A} - (\mathbf{A}\, \text{grad})\mathbf{B} + \mathbf{A}\, \text{div}\, \mathbf{B} - \mathbf{B}\, \text{div}\, \mathbf{A} \tag{A.8}$$

$$\text{div}(f\mathbf{A}) = \mathbf{A} \cdot \text{grad}\, f + f\, \text{div}\, \mathbf{A} \tag{A.9}$$

$$\text{rot}\, \text{grad}\, f = 0 \tag{A.10}$$

$$\text{div}\, \text{rot}\, \mathbf{A} = \mathbf{0} \tag{A.11}$$

$$(\mathbf{A} \cdot \text{grad})\mathbf{A} = \tfrac{1}{2}\, \text{grad}\, \mathbf{A}^2 - \mathbf{A} \times \text{rot}\, \mathbf{A} \tag{A.12}$$

- $(\mathbf{A} \cdot \text{grad})\mathbf{A}$ is short hand writing. For example, the x-component of this vector is

$$\left[(\mathbf{A} \cdot \text{grad})\mathbf{A}\right]_x = A_x \frac{\partial A_x}{\partial x} + A_y \frac{\partial A_x}{\partial y} + A_z \frac{\partial A_x}{\partial z}$$

- Let $\mathbf{A}$ be a vector function over some region G with surface S. Gauss' theorem connects the volume and the surface integral,

$$\int_G \text{div}\,\mathbf{A}\,dV = \oint_S \mathbf{A} \cdot \mathbf{n}\,d\sigma. \tag{A.13}$$

The unit vector $\mathbf{n}$ on the surface element $d\sigma$ on S is directed outwards.

- If one chooses for the vector function in Gauss' theorem

$$\mathbf{A} = u\,\text{grad}\,v$$

where u and v are scalar functions, one arrives at Green's identity,

$$\int_G (u\Delta v - v\Delta u)\,dV = \oint_S \left(u\frac{\partial v}{\partial n} - v\frac{\partial u}{\partial n}\right) d\sigma. \tag{A.14}$$

$\partial u/\partial n = \nabla u \cdot \mathbf{n}$, and likewise $\partial v/\partial n$, denotes the normal derivative taken on the surface S and directed outwards from G.

- Let $\mathbf{A}$ be a vector function over some surface S with boundary B. Stoke's theorem connects the surface integral to a line integral,

$$\int_S \text{rot}\,\mathbf{A} \cdot \mathbf{n}\,d\sigma = \oint_B \mathbf{A} \cdot d\mathbf{s} \tag{A.15}$$

Sums and integrals

$$\sum_{i\geq 0} y^i = \frac{1}{1-y} \implies \sum_{i\geq 1} i y^{i-1} = \frac{d}{dy}\sum_{i\geq 0} y^i = \frac{1}{(1-y)^2} \tag{A.16}$$

$$I(s) = \int_0^\infty \frac{x^{s-1}}{e^x - 1}\,dx = \Gamma(s)\cdot\zeta(s) = (s-1)!\sum_{n\geq 1}\frac{1}{n^s}. \tag{A.17}$$

The last integral is expressed as the product of the gamma-function, $\Gamma(s)$, and Riemann's zeta-function, $\zeta(s)$. In particular,

$$\zeta(4) = \frac{\pi^4}{90} \qquad I(4) = \frac{\pi^4}{15}.$$

$$\int \frac{\sqrt{x^2 - a^2}}{x}\,dx = \sqrt{x^2 - a^2} - a\arccos\frac{a}{x} \tag{A.18}$$

$$\int \frac{\arcsin(ax)}{x^3}\,dx = -\frac{\arcsin(ax)}{2x^2} - a\frac{\sqrt{1 - a^2x^2}}{2x}. \tag{A.19}$$

The integral exponential function is defined by

$$E_n(x) = \int_1^\infty \frac{e^{-xz}}{z^n} \, dz \qquad (x, n \geq 0).$$ (A.20)

Complex integer Bessel functions $J_n(z)$ of the complex variable $z = x + iy$ can be defined through [Abr70, 9.1.21]

$$J_n(z) = \frac{i^{-n}}{\pi} \int_0^\pi e^{iz\cos\varphi} \cos(n\varphi) \, d\varphi.$$ (A.21)

In particular,

$$J_0(z) = \frac{1}{\pi} \int_0^\pi e^{iz\cos\varphi} \, d\varphi.$$ (A.22)

Furthermore [Abr70, 9.1.38]

$$J_0(z) = J_0(-z)$$ (A.23)

$$z \, J_1(z) = \int_0^z J_0(x)x \, dx.$$ (A.24)

The bell curve

For $a > 0$,

$$\int_0^\infty e^{-ax^2} \, dx = \frac{1}{2}\sqrt{\frac{\pi}{a}}$$ (A.25)

$$\int_0^\infty x e^{-ax^2} \, dx = \frac{1}{2a}$$ (A.26)

$$\int_0^\infty x^2 e^{-ax^2} \, dx = \frac{1}{4}\sqrt{\frac{\pi}{a^3}}$$ (A.27)

$$\int_0^\infty x^3 e^{-ax^2} \, dx = \frac{1}{2a^2}$$ (A.28)

$$\int_0^\infty x^4 e^{-ax^2} \, dx = \frac{3}{8}\sqrt{\frac{\pi}{a^5}}$$ (A.29)

$$\int x e^{-ax^2} \, dx = -\frac{1}{2a} e^{-ax^2}$$ (A.30)

$$\int x^2 e^{-ax^2} \, dx = -\frac{x e^{-ax^2}}{2a^2} + \frac{\sqrt{\pi}}{4a^3} \operatorname{erf}(ax)$$ (A.31)

$$\int x^3 e^{-ax^2} \, dx = -\frac{ax^2 + 1}{2a^2} e^{-ax^2}.$$ (A.32)

The time average of an harmonically variable field

Let

$$A_R = A_{0R} \cos \omega t \qquad B_R = B_{0R} \cos \omega t$$

be two *real* physical fields with real amplitudes A_{0R}, B_{0R}. When they are rapidly oscillating, one is mostly interested in mean values over many cycles. For the product

$$S = A_R B_R$$

we get the time average

$$\langle S \rangle = \tfrac{1}{2} A_{0R} B_{0R}$$

or

$$\langle A_R^2 \rangle = \tfrac{1}{2} A_{0R}^2.$$

We may also write the fields in complex form,

$$\mathbf{E} \propto e^{-i\omega t} \qquad \mathbf{H} \propto e^{-i\omega t}$$

which is more convenient when taking derivatives; of course, only the real components have physical meaning. If we express these real quantities A_R, B_R as complex variables,

$$A = A_0 e^{-i\omega t} \qquad \text{and} \qquad B = B_0 e^{-i\omega t}$$

where the amplitudes A_0, B_0 are complex themselves, then $A_R = \mathrm{Re}\{A\}$, $B_R = \mathrm{Re}\{B\}$ and

$$S = \mathrm{Re}\{A\} \cdot \mathrm{Re}\{B\} = \tfrac{1}{4}(A + A^*)(B + B^*) = \tfrac{1}{4}(AB^* + A^*B + AB + A^*B^*).$$

An asterisk denotes the complex conjugate. Taking the mean of S, the third and fourth terms in the last bracket vanish because they contain the factor $e^{-2i\omega t}$. In the remaining expression, the time cancels out, so finally

$$\langle S \rangle = \tfrac{1}{2} \mathrm{Re}\{A_0 B_0^*\} \tag{A.33}$$

and the time average of the square of a complex quantity A is

$$\langle A^2 \rangle = \tfrac{1}{2} |A_0|^2. \tag{A.34}$$

Basic physical and astronomical constants

Table A.1. Cosmic constants.

Bohr radius $\hbar^2/m_e e^2$	a_0	=	5.2918×10^{-8}	cm
Bohr magneton $e\hbar/2m_e c$	μ_B	=	9.2741×10^{-21}	esu cm
Boltzmann constant	k	=	1.3806×10^{-16}	erg K^{-1}
electron rest mass	m_e	=	9.1096×10^{-28}	g
elementary charge	e	=	4.8033×10^{-10}	esu
fine structure constant $e^2/\hbar c$	α	=	$1{:}137.04$	
gravitational constant	G	=	6.673×10^{-8}	cm^3 g^{-1} s^{-2}
Planck's constant	h	=	6.6262×10^{-27}	erg s
proton rest mass	m_p	=	1.6726×10^{-24}	g
Rydberg constant	R_∞	=	$1.097\,373 \times 10^5$	cm^{-1}
velocity of light	c	=	2.9979×10^{10}	cm s^{-1}

astronomical unit	1 AU	=	1.496×10^{13}	cm
parsec	1 pc	=	3.086×10^{18}	cm

Sun	absolute visual magnitude	M_V	=	4.87	mag
	effective temperature	T_{eff}	=	5780	K
	mass	$M_\odot$	=	1.989×10^{33}	g
	luminosity	$L_\odot$	=	3.846×10^{33}	erg s^{-1}
	radius	$R_\odot$	=	6.960×10^{33}	cm
	surface gravity	$g_\odot$	=	2.736×10^4	cm s^{-2}
Earth	mass	$M_\oplus$	=	5.973×10^{27}	g
	equatorial radius	$R_\oplus$	=	6.378×10^8	cm
	equatorial surface gravity	$g_\oplus$	=	9.781×10^2	cm s^{-2}
	tropical year	1 yr	=	3.1557×10^7	s

Appendix B

List of symbols

The notation is intended to be human and is, therefore, not strict. Of course, the velocity of light is always c and the constants of Planck and Boltzmann are h and k. But symbols have multiple meanings. For instance, it is general practice to label the luminosity, the angular momentum and the Lagrange function by the letter L. What a symbol at a given place means is evident from the context, or is stated explicitly, hopefully.

The alternative to our somewhat loose notation is to have strictly one symbol represent one quantity. Then one needs subscripts and superscripts, everything is terribly clear but the formulae look horrible and it takes longer to grasp and memorize them visually. To avoid this, we use, for example, in an equation connecting gas atoms and dust grains, the letter m for the mass of the light atom and a capital M for the mass of the heavy particle, instead of M_{atom} and M_{grain}. Here is a list of selected symbols:

A	area
a	radius of spherical grain or of circular cylinder
a_0	Bohr radius $= 5.29 \times 10^{-9}$ cm
a_-, a_+	lower and upper limit of grain radii in size distribution
A_{ij}	Einstein coefficient for spontaneous emission
A_V	extinction in the visible in magnitudes
$\mathbf{A}$	vector potential of magnetic field
B_{ij}	Einstein coefficient for induced emission
$B_\nu(T)$	Planck function
$\mathbf{B}$	magnetic induction
c	velocity of light
c_n	number density of droplets consisting of n molecules (n-mer)
C^{abs}	absorption cross section of a single grain, likewise C^{sca} ...
C_p, C_v	specific heat at constant pressure and constant volume
C_H, C_M	specific heat at constant magnetic field and constant magnetization
$\mathbf{D}$	displacement in Maxwell's equations

e	charge of a proton but also of an electron
	eccentricity of a spheroid
e	unit vector
E	energy
E	electric field
E_{B-V}	standard color excess
f	number of degrees of freedom
f_{ij}	oscillator strength
f_i	volume fraction of component i
F_ν	electromagnetic flux at frequency ν
F	free energy or Helmholtz potential
F	force
g	asymmetry factor in scattering
	statistical weight
G	free enthalpy or Gibbs potential
	gravitational constant
h	Planck's constant
H	enthalpy
	Hamiltonian
H	magnetic field
I	moment of inertia
I_ν	intensity at frequency ν
J	current density
k	Boltzmann constant
	imaginary part of the optical constant m
k	wavenumber, in vacuo k $= 2\pi/\lambda$, as a vector **k**
K^{abs}	volume or mass absorption coefficient, likewise K^{sca} ...
L	Lagrange function
	luminosity
	angular momentum
	Loschmidt's number $= 6.02 \times 10^{23}$
L_a	shape factor of ellipsoid with respect to axis a, likewise L_b, L_c
m, M	mass
M	magnetization
$m = n + ik$	optical constant
m	magnetic moment
m_e	mass of electron
n	real part of the optical constant m
	number density of particles
n_H	$n(\text{HI}) + n(\text{H}_2)$ = total hydrogen density, atomic plus molecular
N	column density, occasionally also number density
$\mathcal{N}$	occupation number (quantum mecahnics)
p	momentum of a particle
$p(\lambda)$	degree of linear polarization of light at wavelength λ

p	electric dipole moment
P	polarization of matter, related to **D**
p, P	pressure
q	charge of a particle
Q^{abs}	absorption efficiency of a grain, likewise $Q^{sca} \ldots$
Q	heat (thermodynamics)
r	distance or radius
	reflectance
r	position vector
R	Rydberg constant
R_d	dust-to-gas mass ratio
s	saturation parameter of vapor pressure
S	entropy
S	Poynting vector
t	time
T	temperature
	kinetic energy
T_d	dust temperature
u_ν	radiative energy density at frequency ν
U	potential energy
	internal energy (thermodynamics)
v, V	velocity
v_g, v_{ph}	group and phase velocity
v_0	volume of one molecule
V	volume
	potential energy
W	energy loss rate
	work
x	position vector
x	size parameter of grain
X_r	symbol for an element in ionization stage r
α	degeneracy parameter (quantum mechanics)
α_e, α_m	electric, magnetic polarizability
β	$1/kT$ (thermodynamics)
γ	damping constant
δ	penetration depth of electromagnetic field
δA	infinitesimal work (thermodynamics)
ϵ_ν	emissivity at frequency ν
ε	$\varepsilon_1 + i\varepsilon_2$ = complex dielectric permeability
ζ	surface tension
η	efficiency
η_i	sticking probability of gas species i
κ	force constant of a spring
	ratio of specific heats

λ	wavelength
μ	$\mu_1 + i\mu_2 =$ complex dielectric permeability
	damping constant in dissipational motion
μ_{ij}	dipole moment for transition $j \to i$, see (6.20)
ν	frequency
ρ	density or charge density
$\rho(E)$	density of states at energy E
ρ_{gr}	density of grain material
σ	radiation constant
	electric conductivity
	surface charge
$\sigma(\omega)$	cross section of atom at frequency ω
σ_T	Thomson scattering cross section
σ_{geo}	geometrical cross section
τ_ν	optical depth at frequency ν
τ	time scale
ϕ	scalar potential (electrodynamics)
$d\Phi$	size of an infinitesimal cell in phase space
χ_{ad}, χ_T	adiabatic and isothermal (static) susceptibility
χ	$\chi_1 + i\chi_2 =$ complex susceptibility
ψ	quantum mechanical wavefunction
ω	circular frequency
	probability for a configuration of atoms
ω_p	plasma frequency
Ω	thermodynamic probability
$d\Omega$	element of solid angle

References

[Abr74]   Abraham F F 1974 *Homegeneous Nucleation Theory* (New York: Academic)

[Abr70]   Abramowitz M and Stegun I A 1970 *Handbook of Mathematical Functions* (New York: Dover)

[Alt60]   Altenhoff W J, Mezger P G, Wendker H and Westerhout G 1960 *Veröffentlichung Sternwarte Bonn* 59

[And89]   Anders E and Grevasse N 1989 *Geochim. Cosmochim. Acta* **53** 197

[Asa75]   Asano S and Yamamoto G 1975 *Appl. Opt.* **14** 29

[Asp90]   Aspin C, Rayner J T, McLean I S and Hayashi S S 1990 *Mon. Not. R. Astron. Soc.* **246** 565

[Bas95]   Basu S and Mouschovias T C 1995 *Astrophys. J.* **453** 271

[Ber69]   Bertie J, Labbe H and Whalley E 1969 *J. Chem. Phys.* **50** 4501

[Bie80]   Biermann P L and Harwit M 1980 *Astrophys. J.* **241** L107

[Boh83]   Bohren C F and Huffman D R 1983 *Absorption and Scattering of Light by Small Particles* (New York: Wiley)

[Bon56]   Bonnor W B 1956 *Mon. Not. R. Astron. Soc.* **116** 350

[Car89]   Cardelli J A, Clayton G C and Mathis J S 1989 *Astrophys. J.* **245** 345

[Cas38]   Casimir H B G and du Pre F K 1938 *Physica* **5** 507

[Cha39]   Chandrasekhar S 1939 *An Introduction to the Study of Stellar Structure* (Chicago)

[Cha85]   Chase M W, Davies C A, Downey J R, Frurip D J, McDonal R A and Syverud A N 1985 *J. Phys. Chem. Ref. Data* **14** 1

[Chi97]   Chini R *et al* 1997 *Astron. Astrophys.* **325** 542

[Dav51]   Davis L and Greenstein J L 1951 *Astrophys. J.* **114** 206

[Deb29]   Debye P 1929 *Polare Molekeln* (Leipzig: Hirzel)

[Dor82]   Dorschner J 1982 *Astrophys. Space Sci.* **81** 323

[Dra84]   Draine B T 1984

[Dul78]   Debye P 1978 *Astrophys. J.* **219** L129

[Ebe55]   Ebert R 1955 *Z. Astrophys.* **37** 217

[Emd07]   Emden R 1907 *Gaskugeln* (Leipzig: Teubner)

[Far27]   Farkas L 1927 *Z. Phys. Chem.* **125** 236

[Fed66]   Feder J, Russell K C, Lothe J and Pound G M 1966 *Adv. Phys.* **15** 111

[Fie58]   Field G B 1958 *Proc. IRE* **46** 240

[Gil73]   Gillett *et al* 1973 *Astrophys. J.* **183** 87

[Gol52]   Gold T 1952 *Mon. Not. R. Astron. Soc.* **112** 215

[Grü94]   Gr″n E, Gustafson B, Mann I, Baguhl M, Morfill G E, Staubach P, Tayler A and Zook H A 1994 *Astron. Astrophys.* **286** 915

[Guh89]   Guhathakurta P and Draine B T 1989 *Astrophys. J.* **345** 230
[Har70]   Harwit M 1970 *Bull. Astron. Inst. Czech.* **21** 204
[Hel70]   Hellyer B 1970 *Mon. Not. R. Astron. Soc.* **148** 383
[Hol70]   Hollenbach D J and Salpeter E E 1970 *J. Chem. Phys.* **53** 79
[Hul57]   van de Hulst H C 1957 *Light Scattering by Small Particles* (New York: Wiley)
[Joc94]   Jochims H W, Rühl E, Baumgärtel H, Tobita S and Leach S 1994 *Astrophys. J.* **420** 307
[Jon67]   Jones R V and Spitzer L 1967 *Astrophys. J.* **147** 943
[Ker69]   Kerker M 1969 *Scattering of Light* (New York: Academic)
[Kit96]   Kittel C 1996 *Introduction to Solid State Physics* (New York: Wiley)
[Kru53]   Krumhansl J and Brooks H 1953 *J. Chem. Phys.* **21** 1663
[Kuh52]   Kuhrt F 1952 *Z. Phys.* **131** 185
[Lan74]   Landau L D and Lifschitz E M 1974 *Quantum Mechanics* Vol 4, section 81
[Lan80]   Lang K R 1980 *Astrophysical Formulae* (Berlin: Springer)
[Lao93]   Laor and Draine B T 1993 *Astrophys. J.* **402** 441
[Lar69]   Larson R B 1969 *Mon. Not. R. Astron. Soc.* **145** 271
[Leg83]   Leger A, Gauthier S, Defourneau D and Rouan D 1983 *Astron. Astrophys.* **117** 47
[Lin66]   Lind A C and Greenberg J M 1966 *J. Appl. Phys.* **37** 3195
[Lyn74]   Lynden-Bell D and Pringle J E 1974 *Mon. Not. R. Astron. Soc.* **168** 603
[Mas88]   Masevich A G and Tutukov A V 1988 *Stellar Evolution: Theory and Observations* (Moscow: Nauka) (in Russian)
[Mat70]   Mathewson D S and Ford V L 1970 *Mem. R. Astron. Soc.* **74** 139
[Mat77]   Mathis J S, Rumpl W and Nordsieck K H 1977 *Astrophys. J.* **217** 425
[McK87]   McKee C F, Hollenbach D J, Seab C G and Tielens A G G M 1987 *Astrophys. J.* **318** 674
[McK89]   McKee C F 1989 *IAU* **135** 431
[Mie08]   Mie G 1908 *Ann. Phys., Leipzig* **25** 377
[Mil78]   Miller G E and Scalo J M 1978 *Astrophys. J. Suppl.* **41** 513
[Mor65]   Morrish A H 1965 (New York: Wiley)
[Mou91]   Mouschovias T C 1991 *The Physics of Star Formation and Early Stellar Evolution* ed C J Lada and N D Kylafis (Dordrecht: Kluwer Academic) p 61
[Nei90]   Neininger N 1990 *Astron. Astrophys.* **263** 30
[Oll98]   Olling R P and Merrifield M R 1998 *Mon. Not. R. Astron. Soc.* **297** 943
[Pep70]   Pepper S V 1970 *J. Opt. Soc. Am.* **60** 805
[Per87]   Perault M 1987 *Thèse d'Etat* Université de Paris VII
[Pet72]   Petrosian V, Silk J and Field G B 1972 *Astrophys. J.* **177** L69
[Pur56]   Purcell E M and Field G B 1956 *Astrophys. J.* **124** 542
[Pur69]   Purcell E M 1969 *Astrophys. J.* **158** 433
[Pur79]   Purcell E M 1979 *Astrophys. J.* **231** 404
[Ray18]   Rayleigh J W 1918 *Phil. Mag.* **36** 365
[Ryb70]   Rybicki G B 1970 *Spectrum Formation in Stars with Steady-State Extended Atmospheres (NBS Spec. Pub. 332)*
[Sal55]   Salpeter E E 1955 *Astrophys. J.* **121** 161
[Sca87]   Scarrott S M, Ward-Thompson D and Warren-Smith R F 1987 *Mon. Not. R. Astron. Soc.* **224** 299

[Sch93]   Schutte W A, Tielens A G G M and Allamandola L J 1993 *Astrophys. J.* **415** 397

[Sea87]   Seab C G 1987 *Interstellar Processes* ed D J Hollenbach and H A Thronson (Dordrecht: Reidel) p 491

[Sha73]   Shakura N I and Sunyaev R A 1974 *Astron. Astrophys.* **24** 337

[Shu77]   Shu F H 1977 *Astrophys. J.* **214** 488

[Sie99]   Siebenmorgen R, Krügel E and Chini R 1999 *Astron. Astrophys.* **351** 495

[Sie00]   Siebenmorgen R and Krügel E 2000 *Astron. Astrophys.* **364** 625

[Sie01]   Siebenmorgen R, Krügel E and Laureijs R J 2001 *Astron. Astrophys.* **377** 735

[Sob60]   Sobolev V V 1960 *Moving Envelopes of Stars* (Cambridge, MA: Harvard University Press)

[Som94]   Somerville *et al* 1994 *Astrophys. J.* **427** L47

[Spi79]   Spitzer L and McGlynn T A 1979 *Astrophys. J.* **231** 417

[Tie87]   Tielens A G G M and Allamandola L J 1987 *Interstellar Processes* ed D J Hollenbach and H A Thronson (Dordrecht: Reidel) p 397

[Tsc87]   Tscharnuter W M 1987 *Astron. Astrophys.* **188** 55

[Tri89]   Trimble V 1991 *Astron. Astrophys. Rev.* **3** 1

[Vos93]   Voshchinnikov N V and Farafonov V G 1993 *Astrophys. Space Sci.* **204** 19

[Wai55]   Wait J R 1955 *Can. J. Phys.* **33** 189

[Wat72]   Watson W D 1972 *Astrophys. J.* **176** 103

[Wat75]   Watson W D 1975 *Atomic and Molecular Physics and the Interstellar Medium (Les Houches, Session XXVI)* ed R Balian, P Encrenaz and J Lequeux

[Wea77]   Weast R C (ed) 1977 *Handbook of Chemistry and Physics* (Boca Rotan, FL: Chemical Rubber Company)

[Whi63]   Whitten G Z and Rabinovich B S 1963 *J. Chem. Phys.* **38** 2466

[Whi78]   Whitmer J C, Cyvin S J and Cyvin B N 1978 *Z. Naturf.* a **33** 45

[Whi92]   Whittet D C B, Martin P G, Hough J H, Rouse M F, Bailey J A and Axon D J 1992 *Astrophys. J.* **386** 562

[Whi96]   Whittet D C B *et al* 1996 *Astron. Astrophys.* **315** L357

[Win80]   Winkler K-H and Newman M J 1980 *Astrophys. J.* **236** 201

[Wou52]   Wouthuysen S A 1952 *Astrophys. J.* **57** 32

[Yam77]   Yamamoto T and Hasegawa H 1977 *Prog. Theor. Phys.* **58** 816

[Zub96]   Zubko V G, Mennella V, Colangeli L and Bussoletti E 1996 *Mon. Not. R. Astron. Soc.* **281** 1321

Index